IET TRANSPORTATION SERIES 2

# Energy Systems for Electric and Hybrid Vehicles

## Other related titles:

| | |
|---|---|
| Volume 1 | **Clean Mobility and Intelligent Transport Systems** M. Fiorini and J-C. Lin (Editors) |
| Volume 38 | **The Electric Car** M.H. Westbrook |
| Volume 45 | **Propulsion Systems for Hybrid Vehicles** J. Miller |
| Volume 79 | **Vehicle-to-Grid: Linking Electric Vehicles to the Smart Grid** J. Lu and J. Hossain (Editors) |

# Energy Systems for Electric and Hybrid Vehicles

Edited by K.T. Chau

The Institution of Engineering and Technology

Published by The Institution of Engineering and Technology, London, United Kingdom

The Institution of Engineering and Technology is registered as a Charity in England & Wales (no. 211014) and Scotland (no. SC038698).



First published 2016



The Institution of Engineering and Technology
Michael Faraday House
Six Hills Way, Stevenage
Herts, SG1 2AY, United Kingdom

www.theiet.org



**British Library Cataloguing in Publication Data**
A catalogue record for this product is available from the British Library

**ISBN 978-1-78561-008-0 (hardback)**
**ISBN 978-1-78561-009-7 (PDF)**

Typeset in India by MPS Limited
Printed in the UK by CPI Group (UK) Ltd, Croydon

# Contents

# Preface

Electric and hybrid vehicles have been globally identified to be the most environmental friendly road transportation. Technological development of electric and hybrid vehicles covers three main areas: the energy systems, the propulsion systems and the auxiliary systems. Because of the multidisciplinary nature, electric and hybrid vehicle technologies could only be briefly discussed in a book. Over the years, there have been significant advancements in various electric and hybrid vehicle technologies. A single book can hardly be possible to embrace all these technologies without sacrificing in-depth discussions. By the same token, a single author or a few co-authors can hardly be possible to discuss all these technologies without sacrificing insightful discussions.

The purpose of this book is to collect the wisdom of all contributors with expertise in various energy systems for electric and hybrid vehicles. Hence, the book can consolidate the research and development of various energy systems for electric and hybrid vehicles, with emphasis on three main themes – the energy source technology, the battery charging and management technology and the vehicle-to-X technology. Compared with the propulsion systems and the auxiliary systems, the energy systems will be the key to success for widespread use of electric and hybrid vehicles. Since this book starts from general overviews and reviews to new and emerging technologies of various energy systems for electric and hybrid vehicles, it is suitable for a wide range of readers, including engineering students, academic researchers, practicing engineers and relevant administrators.

At the time of harvest, I must express my heartfelt gratitude to all contributors of this book. The Chinese idiom "collective wisdom reaps wide benefits" is so assured that the expertise of various contributors can gather together to derive the complete knowledge of energy systems for electric and hybrid vehicles, which is beneficial not only to technological advancement but also to environmental protection. Without their contributions, this book wouldn't have seen the light of day. Increasingly, without their dedications, this book will not be so comprehensive and coherent. Also, I am grateful to all members of my Electric Vehicle Technologies research group, especially Mr Chaoqiang Jiang and Mr Wei Han, for their contribution to this book.

While electric and hybrid vehicles are the driving force for our better environment, my family is the driving power for my devotion to electric and hybrid vehicles. Without any hesitation, I have to express my special thanks to my son, Aten Man-Ho, and my wife, Joan Wai-Yi, for their hearty support all the way.

HKU, March 2016
K.T. Chau

# Preface

# Organization of this book

The book provides a comprehensive knowledge of energy systems for electric and hybrid vehicles. It is written for a wide coverage of readers including students, researchers, engineers, administrators and general readers, which is organized into three themes:

- The first theme is the knowledge of energy source systems, including the electrochemical energy sources, flywheel energy storage, hybrid energy sources, solar energy harvesting, electromagnetic energy regeneration and thermoelectric energy recovery. It consists of six chapters in which Chapters 2–4 deal with various energy sources, while Chapters 5–7 reveal the on-board energy harvesting for electric and hybrid vehicles.
- The second theme is the knowledge of battery charging and management systems. It spans five chapters in which Chapters 8 and 12 review the battery charging strategies and battery management techniques, respectively, while Chapters 9–11 elaborate emerging wireless charging techniques for electric vehicles and the concept of energy cryptography for secure wireless power transfer.
- The third theme is the knowledge of vehicle-to-X technology, embracing the vehicle-to-home, vehicle-to-vehicle, and vehicle-to-grid energy systems, for electric and hybrid vehicles. It is composed of four chapters in which Chapter 13 brings forward the integration of energy and information, while Chapters 14–16 elaborate various vehicle-to-grid technologies.

In this book, there are totally 16 chapters. Each chapter has different number of sections and subsections. In order to facilitate selection of reading, all chapters are outlined below:

- Chapter 1 gives an overview of energy systems for electric and hybrid vehicles, which covers both existing and emerging technologies of three main areas: energy source systems, battery charging and management systems and vehicle-to-X systems, with emphasis on comparative discussions.
- Chapter 2 gives an overview of electrochemical energy sources for electric vehicles, which include capacitors, batteries and fuel cells. After the description of various candidates, their existing applications and development trend for electric vehicles are discussed.
- Chapter 3 is devoted to presenting the ultrahigh-speed flywheel energy storage system as the main power source or auxiliary energy storage for electric vehicles. Relevant technologies including the system configuration, electric machines and control strategies are discussed.

- Chapter 4 is devoted to presenting the hybridization of multiple energy sources for electric and hybrid vehicles. Various hybrid energy systems including the energy source hybridization in electric vehicles and powertrain hybridization in hybrid electric vehicles are discussed.
- Chapter 5 reviews the existing and emerging technologies of photovoltaic cells. The system configuration, power management and practical application of solar energy harvesting, both on-board and off-board, for electric vehicles are described.
- Chapter 6 reveals on-board electromagnetic energy regeneration for electric vehicles, which includes energy recovery from regenerative braking and suspension systems. System configurations, modelling and control strategies of these systems are discussed.
- Chapter 7 reveals on-board thermoelectric energy recovery for hybrid electric vehicles. The system configuration including series and parallel connections, and the control strategy especially maximum power point tracking for waste heat energy recovery are discussed.
- Chapter 8 reviews the existing and emerging battery charging algorithms for electric vehicles. Starting from the charging algorithms for a single battery, the balancing methods for battery pack charging including passive and active balancing and the charging infrastructure are described.
- Chapter 9 describes the development and basic principles of wireless power transfer. Relevant techniques for inductive power transfer including magnetic resonant, circuit compensation, maximum transmission efficiency and maximum power transfer are discussed.
- Chapter 10 describes the development and trend of dynamic wireless charging for roadway-powered electric vehicles, with emphasis on the emerging concept of on-line electric vehicles. Relevant technologies especially electromagnetic field cancellation are discussed.
- Chapter 11 reveals the concept of energy cryptography – energy transfer with security – for dynamic wireless charging of electric vehicles. The system design and control for realization of energy encryption and decryption techniques are discussed.
- Chapter 12 reviews the battery management systems, including their representation, data management and network as well as standard, for electric vehicles. Particularly, the state-of-charge estimation and battery balancing techniques are elaborated.
- Chapter 13 begins with the introduction of the engineering philosophy of electric vehicle systems. The key is the integration of information flow and energy flow to stimulate synergy, hence offering a win-win ecosystem and deducing the dynamic smart charging.
- Chapter 14 describes the role of information technology for vehicle-to-grid services. Specifically, the optimal scheduling of charging and discharging processes between the utility grid and its coordinated electric vehicles is discussed, hence achieving quality-of-service guarantees.

- Chapter 15 gives an overview of vehicle-to-X technology for electric and hybrid vehicles, which includes the vehicle-to-home, vehicle-to-vehicle and vehicle-to-grid systems. Particularly, the vehicle-to-grid infrastructure, ancillary services and cost-emission optimization are discussed.
- Chapter 16 describes the power interface between the utility grid and electric vehicles. Relevant power converters including the two-stage, three-stage and integrated power interfaces as well as their power flow capability for battery charging and discharging are discussed.

The readers have the flexibility to select reading those chapters that are most interesting to them. The suggestion for reading is as follows:

- Undergraduate students taking a course dedicated to electric vehicles may be particularly interested in Chapters 1, 2, 8, 12 and 13.
- Postgraduate students taking a course dedicated to advanced electric vehicle technologies may be interested in all chapters.
- Researchers in the areas of energy source technology, battery charging and management technology and vehicle-to-X technology may have special interest in Chapters 2–7, Chapters 8–12 and Chapters 13–16, respectively, which involve emerging research topics.
- Practicing engineers for product design and development may be more interested in individual chapters in which new ideas can be triggered and commercial products can be derived.
- Administrators and general readers may be interested in all chapters. They are advised to read the book from the beginning to the end, page by page, which is most enjoyable.

# About the Editor

**K. T. Chau** received his B.Sc. (Eng.) degree with First Class Honors, M.Phil. degree and Ph.D. degree all in Electrical and Electronic Engineering from the University of Hong Kong. He joined the alma mater in 1995, and has served as Director of International Research Centre for Electric Vehicles for over ten years. Currently, he is a professor in the Department of Electrical and Electronic Engineering, and Director of B.Eng. degree in Electrical Engineering.

Professor Chau is Fellow of the Institute of Electrical and Electronics Engineers (IEEE) for Contributions to Energy Systems for Electric and Hybrid Vehicles. He is also a Chartered Engineer, Fellow of the Institution of Engineering and Technology (IET), and Fellow of the Hong Kong Institution of Engineers (HKIE). He has served as editors and editorial board members of various international journals as well as chairs and organizing committee members of many international conferences, especially in the area of Electric Vehicles. He is also an international consultant for electric vehicle technologies.

His teaching and research interests are electric and hybrid vehicles, as well as clean and renewable energies. In these areas, he has published six books, over two hundred refereed journal papers and many industrial reports. His teaching and research philosophy follows the Confucian classic "Li Ji" – Teaching and learning (research) are mutually motivating.

Professor Chau has received many awards: including the Chang Jiang Chair Professorship; the Environmental Excellence in Transportation Award for Education, Training and Public Awareness; the Award for Innovative Excellence in Teaching, Learning and Technology; and the University Teaching Fellow Award.

*Chapter 1*

# Overview of energy systems for electric and hybrid vehicles

*K.T. Chau*[1]

Electric vehicles (EVs) have been identified to be the greenest road transportation, while hybrid vehicles (HVs) have been tagged as the super ultra-low emission vehicles. They have their individual merits and demerits, leading them to have their unique roles in modern society.

In this chapter, the definition and classification of EVs and HVs are first described. Then, their benefits and challenges are discussed. After revealing the multidisciplinary technologies for EVs and HVs, the energy system technologies are discussed, with emphasis on energy source systems, battery charging and management systems, and vehicle-to-X energy systems.

## 1.1 What are EVs and HVs?

The first EV was a tricycle invented by Thomas Davenport in 1834, which was powered by an electric motor with a non-rechargeable battery and operated on a short track (Wakefield, 1994). By the 1880, road EVs began to appear on both sides of the Atlantic Ocean. For instance, the first EV in England was built by William Edward Ayrton and John Perry in 1882, and the first EV in America was built by William Morrison in 1890 (Georgano, 1996). By 1912, nearly 34,000 EVs were registered in America. Nevertheless, their cost was equivalent to a Rolls Royce of today which could only be afforded by the wealthy elite. In 1925, Ford produced the low-cost Model T which was only a fraction of the cost of EVs but could offer a much longer driving range, which eventually finished off EVs by the 1930s. The resurrection of EVs started with the outbreak of energy crisis in the 1970s. Then, the development of EVs was accelerated due to the concern over global warming in the 1980s. Currently, the impetus for the advancement of EVs is the new business opportunities.

The first hybrid EV (HEV) was a series hybrid invented by Ferdinand Porsche in 1898. Then, in 1905, Henri Pieper filed a patent of a parallel hybrid which used an electric motor to boost the acceleration of its heat engine (Wouk, 1995). From 1906 to 1910, some HEVs using this hybrid system were built. By the time this hybrid

[1]Department of Electrical and Electronic Engineering, The University of Hong Kong, Hong Kong, China

went into production, Henry Ford built the first assembly lines to mass-produce the Ford Model T that was powerful enough to accelerate the vehicle unaided. Consequently, by the 1930s, similar to EVs, HEVs had screeched to a halt. Rather than waiting for the breakthrough of EVs, the development of HEVs became widespread beginning in the late 1990s. The first mass-produced HEV was the Toyota Prius which was launched in 1997 (Hermance and Sasaki, 1998). Recently, Toyota has announced that its global hybrid sales surpassed the 7 million unit mark.

There have been various definitions of EVs and HEVs. Some definitions are academic in nature while some are based on general perception, whereas some classifications are quite ad-hoc while some are too general. For instance, EVs are generally classified as the pure EV (PEV) and HEV based on their propulsion systems, whereas they are also classified as the battery EV (BEV), HEV and fuel-cell EV (FEV) based on their energy sources (Chan and Chau, 2001). Sometimes, the BEV is loosely called the EV so that they are also named as EV, HEV and FEV. The HEV may even be extended to embrace the vehicle using any two different energy sources such as the fuel-cell-and-battery hybrid or battery-and-ultracapacitor hybrid (Chau and Wong, 2001). In recent years, there has been a trend that EVs should first be classified by their propulsion devices, and then be further classified by their energy carriers and energy sources (Chau, 2014). So, as shown in Figure 1.1, EVs are first classified as the PEV and HEV families based on their propulsion devices, namely, the PEV solely adopts the electric motor for electric propulsion and the HEV uses both the electric motor and heat engine for hybrid propulsion. It should be noted that the general public prefers to loosely name the PEV as the EV and the HEV as the HV, leading to form the general term of EVs and HVs.

Based on the variation of energy carriers and energy sources, the PEV family can be split into the BEV and FEV due to the use of batteries and fuel cells as their main energy sources, respectively. Taking into account the latest energy sources of capacitors (specifically dubbed ultracapacitors (UCs) or supercapacitors) and

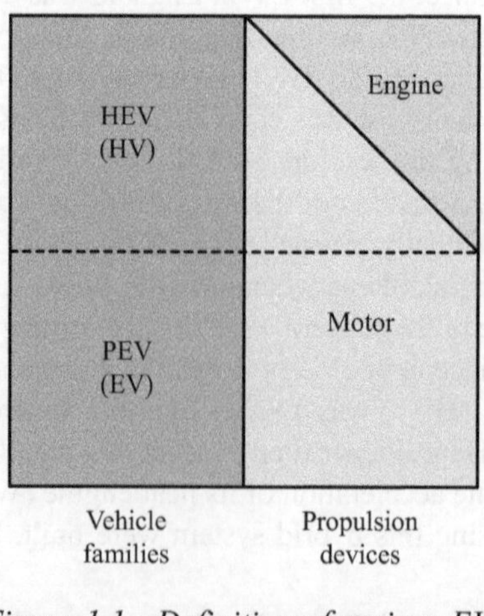


*Figure 1.1 Definition of various EVs*

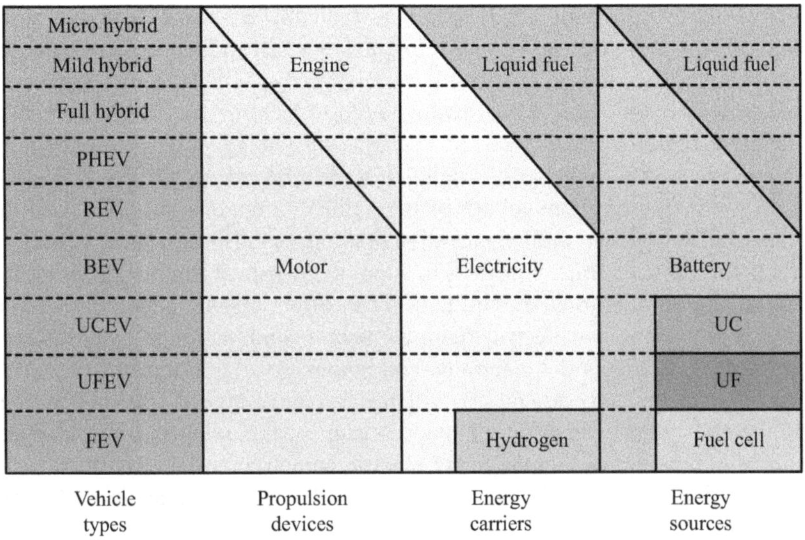


*Figure 1.2 Classification of various EVs*

flywheels (specifically dubbed ultraflywheels (UFs) or ultrahigh-speed flywheels), the ultracapacitor EV (UCEV) and ultraflywheel EV (UFEV) are also members of the PEV family. On the other hand, based on the hybridization level between the electric motor and heat engine, the HEV family consists of five members: the micro hybrid, mild hybrid, full hybrid, plug-in HEV (PHEV) and range-extended EV (REV) in accordance with their increasing contribution from the electric motor for hybrid propulsion. Among them, the micro, mild and full hybrids are termed conventional HEVs which are solely refuelled with liquid fuel in filling stations (Chau and Chan, 2007), whereas the PHEV and REV are called griddable HEVs (GHEVs) which can be recharged by electricity via charging ports or refuelled with liquid fuel in filling stations (Chau, 2010). This classification is depicted in Figure 1.2.

In the PEV family, the BEV is almost exclusively used in real life and is the unique one commercially available. Essentially, all members of the PEV utilize the battery as the sole energy source or one of the energy sources. For instance, the FEV has to incorporate the battery to store the regenerative braking energy, while the UCEV and UFEV should incorporate the battery to provide sufficient energy storage for normal operation because the UC and UF possess relatively low specific energy. From the general public perspective, both the UCEV and UFEC are called an EV using the hybrid energy source of battery and UC and the hybrid energy source of battery and UF, respectively. It should be noted that there are some attempts on using the UC and UF as the sole energy source, the resulting UCEV and UFEV inevitably suffer from the problems of heavy weight, bulky size and very short driving range per charge.

Among those conventional HEVs, the micro, mild and full hybrids offer different capabilities of hybrid features. For the micro hybrid, it is equipped with an integrated-starter-generator (ISG) which is typically 3–5 kW with the system voltage of 14–42 V. Instead of providing power assist to help the heat engine, it offers two important hybrid features: the idle stop-start which shuts down the heat engine whenever the vehicle is at rest so as to reduce the fuel consumption, and regenerative braking which recovers the braking energy during deceleration to regenerate electricity for battery charging. For the mild hybrid, the ISG is increased to 7–15 kW with the system voltage of 100–150 V. It not only provides the hybrid features of idle stop-start and regenerative braking, but also assists the heat engine to propel the vehicle. Thus, it enables to adopt a downsized engine. Differing from the micro and mild hybrids, the full hybrid can offer versatile operations by using the heat engine alone, the electric motor alone or a combination of both. Instead of using the ISG, it adopts the electronic-continuously variable transmission (EVT) system which is typically 50–60 kW with the system voltage of 500–600 V. Using EVT, the full hybrid can offer all hybrid features, including the idle stop-start, regenerative braking, power assist and electric launch. Increasingly, it can enable the heat engine working at its optimal operation line (OOL) to achieve efficiency optimization.

Among those GHEVs, the PHEV and REV are quite similar in terms of their system configurations although the former is derived from the full hybrid while the latter is deduced from the BEV. For the PHEV, it provides all features of the full hybrid, while having an additional feature of plug-in rechargeable. In general, it is equipped with more battery packs than the full hybrid, typically 4–5 kWh, so that it can offer a decent electric range at the pure electric mode. It normally operates at the blended mode in which the electric motor and heat engine complement one another to maximize the fuel economy. Compared with the PHEV, the REV usually installs a smaller heat engine but a larger battery pack. As reflected by its name, it prefers to work as a BEV until the battery capacity drops to a predefined threshold. Then, it works as a series hybrid so that the heat engine drives the generator to charge the battery pack and provide power to the electric motor. At higher speeds and loads, the REV generally operates at the blended mode to maximize the fuel economy. In order to provide the desired electric range, the battery requirement is typically over 16 kWh.

Table 1.1 summarizes the hybrid features of various HEVs, including the conventional HEVs and GHEVs.

*Table 1.1 Hybrid features of HEVs*

| | **Micro** | **Mild** | **Full** | **PHEV** | **REV** |
|---|---|---|---|---|---|
| Idle stop-start | ✓ | ✓ | ✓ | ✓ | ✓ |
| Regenerative braking | ✓ | ✓ | ✓ | ✓ | ✓ |
| Power assist | | ✓ | ✓ | ✓ | ✓ |
| Electric launch | | | ✓ | ✓ | ✓ |
| Efficiency optimization | | | ✓ | ✓ | ✓ |
| Decent electric range | | | | ✓ | ✓ |

## 1.2 Benefits

Internal combustion engine vehicles (ICEVs) are responsible for over 50% of harmful emissions in many cities. In order to solve or at least to alleviate this problem, the use of EVs is the most viable solution. Definitely, the whole PEV family offers zero roadside emissions. When taking into account the emissions to produce petrol for the ICEV and the emissions to generate electricity for the BEV, the harmful emissions of the BEV are still much lower than that of the ICEV as indicated in Figure 1.3, where the carbon monoxide (CO), hydrocarbon (HC), nitrogen oxides ($NO_x$), sulphur oxides ($SO_x$) and particulate matters ($PM_x$) are considered (Chau, 2010). The UCEV and UFEV have the same benefits as both of them are solely fed by electricity. Similarly, when taking account the emissions to produce hydrogen gas, the harmful emissions of the FEV are also much lower than that of the ICEV. On the other hand, the HEV family offers different levels of roadside emissions, but always lower than that of the ICEV. After taking into account the emissions to produce petrol, the harmful emissions of HEVs essentially lie between that of the BEV and the ICEV, with an increasing trend from the REV, then the PHEV, the full hybrid, the mild hybrid to the micro hybrid.

While the above harmful emissions are directly harmful to our health, the carbon dioxide ($CO_2$) emission or loosely called carbon emission is severely detrimental to our planet – the global warming. With an increasing average global temperature, the climate is changing in an unpredictable way, from super hurricanes and typhoons, floods and droughts to deep cold and long heat waves. Differing from the reduction of those harmful emissions, the use of EVs to reduce the carbon emission is controversial. Actually, whether the carbon emission can be reduced or not mainly depends on the fuel mix for electricity generation. When incorporating

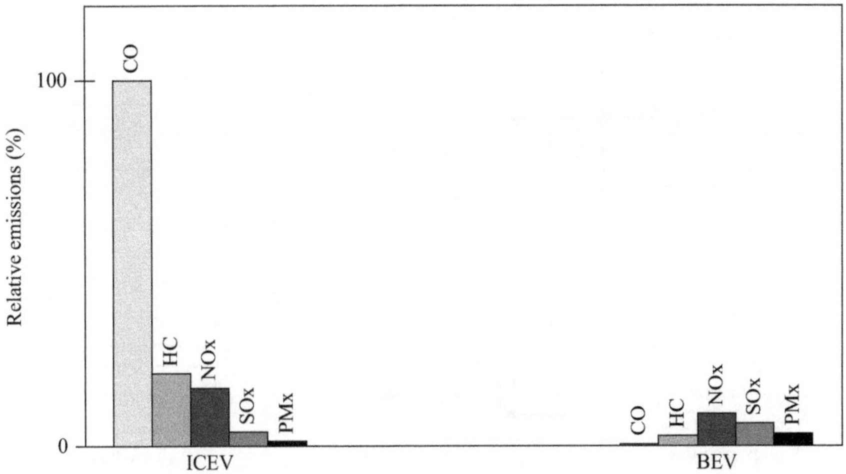


*Figure 1.3 Comparison of harmful emissions between BEV and ICEV*

with higher percentages of clean or renewable power generation, the use of EVs can help reduce the carbon emission. On the contrary, when adopting higher percentages of fossil fuel for electricity generation, the use of EVs may even worsen the carbon emission. For instance, the average carbon emission of EVs in the United Kingdom is about 80 g/km based on the fuel mix of 38% coal, 28% natural gas, 21% nuclear, 11% renewable and 2% other. Meanwhile, the average carbon emission of EVs in Norway is about 3 g/km and in China about 132 g/km, because Norway adopts over 98% hydroelectricity whereas China relies on over 66% coal-fired power generation. With the improvement of heat engine technology, European ICEVs have carbon emission reduced from 172 g/km to 153 g/km over the period from 2000 to 2008, and targeted to reach 130 g/km by 2015. Thus, the use of EVs to improve carbon emission for road transportation desires the use of more clean or renewable power generation; otherwise, the improvement may even be negative when adopting more coal-fired power generation. Consequently, the Chinese government has set a target in its 12th Five-Year Plan to increase the clean or renewable power generation to 15% and reduce the coal-fired generation to 63% by 2020.

ICEVs adopt various liquid fuels or gaseous fuels as their energy carriers which are mainly derived from oil and natural gas, respectively. Differing from ICEVs, PEVs take the definite advantage that they adopt electricity or hydrogen as their energy carriers which can be derived from almost all kinds of energy resources. Figure 1.4 depicts the energy flow from various energy resources including the oil,

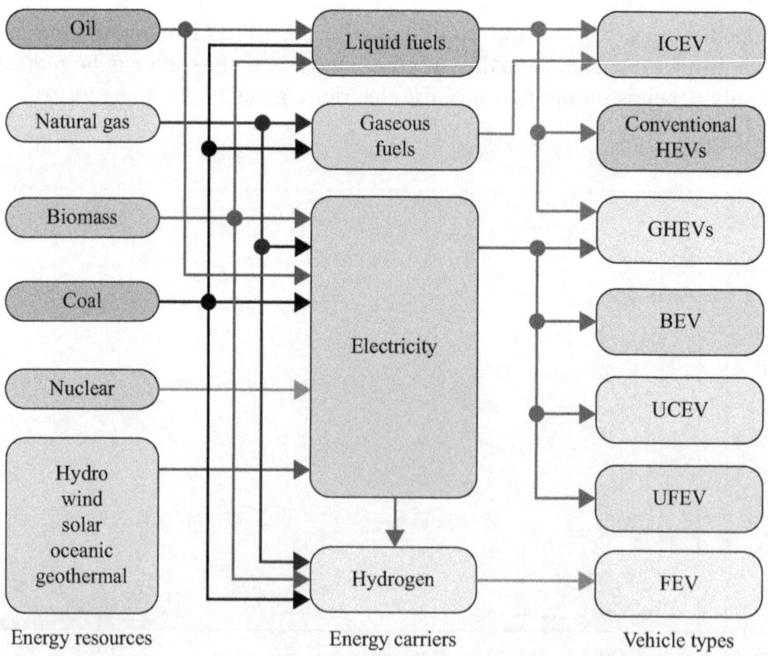


*Figure 1.4 Comparison of energy diversification between various EVs and ICEV*

natural gas, biomass, coal, nuclear, hydro, wind, solar, oceanic, geothermal to various vehicle types including the ICEV, conventional HEVs, GHEVs, BEV, UCEV, UFEV and FEV via various energy carriers including the liquid fuels, gaseous fuels, electricity and hydrogen. It illustrates that EVs take the definite merit of energy diversification. Among various types of EVs, the GHEVs are most energy diversified because they can be refuelled by two major energy carriers – liquid fuels and electricity which have excellent infrastructure support.

While EVs have the definite advantage of energy diversification over the ICEV, they can offer another advantage of higher energy efficiency. Figure 1.5 compares the tank-to-wheels energy efficiencies of the BEV, FEV and ICEV where their tanks of fuel refer to electricity, hydrogen and petrol, respectively. It can be observed that the tank-to-wheels energy efficiency of the BEV is much better than the others, namely, 1.8 and 4.5 times higher than that of the FEV and ICEV, respectively. This comparison may be a bit unfair because the major loss of the BEV occurs at the power generation plant which is not included in the calculation of its tank-to-wheels energy efficiency, whereas the major losses of the FEV and

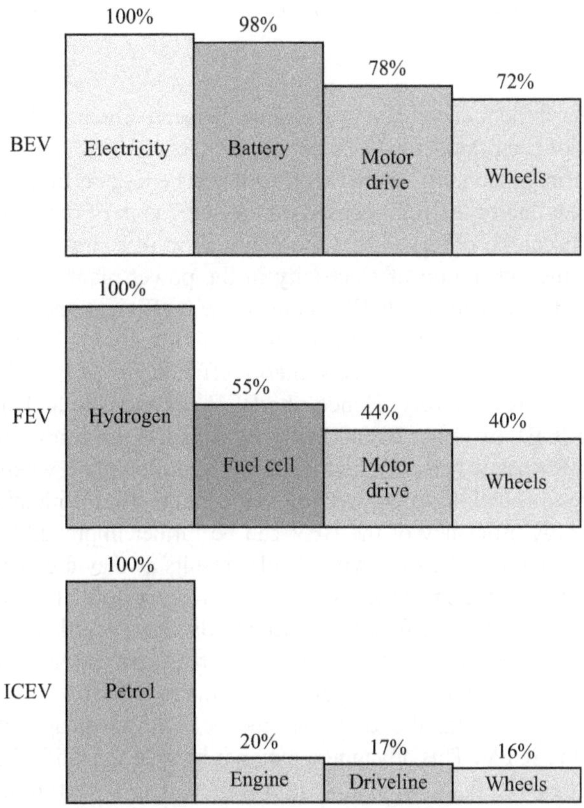


*Figure 1.5 Comparison of tank-to-wheels efficiencies of BEV, FEV and ICEV*

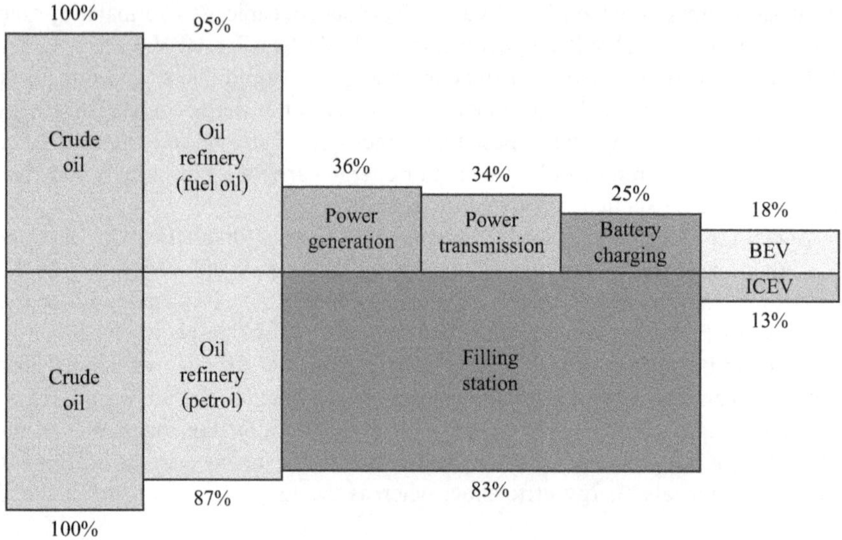


*Figure 1.6 Comparison of well-to-wheels efficiencies of BEV and ICEV*

ICEV that occur at the fuel cell and heat engine, respectively, have been included in the calculation of their tank-to-wheels energy efficiencies.

For a fair comparison, the overall well-to-wheels energy efficiency of the BEV is compared with that of the ICEV based on the same tank of crude oil as depicted in Figure 1.6. It can be observed that the major drop in energy efficiency for the BEV is due to the generation of electricity in the power plant whereas the major loss in energy efficiency for the ICEV is due to the inefficient operation of the heat engine. Since the electric motor can offer much higher energy efficiency than the heat engine, the overall well-to-wheels energy efficiencies of the BEV and ICEV are 18% and 13%, respectively. Hence, the BEV can provide higher energy efficiency than the ICEV by about 38%. Moreover, the BEV can readily regenerate the braking energy for battery charging during braking or downhill, whereas the ICEV simply dissipates this braking energy as heat in the brake discs. With this regenerative braking, the energy efficiency of the BEV can be further improved by about 10%. Meanwhile, the UCEV and UFEV have similar results as they are also fed by electricity. On the other hand, since the HEV family involves both of the electric motor and heat engine, it offers different well-to-wheels energy efficiencies which lie between that of the BEV and the ICEV, with a decreasing trend from the REV, then the PHEV, the full hybrid, the mild hybrid to the micro hybrid.

There is no doubt that the BEV has the definite advantage of higher fuel economy than the ICEV. This advantage may not be able to justify that the BEV is much more expensive than the ICEV, probably two to four times, because the battery cost dominates the overall cost. Nevertheless, due to use of lower battery capacities, the conventional HEVs are well justified to improve the fuel economy.

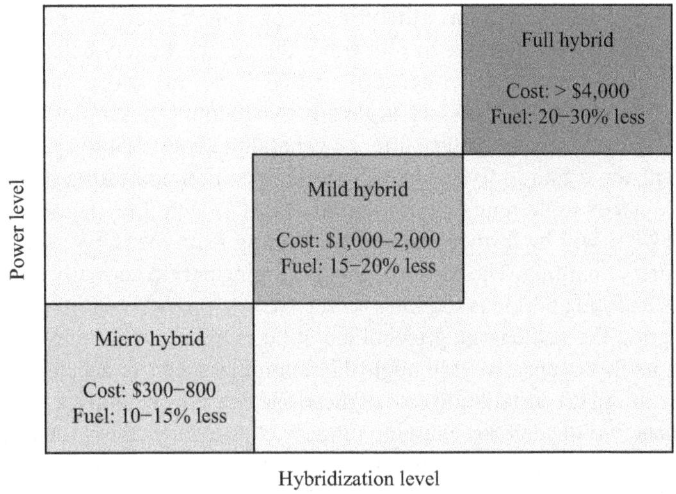


*Figure 1.7 Comparison of cost premium and fuel economy of conventional HEVs*

Among them, they need different price premiums (a few hundred to several thousand dollars) so as to achieve 10–30% improvement in fuel economy. As depicted in Figure 1.7, the micro hybrid takes the advantage of cost-effectiveness in fuel economy, the full hybrid enjoys the highest fuel economy, and the mild hybrid offers a compromise between the cost premium and fuel economy.

Compared with the BEV and conventional HEVs, the GHEVs have the following definite advantages:

- They are equipped with a petrol tank so that the problem of range anxiety can be avoided which generally occurs at the BEV.
- They can be externally charged by the power grid which is more effective and efficient than internal charging in the conventional HEVs.
- They can offer a decent electric range with zero-emission operation which cannot be provided by the conventional HEVs.

## 1.3 Challenges

### *1.3.1 Pure electric vehicles*

The pure electric vehicle (PEV), loosely termed the EV, consists of the BEV, UCEV, UFEV and FEV types in which their major energy sources are the battery, UC, UF and fuel cell, respectively. Because of different features and properties of these energy sources, they have their individual challenges.

The BEV is the most mature PEV, and is the only one commercially available. Its major challenges are the limited driving range, high initial cost and lack of charging infrastructure. Currently, a typical BEV can only travel about 120 km per charge while a typical ICEV can run over 500 km per refuel. With such a short driving range, the BEV will have the problem of range anxiety. Although there are some BEV models having three to four times the general battery capacity to extend the driving range comparable with that of the ICEV, they are two to three times more expensive than a typical BEV which is already two to four times more expensive than an ICEV. Moreover, the battery life is generally about 1,500 cycles which is equivalent to about four to five years of vehicle operation, indicating that the battery of the BEV needs to be renewed in the midway of vehicle life. Thus, the effective cost of the BEV will be further increased. Differing from the ICEV, the BEV takes time for battery charging. The corresponding charging period normally ranges from 5 to 8 h. This charging period is too long for the BEV to provide continuous operation. When adopting the fast charging technique, it takes about 20–30 min to charge the battery up to 80% capacity. Although this charging speed is acceptable for continuous operation, the installation cost of these fast charging stations is high while the establishment cost of such fast charging infrastructure is huge. Also, the fast charging process inevitably burdens the peaking power of the power system. In case the BEV allows for battery swapping, it takes only a few minutes to mechanically refuel the battery. However, the necessary space for each swapping station is much larger than charging stations. In addition, it involves two great challenges, namely, the battery size and location inside the BEV have to be standardized, and the single ownership of the battery of those BEVs needs a new business model.

The FEV, loosely termed the fuel-cell vehicle, offers the same environmental advantages as the BEV. Additionally, it can offer the driving range comparable to that of the ICEV. Its major challenges are the high initial cost and lack of hydrogen refuelling infrastructure. The high initial cost is due to the use of expensive fuel cells. The hydrogen refuelling infrastructure is generally absent in our society, and the establishment of such infrastructure involves a huge investment cost. There are three practical ways to store hydrogen in the FEV, namely, the compressed hydrogen gas (CHG), liquid hydrogen (LH) and metal hydride (MH). When adopting the CHG for the FEV, the infrastructure is similar to that of compressed natural gas for some alternative fuel vehicles. When adopting the LH, the infrastructure is very demanding since the hydrogen needs to be cooled to about −253 °C while still pressurized. When adopting the MH, it needs to have a similar infrastructure as battery swapping to mechanically replace the discharged MH with the fully charged MH. Both the CHG and LH enjoy the merit of high specific energy which is desirable for the FEV, but face the same safety concern which can be an explosion hazard. In contrast, the MH takes the merit of safety which is essential for the FEV, but suffers from the problem of low specific energy which deteriorates the driving range. In the coming future, the commercialization of the FEV depends on whether the cost of fuel cells can be significantly reduced and whether there will be a policy to establish the hydrogen refuelling infrastructure.

The UCEV and UFEV are similar to the BEV because they are all fed by electricity. Although the UC and UF take the definite advantage of practically unlimited cycle life, they suffer from two fundamental challenges: very expensive and low specific energy. These challenges essentially limit their application as a sole energy source; otherwise, the UCEV and UFEV have to be equipped with many UCs and UFs, respectively, or need to be frequently recharged. In addition, the UFEV has been challenged that the UF might exhibit flywheel explosion in a car accident.

### *1.3.2 Hybrid electric vehicles*

The hybrid electric vehicle (HEV), loosely termed the HV, consists of the conventional HEV types and the GHEV types, which depend on whether their batteries are only recharged internally by the heat engine or can also be recharged externally by the power grid. The conventional HEVs include the micro, mild and full hybrids, while the GHEVs include the PHEV and REV. Various HEVs have their individual challenges.

For the micro hybrid, it serves to provide two basic hybrid features: the idle stop-start and regenerative braking. The key challenge is how to reduce the cost premium to realize these basic hybrid features. For the full hybrid, it aims to provide all hybrid features: the idle stop-start, regenerative braking, power assist and electric launch. Its key challenges are how to reduce the system complexity which involves both the electric motor and heat engine for propulsion, and how to coordinate these two propulsion devices to achieve optimal efficiency operation. Also, it generally suffers from the reliance on planetary gearing which involves transmission loss, gear noise and regular lubrication. For the mild hybrid, it can provide three hybrid features: the idle stop-start, regenerative braking and power assist. Nevertheless, its role is relatively vague because the desire of simple configuration and low-cost premium will certainly go for the micro hybrid whereas the demand of versatile operation and efficiency optimization will prefer to select the full hybrid.

The key challenges of the GHEVs are the system complexity and initial cost. Their system complexity is similar to that of the conventional HEVs, mainly due to the use of both the electric motor and engine. Differing from the conventional HEVs, they need to install the on-board charger to plug in the power grid for battery charging. Their initial cost is much higher than that of the conventional HEVs due to the use of a large number of batteries for the pure electric mode. Of course, when the PHEV operates at the blended mode or the REV operates at the extended mode, they lose the merit of zero roadside emissions.

## 1.4 Multidisciplinary technologies

Although EVs have definite environmental advantages over the ICEV, there are many challenges that need to be solved. The key is to advance our technologies for EVs (Chau, 2012). EVs are an integrated system involving multidisciplinary technologies such as electrical engineering, mechanical engineering and chemical engineering. The electrical discipline includes the electric machine, power converter, electronic controller, energy management unit, energy refuelling unit, auxiliary power supply, temperature control unit, power steering unit and so on. The mechanical discipline includes the car body, heat engine, mechanical transmission, chassis system, suspension system, braking system, steering mechanism and so on. The chemical discipline includes the battery, fuel cell, liquid fuel, gaseous fuel and so on.

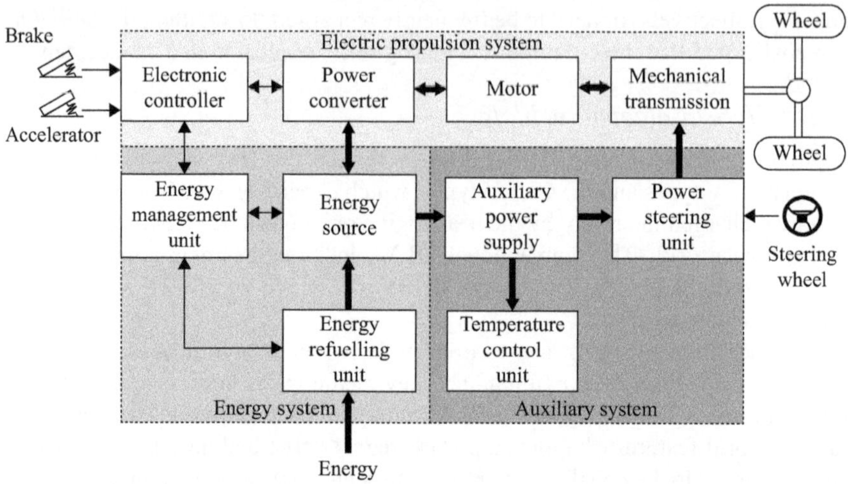


*Figure 1.8 System configuration of PEV*

Figure 1.8 depicts the system configuration of a typical PEV where the thick and thin arrows denote the power and signal flows, respectively, which can be split into three subsystems: energy system, electric propulsion system and auxiliary system. Among them, the energy system is most actively developed in recent years where there are many innovations and advancements in the areas of energy sources, energy management and energy refuelling. The development trends are to reduce the energy cost or increase the cost-effectiveness, to improve the energy storage or generation capacity, and to automate the recharging or refuelling process. Meanwhile, the electric propulsion system is also actively developed in recent years, focusing on the advancement of electric motors, power converters and electronic controllers (Chau, 2015). It aims to further improve the efficiency, torque density, power density and controllability of electric motor drives (Chau, 2009). In addition, the auxiliary system is developed with emphasis on the areas of auxiliary power supply, power steering and temperature control, aiming to reduce the power consumption of vehicular electronics and enhance the vehicular manoeuvrability. This auxiliary system can readily be extended to the conventional ICEV.

Figure 1.9 depicts the system configuration of a typical HEV, namely, the PHEV or REV, where the external energy includes the electricity and petrol, the energy refuelling unit consists of the battery charger and petrol pump, and the internal energy source is the battery. The electronic controller, the energy management unit and the whole auxiliary system that are common with the PEV are ignored. Different kinds of HEVs have similar system configurations but with some modifications. For instance, the electricity source and battery charger are absent in the full hybrid, whereas the electricity source, battery charger and generator are all absent in the mild hybrid. It can be observed that both the energy system and propulsion system of the HEV are more complicated than those of the PEV. Particularly, the development

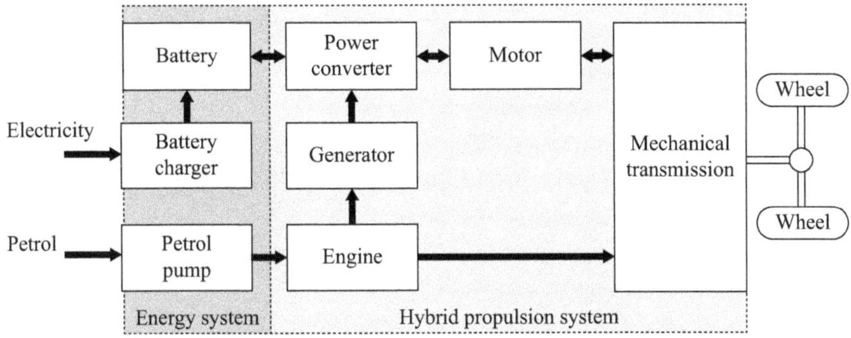


*Figure 1.9 System configuration of HEV*

trends are to optimally utilize the hybrid energy sources involving the battery and petrol (Chau and Wong, 2002), and to optimally coordinate the hybrid propulsive powers created by the motor and engine (Chau, 2015). Meanwhile, the corresponding auxiliary system including the auxiliary power supply, power steering unit and temperature control unit is essentially the same as that of the PEV.

## 1.5 Energy system technologies

The energy system is the core of EVs, including the PEV and HEV. The corresponding technologies are most actively developed, especially the energy source technology, the battery charging and management technology, and the latest vehicle-to-X (V2X) technology. The first one serves to reduce the energy cost and increase the energy capacity, the second one functions to facilitate the energy refuelling and enhance the energy management, and the third one aims to expand the role and increase the cost-effectiveness of EVs.

### *1.5.1 Energy source systems*

Energy sources are the core technology which provides on-board electrical energy for EVs. Currently, actually over the last two decades, there are four viable EV energy sources: electrochemical batteries (normally termed batteries), UCs, UFs and fuel cells (Chau *et al.*, 1999). The batteries are electrochemical devices which store electrical energy during charging and produce electricity during discharging. The UCs are essentially capacitors with ultrahigh capacitances which store and produce electrical energy by electrostatic means. The UFs are essentially electric machines spinning at ultrahigh speeds which store and produce electrical energy by electromechanical means. Namely, they work as motors during charging and serve as generators during discharging. Differing from the batteries, UCs and UFs that are energy storage devices with the capability of bidirectional energy conversion, the fuel cells are electrochemical energy generation devices which directly convert chemical fuels into electricity. As their energy conversion is unidirectional, they cannot accept electricity.

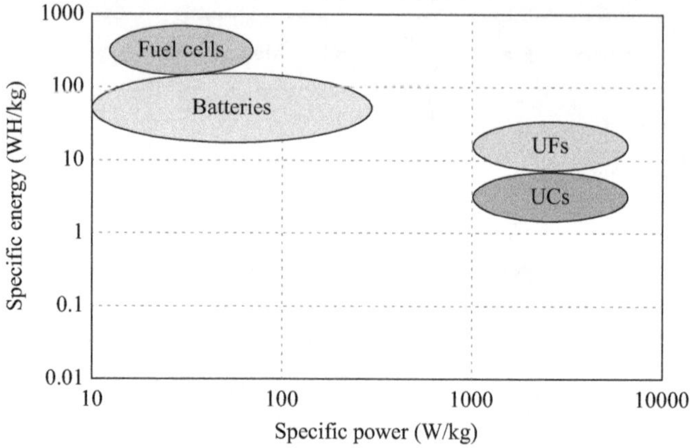


*Figure 1.10 Ragone chart of various energy sources*

A Ragone chart of those viable energy sources is shown in Figure 1.10, which indicates that none of them can simultaneously offer high specific energy and high specific power, analogous to none of athletes can simultaneously be suitable for marathon running and 100-m sprint. Mencius said, 'Fish is what I want; bear's palm is also what I want. If I cannot have both, I would rather take the bear's palm than fish' (Chan, 2002). For the BEV, the bear's palm is the specific energy which governs the driving range per charge while the fish is the specific power which determines the acceleration rate. In general, a compromise between the specific energy and specific power is necessary, but with more preference on the specific energy. If allowable in terms of cost, space and complexity, a hybridization of two energy sources (one with high specific energy and another with high specific power) can be realized (Chau and Wong, 2001; Wong and Chan, 2012). In particular, the FEV usually adopts the hybridization of fuel cells and batteries because the former can offer excellent specific energy and the latter can provide very good specific power. Also, the batteries can complement the fuel cells to accept the recovered electricity during regenerative braking. Alternatively, the hybridization can be realized within a battery to form the so-called hybrid battery (Li *et al.*, 2011).

In the foreseeable future, batteries are still the major energy source for EVs. Over the last two decades, there have been many types of batteries that have been developed for EVs. The viable EV batteries consist of the valve-regulated lead acid (VRLA), nickel-cadmium (Ni-Cd), nickel-zinc (Ni-Zn), nickel-metal hydride (Ni-MH), zinc/air (Zn/air), aluminium/air (Al/air), sodium/sulphur (Na/S), sodium/nickel chloride ($Na/NiCl_2$) and lithium-ion (Li-ion) types. Among them, the VRLA battery is accepted for low-cost low-end EVs, the Ni-MH battery is preferred for well-performed EVs, the Li-ion battery is attractive for high-performance EVs, and the Zn/air battery is unique for EVs using battery swapping. Figure 1.11 compares their performances in terms of the specific energy which dictates the driving range per charge, the specific power which governs the acceleration rate, the cycle life

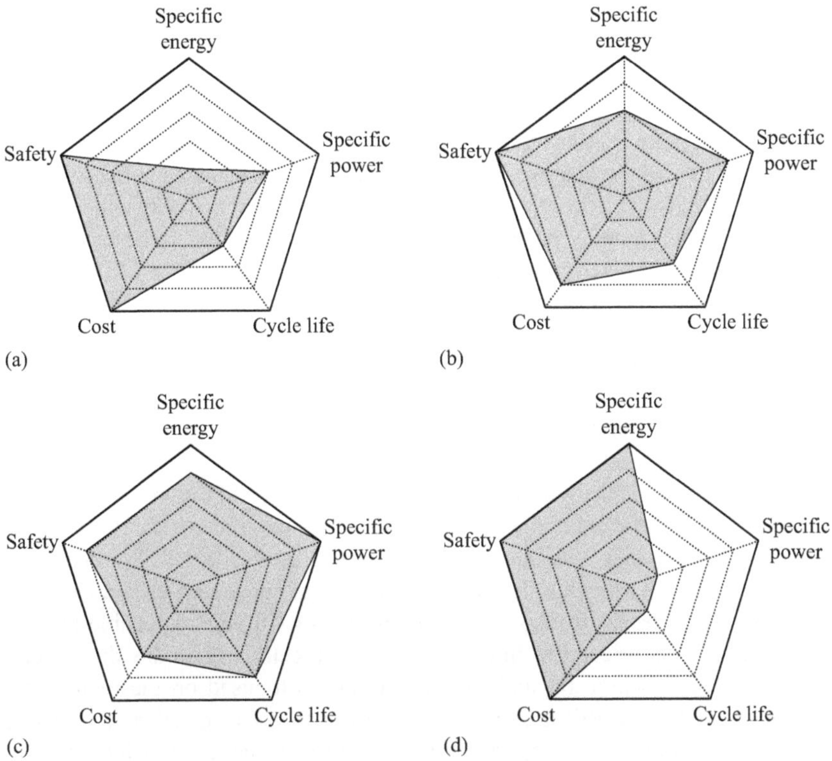


*Figure 1.11 Comparison of battery performances: (a) VRLA battery, (b) Ni-MH battery, (c) Li-ion battery and (d) Zn/air battery*

which represents the longevity, the cost which denotes the frugality, and the safety which indicates the risk of flame or explosion. It can be observed that the Li-ion battery is relatively most preferable. At the present status of battery technology, the BEV can only offer acceptable driving range with affordable price. In order to enable the BEV offering comparable price and driving range as the ICEV, the battery specific energy and cycle life need to be greatly increased while the battery cost needs to be significantly reduced. Currently, the main research on battery technology is being focused on the development of various Li-ion battery types (Omar *et al.*, 2012; Giuliano *et al.*, 2011), such as the lithium manganese oxide (LMO), lithium nickel manganese cobalt oxide (NMC), lithium iron phosphate (LFP), lithium nickel cobalt aluminium oxide (NCA) and lithium titanate (LTO). Meanwhile, another key research direction is to develop the lithium/air (Li/air) battery and lithium/sulphur (Li/S) battery (Christensen, *et al.*, 2012; Zhang, 2013).

The UC technology is promising for EVs since it offers exceptionally high specific power and practically unlimited cycle life. Nevertheless, the UC needs significant improvement before practically applicable as the sole energy source for EVs, namely, the specific energy has to be greatly increased and the initial cost has to be greatly reduced. Current research on UC technology is being focused on using graphene and carbon nanotubes to increase the usable surface area and hence the energy storage capacity (Liu *et al.*, 2010; Du and Pan, 2006). In addition, the UF technology exhibits potentiality for EVs. By providing vacuum environment to remove the air friction and magnetic bearings to eliminate the bearing loss, the flywheel can spin to ultrahigh speeds so as to achieve high specific energy, excellent specific power and high energy efficiency. However, it suffers from the problem of safety concern when the flywheel is accidentally damaged; otherwise, the casing has to be significantly thickened which will deteriorate the merit of high specific energy. Current research activities are focused on improving the reliability of the flywheel, and extending its application to off-board energy storage for EV charging stations where the whole UF can be embedded in the ground (Strasik *et al.*, 2007; Li *et al.*, 2015).

The fuel-cell technology is one of the most actively researched topics for EVs, which is the core technology for the FEV. The leading types of fuel cells include the direct methanol fuel cell (DMFC), alkaline fuel cell (AFC), proton exchange membrane fuel cell (PEMFC), phosphate acid fuel cell (PAFC), molten carbonate fuel cell (MCFC) and solid oxide fuel cell (SOFC). Among them, the PEMFC, also called the solid polymer fuel cell (SPFC), is most popular for the FEV because of its solid electrolyte nature, low-temperature operation, quick start-up, proper power level, high power density and good system efficiency. In order to enable the FEV offering affordable price, the initial cost of fuel cells has to be dramatically reduced. Current research on fuel-cell technology is focused on the reduction of platinum usage for the PEMFC which requires such noble metal as the electrocatalyst, and the reduction of operating temperature for the SOFC which does not desire a noble metal as the electrocatalyst (Martin *et al.*, 2010; Wang *et al.*, 2011).

Figure 1.12 compares the performances of the Li-ion battery, UC, UF and PEMFC in terms of the specific energy, specific power, cycle life (equivalent to service life for fuel cells), cost and safety. It can be observed that the Li-ion battery is still most preferable, holistically.

In recent years, the concept of on-board renewable energy systems has been very attractive for EVs. As illustrated in Figure 1.13, there are four viable renewable energies that can be harnessed:

- Braking energy
- Solar energy
- Waste heat energy
- Vibration energy

First, the conversion of braking energy into electrical energy is well known, dubbed the regenerative braking, which is widely used in all types of EVs. Whenever an EV performs braking for speed reduction, downhill or even cruising on highways, the propulsion motor will work as a generator to produce electricity for battery charging. It should be noted that the regenerative braking needs to cooperate with the traditional hydraulic braking so that good pedal feel and safe braking can be ensured. Second, similar to the Solar Impulse (the first solar-powered plane), an EV can

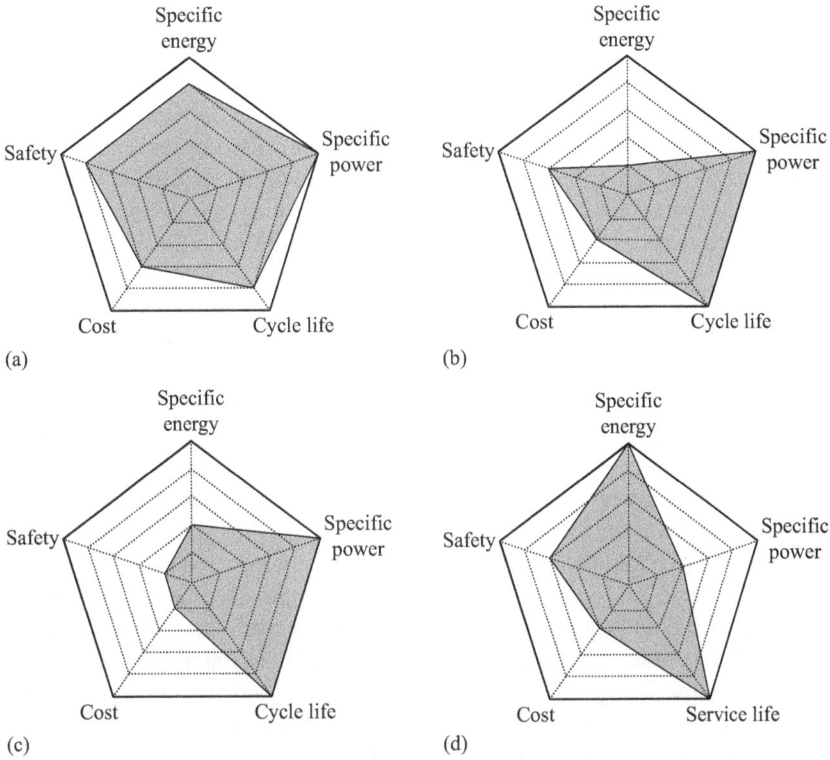


*Figure 1.12 Comparison of energy source performances: (a) Li-ion battery, (b) UC, (c) UF and (d) PEMFC*

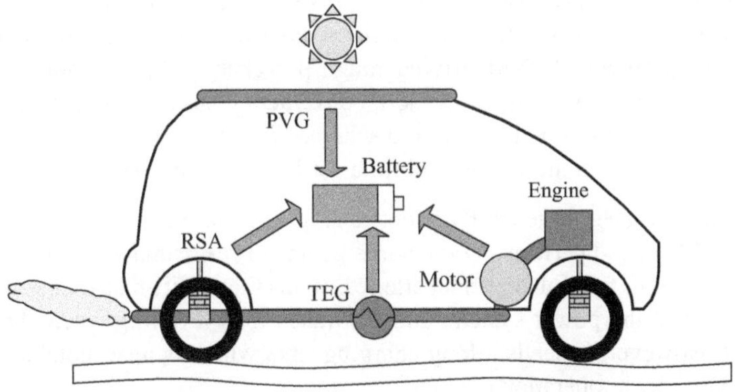


*Figure 1.13 On-board renewable energy systems*

mount the photovoltaic generator (PVG) or called the solar panel on the roof so that solar energy can readily be harnessed to help charge the batteries. While monocrystalline and polycrystalline silicon modules have been used in most of the PVG currently available, copper indium gallium diselenide modules are promising thin-film technology to improve the flexibility, increase the efficiency and reduce the cost (Lam *et al.*, 2004). Differing from the Solar Impulse which can provide a very large surface area to capture sufficient solar energy for travelling, an EV can only harness a few hundred watts from solar radiation. Third, since the fuel efficiency of the heat engine of HEVs is only around 25% and about 40% is lost in the form of heat energy in exhaust gas, the thermoelectric generator (TEG) can be mounted at the exhaust pipe to recover the waste heat and help charge the batteries (Yu and Chau, 2009). Similar to the PVG, the TEG can harness a couple of hundred watts from the waste heat of an HEV. Of course, this type of on-board renewable energy is absent in the PEV. It should be noted that both the PVG and TEG exhibit nonlinear characteristics at different irradiances and temperatures, respectively, the power converter with maximum power point tracking should be incorporated to maximize energy harvesting (Zhang *et al.*, 2010; Zhang and Chau, 2011). Fourth, as the vehicle suspension serves to damp the vibration disturbance from road roughness and vehicle dynamics for better ride comfort and road handling, shock absorbers are widely used to absorb the vibration energy and dissipate it as waste heat. The regenerative shock absorber (RSA), which is generally an electromagnetic generator, is able to not only perform the vibration damping, but also harness the vibration energy for battery charging (Li *et al.*, 2013c). Typically, an EV with four RSAs can collect a few hundred watts when running on a fairy smooth road. In order to improve the efficiency and reduce the size, some advanced permanent-magnet linear generators are developed, such as the vernier permanent-magnet topology (Li *et al.*, 2013a) and transverse-flux permanent-magnet topology (Li *et al.*, 2013b), which are promising to serve as the RSA.

### *1.5.2 Battery charging and management systems*

In the foreseeable future, batteries are the main energy source of the PEV. Consequently, in recent years, many researchers have developed various methods to alleviate the problem of short driving range per charge. Apart from developing better batteries, the development of more convenient chargers has attracted a lot of attention. There are four major charging schemes for the BEV, which vary with the allowable voltage, current, power and time for battery charging:

- *Normal charging*. It generally adopts a single-phase AC, 110–240 V, 13–20 A and 2–4 kW, and performs at the home garage or residential car park. It usually takes 5–8 h to fully charge the batteries. It has the merits that it does not cause a burden to the power system, and facilitates load levelling when charging at night. However, it needs a long charging time which is unacceptable for continuous vehicle operation.
- *Opportunity charging*. It typically uses a three-phase AC, 110–240 V, 32–80 A and 8–20 kW, and conducts at the public car park whenever possible. It does

not care too much about how many times of charging per day and how long of the duration per charge, provided that the batteries have not been fully charged. It is a good choice for extended shift operations without causing a significant burden to the power system.

- *Fast charging*. It generally adopts DC, 200–450 V, 80–200 A, 36–90 kW, and performs at the dedicated fast charging station. It only takes 20–30 min to charge 80% of the battery usable capacity. However, it causes an additional burden to the power system, especially during daytime peak hours. Also, it needs dedicated hardware and safety precaution.
- *Battery swapping*. It mechanically swaps the discharged and charged batteries within 5 min so that the discharged batteries can be recharged by normal charging during off-peak hours. As aforementioned, it suffers from the problem of standardization of batteries used for various BEV models, and the desire of a large area to conduct battery swapping.

On the basis of these four charging schemes, there are various charging algorithms such as the constant-current charging, constant-voltage charging, multistage charging, pulse charging and trickle charging for various batteries, aiming to improve the charging efficiency, reducing the charging time and enhancing the battery life (Shen *et al.*, 2012). In addition, the charging process can be further improved by searching the optimal charging current pattern and terminating the process based on state-of-charge (SOC) estimation (Vo *et al.*, 2015). On the other hand, all these four charging schemes can be performed via wire or wireless. It is obvious that the use of conductive wires for such power transfer is simple, low cost and high efficiency, but inevitably suffers from the lack of convenience and risk of electrocution. In contrast, the use of wireless technology for battery charging takes the definite advantages of convenience and free from electrocution, but suffers from inevitable power transfer loss and high installation cost (Hui *et al.*, 2014).

In recent years, the development of wireless power transfer technology has been accelerated, including the far-field strategies such as the microwave radiation and laser, and the near-field strategies such as the capacitive power transfer and inductive power transfer (IPT). Among them, the IPT has been identified to be the most viable technology for EV battery charging (Qiu *et al.*, 2014a; Musavi and Eberle, 2014). The basic principle of inductive charging for EVs is illustrated in Figure 1.14, which is based on the magnetic coupling between two coils of a high-frequency transformer. One of the coils is installed in the charger coupler while the other is embedded in the vehicle inlet. Firstly, the main AC supply with a frequency of 50 or 60 Hz is rectified and converted to a high-frequency AC power of about 80 kHz within the charger module, then the high-frequency AC power is transferred to the EV side by induction, and finally this AC power is converted into DC power for battery charging. This IPT can operate over a wide frequency range, and may readily be scaled up to meet various power levels for EV charging.

In order to facilitate the park-and-charge process for EVs, the IPT technology is extended to be plugless, in which the primary coil is installed on the floor of a garage or parking lot and the secondary coil is installed in the vehicle. The driver needs no

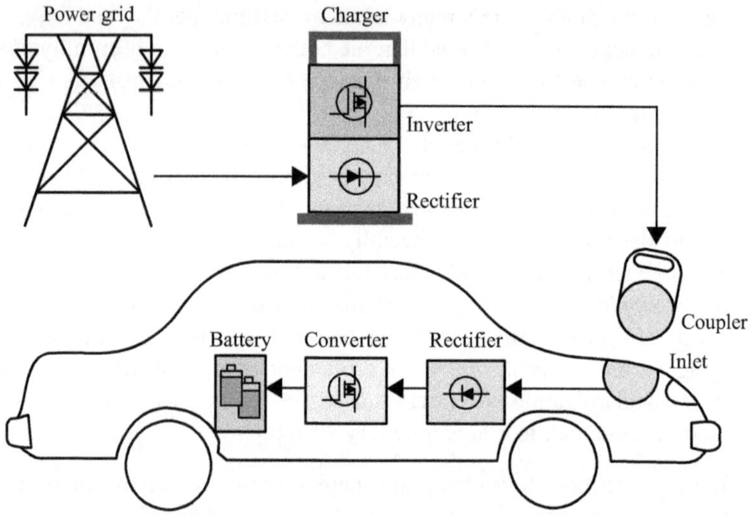


*Figure 1.14 Inductive charging for EVs*

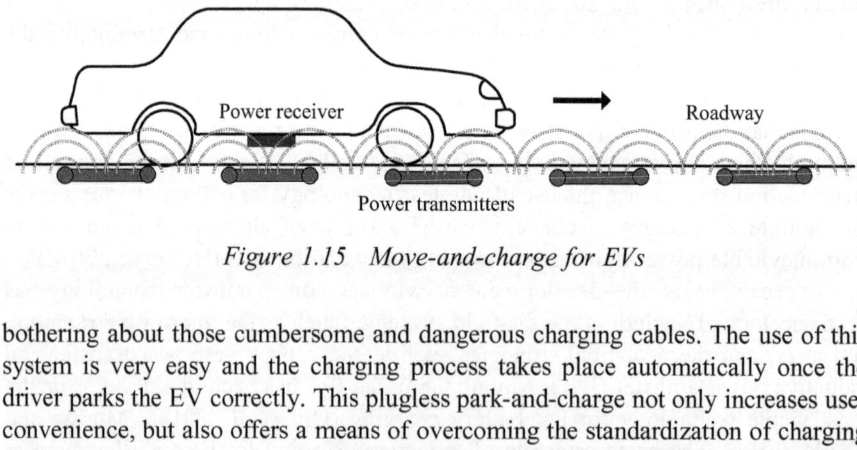


*Figure 1.15 Move-and-charge for EVs*

bothering about those cumbersome and dangerous charging cables. The use of this system is very easy and the charging process takes place automatically once the driver parks the EV correctly. This plugless park-and-charge not only increases user convenience, but also offers a means of overcoming the standardization of charging plugs. Due to the existence of a large air-gap or clearance between the primary and secondary coils, the magnetic resonant coupling technique is usually adopted in which the primary and secondary resonant coils having the same resonant frequency can wirelessly transfer power efficiently with high power density, while dissipating relatively little energy in non-resonant objects such as vehicle bodies or drivers (Cannon *et al.*, 2009; Imura and Hori, 2011; Zhang *et al.*, 2014a).

Rather than stopping or parking, EVs prefer to be wirelessly charged during moving. Namely, an array of power transmitters are embedded beneath the roadway (so-called the charging zone or lane) while a receiver is mounted at the bottom of an EV as depicted in Figure 1.15. This move-and-charge technology has high potentiality to fundamentally solve the long-term problems of the BEV – high initial cost and short driving range. Namely, there is no need to install so many batteries in the BEV, hence dramatically cutting its initial cost; and the BEV can be conveniently charged at the charging zone during driving, hence automatically extending the driving range (Huh *et al.*, 2011). However, there are many challenges to be tackled before the realistic application to EVs. First, the efficiency of wireless power transfer heavily depends on the vertical distance and horizontal misalignment between the transmitter of the array and the receiver of the vehicle (Qiu *et al.*, 2014b). Since such distance and misalignment are inevitably time-varying and significantly affected by the road condition and vehicle payload, the resonant frequency of wireless power transfer is not constant, termed the resonance shifting. Thus, the power converter that excites the power transmitter needs to be dynamically controlled to maintain high-efficiency power transfer. Second, the effectiveness of move-and-charge operation heavily depends on the coverage of wireless power transfer as well as the position and speed of vehicles running on the charging zone (Zhang and Chau, 2015). The location of power transmitters needs to be optimized in such a way that the electromagnetic field intensities at different locations over the charging zone are uniform. Third, as there are many EVs running on the electrified roadway, the move-and-charge operation needs to distinguish which EVs are authorized to retrieve wireless power or to prevent unauthorized vehicles from stealing the energy (Zhang *et al.*, 2014b, 2015).

At the present status of battery technology, the battery storage capacity is quite limited so that the battery management system (BMS) is highly desirable. Figure 1.16 depicts the functional blocks of BMS for EVs. Essentially, on the basis of the real-time measured battery parameters, namely, the voltage, current and temperature ($V$, $I$, $T$), the central processor will perform the cell balancing and thermal

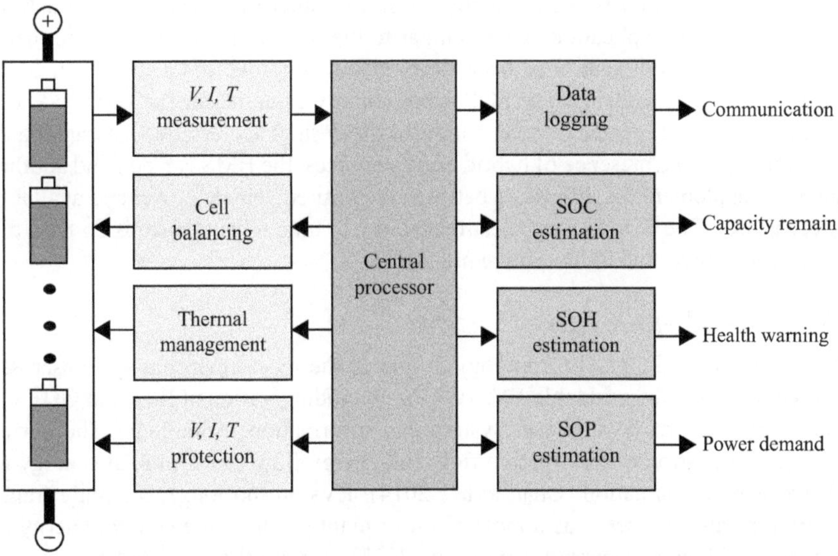


*Figure 1.16 Battery management system for EVs*

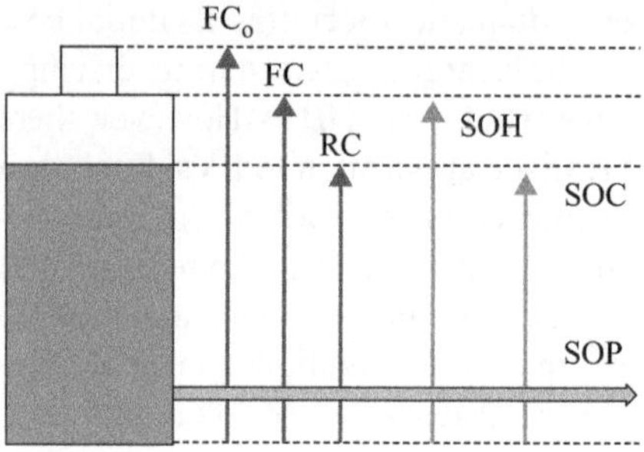


*Figure 1.17 Battery informative parameters*

management of the batteries as well as the overvoltage, overcurrent and overheat protections. Meanwhile, it will log the data for further communication, and estimate three important informative parameters, namely, the SOC, state-of-health (SOH) and state-of-power (SOP), for the EV driver. As shown in Figure 1.17, the SOC is defined as the ratio of the remaining capacity (RC) to the full-charge capacity (FC); the SOH is defined as the ratio of FC to the original full-charge capacity ($FC_0$); and the SOP refers to the capability to provide power under voltage and/or current limits. The SOC is particularly important because it directly provides instant information to the driver the amount of energy that is still available for use by the EV. The corresponding range remaining prediction relies greatly on the SOC. Consequently, many sophisticated methods have been developed to improve the accuracy of SOC estimation (Shen *et al.*, 2002, 2005; Chau *et al.*, 2003, 2004; Cheng *et al.*, 2011). The SOH is related to the loss of battery capacity due to charge–discharge cycling and aging. When the SOH drops below a threshold (typically 70%), the battery is considered to reach its end of life for EV application. Thus, similar to the SOC estimation, there are many sophisticated methods that have been developed to improve the accuracy of SOH estimation (Hu *et al.*, 2014). The SOP is particularly essential for the BMS to know whether the available battery power can fulfil the desired acceleration, overtaking or hill climbing. In the presence of hybrid energy sources, the BMS can demand another source to supplement the difference between the desired vehicle power and available battery power. The SOP has to be estimated in real time which depends on both the instantaneous SOC and SOH (Sun *et al.*, 2014).

### *1.5.3 Vehicle-to-X energy systems*

The vehicle-to-grid (V2G) technology is one of the most emerging system-crossover technologies for griddable EVs (GEVs), including various PEVs and GHEVs. It is a crossover of EVs, power system and information technology. The corresponding integration of information flow and energy flow can stimulate synergy to achieve win–win situation (Chan *et al.*, 2014). EVs are no longer a simple transportation means, but serve as a mobile power plant generating electrical energy to the power grid when necessary (Liu *et al.*, 2013). Among those major EV types, as shown in Figure 1.18, the BEV and GHEVs are fully griddable where bidirectional

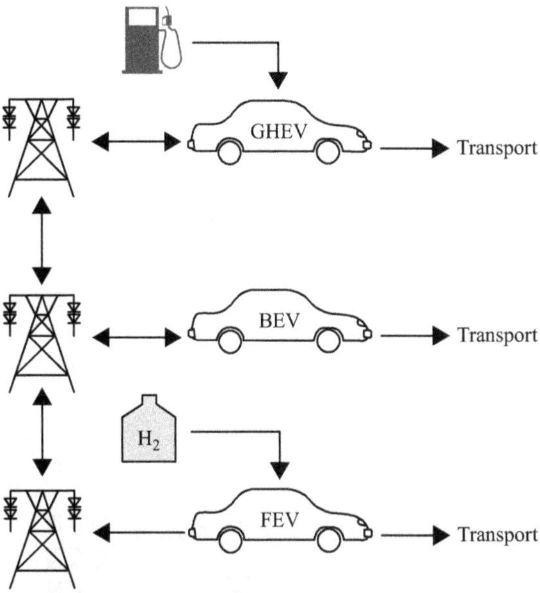


*Figure 1.18 EVs for V2G*

power flow between the vehicle and grid can be realized whereas the FEV is partially griddable where there is unidirectional power flow only from the vehicle to the grid. Thus, the BEV and GHEVs (including the PHEV and REV) are considered as fully functional GEVs for V2G operation, whereas the FEV is generally excluded from such operation.

The essence of V2G operation is that EVs communicate with the power grid to sell services by delivering electricity into the grid or by controlling the charging rate for GEVs. Since most GEVs are parked with an average of 95% of the time, their batteries can be used to let electricity flow between the vehicles and the grid. When there is a reasonable penetration rate of GEVs (such as 20–40% vehicles are griddable) and each GEV can store or generate electrical energy of 4–80 kWh, the V2G operation will have a significant impact on power system. Economically, the V2G service will be a new business, namely, the energy arbitrage between the power utilities and the GEV drivers.

Since a GEV can only store a limited capacity, from 4 kWh in a typical PHEV to 85 kWh in a high-performance BEV, individual V2G operation of each GEV with the power grid is ineffective and inefficient. So, an aggregator is necessary which is responsible for gathering a number of GEVs (Guille and Gross, 2009). The aggregator controls a collection of homes performing vehicle-to-home (V2H) operation or a building with many parking lots performing vehicle-to-building (V2B) operation, and then communicates with the power grid to perform V2G operation. Based on the willingness of drivers and the battery capacity of GEVs, the aggregator controls proper GEVs to achieve smart charging and discharging

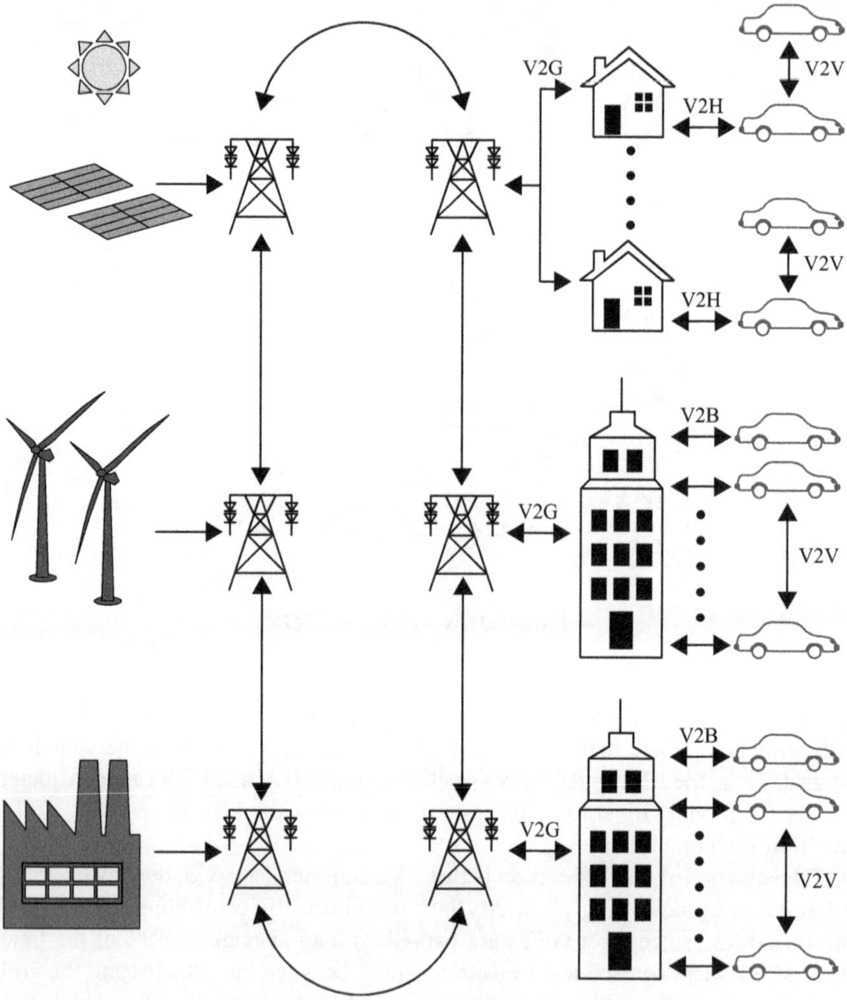


*Figure 1.19 V2X system framework*

(Chan *et al.*, 2014). Moreover, the corresponding energy arbitrage can be performed internally, so-called the vehicle-to-vehicle (V2V) operation, by the aggregator via V2H or V2B. For instance, when some BEVs preferably perform fast charging for instant usage and some PHEVs preferably sell electricity for profit making, the aggregator can realize internal V2V operation for energy arbitrage. The whole V2X system framework, including the V2G, V2H, V2B and V2V, is depicted in Figure 1.19 in which the aggregator functions to aggregate the GEVs to deal with the energy service provider to buy power and energy, as well as deal with the independent system operator or regional transmission organization to sell power and energy, thus maximizing the overall profit of V2X operation.

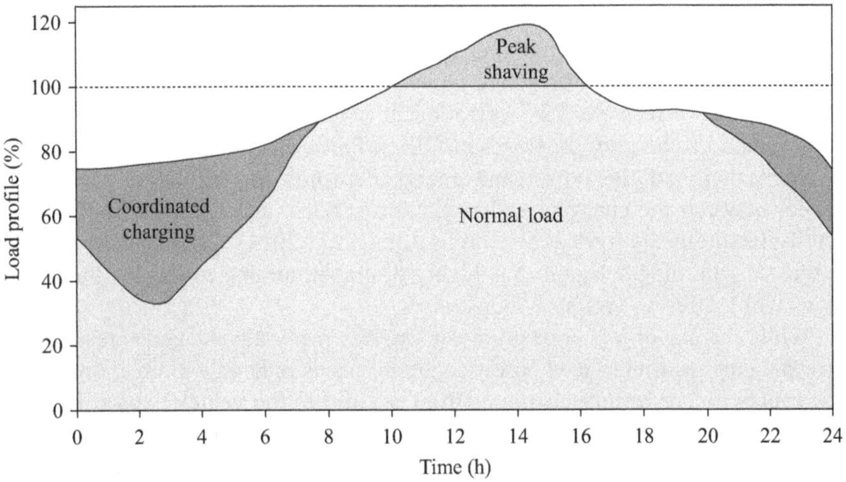


*Figure 1.20 V2G coordinated charging and peak shaving*

The V2G operation has been identified to have various potential applications such as charging coordination, peak shaving, active regulation, spinning reserve, motor starting, reactive regulation and renewable transients (Ehsani *et al.*, 2012). Since the power generation capacity has to match with the load demand, a large fluctuation of load demand will significantly increase the capital cost and operating cost of the power system. As shown in Figure 1.20, the V2G operation preferably performs EV battery charging to absorb or buy electrical energy from the grid during the off-peak period (so-called coordinated charging), and to generate or sell electrical energy to the grid during peak period (so-called peak shaving). Also, the corresponding charging and discharging processes of EVs are much faster than the shutoff and start-up processes of standby generators (Wu *et al.*, 2010, 2012). However, at the present status of battery technology, the EV batteries still suffer from a limited cycle life. Deep discharging of EV batteries for peak shaving will inevitably degrade their life for normal vehicular operation (Gao *et al.*, 2011).

Since renewable power generations such as wind power and solar power are intermittent in nature, the use of standby generators to back up the intermittent power outage is expensive, inefficient and sluggish. Although the battery energy storage system can provide the desired efficient and fast backup, it is too expensive and bulky. The V2G operation can fully utilize the EV batteries to complement the intermittent outage of power grid with renewables (Gao *et al.*, 2012, 2014). Differing from peak shaving, the renewable transients will not cause deep discharging of EV batteries or noticeable degradation of battery life.

Apart from using EV batteries for V2G operation, the DC-link capacitors of the off-board bidirectional EV chargers can be fully utilized to perform reactive power compensation for the power grid. In general, the DC-link capacitors are sufficient enough to supply reactive power to the grid even without engaging the EV batteries. Thus, it causes no degradation on the battery life. This feature makes the EV chargers very attractive for V2G reactive regulation (Kesler *et al.*, 2014). Similar to the V2G active regulation, an aggregation of EV chargers is necessary to provide the desired level of reactive power compensation.

In order to effectuate V2G operation, it requires sophisticated control of the charging and discharging processes of GEVs. Particularly, there are challenges of the stochasticity of the power and energy demands for regulation service, the conflict between the charging and regulation service, and the computational and security requirements for V2G operation (Lin *et al.*, 2014). Consequently, optimal charging/discharging scheduling is highly desirable, aiming to resolve these challenges while offering various V2G services.

While the use of V2G operation can provide many advantageous features, the corresponding grid-interfaced power converter needs to handle bidirectional power flow capability for battery charging (from the grid to the vehicle) and battery discharging (from the vehicle to the grid). Consequently, harmonic contaminations to both the grid and the load are inevitable. In order to provide effective power control and compensate the harmonics simultaneously, sophisticated control and filtering are necessary (Zou *et al.*, 2014).

## Acknowledgements

The author is grateful to all members of Electric Vehicle Technologies research group for their contributions to the preparation of this chapter. He expressed his indebtedness to Joan Wai-Yi and Aten Man-Ho for their hearty support all the way.

## References

Cannon, B.L., Hoburg, J.F., Stancil, D.D., and Goldstein, S.C. (2009) Magnetic resonant coupling as a potential means for wireless power transfer to multiple small receivers. *IEEE Transactions on Power Electronics*, 24, 1819–1825.

Chan, A.K.L. (2002) *Mencius: Contexts and Interpretations*, Honolulu: University of Hawaii Press.

Chan, C.C. and Chau, K.T. (2001) *Modern Electric Vehicle Technology*, Oxford: Oxford University Press.

Chan, C.C., Jian, L., and Tu, D. (2014) Smart charging of electric vehicles – integration of energy and information. *IET Electrical Systems in Transportation*, 4, 89–96.

Chau, K.T. (2009) Electric motor drives for battery, hybrid and fuel cell vehicles. In *Electric Vehicles: Technology, Research and Development*, ed. G.B. Raines, New York, NY: Nova Science.

Chau, K.T. (2010) Hybrid vehicles. In *Alternative Fuels for Transportation*, ed. A.S. Ramadhas, Boca Raton, FL: CRC Press.

Chau, K.T. (2012) Research on electric vehicles: challenges, opportunities and emerging technologies. *Studies in Science and Technology*, 1, 13–24.

Chau, K.T. (2014) Pure electric vehicles. In *Alternative Fuels and Advanced Vehicle Technologies for Improved Environmental Performance*, ed. R. Folkson, Amsterdam: Woodhead Publishing.

Chau, K.T. (2015) *Electric Vehicle Machines and Drives – Design, Analysis and Application*, Chichester, West Sussex: Wiley-IEEE Press.

Chau, K.T., and Chan, C.C. (2007) Emerging energy-efficient technologies for hybrid electric vehicles. *Proceedings of IEEE*, 95, 821–835.

Chau, K.T., and Wong, Y.S. (2001) Hybridization of energy sources in electric vehicles. *Energy Conversion and Management*, 42, 1059–1069.

Chau, K.T., and Wong, Y.S. (2002) Overview of power management in hybrid electric vehicles. *Energy Conversion and Management*, 43, 1953–1968.

Chau, K.T., Wong, Y.S., and Chan, C.C. (1999) An overview of energy sources for electric vehicles. *Energy Conversion and Management*, 40, 1021–1039.

Chau, K.T., Wu, K.C., and Chan, C.C. (2004) A new battery capacity indicator for lithium-ion battery powered electric vehicles using adaptive neuro-fuzzy inference system. *Energy Conversion and Management*, 45, 1681–1692.

Chau, K.T., Wu, K.C., Chan, C.C., and Shen, W.X. (2003) A new battery capacity indicator for nickel–metal hydride battery powered electric vehicles using adaptive neuro-fuzzy inference system. *Energy Conversion and Management*, 44, 2059–2071.

Cheng, K.W.E., Divakar, B.P., Wu, H., Ding, K., and Ho, H.F. (2011) Battery-management system (BMS) and SOC development for electrical vehicles. *IEEE Transactions on Vehicular Technology*, 60, 76–88.

Christensen, J., Albertus, P., Sanchez-Carrera, R.S., et al. (2012) A critical review of Li/air batteries. *Journal of the Electrochemical Society*, 159, R1–R30.

Du, C., and Pan, N. (2006) Supercapacitors using carbon nanotubes films by electrophoretic deposition. *Journal of Power Sources*, 160, 1487–1494.

Ehsani, M., Falahi, M., and Lotfifard, S. (2012) Vehicle to grid services: potential and applications. *Energies*, 5, 4076–4090.

Gao, S., Chau, K.T., Chan, C.C., Liu, C., and Wu, D. (2011) Optimal control framework and scheme for integrating plug-in hybrid electric vehicles into grid. *Journal of Asian Electric Vehicles*, 9, 1473–1481.

Gao, S., Chau, K.T., Liu, C., Wu, D., and Chan, C.C. (2014) Integrated energy management of plug-in electric vehicles in power grid with renewables. *IEEE Transactions on Vehicular Technology*, 63, 3019–3027.

Gao, S., Chau, K.T., Liu, C., Wu, D., and Li, J. (2012) SMES control for power grid integrating renewable generation and electric vehicles. *IEEE Transactions on Applied Superconductivity*, 22, 5701804:1–4.

Georgano, N. (1996) *Electric Vehicles*, Buckinghamshire: Shire Publications.

Giuliano, M.R., Advani, S.G., and Prasad, A.K. (2011) Thermal analysis and management of lithium-titanate batteries. *Journal of Power Sources*, 196, 6517–6524

Guille, C., and Gross, G. (2009) A conceptual framework for the vehicle-to-grid (V2G) implementation. *Energy Policy*, 37, 4379–4390.

Hermance, D., and Sasaki, S. (1998) Hybrid electric vehicles take to the streets. *IEEE Spectrum*, 35, 48–52.

Hu, X., Li, S.E., Jia, Z., and Egardt, B. (2014) Enhanced sample entropy-based health management of Li-ion battery for electrified vehicles. *Energy*, 64, 953–960.

Huh, J., Lee, S.W., Lee, W.Y., Cho, G.H., and Rim, C.T. (2011) Narrow-width inductive power transfer system for online electrical vehicles. *IEEE Transactions on Power Electronics*, 26, 3666–3679.

Hui, S.Y.R., Zhong, W., and Lee, C.K. (2014) A critical review of recent progress in mid-range wireless power transfer. *IEEE Transactions on Power Electronics*, 29, 4500–4511.

Imura, T., and Hori, Y. (2011) Maximizing air gap and efficiency of magnetic resonant coupling for wireless power transfer using equivalent circuit and Neumann formula. *IEEE Transactions on Industrial Electronics*, 58, 4746–4752.

Kesler, M., Kisacikoglu, M.C., and Tolbert, L.M. (2014) Vehicle-to-grid reactive power operation using plug-in electric vehicle bidirectional offboard charger. *IEEE Transactions on Industrial Electronics*, 61, 6778–6784.

Lam, K.H., Close, J., and Durisch, W. (2004) Modelling and degradation study on a copper indium diselenide module. *Solar Energy*, 77, 121–127.

Li, H., Weng, G., Li, C.Y.V., and Chan, K.Y. (2011) Three electrolyte high voltage acid–alkaline hybrid rechargeable battery. *Electrochimica Acta*, 56, 9420–9425.

Li, W., Chau, K.T., Ching, T.W., Wang, Y., and Chen, M. (2015) Design of a high-speed superconducting bearingless machine for flywheel energy storage systems. *IEEE Transactions on Applied Superconductivity*, 25, 5700204:1–4.

Li, W., Chau, K.T., Liu, C., Gao, S., and Wu, D. (2013a) Analysis of tooth-tip flux leakage in surface-mounted permanent magnet linear vernier machines. *IEEE Transactions on Magnetics*, 49, 3949–3952.

Li, W., Lee, C.H.T., and Ching, T.W. (2013b) A transverse flux permanent magnet linear generator for hybrid electric vehicles. Proceedings of IEEE International Symposium on Industrial Electronics, Taipei, Taiwan, pp. 1–6.

Li, Z., Zuo, L., Luhrs, G., Lin, L., and Qin Y. (2013c) Electromagnetic energy-harvesting shock absorbers: design, modeling, and road tests. *IEEE Transactions on Vehicular Technology*, 62, 1065–1074.

Lin, J., Leung, K.C., and Li, V.O.K. (2014) Optimal scheduling with vehicle-to-grid regulation service. *IEEE Internet of Things Journal*, 1, 556–569.

Liu, C., Chau, K.T., Wu, D., and Gao, S. (2013) Opportunities and challenges of vehicle-to-home, vehicle-to-vehicle, and vehicle-to-grid technologies. *Proceedings of IEEE*, 101, 2409–2427.

Liu, C., Yu, Z., Neff, D., Zhamu, A., and Jang, B.Z. (2010) Graphene-based supercapacitor with an ultrahigh energy density. *Nano Letters*, 10, 4863–4868.

Martin, S., Garcia-Ybarra, P.L., and Castillo, J.L. (2010) Electrospray deposition of catalyst layers with ultra-low Pt loadings for PEM fuel cells cathodes. *Journal of Power Sources*, 195, 2443–2449.

Musavi, F., and Eberle, W. (2014) Overview of wireless power transfer technologies for electric vehicle battery charging. *IET Power Electronics*, 7, 60–66.

Omar, N., Daowd, M., Hegazy, O., et al. (2012) Standardization work for BEV and HEV applications: critical appraisal of recent traction battery documents. *Energies*, 5, 138–156.

Qiu, C., Chau, K.T., Ching, T.W., and Liu, C. (2014a) Overview of wireless charging technologies for electric vehicles. *Journal of Asian Electric Vehicles*, 12, 1–7.

Qiu, C., Chau, K.T., Liu, C., Li, W., and Lin, F. (2014b) Quantitative comparison of dynamic flux distribution of magnetic couplers for roadway electric vehicle wireless charging system. *Journal of Applied Physics*, 115, 17A334:1–3.

Shen, W.X., Chan, C.C., Lo, E.W.C., and Chau, K.T. (2002) A new battery available capacity indicator for electric vehicles using neural network. *Energy Conversion and Management*, 43, 817–826.

Shen, W.X., Chau, K.T., Chan, C.C., and Lo, E.W.C. (2005) Neural network based residual capacity indicator for nickel–metal hydride batteries in electric vehicles. *IEEE Transactions on Vehicular Technology*, 54, 1705–1712.

Shen, W.X., Vo, T.T., and Kapoor, A. (2012) Charging algorithms of lithium-ion batteries: an overview. Proceedings of IEEE Conference on Industrial Electronics and Applications, Singapore, pp. 1567–1572.

Strasik, M., Johnson, P.E., Day, A.C., et al. (2007) Design, fabrication, and test of a 5-kWh/100-kW flywheel energy storage utilizing a high-temperature superconducting bearing. *IEEE Transactions on Applied Superconductivity*, 17, 2133–2137.

Sun, F., Xiong, R., and He, H. (2014) Estimation of state-of-charge and state-of-power capability of lithium-ion battery considering varying health conditions. *Journal of Power Sources*, 259, 166–176.

Vo, T.T., Chen, X., Shen, W.X., and Kapoor, A. (2015) New charging strategy for lithium-ion batteries based on the integration of Taguchi method and state of charge estimation. *Journal of Power Sources*, 273, 413–422.

Wakefield, E.H. (1994) *History of the Electric Automobile: Battery-Only Powered Cars*, Warrendale, PA: Society of Automotive Engineers.

Wang, X., Ma, Y., Li, S., Kashyout, A.H., Zhu, B., and Muhammed, M. (2011) Ceria-based nanocomposite with simultaneous proton and oxygen ion conductivity for low-temperature solid oxide fuel cells. *Journal of Power Sources*, 196, 2754–2758.

Wong, Y.S., and Chan, C.C. (2012) Vehicle energy storage: batteries. In *Encyclopedia of Sustainability Science and Technology*, ed. R.A. Meyers, New Yoke: Springer.

Wouk, V. (1995) Hybrids: then and now. *IEEE Spectrum*, 32, 16–21

Wu, D., Chau, K.T., and Gao, S. (2010) Cost-emission analysis of vehicle-to-grid system. *World Electric Vehicle Journal*, 4, 767–773.

Wu, D., Chau, K.T., Liu, C., Gao, S., and Li, F. (2012) Transient stability analysis of SMES for smart grid with vehicle-to-grid operation. *IEEE Transactions on Applied Superconductivity*, 22, 5701105:1–5.

Yu, C., and Chau, K.T. (2009) Thermoelectric automotive waste heat energy recovery using maximum power point tracking. *Energy Conversion and Management*, 50, 1506–1512.

Zhang, S.S. (2013) Liquid electrolyte lithium/sulfur battery: fundamental chemistry, problems, and solutions. *Journal of Power Sources*. 231, 153–162.

Zhang, X., and Chau, K.T. (2011) An automotive thermoelectric–photovoltaic hybrid energy system using maximum power point tracking. *Energy Conversion and Management*, 52, 641–647.

Zhang, X., Chau K.T., and Chan, C.C. (2010) A multi-hybrid energy system for hybrid electric vehicles. *World Electric Vehicle Journal*, 4, 505–510.

Zhang, Z., and Chau, K.T. (2015) Homogeneous wireless power transfer for move-and-charge. *IEEE Transactions on Power Electronics*, 30, 6213–6220.

Zhang, Z., Chau, K.T., Liu, C., Li, F., and Ching, T.W. (2014a) Quantitative analysis of mutual inductances for optimal wireless power transfer via magnetic resonant coupling. *IEEE Transactions on Magnetics*, 50, 8600504:1–4.

Zhang, Z., Chau, K.T., Liu, C., Qiu, C., and Lin, F. (2014b) An efficient wireless power transfer system with security considerations for electric vehicle applications. *Journal of Applied Physics*, 115, 17A328:1–3.

Zhang, Z., Chau, K.T., Qiu, C., and Liu, C. (2015) Energy encryption for wireless power transfer. *IEEE Transactions on Power Electronics*, 30, 5237–5246.

Zou, Z., Wang, Z., and Cheng, M. (2014) Modeling, analysis, and design of multifunction grid-interfaced inverters with output LCL filter. *IEEE Transactions on Power Electronics*, 29, 3830–3839.

# *Chapter 2*

# Overview of electrochemical energy sources for electric vehicles

*Christopher H.T. Lee*[1] *and K.Y. Chan*[2]

In this chapter, the electrochemical energy sources for electric vehicles (EVs) are discussed. For those energy sources, their backgrounds, operating principles, performances, and applications are presented. First, the high power density power devices, the capacitors, are described. Next, the most commonly available energy sources, the batteries, are elaborated. Then, the fuel cells, which convert the chemical energy into the electrical energy, are discussed. After the description of various candidates, their existing applications to EVs and the development trend of electrochemical energy sources are given.

## 2.1 Types of electrochemical cells for electric vehicles

The EV technology has been regarded as one of the most promising solutions to resolve the problems of the environmental pollution and energy crisis (Chan and Chau, 2001). As the major component of the EV technology, the development of the electric motors has drawn many attentions (Chau, 2015). Nowadays, the electric motors can fulfil most of the requirements for the standard vehicular driving, such as high efficiency, high torque density, high power density, and maintenance-free operation. However, because of the bottle neck in the electrochemical energy sources, the EV models have not generally penetrated into the current market except in ancillaries (Chau *et al.*, 1999; Burke, 2007).

The electrochemical energy sources are used to power the electric motor as well as the EV ancillaries. All the developed electrochemical energy sources can only provide either high specific energy or high specific power, but not both simultaneously (Khaligh and Li, 2010). Hence, the existing EV model cannot compete with the gasoline combustion engine vehicles at the moment (Chau, 2012). To improve the situation, the development of electrochemical energy sources has become one of the hottest research topics recently (Whittingham, 2012). There are various electrochemical energy sources available in the existing market, and they

[1]Department of Electrical and Electronic Engineering, The University of Hong Kong, Hong Kong, China
[2]Department of Chemistry, The University of Hong Kong, Hong Kong, China

can be classified into three major groups, namely, the capacitor, the battery, and the fuel cell (Lukic *et al.*, 2008).

### 2.1.1 Basic differences among electrochemical cells

The three major electrochemical energy sources can be further classified based on two different electrode processes, namely, the non-Faradaic and Faradaic processes. The non-Faradaic process means there is no chemical reaction associated with the cycling processes and the charges are distributed only by the physical mean without any formation of the chemical bonds. On the other hand, the Faradaic process means the chemical reactions, such as the oxidation–reduction reactions, are associated to transfer the charges. Typically, the capacitors (except the pseudocapacitor) consist of the non-Faradaic process, while the batteries and fuel cells consist of the Faradaic processes. Since the capacitors that consist of the non-Faradaic process involve only the physical process of charges movement, its overall charging and discharging mechanisms are faster. Hence, these types of energy sources are more suitable for the power devices. On the other hand, the batteries and the fuel cells involve the Faradaic processes so that higher energy can be stored and released. The graphical comparisons among the representative energy sources are shown in Figure 2.1. Unfortunately, in the current situation, none of the single energy sources can fulfil the desired expectation for EV applications, so that the development of electrochemical energy sources has been a very hot research topic since the last century.

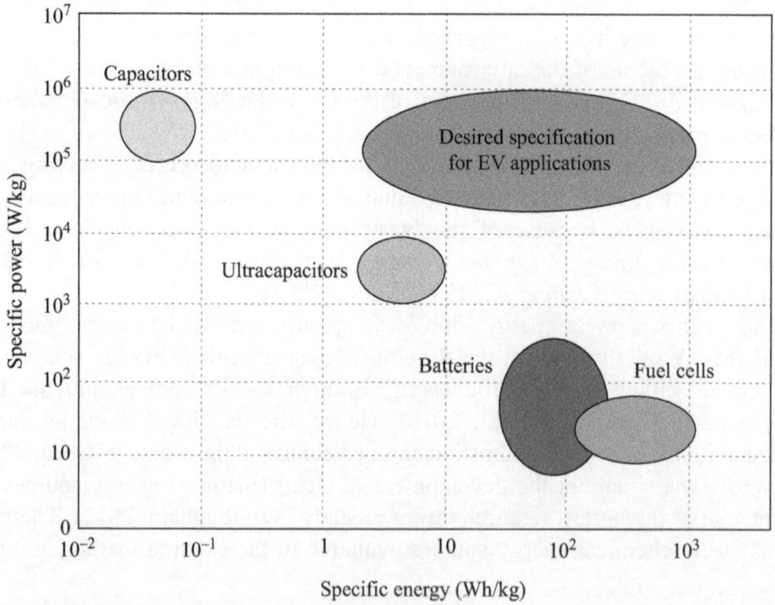


*Figure 2.1 Comparison of the electrochemical energy sources*

As the common candidates for EV applications, the batteries and fuel cells can be further distinguished by the types of chemical reactions. Fuel cells consist of combustion of a fuel as the overall reaction, and hence it involves oxygen (the most common and generally available oxidant) at the cathode (positive electrode). Because of the difficulty of a three-phase reaction, i.e., phase of gas/solid/electrolyte form, the fuel cell electrochemical reactions are relatively slower with poorer power density. Yet, it involves oxygen as the active material so that the energy density is higher.

A large variety of batteries provides the variety choices of active materials, voltage, current, electrolyte, reversibility, and safety. The variations stem from reactions at the negative and positive electrodes involving different reactants and products. Kinetics and reversibility are generally limited by mass and energy transfer across phase boundaries and the solid phase morphology change. Since a battery consists of a batch reaction, its capacity is limited by changes in solid phase of electrode. On the other hand, a fuel cell has continuous supply of active material (fuel) and therefore offering superior capacity.

## *2.1.2 Specific energy of electrochemical cells*

The driving range has been one of the major considerations for EV applications, although it can be effectively determined by the specific energy of electrochemical cells. Theoretically, the specific energy can be calculated based on the following simple relationship as *Cell capacity* × *Cell voltage*. Some examples are shown in Figure 2.2 which includes the nickel-metal hydride (Ni-$MH_x$) battery, all-vanadium redox flow (VRF) battery, and hybridizing positive electrode of VRF with either

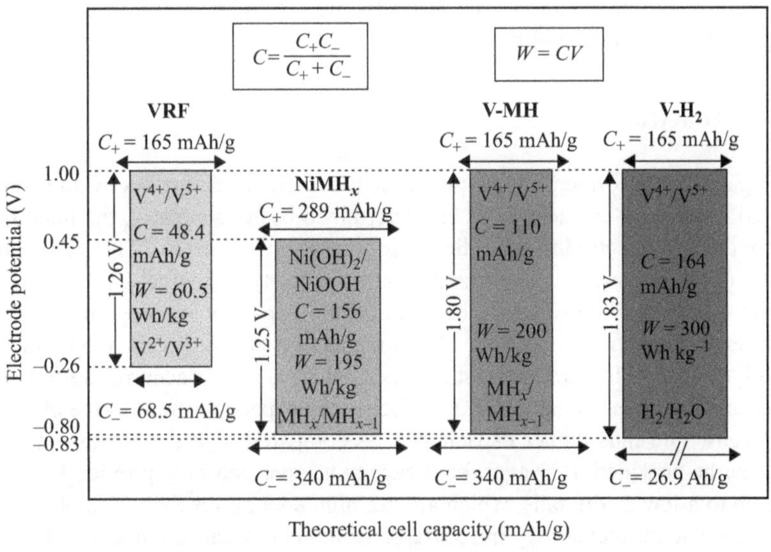


*Figure 2.2 Theoretical specific energies among various batteries*

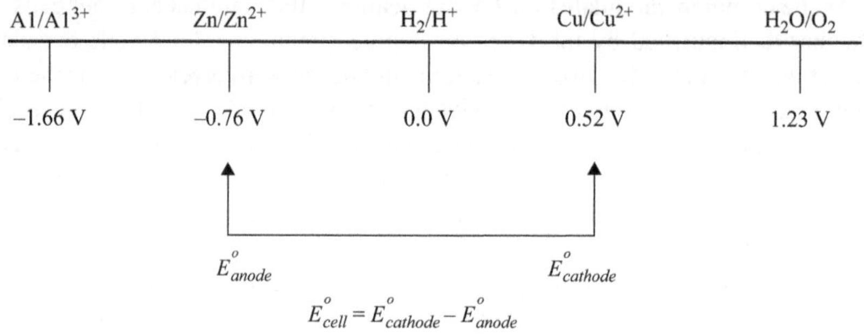


*Figure 2.3 Theoretical electrochemical activity series among various materials*

MH or hydrogen. In the VRF battery, $VOSO_4$ and $V_2(SO_4)_3$ are the active material for $V^{4+}/V^{5+}$ and $V^{2+}/V^{3+}$ redox couples, respectively.

The theoretical cell capacity is calculated by $C = C_+C_-/(C_+ + C_-)$, where $C_+$ and $C_-$ represent the theoretical capacity of positive electrode/electrolyte material (width of bar at the top) and negative electrode material (width of bar at the bottom), respectively (Li *et al.*, 2009, 2011). $C_+$ and $C_-$ are scaled by Faraday's law to their individual active material, which states that capacity is proportional to the maximum number of charge transferred and inversely proportional to mass of the active material.

The theoretical cell voltage (or known as the electrochemical window), which serves as an important factor determining the key performances of the electrochemical cells, can be described by the pairing of negative and positive electrode reactions and their corresponding standard potential, i.e., the position in the electrochemical activity series, as shown in Figure 2.3. As illustrated, the theoretical maximum cell voltage can be calculated by the difference between the individual voltages of the cathode and the anode.

## 2.2 Capacitors

The capacitors are formed by separating the two metallic electrodes by an insulating dielectric, while its characteristics are determined by its geometry, the materials of the conducting electrodes and dielectric (Deshpande, 2015). The conventional capacitors can be roughly categorized into two major types, namely, the electrostatic and the electrolytic capacitors. The electrostatic capacitor employs the insulating materials as dielectric to separate the two metallic electrodes. It adopts the non-electrolytic materials, such as plastic, paper, mica, or even vacuum, to serve as the dielectric component. Meanwhile, the electrolytic capacitor employs the electrolyte as dielectric and thus higher capacitance, as compared with the electrostatic capacitor, can be provided. However, both capacitor types can only provide the capacitance up to a few millifarads, which are not high enough for the EV applications.

The capacitor technology has suffered from the low capacitance over the years, although there is a turning point upon the invention of the ultracapacitor. The ultracapacitor is given the adjective "ultra" because of its supreme capacitance levels as compared with the conventional one. There are fundamentally two types of ultracapacitor technologies available, namely, the double layer capacitor and the pseudocapacitor, in the commercial market. Both of them store the energy based on their unique principles, where the doubly layer capacitor involves the non-Faradaic process and the pseudocapacitor involves the Faradaic process. In general, the pseudocapacitor is less common as compared with the double layer capacitor, yet both of them have its own pros and cons. The incorporation between the two ultracapacitors has been proposed recently, and this leads to form the new type of ultracapacitor, known as the hybrid capacitor.

### *2.2.1 Double layer capacitor*

The idea of the double layer capacitor, which stored the charge on the electrode surface that separating the electrode and the electrolyte, was invented by von Helmholtz in the late 1800s (Deshpande, 2015). The development of the double layer capacitor was not very astonishing until the breakthrough of the nanotechnology. The highly porous materials with the nanometer scale were employed to make the capacitor electrodes. Figure 2.4 shows the basic structure of the double layer capacitor. Upon the application of the voltage, the very thin double dielectric layer was then naturally formed at the interface of the conducting surface and the electrolyte, while no more dielectric material was needed. Hence, this new capacitor is known as the double layer capacitor, as one type of the ultracapacitor.

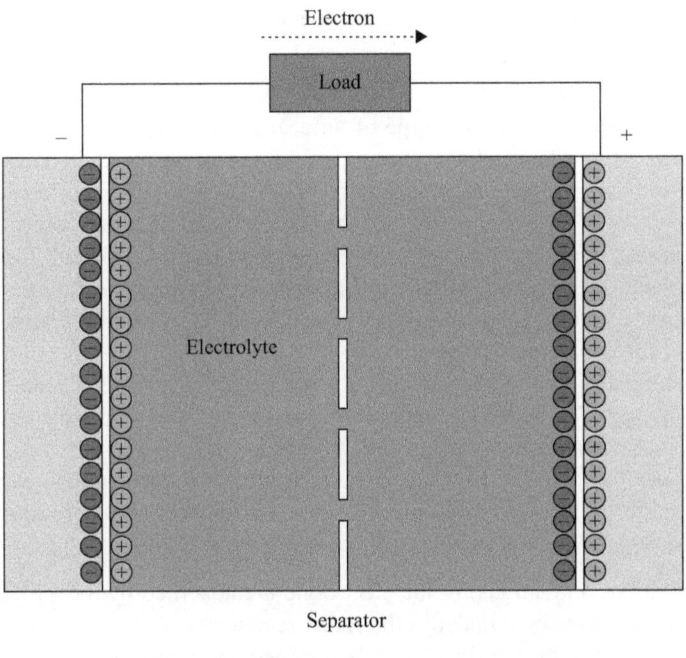


*Figure 2.4 Basic structure of double layer capacitor*

The electrodes of the double layer capacitor are made by the porous activated carbon. The charged ions are mobilized but not electrically contacted. Same as the conventional capacitors, the double layer capacitor stores the energy by the electrostatic mechanism. The ultra large capacitance value is achieved by pushing the two particular physical values, namely, the surface area and the charge separation distance, into the extreme. With the nanometer scale dielectric thickness and the large surface area, the double layer capacitor can provide unexpected huge amount of energy, as compared with the traditional capacitors. It is the nanotechnology which provides the material with the high porosity property, enriches the potential of the ultracapacitor.

Unlike the conventional capacitor, the double layer capacitor employs no dielectric material, while an extremely thin layer is automatically formed at the two electrode surfaces as the dielectric. By the attractions of the unlike polarities, the charged ions in the electrolyte diffuse across the separator and form the double layer charge at each electrode. Due to the highly porous property, the effective contact area is increased drastically. In the meantime, its low barrier thickness also offers the positive effect to the capacitance value and hence higher energy density is achieved. These two phenomena enable the double layer capacitor to reach the extremely high capacitance value, up to several thousand farads.

The operating mechanism of the double layer capacitor is regarded as the non-Faradaic process, i.e., there is no chemical reaction associated with the charging or discharging processes. Without associating with any chemical reaction, the charge storage is very stable with high cycling stability. Hence, the double layer capacitor suffers very little from the aging effect problem, so that it can provide very stable performances even after numerous operating cycles.

### 2.2.2 *Pseudocapacitor*

The pseudocapacitor, as another type of ultracapacitor, adopts the similar concept and structure of the double layer capacitor. Yet, the pseudocapacitor involves the oxidation–reduction reactions at an interface and the operating mechanism of the pseudocapacitor is regarded as the Faradaic process, i.e., there are chemical reactions associated with the charging or discharging processes. Figure 2.5 shows the charge transfer of the pseudocapacitor. Because of the Faradaic process, the pseudocapacitor can offer higher capacitance as well as energy densities than the double layer capacitor counterpart does.

The electrodes of the pseudocapacitor are made by the metal oxides or conducting polymer. The pseudocapacitor employs the fast and reversible reactions to store the energy faradically, where the charges are transferred between the electrode and electrolyte. The Faradaic process of the pseudocapacitor involves the electrosorption, reduction–oxidation reactions, and intercalation processes, while major processes of the electrochemical reactions are as follows:

- Ions (cations and anions) in the electrolyte are adsorbed by the surface.
- Ions in the electrolyte undergo the redox reactions.
- Conducting polymer in the electrode is doped and undoped.

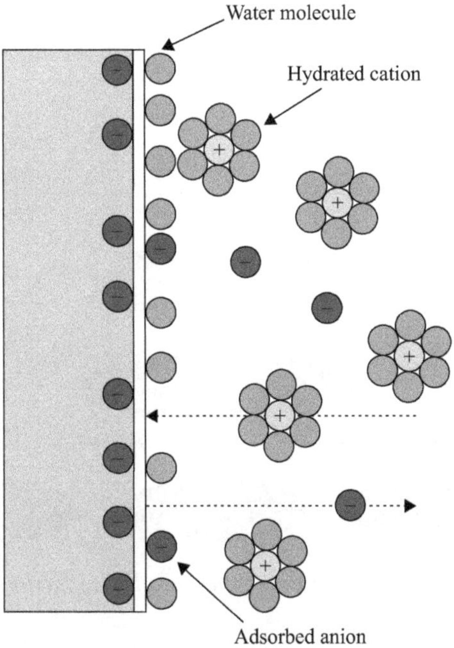


*Figure 2.5 Charge transfer of pseudocapacitor*

Similar as the double layer capacitor, the pseudocapacitor relies the nanotechnology to prosper because the micro-pores can provide higher surface area to promote the accessibility of the electrode materials. Even the pseudocapacitor employs similar concept as the double layer capacitor does, it enjoys extra benefit from the oxidation–reduction at the interface. Hence, the pseudocapacitor can provide better performance in terms of the capacitances as well as the energy densities. However, because of the intrinsic electrode-kinetic by the Faradaic leakage resistance, the pseudocapacitor suffers from much shorter cycle life than the double layer capacitor.

### *2.2.3 Hybrid capacitors*

By incorporating both the Faradaic and non-Faradaic processes to store the electric charges, the hybrid capacitor is formed. With the implementation of the characteristics of the two ultracapacitors, namely, the double layer capacitor and the pseudocapacitor, the hybrid capacitor can enjoy the high energy as well as high power density while maintaining the satisfactory cycle life. In other words, the hybrid capacitor offers the characteristics in between the two ultracapacitors. So the ultracapacitors can be classified into three classes as shown in Figure 2.6.

The hybrid capacitors can be further categorized as three major sub-classes, namely, the composite type, the asymmetric type, and the battery type. The composite type purposely integrates the carbon-based materials with other materials,

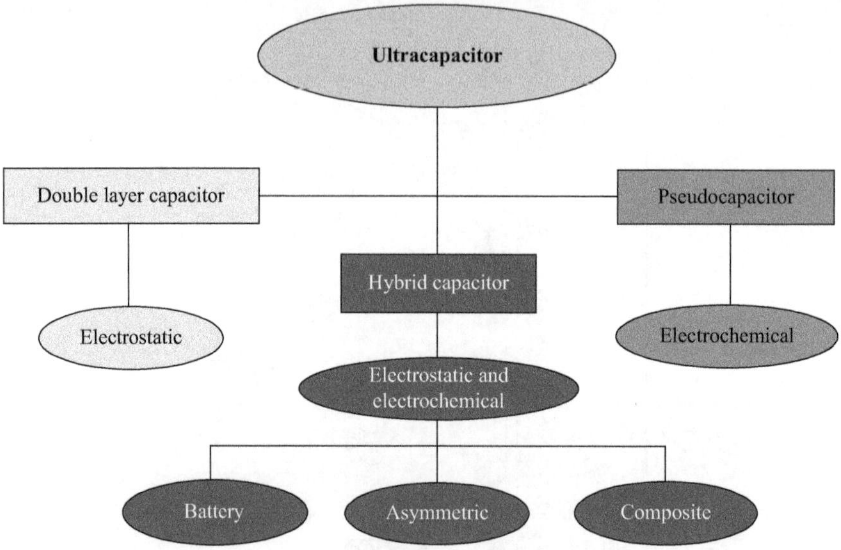


*Figure 2.6 Classification of ultracapacitors*

such as the polymer material or the metal oxide material, to form as a single electrode. The composite electrode can serve similarly as both the double layer capacitor and the pseudocapacitor. On the other hand, the asymmetric type employs the asymmetric electrodes to facilitate the Faradaic and non-Faradaic processes that exist in the double layer capacitor and the pseudocapacitor. In particular, the hybrid capacitor purposely couples the double layer capacitor electrode with the pseudocapacitor electrode to inherit the characteristics from its precedents. In the meantime, the battery type instead utilizes two different electrodes from two technologies to couple with the battery electrodes.

## 2.3 Batteries

Battery is an electrochemical interchange system that converts the active materials into electrical energy (Reddy, 2011). The battery basically consists of three major components, namely, the positive electrode, the negative electrode, and the electrolyte, as shown in Figure 2.7. In order to fulfil the criteria for different applications, many different types of batteries with different characteristics are invented. Yet, the batteries can be normally categorized into three major classes, namely, the primary batteries, the reserve batteries, and the secondary batteries. The primary batteries refer to the batteries with non-rechargeable properties so that they can only be discharged once. These types of batteries sometimes are also known as the "dry cells." The reserve batteries refer to the batteries with the artificial activation properties so that they can only be activated to discharge if and only if the key

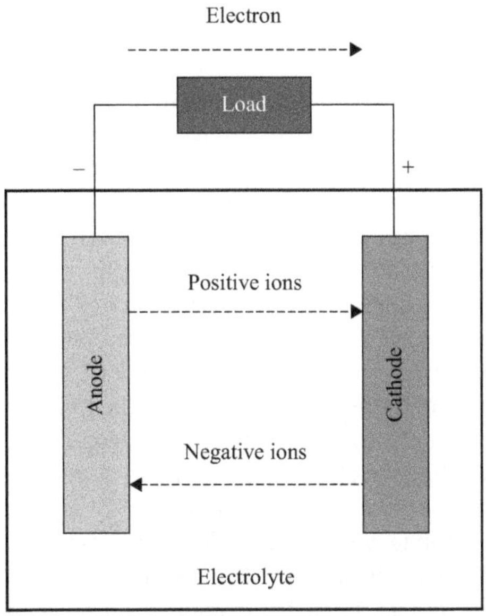


*Figure 2.7 Basic structure of battery system*

component of the batteries is given or launched. For instance, the gas-activated batteries, as one candidate of the reserve batteries, can only be activated when the gas is introduced to the battery system. The reserve batteries can then minimize the problem of self-discharging such that they can reserve the energy storage for longer term than its counterparts. In the meantime, the secondary batteries refer to the batteries with the rechargeable properties so that they can be recharged to their original conditions after discharged.

As the energy sources of the EVs or hybrid EVs (HEVs), the batteries have to offer the rechargeable properties with the satisfactory performances, including the high specific energy for sufficient driving range, the high specific power for powerful acceleration, the long cycle life for minimized maintenance, the non-artificial activation properties for easy adoption, and the low cost for better market penetration. Hence, the secondary batteries that fulfil most of the criteria are extensively adopted in EV applications.

The first secondary battery, the lead-acid battery, was invented in 1859 by the French physicist, Gaston Plante (Reddy, 2011). After numerous modifications and improvements, the lead-acid battery has been widely used in many areas, including the EV applications. The development of secondary batteries has continued over 100 years, while there are many different types of secondary batteries available in the commercial market. Some of the secondary batteries have been adopted in EV applications as the energy sources for the propulsion system. The representative candidates include the lead-acid batteries, the nickel-based batteries, and the ambient-temperature lithium batteries.

### 2.3.1 Lead-acid battery

The lead-acid battery, which is renowned as the first commercial secondary battery, has been successfully penetrated the battery market for more than a century. The lead-acid battery has been employed to numerous applications, such as the communication systems, the portable devices, and the power sources for heavy industry. In the original design of the lead-acid battery, there is no intention to replace the electrolyte and there is no true hermetic seal installed. To limit the interflow of the gas, only a pressure release valve, which is sensitive to the change of the pressure values, is introduced. To improve the performances of the conventional design, the valve regulated lead-acid (VRLA) battery has been proposed.

Unlike the conventional lead-acid battery, the VRLA purposely contains the limited electrolyte as two forms, namely, the absorbed electrolyte and the gelled electrolyte, in which both of them are immobilized electrolytes. The absorbed electrolyte is achieved by filling the electrolyte to a highly porous materials that serve as the separator. On the other hand, the gelled electrolyte is achieved by adding the fumed silica to the electrolyte, and hence hardening the mixture into the gelled form. With the immobilized electrolyte, the VRLA can reduce the evaporation, leakage, and vibration problems so that it can be operated in different harsh conditions.

Similar as the conventional lead-acid battery, the VRLA employs the lead dioxide as the positive electrode, while the metallic lead as the negative electrode. To improve the mechanical strength and electrical properties, sometimes the lead alloy, such as lead-tin is employed instead of pure lead. Even though the lead-acid battery and the VRLA have different structures, both of them share the same chemical reactions. The chemical reaction at positive electrode is:

$$PbO_2 + 3H^+ + HSO_4{}^- + 2e^- \leftrightarrow PbSO_4 + 2H_2O \tag{2.1}$$

and at the negative electrode:

$$Pb + HSO_4{}^- \leftrightarrow PbSO_4 + H^+ + 2e^- \tag{2.2}$$

Hence, the overall reaction is:

$$Pb + PbO_2 + 2H_2SO_4 \leftrightarrow 2PbSO_4 + 2H_2O \tag{2.3}$$

According to (2.3), it can be seen that the electrolyte losses its sulfuric acid and eventually becomes water, during the discharge process. The specific gravity of the electrolyte can then be used to determine the state-of-charge of the battery. On the other hand, the $PbSO_4$ is converted to the lead dioxide at the positive electrode, while the sponge lead at the negative electrode, during the recharging process. Upon the point when almost all the $PbSO_4$ is converted, the overcharge state starts to begin. Same as the conventional lead-acid battery, the VRLA produces the hydrogen and oxygen gases, during the overcharge state. Yet, the VRLA can allow the oxygen to react with the lead to produce the water within the battery as:

$$Pb + HSO_4{}^- + H^+ + \frac{1}{2}O_2 \leftrightarrow PbSO_4 + H_2O \tag{2.4}$$

In the meantime, the generated hydrogen is released through the vent. Hence, the water, instead of losing from the cell, can be recycled during the overcharge state so that higher recharging rates can be achieved.

## *2.3.2 Nickel-based batteries*

The nickel-based rechargeable batteries, which employ the nickel oxide as the positive electrode, were developed more than 100 years ago. These rechargeable batteries have been employed in many different applications, such as portable devices, telecommunication systems, heavy industry, military, and EVs. The nickel-based batteries include the nickel-iron (Ni-Fe) battery, the nickel-cadmium (Ni-Cd) battery, the nickel-metal hydride (Ni-MH) battery, and the nickel-zinc (Ni-Zn) battery.

### **2.3.2.1 Ni-Fe battery**

The Ni-Fe battery, which is highly ruggedness for harsh physical and electrical conditions, is the most common secondary battery among its iron electrode group. The Ni-Fe battery employs the nickel oxyhydroxide as the positive electrode, the metallic iron as the negative electrode, and a potassium hydroxide solution mixing with lithium hydroxide as the electrolyte. The Ni-Fe battery transfers the oxygen from one electrode to another, while the reactions are quite complex due to the transitory existence. The overall reaction can be described as:

$$3Fe + 8NiOOH + 4H_2O \leftrightarrow 8Ni(OH)_2 + Fe_3O_4 \tag{2.5}$$

According to (2.5), it can be seen that the electrolyte gets no involvement during the charging and discharging processes, so that the specific gravity of the electrolyte cannot be used to determine the state-of-charge of the battery, which is employed by the lead-acid battery.

Even though the Ni-Fe battery is renowned for its ruggedness and long life, extra care and special handling procedures should be provided. The Ni-Fe battery can produce a thermal runaway easily under the constant-voltage charging condition, as the result of the increased charging current at the fully charged state. Hence, the electrolyte management implementing with the battery system should be employed to improve its safety performance.

### **2.3.2.2 Ni-Cd battery**

Even though the cadmium results adverse effect to human being and environment, the development of the Ni-Cd battery has begun long time ago. There are different types of Ni-Cd battery available in the commercial market, such as the pocket-plate type, the sintered-plate type, and the fiber-plate type. The pocket-plate type is the most mature Ni-Cd battery that offers very reliable performance. However, this type of battery suffers from the heavy weight so that it is not applicable to many operations. The sintered-plate type is constructed in a thinner form so that the internal resistance of the battery can be reduced, hence achieving the higher charging and discharging rates. Yet, the sintered-plate type suffers from the manufacturing complexity and the cost problem. By employing the polymer materials and plating techniques, the fiber-plate type can eliminate most of the drawbacks than its predecessors.

The Ni-Cd battery, for any type, employs the nickel oxyhydroxide as the positive electrode, the cadmium as the negative electrode, and the potassium hydroxide solution as the electrolyte. The overall reaction can be described as:

$$2NiOOH + 2H_2O + Cd \leftrightarrow 2Ni(OH)_2 + Cd(OH)_2 \tag{2.6}$$

According to (2.6), the metallic cadmium is oxidized to produce the $Cd(OH)_2$, while the NiOOH is reduced to produce the $Ni(OH)_2$, during the discharging process. Unlike the sulfuric acid in the lead-acid battery, the potassium hydroxide electrolyte in the Ni-Cd battery has no significant change during the charging and discharging processes.

#### 2.3.2.3 Ni-MH battery

The Ni-MH battery shares similar principle as the Ni-Cd battery, yet the former one employs metal alloy to absorb the hydrogen as its negative electrode. Without using any cadmium, the Ni-MH battery is safer and more environmental friendly than the Ni-Cd battery. As compared with the Ni-Cd battery, the Ni-MH battery can provide higher energy density and longer cycle life, yet it suffers from relatively lower rate capability.

The Ni-MH battery employs the nickel oxyhydroxide as the positive electrode, the metal hydride as the negative electrode and the potassium hydroxide solution as the electrolyte. There are numerous choices of metal hydride available, while the most common type are the $AB_2$ alloy, a metal alloy of vanadium-titanium-zirconium-nickel, and the $AB_5$ alloy, a rare-earth metal alloy of lanthanum-nickel. The overall reaction can be described as:

$$MH + NiOOH \leftrightarrow M + Ni(OH)_2 \tag{2.7}$$

The oxygen-recombination mechanism exists in the Ni-MH battery so that the generated pressure by the released gases can be minimized. To be specific, the oxygen reacts with the hydrogen electrode to produce the water, hence releasing the pressure. In the meantime, the charge rate should be controlled to ensure that the generated oxygen does not exceed the recombination rate.

#### 2.3.2.4 Ni-Zn battery

The original aim of the Ni-Zn battery is to implement the Ni-Cd battery to form a new battery type. The Ni-Zn battery was first patented in 1901 in Russia while further studies were started by the Irish chemist, James J. Drumm in the 1930s (Reddy, 2011). The Ni-Zn battery is an alkaline battery that combines the nickel electrode and the zinc electrode. It should be noted that the nickel electrode is commonly employed in the Ni-Fe, the Ni-Cd, and the Ni-MH batteries. Upon its outstanding performances such as fast recharging capability, high specific energy, environmental friendly, and cost-effectiveness, the Ni-Zn battery has gained many attentions for different applications.

The Ni-Zn battery employs the nickel oxyhydroxide as the positive electrode, the zinc as the negative electrode, and the alkaline solution as the electrolyte. The chemical transitions of zinc in the alkaline solution are very complex so that the overall reaction is only for illustration:

$$2NiOOH + 2H_2O + Zn \leftrightarrow 2Ni(OH)_2 + Zn(OH)_2 \tag{2.8}$$

The nickel oxyhydroxide is reduced to form the nickel hydroxide, while the metallic zinc is oxidized to form the zinc hydroxide, during the discharging process. The oxygen gas is produced at the nickel electrode and the hydrogen gas at the zinc electrode, if the battery is at the undercharged state. Yet, the oxygen and hydrogen gases can then be recombined to produce water.

## *2.3.3 Ambient-temperature lithium batteries*

Although the ambient-temperature lithium batteries suffer from lower charging and discharging rates than the high-temperature lithium batteries, they enjoy the definite advantage of convenient operating temperature ranges. Meanwhile, they can generally provide relatively higher specific energy as compared with other rechargeable batteries.

The ambient-temperature lithium batteries employ different types of materials, such as the intercalation compounds, inorganic cathodes, and polymers, as the positive electrode. On the other hand, they generally employ the lithium metal as the negative electrode, which unavoidably creates a safety problem. To improve the situation, the intercalation material is purposely employed to replace the lithium metal as the negative electrode. This improved lithium battery is known as the lithium-ion (Li-Ion) battery. Hence, the ambient-temperature lithium batteries can be roughly categorized into two classes, namely, the metallic lithium (Li-Metal) battery and the Li-Ion battery.

### 2.3.3.1 Li-Metal battery

The Li-Metal battery undoubtedly enjoys the definite advantage of highest specific energy among its counterparts, yet it suffers from the reactivity problem. Unlike other types of rechargeable batteries, the electrochemical characteristics of the Li-Metal battery are quite unique. Upon the application of the lithium anode as the conjunction for the intercalation cathode, the Li-Metal battery can provide very attractive performances, including high specific energy, good storage characteristic, and satisfactory cycle life.

Different materials have been adopted as the positive electrode and the electrolyte. For instance, the intercalation compounds or the polymer are the typical choices for the positive electrode. Meanwhile, the inorganic electrolyte, the polymer electrolyte, and the solid electrolyte are suitable to match with different positive electrodes. The corresponding chemical reactions for different combinations are different, while the most common one is:

$$x\mathrm{Li} + \mathrm{M}_y\mathrm{B}_z \leftrightarrow \mathrm{Li}_x\mathrm{M}_y\mathrm{B}_z \tag{2.9}$$

where $\mathrm{M}_y\mathrm{B}_z$ is the transition metal compound. With different combinations of positive electrode and electrolyte, different performances can be achieved.

#### 2.3.3.2 Li-Ion battery

The Li-Ion battery employs the lithium intercalation compounds as the positive and negative electrodes. It inserts and removes the lithium ions between the positive and negative electrodes during the charging and discharging processes. During the movement of the lithium ions, there is no significant change among the electrodes, and the movement of the lithium ions is known as the intercalation processes as shown in Figure 2.8.

The Li-Ion battery employs the materials with two structures, namely, the layered structure and the three-dimensional framework structure, as the positive electrode. The lithium cobalt oxide ($LiCoO_2$) and the lithium nickel oxide ($LiNiO_2$) are the typical candidates as the layered structure, while the lithium manganese oxide ($LiMn_2O_4$) as the three-dimensional framework structure. In general, the Li-Ion battery employs the carbon materials, which consist of different precursor phases, as the negative electrode. However, there is an exceptional case where the negative electrode adopts the lithium titanate ($Li_4Ti_5O_{12}$) and this battery type is known as the $Li_4Ti_5O_{12}$ battery. The surface area of the negative electrode becomes much larger than that using carbon so that the $Li_4Ti_5O_{12}$ battery allows the electrons to interchange quickly and hence resulting faster charging rate than other types of Li-Ion battery.

Not surprisingly, different types of carbon electrodes result with different lithium intercalation capacity as well as performances. In the meantime, four major types of electrolytes, namely, the liquid, the gel, the polymer, and the ceramic electrolytes, have been adopted for the Li-Ion battery. Similar as the Li-Metal

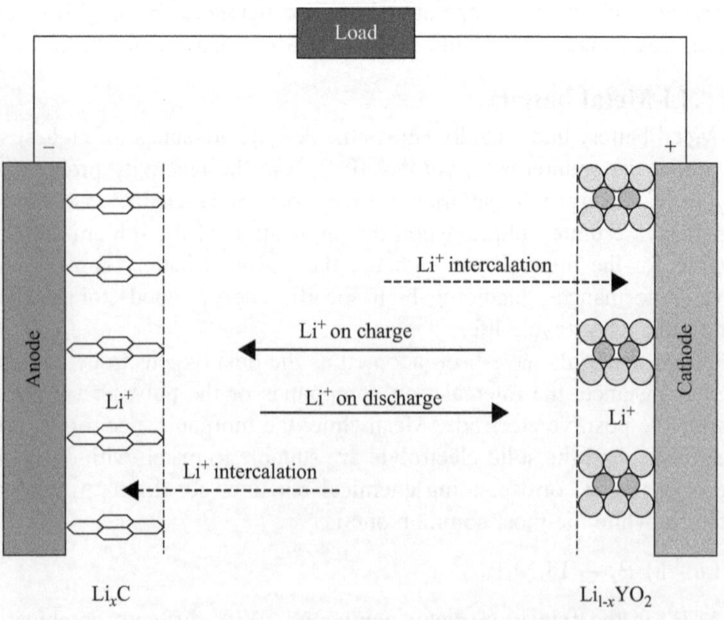


*Figure 2.8 Intercalation processes of Li-Ion battery*

battery, the Li-Ion battery consists of slightly different chemical reactions upon different combinations of adopted materials, while the general one is:

$$LiYO_2 + C \leftrightarrow Li_xC + Li_{1-x}YO_2 \tag{2.10}$$

As compared with the Li-Metal battery, the Li-Ion battery purposely replaces the lithium metal by the carbon material so that it suffers from slightly lower specific capacity. Yet, the Li-Ion battery can instead take the definite superiority over the Li-Metal battery regarding to the safety issue.

## 2.4 High-temperature batteries

While the ambient-temperature batteries enjoy the merit of effective thermal management, some researches have been focused on the development of the high-temperature batteries where these candidates enjoy the advantages of potentially higher power and energy densities. In particular, there are two high-temperature batteries available for EV applications, namely, the sodium-beta batteries and the high-temperature lithium batteries.

### *2.4.1 Sodium-beta batteries*

Upon the employment of the metallic molten sodium as the reactants, a new type of battery family for high energy density applications has been generated. There are many sodium-based technologies explored over the years, while only two of them have been successfully produced as the sodium-beta batteries. These two technologies are the sodium/sulfur (Na/S) and the sodium/metal-chloride ($Na/MCl_2$) batteries. Because both of them employ the liquid sodium as the negative electrode, while the ceramic beta″-alumina as the electrolyte, these battery types are known as the sodium-beta batteries. In order to provide the function of ionic conductivity, the sodium-beta batteries, including the Na/S and the $Na/MCl_2$ batteries, have to be operated at relatively high temperature from 270 °C to 350 °C.

#### 2.4.1.1 Na/S battery

The Na/S battery employs the sulfur as the positive electrode, the sodium as the negative electrode, and the ceramic beta″-alumina as the electrolyte. The sodium at the negative electrode is oxidized to form the sodium ion, while the produced ion reacts with the sulfur at the positive electrode to form the sodium pentasulfide. Two-phase liquid mixture is then formed because the sodium pentasulfide is non-miscible with the sulfur. The chemical reaction at the positive electrode is:

$$xS + 2e^- \leftrightarrow S_x{}^{-2} \tag{2.11}$$

and at the negative electrode:

$$2Na \leftrightarrow 2Na^+ + 2e^- \tag{2.12}$$

Hence, the overall reaction is:

$$2Na + xS \leftrightarrow Na_2S_x \tag{2.13}$$

The operating voltages of the Na/S battery relate to the depth of discharge, which should be limited to far below 100% of the expected value, in order to prevent the over-discharging and the corrosive consequences. When the discharging process exceeds a certain state, an additional two-phase mixture is formed again. This phenomenon increases the internal resistance, hence degrading the battery performance.

#### 2.4.1.2 Na/$MCl_2$ battery

The major distinction between the Na/$MCl_2$ and the Na/S batteries comes from the material adoption of the positive electrode. The Na/$MCl_2$ battery employs the transition metal chlorides as the positive electrode to react with the molten secondary electrolyte, $NaAlCl_4$. The parent metal and sodium chloride crystals are formed from the solid metal chloride during the discharging process. The Na/$MCl_2$ battery has two main alternatives which are based on nickel and iron to form the Na/$NiCl_2$ and Na/$FeCl_2$ types. The overall chemical reactions of these two types of Na/$MCl_2$ battery are described as:

$$NiCl_2 + 2Na \leftrightarrow Ni + 2NaCl \tag{2.14}$$

$$FeCl_2 + 2Na \leftrightarrow Fe + 2NaCl \tag{2.15}$$

The operating voltage drops drastically upon the point when all the metal chloride is used up. The secondary electrolyte is then reduced to form the aluminum as:

$$NaAlCl_4 + 3Na \leftrightarrow Al + 4NaCl \tag{2.16}$$

Since the metallic aluminum is electrically conductive, the Na/$MCl_2$ battery continues to maintain its conductivity even all the sodium is depleted.

The Na/$NiCl_2$ battery resolves most of the problems that exist in the Na/S battery, including the limited operating temperature range, severe metallic material corrosion, irritating side-products, and relative low power density. Hence, the Na/$NiCl_2$ battery can provide greater potential to serve for commercial applications.

### *2.4.2 High-temperature lithium batteries*

Implementation of the alkali-metal electrodes and the molten-salt electrolytes more than 50 years ago was followed by the invention of high-temperature lithium batteries. The high-temperature lithium batteries consist of the solid porous electrodes and the molten-salt electrolyte with an operating temperature of over 400 °C. In the very beginning, the lithium/sulfur (Li/S) battery has been considered as a suitable candidate for the high energy demanding applications. However, the Li/S battery suffers from the decline of capacity due to non-sustainable reactions at the active electrodes and thus the development of this battery type has been severely hindered. Lithium alloys and metal sulfides are introduced to replace the elemental lithium and sulfur as the negative and positive electrodes, respectively.

Lithium alloys that consist of relatively lower reactivity than the metal lithium are purposely adopted as the negative electrode so that the reaction rate of the lithium batteries can be limited to within the manageable range. The two-component lithium alloys, such as $\alpha + \beta$ Li-Al and $Li_5Al_5Fe_2$, are adopted as the negative electrode materials. These lithium alloys not only can provide the overcharge tolerability, but also increase the reactivity, as compared with other lithium alloy counterparts. On the other hand, different types of metal sulfides, such as iron, nickel, and cobalt, have been employed as the positive electrodes. All the chosen metal sulfides can minimize the vapor pressure as well as the corrosion problems, yet the FeS and $FeS_2$ are commonly selected due to the cost consideration. The overall reactions for the Li/FeS and Li/$FeS_2$ batteries can be described as:

$$2Li - Al + FeS \leftrightarrow Li_2S + Fe + 2Al \quad (2.17)$$

$$2Li - Al + FeS_2 \leftrightarrow Li_2FeS_2 + 2Al \quad (2.18)$$

The Li/FeS and Li/$FeS_2$ batteries employ the LiCl-rich electrolyte, which offers the advantages of low melting point and high conductivity. Upon the application of LiCl-rich electrolyte, the operating temperatures of the Li/FeS and Li/$FeS_2$ batteries are reduced to the range around 400 °C. Yet, the thermal management system is still needed for normal operations. Complex multi-step reactions at the positive electrode, while intermediate sulfur complex are soluble and their movement to the negative electrode may end up some problems.

## 2.5 Metal/air batteries

The metal/air batteries purposely employs the air as the cathode to couple electrochemically with the metallic reactive anode. With the application of an air electrode as cathode, the reactant can then be inexhaustible and hence very high specific energy can be purposely achieved. Upon the air electrode mechanism, the energy capacity of the battery is limited by the capacity of the anode and the handling procedures of the reactants and the by-products. It is believed that these batteries can offer the excellent performances with reasonable cost, upon the appropriate selections of the metallic reactants.

There are different battery configurations available for the metal/air batteries, namely, the primary, the reserve, the electrically rechargeable, and the mechanically rechargeable types. For the mechanically rechargeable type, its recharging mechanism involves only the replacement of the discharged metal electrode. It functions similarly as the primary battery does, where the air electrode has to be operated only during the discharging process. Hence, the mechanical rechargeable system has great potential for the high specific energy applications.

Apart from the classifications by their configurations, the metal/air batteries can also be categorized by the metal types that are employed as the reactive anode. Various types of metals, such as the calcium, magnesium, iron, aluminum, zinc, and lithium, have been explored to employ as the reactants for the metal/air batteries. In particular, the calcium/air and magnesium/air batteries have limited development due to the high cost, instability, corrosion, unsteady dissolution, and handling problems. In the meantime, the iron/air (Fe/Air) battery which enjoys the long cycle life and high recharging rate, has been developed for the rechargeable system. Due to the advantages of rich abundance and low cost, the aluminum has gained many attentions to adopt as a metal/air battery. Yet, the aluminum/air (Al/Air) battery offers far too high charging potential for the aqueous system, so that it can be only adopted for the reserve and mechanically rechargeable systems. With the relatively high stability in alkaline electrolytes and as the most active metal in an aqueous electrolyte, the zinc has been confirmed to be one of the best candidates for the electrically rechargeable system. It is the fact that there are still problems awaited to be solved, while the potential of the zinc/air (Zn/Air) battery is quite promising. In the meantime, the lithium/air (Li/Air) battery, which has been proposed in the past decade, has been widely accepted as the most promising candidate for the high specific energy applications among the metal/air battery group.

The metal/air batteries employ the air as the positive electrode, the metal as the negative electrode, and the potassium hydroxide solution as the electrolyte. The chemical reaction at the positive electrode is:

$$O_2 + 2H_2O + 4e^- \leftrightarrow 4OH^- \tag{2.19}$$

and at the negative electrode:

$$M \leftrightarrow M^{n+} + ne^- \tag{2.20}$$

Hence, the overall reaction is:

$$4M + nO_2 + 2nH_2O \leftrightarrow 4M(OH)_n \tag{2.21}$$

where M is the metal adopted in the metal/air batteries, and $n$ is the valence change of the oxidization of the metal. In the aqueous electrolyte, the metals become unstable and hence generating hydrogen as:

$$M + nH_2O \leftrightarrow M(OH)_n + \frac{n}{2}H_2 \tag{2.22}$$

This self-discharging mechanism degrades the battery performances and thus must be handled carefully.

## 2.6 Fuel cells

Fuel cells are the electrochemical energy sources, which convert the chemical energy to the electrical energy (Revankar and Majumdar, 2014). The fuel cells have been regarded as one of the most promising candidates that can eventually achieve the objective of zero emission with satisfactory specific energy and specific power performances. Even though the fuel cells sound to be the advancing technologies to most of the people, these types of energy sources were proposed long time ago. Yet, the development of the fuel cells is generally very slow as compared with other energy sources.

The fuel cells share some similarities as well as dissimilarities with the batteries do. For instance, both of the two energy sources are the electrochemical devices that consume the fuels and oxidants for the chemical reactions to generate electric power. In addition, both of them share similar structures where the electrodes and

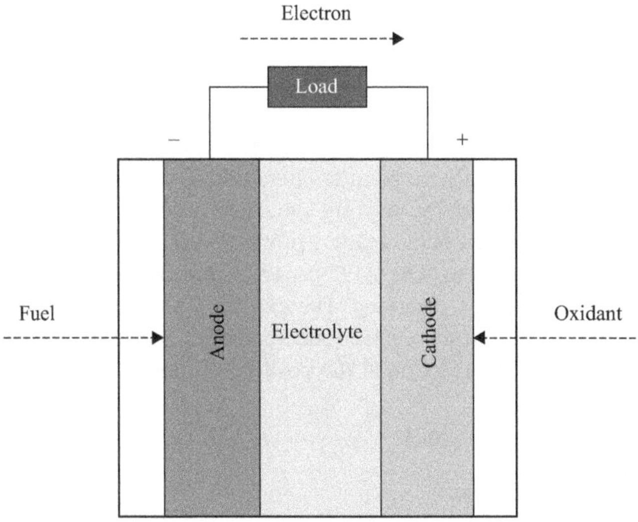


*Figure 2.9 Basic structure of fuel cell system*

electrolytes are installed. Nevertheless, the fuel cells distinguish themselves by the unique operating mechanism. In particular, the fuel cells purposely store the fuels and oxidants outside the system, so that the fuel cells can generate the electric power as long as the external fuels are available as shown in Figure 2.9. On the other hand, the batteries store all the fuels and the oxidants within the system, and hence the specific energy is limited upon the availability of the fuels. Hence, in principle, the fuel cells can get rid of the ranging problem for EV applications.

The first fuel cells were invented by the Welsh judge and physical scientist, William Grove in 1839, yet the performances of the fuel cells were not satisfactory. The fuel cell technologies have drawn very little attention while the situation has been changed upon the invention of the alkaline fuel cell (AFC). After the success of the AFC, the potentiality of the fuel cell technologies have been figured out and thus intensive researches have been started since several decades ago. Until now, there are several promising fuel cell candidates, including the AFC, the phosphoric acid fuel cell (PAFC), the proton exchange membrane fuel cell (PEMFC), the molten carbonate fuel cell (MCFC), the solid oxide fuel cell (SOFC), and the direct methanol fuel cell (DMFC).

### *2.6.1 Alkaline fuel cell*

The AFC has been widely recognized as the first practical fuel cell that consumes the hydrogen fuel to produce the electric energy. It was first proposed by Sir William Robert Grove in the mid of 1800s (Revankar and Majumdar, 2014), yet the prototype was not feasibly applicable for the practical works. After numerous developments, the functional and applicable AFC was released in 1930s. The AFC is operated with the pure hydrogen, the pure oxygen, and the circulating electrolyte. It enjoys the advantages of high efficiency, mature technology, and cost-effective. Hence, the AFC has been adopted in many applications, including the EV applications.

The AFC consists of two porous electrodes, separated by the porous matrix saturated alkaline solution, while the potassium hydroxide (KOH) is a common choice. The KOH is chosen because it can effectively transfer the heat and hence the AFC can even work at the ambient temperature. The carbon dioxide ($CO_2$) can contaminate the fuel cell by converting the KOH into the potassium carbonate ($K_2CO_3$). To minimize the contamination problem, the AFC purposely adopts the circulating electrolyte system. The AFC generates electricity via a redox reaction, i.e., oxidation and reduction reactions. The oxygen from the air and the hydrogen fuel are inputted to the positive electrode (cathode) and the negative electrode (anode), respectively. The reaction at the positive electrode (cathode) is:

$$\frac{1}{2}O_2 + H_2O + 2e^- \rightarrow 2OH^- \tag{2.23}$$

and at the negative electrode (anode):

$$H_2 + 2OH^- \rightarrow 2H_2O + 2e^- \tag{2.24}$$

Hence, the overall reaction is:

$$H_2 + \frac{1}{2}O_2 \rightarrow H_2O \tag{2.25}$$

The schematic operation of the AFC is shown in Figure 2.10. The AFC combines the hydrogen and oxygen through the redox reaction to produce the water, the

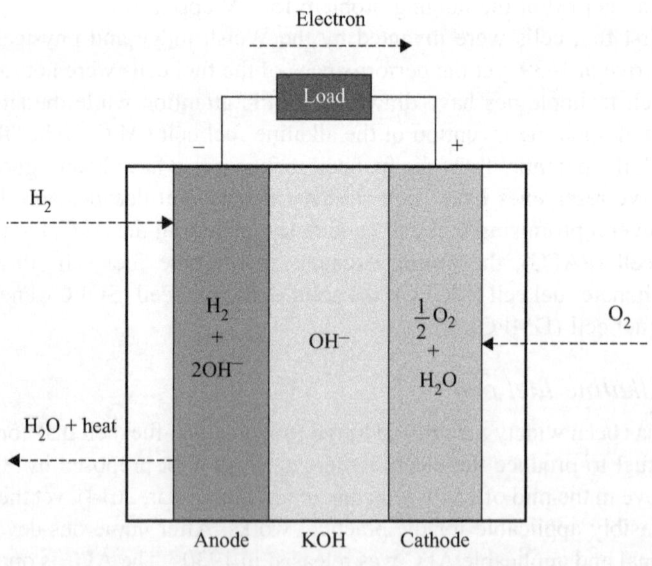


*Figure 2.10 Schematic operation of AFC system*

heat, and the electrical energy. The AFC needs to be operated at the low $CO_2$ environment, otherwise the alkaline electrolyte will be poisoned:

$$CO_2 + 2KOH \rightarrow K_2CO_3 + H_2O \tag{2.26}$$

The formed $K_2CO_3$ reduces the concentration of the hydroxyl ions and hence the fuel cell performances are degraded. In addition, the formed $K_2CO_3$ can block the electrolyte pathways so that the fuel cell may be malfunctioned.

### *2.6.2 Phosphoric acid fuel cell*

The PAFC is one of the most mature fuel cell technology that employed in many applications and in particular, it has been generally accepted as the first commercialized fuel cell across the electric power industry. It employs the liquid phosphoric acid as the electrolyte, and the carbon paper that coated with platinum catalyst as the electrodes. The positively charged hydrogen ions can migrate from the positive electrode (cathode) to the negative electrode (anode) within the acidic electrolyte, while the electrons instead move from the anode to cathode through the external circuit. The electrical energy and the bi-products, such as the water and heat, are then generated.

The PAFC has to operate in the temperature range higher than the boiling water, so that the phosphoric acid is chosen as its electrolyte. Due to the adoption of platinum as catalyst, the PAFC suffers from higher material costs than its counterparts. Yet, the PAFC can offer high power density and good quality of energy, and it is particularly attractive for some premium applications. Different acidic solutions have been attempted to serve as the electrolyte, while the phosphoric acid enjoys the advantage of stability as compared with its counterparts. However, the acidic solutions suffer the major disadvantage of lower oxygen reducing rate as compared with the alkaline electrolytes.

The PAFC employs the platinum-based catalysis to trigger the reaction, while the oxygen from the air and the fuel are inputted to the positive electrode (cathode) and the negative electrode (anode), respectively. The reaction at the positive electrode (cathode) is:

$$\frac{1}{2}O_2 + 2H^+ + 2e^- \rightarrow H_2O \tag{2.27}$$

and at the negative electrode (anode):

$$H_2 \rightarrow 2H^+ + 2e^- \tag{2.28}$$

Hence, the overall reaction is:

$$H_2 + \frac{1}{2}O_2 \rightarrow H_2O \tag{2.29}$$

The schematic operation of the PAFC is shown in Figure 2.11. The PAFC is suitable to operate with natural gas, and provides the potential to develop high energy efficiency accordingly. As compared with other fuel cells, the PAFC can

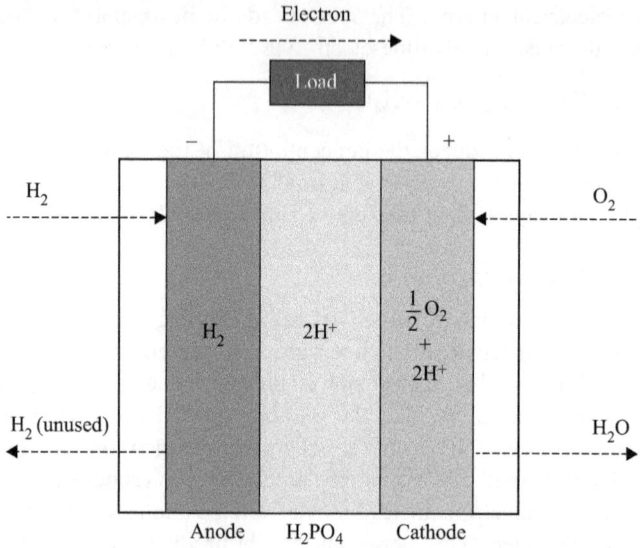


*Figure 2.11 Schematic operation of PAFC system*

effectively reject the $CO_2$ from the electrolyte, so that the contamination problem that happened in the AFC can be minimized.

## 2.6.3 Proton exchange membrane fuel cell

By using the proton conducting membrane in the solid polymer form as the electrolyte, the PEMFC is sometimes known as the ion exchange membrane fuel cell or the solid polymer electrolyte fuel cell. With the employment of the proton conducting membrane, the PEMFC can provide definite merit of high power density so that the PEMFC was the first fuel cell to serve as the energy source for NASA Gemini space project in 1960s (Revankar and Majumdar, 2014). Even though the PEMFC inherits the advantage of high power potential, the operating temperature of the PEMFC is relatively low so that the expensive platinum metal has to be adopted as the catalyst. The high material cost has hindered the development pace of this type of fuel cell. Until recently, with the technology advancement, the platinum loading has been reduced so that the PEMFC has once again become the very hot research topic.

Upon the application of the solid electrolyte, the PEMFC enjoys the extra benefits such as the simple topology, robust operation, and ease of manufacturing process. In the meantime, the internal cell resistance can be effectively minimized as the solid electrolyte can be constructed as very thin sheets. Hence, the high efficiency and high power density can be achieved. The solid membrane gives extra support for the PEMFC to withstand the high pressure condition. All these characteristics enable the PEMFC as a good candidate to serve for mobile applications.

To improve the stability and to extend the lifetime of the membrane, the operating temperature of the PEMFC is purposely pulled down. Upon the low operating temperature, the thermal integration between the waste heat and the fuel reforming processor is not very effective. In addition, the acidic electrolyte in the low operating temperature has to adopt the expensive noble metals as the catalyst, which can be poisoned by the presence of carbon monoxide. Meanwhile, the adoption of the expensive metallic catalysts undoubtedly hinders the market penetration of the PEMFC.

The PEMFC employs the carbon as the electrodes, with the metallic platinum as the catalyst, and the solid polymer membrane as the electrolyte. To enable the polymer membrane to operate normally, the complete hydration environment must be provided. Hence, the fuel and oxidant are purposely as the form of fully humidified streams. The reaction at the positive electrode (cathode) is:

$$\frac{1}{2}O_2 + 2H^+ + 2e^- \rightarrow H_2O \tag{2.30}$$

and at the negative electrode (anode):

$$H_2 \rightarrow 2H^+ + 2e^- \tag{2.31}$$

Hence, the overall reaction is:

$$H_2 + \frac{1}{2}O_2 \rightarrow H_2O \tag{2.32}$$

The schematic operation of the PEMFC is shown in Figure 2.12. Due to the adoption of the sulfonic acid with the solid polymer form as the electrolyte, the PEMFC shares the same electrochemical reaction with the PAFC. Both the PEMFC

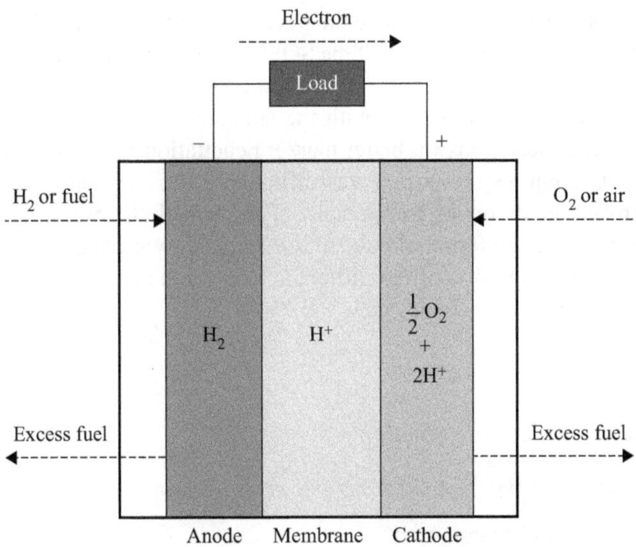


*Figure 2.12 Schematic operation of PEMFC system*

and PAFC employ the platinum metals as the catalyst that expose to the risk by the carbon monoxide poisoning. Because the PEMFC is operated with relatively lower temperature, the poisoning effect in the PEMFC is even more severe than that happened in the PAFC.

### *2.6.4 Molten carbonate fuel cell*

The MCFC was first patented by W.D. Treadwell in 1910s, while this technology has not drawn many attentions until the successful conversion from coal to electrical energy with the carbonate media in 1940s (Revankar and Majumdar, 2014). Unlike the PAFC that employed the natural gas as its major fuel, the MCFC was originally proposed to operate with the coal as the major fuel. However, the adoption of the coal as major fuel was not realized, yet the deviated fuels from the coal and natural gas were instead employed. Even though the MCFC has led great discussions in the 1960s, its development kept stagnated as compared with the market-penetrated PAFC counterpart. Yet, the development of the MCFC technology has not slowed down. According to the common expectations, the MCFC is now in the developing stage that awaits to reach the commercialization stage.

The normal operating temperature range of the MCFC is around 650 °C, which is the highest temperature value among the previously discussed candidates, namely, the AFC, the PAFC, and the PEMFC. As mentioned before, the carbon dioxide and carbon monoxide can inevitably poison the AFC and the acidic electrolyte fuel cells. In contrast, the relatively high operating temperature allows the MCFC to enjoy the definite merits of tolerability towards both carbon monoxide and carbon dioxide. In additional to the tolerability, the high operating temperature can assist the co-generation process of the wasted heat so that the overall system efficiency can be improved. Furthermore, this high operating temperature provides the potential to utilize the carbonaceous fuel to achieve the internal reforming by the electrochemical reactions. The operating temperature of the SOFC, which will be discussed in the next section, is about 1,000 °C. As compared with the SOFC, the relatively lower operating temperature grants the MCFC with the simpler construction structure, hence leading to higher expectations for better market penetration and commercialization.

The MCFC employs the porous gas-diffusion electrodes and the molten carbonate electrolyte as its major components. The formed electrons are transported from the anode, via an external circuit, to the cathode, while the electric circuit is completed upon the migration of the carbonate ions via the electrolyte. The reaction at the positive electrode (cathode) is:

$$\frac{1}{2}O_2 + CO_2 + 2e^- \rightarrow {CO_3}^{2-} \tag{2.33}$$

and at the negative electrode (anode):

$$H_2 + {CO_3}^{2-} \rightarrow H_2O + CO_2 + 2e^- \tag{2.34}$$

Hence, the overall reaction is:

$$H_2 + \frac{1}{2}O_2 \rightarrow H_2O \tag{2.35}$$

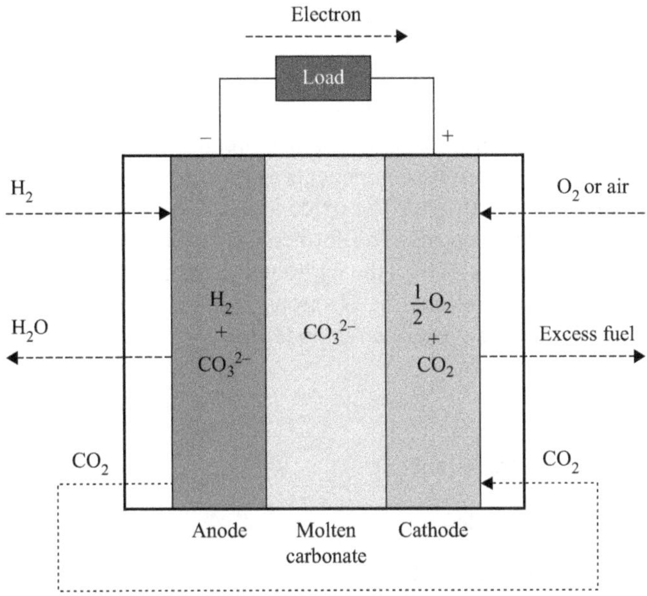


*Figure 2.13 Schematic operation of MCFC system*

The schematic operation of the MCFC is shown in Figure 2.13. Similar as other fuel cells, the MCFC generally produces the water, heat, and electrical energy via the electrochemical reactions. Even though the reaction rate is not very fast, any other fuel gases, such as carbon monoxide or higher hydrocarbons, can be oxidized into hydrogen via the direct electrochemical oxidation.

## *2.6.5 Solid oxide fuel cell*

As compared with other representative fuel cells, the SOFC exhibits the highest operating temperature, up to around 1,000 °C. The first solid-state oxygen ion conductor, as the solid electrolyte for the SOFC, was first suggested by Nernst in 1899. Yet, the first SOFC was proposed by the Nernst's student, Schotkky, in 1930s (Revankar and Majumdar, 2014). Similar as the MCFC, the SOFC was originally designed to operate with the coal as the major fuel. The SOFC was also known as a coal gas cell, indicating that fuels derived from the coal would be employed as its major fuel. Nowadays, both natural gas and derived gas are employed as its major fuel.

To guarantee the ionic and electronic conductivity for normal operation, high operating temperature must be adopted for this type of fuel cell. In the current development, the SOFC normally employs the oxygen ion conductor as its electrolyte. Unlike other the fuel cells with the liquefied electrolyte, the SOFC consists of all solid components so that it enjoys the definite advantage of simpler structure and construction. Since the SOFC has to operate with relatively high temperature, the reaction rate at the electrodes is fast enough to activate the chemical reactions without the presence of the noble metals as catalysts. In other words, no expensive metallic catalysts are needed and hence the materials costs of the SOFC are relatively cheaper. In addition, the high operating temperature also offers the extra tolerant capability when it comes to the presence of the gaseous impurities. Even though there are many advantages of the high operating temperature, there are very few materials available to withstand the normal operation. Hence, the development of the low-temperature SOFC has been the new trend recently.

The SOFC employs the solid components as its major substances, including the oxide ion conductor as electrolyte. The oxide ions are formed by the reaction from the oxygen gas and the electrons. The formed oxide ions are then transported to the anode via the electrolyte. Then, the hydrogen gas at the anodes reacts with the transported oxide ions to release the electrons as well as the energy and water. The reaction at the positive electrode (cathode) is:

$$\frac{1}{2}O_2 + 2e^- \rightarrow O^{2-} \tag{2.36}$$

and at the negative electrode (anode):

$$H_2 + O^{2-} \rightarrow H_2O + 2e^- \tag{2.37}$$

Hence, the overall reaction is:

$$H_2 + \frac{1}{2}O_2 \rightarrow H_2O \tag{2.38}$$

On the other hand, if the carbon monoxide is present to replace the hydrogen gas, the overall reaction becomes:

$$CO + \frac{1}{2}O_2 \rightarrow CO_2 \tag{2.39}$$

The schematic operation of the PAFC is shown in Figure 2.14. Unlike some other fuel cells, such as the AFC or PEMFC, where the presence of the carbon monoxide will contaminate the cell itself, the carbon monoxide instead can be used as the fuel for the SOFC to produce the electrical energy. Hence, the SOFC enjoys the definite advantage of high tolerability towards the carbon monoxide.

### 2.6.6 Direct methanol fuel cell

In the previous sections, as the common trend, all the discussed fuel cells have adopted the hydrogen as the direct fuel. However, the application of hydrogen suffers from many disadvantages. Due to the fact that hydrogen gas is not immediately available on Earth, the hydrogen fuel has to be generated by the primary fuel through some complicated processes. Yet, the formation process of the hydrogen gas unavoidably produces a small portions of carbon monoxide and carbon dioxide, where both of these by-products have poisoning effects to some of the fuel cells. In the meantime, if the hydrogen is in the liquid form, the overall system can become lighter and more compact. Nevertheless, the storage and liquefaction of the hydrogen fuel needs extra equipment, such as the pressure vessel and cryogenic system, which increases the overall weight and degrades the system

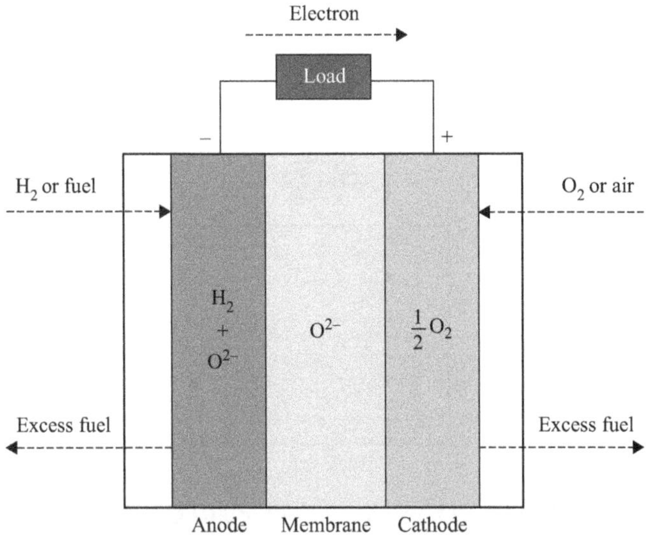


*Figure 2.14 Schematic operation of SOFC system*

performances. Hence, the research on the liquid fuel at ambient temperature and atmospheric pressure, as the replacement of the hydrogen fuel, has drawn many attentions. After various researches on this issue, the methanol has been confirmed as one of the most promising candidates to cope with the task.

The DMFC employs the methanol as its direct fuel to generate the electrical energy through the electrochemical reactions. Instead of being employed directly, the methanol can also be transformed as hydrogen reformate gas mixture within the fuel cell for energy production. This approach is known as the indirect methanol fuel cell (IMFC), and the system layout is shown in Figure 2.15. The IMFC acts similarly as other fuel cells, where the hydrogen gas is employed directly. In terms of performances, the cell potential of the hydrogen fuel is higher than that of the methanol fuel. Hence, it can be expected that the power density of IMFC is higher than that of the DMFC. In contrast, as compared with the IMFC, the DMFC consists of a relatively simpler structure, smaller size, and more cost-effective. Taking into consideration both the performances and costs, the DMFC has accrued more attentions than the IMDC. The schematic structure of the DMFC is shown in Figure 2.16.

Theoretically, with the employment of the methanol as direct fuel, the DMFC can apply any candidates of electrolytes to form many DMFC derivatives, such as the alkaline DMFC (A-DMFC), the phosphoric acid DMFC (PA-DMFC), the proton exchange membrane DMFC (PEM-DMFC), and the molten carbonate DMFC (MC-DMFC). As expected, all of these candidates result with different chemical reactions as well as performances. Due to the consideration of stability, power density, operating temperature range, and technology maturity, the PEM-DMFC has

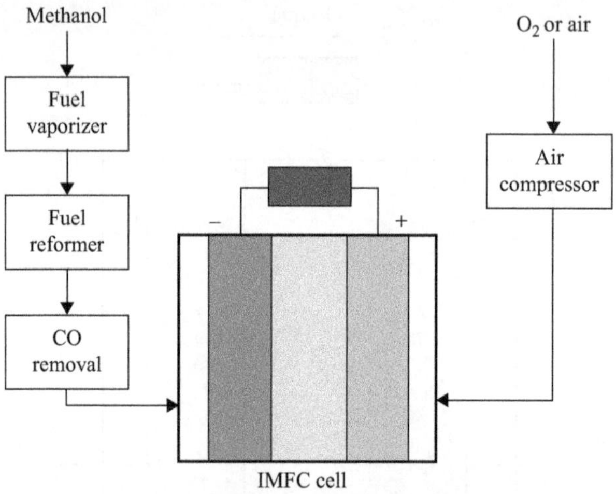


*Figure 2.15 System layout of IMFC system*

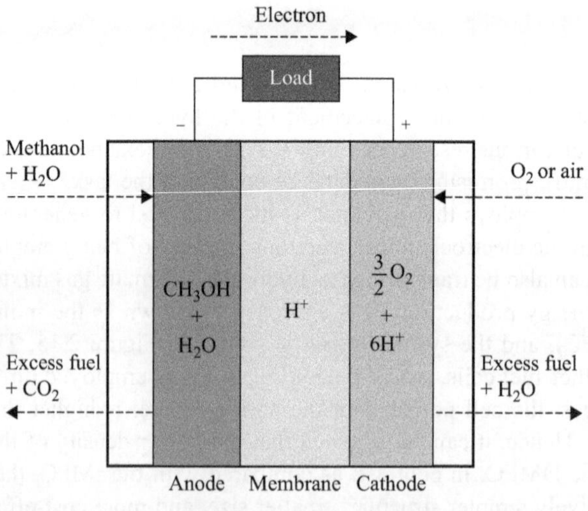


*Figure 2.16 Schematic operation of DMFC system*

been regarded as the most promising candidate for various applications. The reaction of the PEM-DMFC at the positive electrode (cathode) is:

$$\frac{3}{2}O_2 + 6H^+ + 6e^- \rightarrow 3H_2O \tag{2.40}$$

and at the negative electrode (anode):

$$CH_3OH + H_2O \rightarrow 6H^+ + CO_2 + 6e^- \tag{2.41}$$

Hence, the overall reaction is:

$$CH_3OH + \frac{3}{2}O_2 \rightarrow CO_2 + 2H_2O \tag{2.42}$$

It is shown that water is needed for the oxidation at the anode and extra water is produced at the cathode. However, the water is also transported from the anode to the cathode due to the drag effect so that it is more preferable to supply the mixture of the methanol and water to the anode.

## 2.7 Flow batteries

The flow batteries, or sometimes also known as the redox flow batteries, can be regarded as the rechargeable battery with two chemical substances dissolved in liquid form as separated by the membrane. Even though they are commonly called as a "battery," the flow batteries perform similarly as the fuel cells do. There are two major types of flow batteries available, namely, the zinc/chlorine ($Zn/Cl_2$) battery and the zinc/bromine ($Zn/Br_2$) battery. Due to the high specific energy and high efficiency under ambient-temperature operation, the zinc/halogen batteries have ever been considered as the most promising candidates for the utility energy storage and EV applications. Yet, the operation of the zinc/halogen batteries needs extensive plumbing and frequent maintenances of the electrolyte. In particular, the $Zn/Cl_2$ battery suffers from the relatively low specific energy with high costs and thus its development for the utility energy storage and EV applications is severely hindered. Therefore, only the $Zn/Br_2$ battery will be focused for further discussions.

The $Zn/Br_2$ battery employs the bipolar electrodes composed of carbon plastic, the aqueous electrolyte stored in the reservoir, the separator, and the pumps as its major components. The chemical substances of the electrolyte are very complex with polybromide ions presented. The chemical reaction at the positive electrode is:

$$Br_2 + 2e^- \leftrightarrow 2Br^- \tag{2.43}$$

and at the negative electrode:

$$Zn \leftrightarrow Zn^{2+} + 2e^- \tag{2.44}$$

Hence, the overall reaction is:

$$Zn + Br_2 \leftrightarrow Zn^{2+} + 2Br^- \tag{2.45}$$

The zinc and bromide ions are produced during the discharging process, while the zinc and bromine are formed at the respective electrodes during the charging process. The zinc dendrites are formed during the chemical reaction which degrades the battery performances. Hence, the electrolyte circulation system is adopted as shown in Figure 2.17.

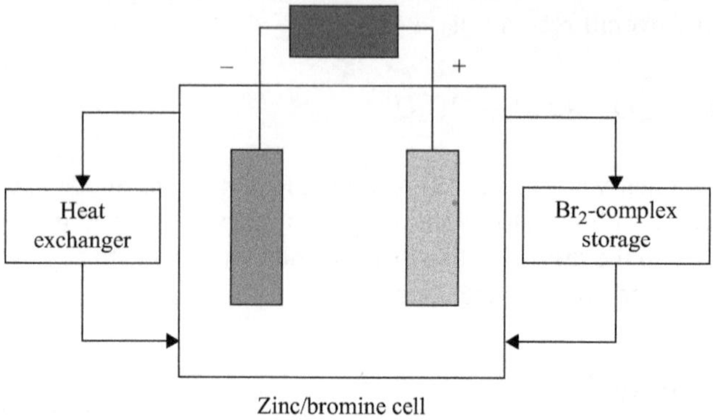


*Figure 2.17 Basic structure of zinc/bromine battery*

## 2.8 Characteristics and EV applications

After the comprehensive discussions on various electrochemical energy sources, the individual characteristics and EV applications are also provided in the following section. Generally speaking, the high power density ultracapacitors are still in the early developing stage where this type of technology normally serves as the auxiliary energy source for EV applications. As compared with the ultracapacitors, the fuel cell technology has provided relatively promising potential to serve as the main EV energy sources, and hence it has drawn many attentions in the past few decades. In the meantime, the batteries have already been adopted as the major energy sources for the existing EV applications.

### *2.8.1 Ultracapacitor characteristics*

The double layer capacitor is more common and representative than the pseudocapacitor and the hybrid capacitor, and hence the ultracapacitor in the following discussion purposely refers to the former one. Operating similarly as the conventional capacitor, the ultracapacitor stores the energy based on the standard capacitor law as:

$$E = \frac{1}{2}CV^2 \tag{2.46}$$

where $E$ is the stored energy, $C$ is the capacitance value, and $V$ is the applied voltage. The ultracapacitor utilizes the nanoporous carbon material as the electrodes to increase the effective surface area, hence improving the capacitance value and the energy density, as compared with the conventional capacitor. The typical ultracapacitor normally provides the specific energy up to 10 Wh/kg, which is 100 times larger than the conventional capacitor does. However, the ultracapacitor still stores much less energy than the batteries and the fuel cells, where the specific energy of the batteries and fuel cells are up to 300 Wh/kg and 1,000 Wh/kg, respectively.

The ultracapacitor, same as other energy sources, follows the law of diminishing return, where there is a trade-off between the amount of energy and power. Hence, it can be expected that the ultracapacitor can be charged up and deliver the energy extremely fast, as compared with other energy sources. Typical ultracapacitors, batteries and fuel cells can provide the specific power around 10 kW/kg, 0.5 kW/kg and 0.08 kW/kg, respectively.

Apart from the high specific power performance, the ultracapacitor outperforms its counterparts in terms of the cycle life, where one life cycle represents one full charge and one full discharge. The ultracapacitor can offer a minimum of 100,000 cycles, and there are no obvious deterioration after 1,000,000 cycles. On the other hand, the batteries and fuel cells can sustain their performances not great than 2,500 cycles and 40,000 h, respectively.

### *2.8.2 Ultracapacitors for EV applications*

Due to its high specific power characteristic, the ultracapacitor nowadays serves as the supplement to another energy source with high specific energy characteristic, including the petrol, the battery, and the fuel cell. The ultracapacitor can store and deliver energy very effectively, while the amount of stored energy is not enough for normal EV applications. The ultracapacitor can be used to handle the peak load, to provide the instant acceleration, and to capture the regenerative braking energy in EVs or HEVs. Because of its low specific energy, the ultracapacitor seldom serves as a sole energy source in EVs.

Unlike the conventional vehicles that employ the gasoline or diesel as the only energy source, the HEVs adopt at least two energy sources for propulsion. Hence, the ultracapacitor can be utilized to incorporate with other energy sources to form different types of HEVs. For instance, the racing car, Toyota Supra-HV-R hybrid car, which is powered by the gasoline internal combustion engine, utilizes the ultracapacitor to capture the regenerative braking energy to provide the starting torque and boosting acceleration from the electric motor. Hence, the overall performances, including the fuel efficiency, can be improved.

By taking the advantages over its competitors, the ultracapacitor has been considered as the replacement of the existing energy sources. For instance, the BMW X3 originally employed the Ni-MH battery as the hybrid energy sources. In contrast, the Ni-MH battery suffered from the problems of relatively high operating temperature and low efficiency. With the replacement of the Ni-MH battery by the ultracapacitor, the new BMW X3 has been released. The new BMW X3 model can be operated at the lower temperature environment with higher efficiency, as compared with the Ni-MH battery–based version. Hence, the problems of thermal runway and fire hazard can be minimized accordingly.

### *2.8.3 Battery characteristics*

The maximum energy available or the theoretical energy capacity of batteries can be calculated by the total amount of the active materials stored within the cell. In general, the specific energy (Wh/kg) is the most commonly adopted quantity to determine the energy capacities. In terms of the energy capacity performances, the batteries can be roughly classified as three groups, namely, the high energy capacity, the fair energy capacity, and the low energy capacity groups. The high energy capacity candidates include the metal/air batteries, and the high-temperature lithium batteries. The fair energy capacity candidates include the sodium-beta batteries, and the ambient-temperature lithium batteries. The low energy capacity candidates include the lead-acid batteries, the nickel-based batteries, and the zinc/halogen batteries. Among all the battery candidates, the metal/air batteries can provide the highest specific energy up to 250 Wh/kg. In particular, these values are almost 5 times larger than those produced by the lead-acid battery.

Apart from the energy capacity, the specific power (W/kg) is another important factor to determine the performances of any energy sources. The specific power is derived as how much energy can the energy sources deliver in a particular period of time. In other words, the specific power means how quick the energy sources can provide the necessary energy to the system. Generally speaking, there is a trade-off between the amount of energy and power, and it is known as the law of diminishing return. Again, the batteries can be classified into three classes based on their specific power performances. The high power candidates include the nickel-based batteries, the high-temperature lithium batteries, and the ambient-temperature lithium batteries. The fair power candidates include the lead-acid batteries, and the sodium-beta batteries. The low power candidates include the zinc/halogen batteries, and the metal/air batteries. Among all the battery candidates, the high-temperature lithium batteries can provide highest specific power up to 400 W/kg.

The rechargeable batteries can undergo the charging and recharging processes throughout the operations, while they suffer from the aging effect after a number of cycles. In the meantime, the maximum number of cycles that the batteries can provide without severe degraded performances is known as the cycle life. Different batteries provide different values of cycle life, and the average number is around 1,200 cycles. Typical characteristics of the batteries are summarized in Table 2.1.

*Table 2.1 Typical characteristics of batteries*

| | **Specific energy (Wh/kg)** | **Specific power (W/kg)** | **Cycle life (cycles)** |
|---|---|---|---|
| Lead-acid | 30–50 | 150–200 | 400–800 |
| Nickel-based | 35–75 | 150–450 | 800–2,000 |
| Ambient-temperature lithium | 120–150 | 200–350 | 600–2,500 |
| Sodium-beta | 115–200 | 120–250 | 800–1500 |
| High-temperature lithium | 130–180 | 240–400 | 1,000–1,200 |
| Metal/air | 75–250 | 100–200 | 300–800 |
| Zinc/halogen | 65–75 | 60–110 | 200–400 |

### *2.8.4 Batteries for EV applications*

All the mentioned batteries have ever been employed in practical EVs, or at least in some concept EVs. The batteries can serve as the energy source to power the pure EVs, while they can also incorporate with gasoline/diesel internal combustion engine to power the HEVs. Due to various considerations, three particular battery types, namely, the VRLA battery, the Ni-MH battery, and the Li-Ion battery, are the most commonly adopted candidates for EV applications. As one of the earliest developed battery, the VRLA battery has been utilized to power the commercial EVs, such as the General Motors EV1. The EV1 was released in the late 1990s, while it was renowned as the first mass-produced EV of the modern era.

Even the Ni-MH battery is not as old as the VRLA battery, it has been regarded as the relatively mature technology, as compared with the remaining battery types. The Ni-MH battery was employed in the General Motors EV1, as an alternative choice of the VRLA battery. Hence, it could run more than the double range than the original EV1 did. After the success of the EV1, the Toyota RAV4 EV that adopted the Ni-MH battery has also drawn some of the attentions.

The Li-Ion battery has been renowned as the most promising candidate for the EV applications, and it was adopted in the second-generation RAV4 EV to replace the Ni-MH battery. The Li-Ion provided better performances than the Ni-MH, especially in terms of specific energy and cycle life. After many applications of the Li-Ion battery to EVs, it came to the remarkable model that broke the selling record of the EV market – the legendary Tesla Model S. The Model S has become the first EV that could rank the top new car sales among the world, where it could even lead the first selling places in some countries. Table 2.2 summarizes typical battery suppliers and their EV applications.

### *2.8.5 Fuel cell characteristics*

Unlike the batteries that take the specific energy as their major concern, the fuel cells instead pay more attentions on their power density. In terms of the power

*Table 2.2 Battery suppliers and EV applications*

| | **Battery suppliers** | **EV applications** |
|---|---|---|
| VRLA | GS, Horizon, Panasonic, Sonnenschein, YUASA | Chrysler Voyager, Daihatsu Hijet, Ford Ranger, GM EV1, Mazda Bongo Friendee, Suzuki Alto |
| Ni-MH | GP, GS, Ovonic, Panasonic, SAFT, Varta, YUASA, Sanyo | Honda EV Plus, Mazda Demio, Peugeot 106, Solectria Force, Toyota RAV4L |
| Li-Ion | GS, SAFT, Sony, Varta, Bosch, Samsung, Panasonic, Johnson Controls | Nissan Prairie Joy, Tesla Model S, BMW i3, Mitsubishi i-MiEV, Nissan Leaf, Ford Focus Electric, Chevrolet Spark EV |
| $Na/NiCl_2$ | Zebra | BMW AG, Mercedes-Benz Vito |
| Zn/Air | Electric Fuel | GM-Opel Corsa Combo, Mercedes-Benz MB410 |

density performances, the fuel cells can be roughly classified into three groups, namely, the high power density, the fair power density, and the low power density groups. The high power density candidates include the PEMFC and the DMFC. The fair power density candidates include the AFC and the SOFC. The low power density candidates include the PAFC and the MCFC. Among all the fuel cell candidates, the PEMFC can provide the highest power density up to 6,500 W/m$^3$. In particular, these values are almost 2.5 times larger than those produced by the PAFC or the MCFC.

Apart from the power density, the operating temperature is another key factor to determine the performances of any energy sources. The operating temperature is important because the high temperature can assist the fuel cell reactions in the absence of noble metal catalysts. However, it suffers from the requirement of higher cost for the thermal management. The fuel cells can be classified into two classes based on their operating temperatures, namely, the high-temperature candidates and the fair-temperature candidates. The high-temperature candidates include the MCFC and the SOFC. The fair-temperature candidates include the AFC, the PAFC, the PEMFC, and the DMFC. In particular, the AFC, the PEMFC, and the DMFC can be operated without any thermal management, unless there are any further purposes to be achieved.

For the fuel cell performances, the fuel efficiency is one of the major criteria to be considered. The fuel efficiency shows the effectiveness of the chemical reaction associated within the fuel cells. It should also be noted that hydrogen can be employed as the major fuel for most of the fuel cells, except the DMFC. Unlike other fuel cells, the DMFC employs the methanol as its major fuel. All the fuel cell candidates perform similarly in terms of the fuel efficiency, with the efficiency around 40–60%. Yet, the DMFC suffers from the lowest value among its counterparts, with the efficiency around 30–40%. Typical characteristics of the fuel cells are summarized in Table 2.3.

### *2.8.6 Fuel cells for EV applications*

Unlike the relatively well developed battery EVs, the fuel cell EVs are still in the developing stage. Yet, they enjoy the advantage of unlimited range, and hence have drawn many attentions since the past few decades. The first fuel cell road vehicle,

*Table 2.3 Typical characteristics of fuel cells*

| | **Primary fuel** | **Power density (W/m$^3$)** | **Operating temperature (°C)** | **Fuel efficiency (%)** |
|---|---|---|---|---|
| AFC | $H_2$ | 1,000–3,000 | 50–150 | 40–60 |
| PAFC | $H_2$ | 800–2,500 | 150–220 | 40–55 |
| PEMFC | $H_2$ | 3,500–6,500 | 50–100 | 45–60 |
| MCFC | $H_2$, CO, $CH_4$ | 1,500–2,600 | 500–700 | 45–65 |
| SOFC | $H_2$, CO | 1,500–3,000 | 700–1,000 | 45–65 |
| DMFC | Methanol | 1,500–3,500 | <100 | 30–40 |

*Table 2.4 Fuel cell EV manufacturers and their EV applications*

| Manufacturers | EV applications |
|---|---|
| General Motors | GM Electrovan, HyrdoGen3, Chevrolet Equinox Fuel Cell, HydroGen4 |
| Toyota | Toyota RAV4 FC EV, Toyota FCHV-3, Toyota Mirai |
| Daimler | Mercedes-Benz NeCAR 1, Mercedes-Benz F-Cell Roadster |
| Ford | Ford Focus FCV, F-250 Super Chief |
| Honda | Honda FCX Clarity |
| Mazda | Mazda HR-X, Mazda Capella Cargo, Mazda 5 Hydrogen RE Hybrid |

the GM Electrovan, was developed by General Motors in 1966. It employed the AFC with liquid hydrogen and oxygen fuel tanks, which ranged around 190 km with top speed of 70 mph. The GM Electrovan demonstrated the potential of the fuel cell EV, yet there were many technical problems awaited to solve before it could be employed for commercial applications. General Motors and some other famous companies, such as Toyota and Mazda, have continued their work on the fuel cell EVs, while most of the attentions are drawn by the battery EVs and HEVs.

The development of fuel cell EVs was slow in the past few decades, while there were still some well-known companies have kept the projects for the fuel cell EVs. Toyota Motor was one of the largest vehicle manufacturers in Japan that has continued their R&D on fuel cell EVs. One of the representative fuel cell EVs developed by Toyota was the FCHV-3, which was released in 2001. The FCHV-3 employed the PEMFC, with the direct hydrogen as the major fuel. It purposely installed the separated Ni-MH battery sets to capture the power from the regenerative braking and to provide some other features for high-efficiency driving performances. With the combination of the hybrid features, it could provide the range around 300 km.

Some other companies, such as Honda, Mazda, or BMW, have also launched the fuel cell EV projects in the past few decades, yet most of them offered only the limited commercial releases or even the concept car models. The fuel cell EV industry has been waiting to have a breakthrough while it might happen soon. According to the reports by US Environmental Protection Agency, Hyundai and Toyota will release their new models in 2016, while each of them is estimated to offer the range over 400 km per refuel. Table 2.4 summarizes some fuel cell EV manufacturers and their EV applications.

## 2.9 Trend

Due to the consideration of the technology maturity, cost-effectiveness and overall performances, the battery EVs have been dominating the current EV market. In particular, the Li-Ion batteries have been widely employed as the major candidate for EV applications. Theoretically, the Li/Air battery and the Li/S battery can provide higher specific energy or higher specific power than the Li-Ion batteries.

Once these batteries can resolve the corresponding problems, such as the high reactive rate problem, the thermal management, and the cost issue, it is expected that the Li/Air or the Li/S battery will take over the dominating position.

Unlike the battery candidates, the fuel cells and ultracapacitors serve instead as the supplementary roles. Nevertheless, the fuel cells enjoy the exceptionally high specific energy, while the ultracapacitors enjoy the extremely high specific power. Hence, more attentions have been recently shifted to the development of the fuel cell EVs and the hybrid EVs installed with ultracapacitors. Based on the considerations of various factors, it is anticipated that the PEMFC and the SOFC will be the major trend for the fuel cell EVs.

The pure EV enjoys the definite advantages of zero emission, silent operation, and high fuel efficiency, yet it suffers the major problem of the short driving range. As compared with other parts of EV technologies, the EV energy sources are relatively immature and not satisfactory enough for daily applications. In the foreseeable future, unless there is any significant breakthrough, none of the single EV energy source can carry out all the vehicular requirements to enable the pure EV competing with the diesel or gasoline internal combustion engine vehicles. Instead, some alternative ways that can improve the deficiencies of the EV energy sources should be developed. In particular, the concept of energy hybridization is accepted as one of the best solutions to reframe the current situation.

Due to the law of diminishing return, any particular EV energy source can only provide either the high specific energy or high specific power, but not both simultaneously. As stated in the earlier section and depicted in Figure 2.1, it is clear there is no single energy source that can carry out all the requirements for EV applications. Even though there are derivatives among each group of technologies, the batteries and the fuel cells can typically offer relatively higher specific energy performances, while the ultracapacitors can instead offer the relatively higher specific power performances. Hence, the implementation of multiple energy sources can definitely cope with the existing problems. To relieve the problem, the EVs, which incorporate multiple energy sources with different characteristics, seem to be the best solution for the current situation. Actually, HEVs, in which the liquid fuel and batteries serve as the multiple energy sources, are a success to solve the problem of short driving range even though they violate the concept of zero emission. It is anticipated that pure EVs with energy hybridization and HEVs will be the major trend in the coming future.

## References

Chan, C. C. and Chau, K. T. (2001) *Modern Electric Vehicle Technology*, Oxford: Oxford University Press.

Chau, K. T. (2012) Research on electric vehicles: challenges, opportunities and emerging technologies. *Studies in Science and Technology*, 1, 13–24.

Chau, K. T. (2015) *Electric Vehicle Machines and Drives – Design, Analysis and Application*, Chichester, West Sussex: Wiley-IEEE Press.

Chau, K. T., Wong, Y. S., and Chan, C. C. (1999) An overview of energy sources for electric vehicles. *Energy Conversion and Management*, 40, 1021–1039.

Burke, A. F. (2007) Batteries and ultracapacitors for electric, hybrid, and fuel cell vehicles. *Proceedings of the IEEE*, 95, 806–820.

Deshpande, R. P. (2015) *Ultracapacitors*, New York: McGraw-Hill Education.

Khaligh, A., and Li, Z. (2010) Battery, ultracapacitor, fuel cell, and hybrid energy storage systems for electric, hybrid electric, fuel cell, and plug-in hybrid electric vehicles: state of the art. *IEEE Transactions on Vehicular Technology*, 59, 2806–2814.

Li, H. Q., Wang, Y. G., Na, H. T., Liu, H. M., and Zhou, H. S. (2009) Rechargeable Ni-Li battery integrated aqueous/nonaqueous system. *Journal of the American Chemical Society*, 131, 15098–15099.

Li, H. Q., Weng, G. M., Li, C. Y. V., and Chan, K. Y. (2011) Three electrolyte high voltage acid-alkaline hybrid rechargeable battery. *Electrochimica Acta*, 56, 9420–9425.

Lukic, S. M., Cao, J., Bansal, R. C., Rodriguez, F., and Emadi, A. (2008) Energy storage systems for automotive applications. *IEEE Transactions on Industrial Electronics*, 55, 2258–2267.

Reddy, T. (2011) *Linden's Handbook of Batteries*, New York: McGraw-Hill Education.

Revankar, S., and Majumdar, P. (2014) *Fuel Cells: Principles Design and Analysis*, Boca Raton: CRC Press, Taylor & Francis Group.

Whittingham, S. (2012) History, evolution, and future status of energy storage. *Proceedings of the IEEE*, 100, 1518–1534.

*Chapter 3*

# Ultrahigh-speed flywheel energy storage for electric vehicles

*Wenlong Li*[1] *and T.W. Ching*[2]

Flywheel energy storage systems (FESSs) have been investigated in many industrial applications, ranging from conventional industries to renewables, for stationary emergency energy supply and for the delivery of high energy rates in a short time period. FESSs can be used for industrial applications ranging from aerospace stations and railway trains to electric vehicles (EVs). They have their own individual advantages and disadvantages, leading them to have their own unique roles for energy storage applications. Compared to the limitation of an electrochemical battery imposed by its inherent features, such as low power density, short duration of service, limited charge-discharge cycles and being environmentally unfriendly, FESSs exhibit some distinctive merits, such as high energy density, low cost, high reliability, high dynamics, long lifetime, high efficiency, environmental friendliness and easy monitoring of the state of charge.

In this chapter, FESSs applied in EVs are discussed, with an emphasis on those operating at speeds over 10,000 rpm. The organization of this chapter is as follows. In section 3.1, a brief introduction of FESSs is presented. In section 3.2, the configuration of an FESS, including a flywheel, a motor/generator, a bearing, a power converter and an enclosure, is described. Then, in section 3.3, possible candidates for ultrahigh-speed motors/generators for FESSs are reviewed. Lastly, in section 3.4, control strategies for motor/generator control, flywheel control and power flow control are discussed.

## 3.1 Flywheel energy storage

An FESS stores energy in the form of kinetic energy based on a rotating mass driven by an electric machine. An FESS can be seen as including two fundamental parts: an energy storage part and a power transformation part. Since it is stored as kinetic energy in the rotating mass, the stored mechanical energy is proportional to the weight of the rotating mass and the square of the rotating speed. Therefore, the higher

[1]Department of Electrical and Electronic Engineering, The University of Hong Kong, Hong Kong, China
[2]Department of Electromechanical Engineering, University of Macau, Macao, China

*Table 3.1 Performance comparison for different energy storage devices*

| | FESS | Battery | Fuel cell | Supercap | SMES |
|---|---|---|---|---|---|
| Power density | 4 | 2 | 1 | 5 | 3 |
| Energy density | 3 | 4 | 5 | 1 | 2 |
| Duration per charge | 2 | 4 | 5 | 3 | 1 |
| Lifetime | 5 | 1 | 3 | 2 | 4 |
| Environmental-friendliness | 4 | 1 | 5 | 2 | 3 |
| Price | 3 | 1 | 5 | 2 | 4 |

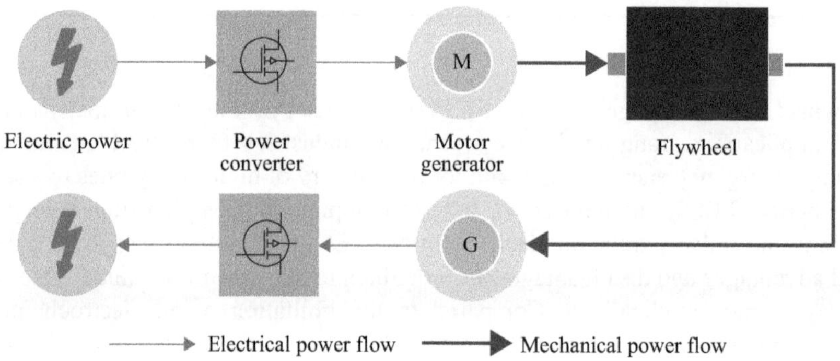


*Figure 3.1 Flywheel energy storage: operating principle*

the speed at which it rotates, the more mechanical energy can be stored. Compared with other energy storage devices, FESSs have many distinct advantages, such as high specific energy and power, a long cycle life and high energy efficiency, along with a reduction of maintenance, contamination and system overheads. A comparison with other energy storage technologies including the battery, fuel cell, supercapacitor (supercap) and superconducting magnetic energy storage (SMES) is shown in Table 3.1 (Hebner *et al.*, 2002; Chen *et al.*, 2009; Vazquez *et al.*, 2010), where a point grading system ranging from 1 to 5 points is adopted, with 1 point being the worst and 5 points the best. As a core component in an FESS, an electrical machine realizes energy interconversion between kinetic energy and electric energy. Therefore, a suitable candidate is a machine that can offer high power density, high efficiency, high reliability and a very wide speed range.

In recent decades, FESSs have been mostly mechanically based. The classic one was developed by General Motors, and required continuous variable ratio transmission and lots of clutches (Acarnley *et al.*, 1996). Due to mechanical power transmission problems caused by mechanical components, FESSs of this kind have been abandoned. By introducing an electric motor, mechanically complex arrangements can be avoided. The modern electrically based FESS mainly consists of a flywheel for storing kinetic energy, a motor/generator and power electronic converters, as shown in Figure 3.1. To improve the energy density, the operating

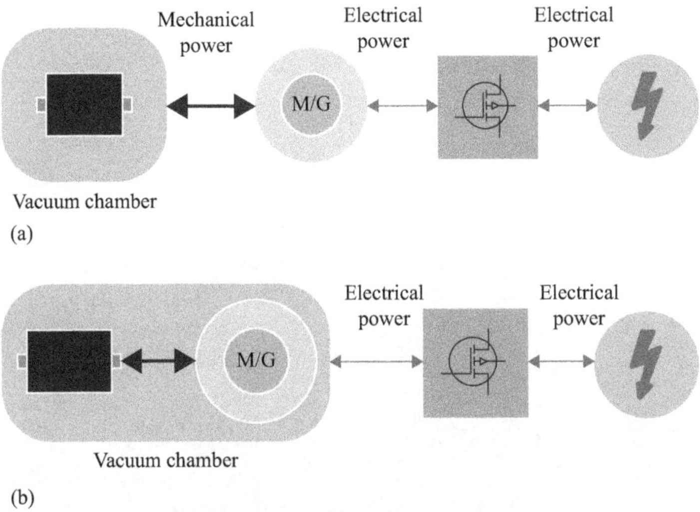


*Figure 3.2 Flywheel and machine arrangements: (a) separate flywheel and machine, and (b) integrated flywheel and machine*

speed should be as high as possible. However, at these high speeds, the windage loss is excessive. In order to reduce the windage loss, the flywheel is placed in an evacuated container. As shown in Figure 3.2, the flywheel rotates in a vacuum and is supported by magnetic bearings while the magnetic bearings are aided by mechanical bearings for fault-tolerant purposes. However, the use of auxiliary mechanical bearings can significantly reduce the system's life.

Due to the distinct advantages provided by the FESS, it is applied in a wide range of applications, such as uninterruptible power supply (Lawrence *et al.*, 2003), the railway (Lee *et al.*, 2013), a space station (Cosgriff and Baaklini, 2003), wind power generation (Diaz-Gonzalez *et al.*, 2014), and EVs (Dhand and Pullen, 2015). In this chapter, only the application for EVs, including hybrid EVs (HEVs) and battery EVs (BEVs), is presented in details.

For EVs, the FESS can be applied as a main power source for driving the electric motor or as an auxiliary energy storage for assisting the power source and absorbing the power recovered from regenerative braking.

### *3.1.1 FESSs as the main power source*

The first EV entirely powered by an FESS was developed in the 1940s by Oerlikon, a company based in Switzerland. Compared to battery-powered EVs, this vehicle is much quieter and requires limited overhead charging wires, as shown in Figure 3.3. The first EV of this kind was commercialized in 1953, and was an electric bus called a 'gyrobus'. It adopted an induction motor to drive a large flywheel rotating up to 3,000 rpm. Power for charging the flywheel was fed from three booms mounted on the roof of the bus, which contacted charging points located as desired. To obtain tractive power, capacitors excited the electric machine which operated as

*Figure 3.3 An FESS-powered electric bus being charged (Courtesy Smiley.toerist, Wikimedia Commons, https://upload.wikimedia.org/wikipedia/commons/3/32/Gyrobus_bij_oplaadpunt.jpg)*

a generator and converted the kinetic energy stored in the flywheel back into electricity. The vehicle braking was electric, and some of the energy was recuperated back into the flywheel and extended its range.

One of the major problems with gyrobuses was the excessive wear and tear. Other problems included breakage of gyro ball bearings, and high humidity resulting in traction motor overload making the gyrobuses unreliable. Also, the heavy weight of gyrobuses damaged road surfaces. For these reasons, most of the gyrobus services were eliminated from the market.

### *3.1.2 FESSs as auxiliary energy storage*

FESSs can also act as an auxiliary energy storage for both HEVs and BEVs for recovering braking power and providing power enhancement.

When starting up and accelerating, the FESS can provide a large amount of energy in a short time to meet the abrupt power demand. For retrieving the power from regenerative braking, the FESS is capable of enhancing the fuel economy of a HEV and extending the driving range of a BEV.

HEVs are usually powered by more than one source, and the scheme for integration with FESSs varies. Figure 3.4 shows an FESS applied in HEVs. Conventionally, they can be classified into two categories: one scheme uses a mechanical gear and clutch for integration, and the other uses one or several planetary gears for integration, such as the transmotor systems proposed by Texas A&M

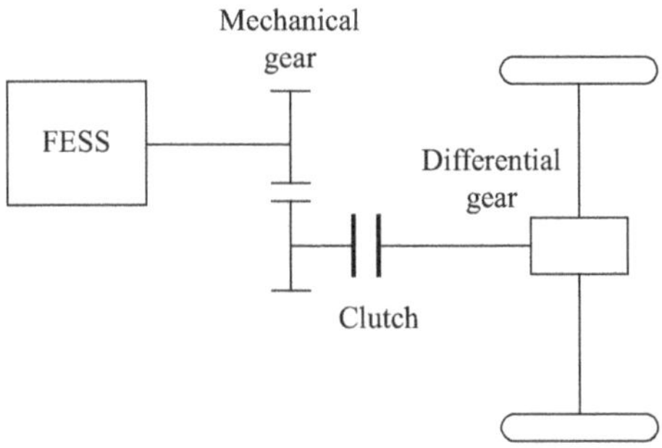


(a)

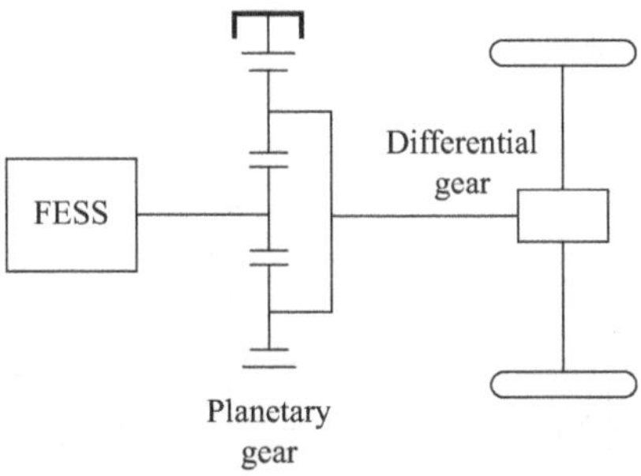


(b)

*Figure 3.4 An FESS in an HEV as an auxiliary energy source: (a) coupled with mechanical gear and clutch, and (b) coupled with planetary gear*

University, the petrol electric drivetrain proposed by Jeffries, and the Swiss Federal Institute flywheel concept (Miller, 2004; Read *et al.*, 2015).

FESSs utilized as an auxiliary energy storage for BEVs can be dated back to the early 1970s. By using an FESS to compensate for the battery, the battery capacity and life can be extended due to the avoidance of overloading and drawing of peak current. As the flywheel accelerates when the vehicle is decelerating and decelerates when the vehicle is accelerating, the transmission is continuously variable. Several transmissions, namely, hydromechanical, electromechanical, mechanical-belt, mechanical-toroidal, mechanical-cone, planetary-gear and electrically integrated, are adopted for integration with the FESSs (Dhand and Pullen, 2015). Figure 3.5 illustrates the most common schemes: hydromechanical, planetary gear and electrical fully integrated styles.

Since the on-board flywheel acts as a gyroscope, it resists changes in orientation. If the flywheel is mounted horizontally with a horizontal rotation axis, it may cause the vehicle to tilt while the vehicle makes a turn. In order to alleviate the gyroscopic effect, the flywheel can be implemented into a gimbal system, which can mitigate the gyroscopic torque (Bolund *et al.*, 2007). Therefore, when the vehicle is turning or leaning, the position of the flywheel can be kept unchanged and the gyroscopic effect can be eliminated.

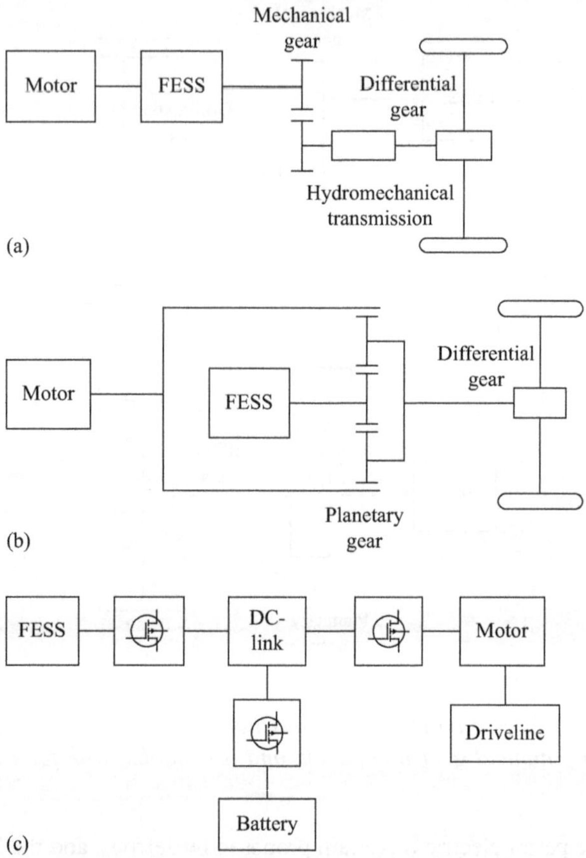


*Figure 3.5 An FESS in a BEV as an auxiliary energy source: (a) coupled with hydromechanical transmission, (b) coupled with planetary gear and (c) electrical fully integrated*

## 3.2 System configuration

As discussed in the previous section, an FESS consists of a flywheel, a motor/generator, a power converter and an enclosure. The flywheel stores the energy in a mechanical form. The motor and generator perform the electromechanical energy interconversion between the flywheel rotor and the power source. The power converter and the controller serve both the motoring and generating operations and realize the power flow control between the power source, the flywheel rotor and the load. In this section, every component of FESSs will be briefly presented.

### *3.2.1 Flywheel*

Due to the ultrahigh speed of the flywheel, a huge stress is exerted on the rotating flywheel that may make the rotor disintegrate. For safety reasons, the selection of

*Table 3.2 Properties for different flywheel materials*

| | Density (kg/m$^3$) | Tensile strength (MPa) | Energy density (MJ/kg) |
|---|---|---|---|
| 4340 Steel | 7,700 | 1520 | 0.19 |
| composites | | | |
| E-glass | 2,000 | 100 | 0.05 |
| S2-glass | 1,920 | 1470 | 0.76 |
| Carbon T1000 | 1,520 | 1950 | 1.28 |
| Carbon AS4C | 1,510 | 1650 | 1.1 |

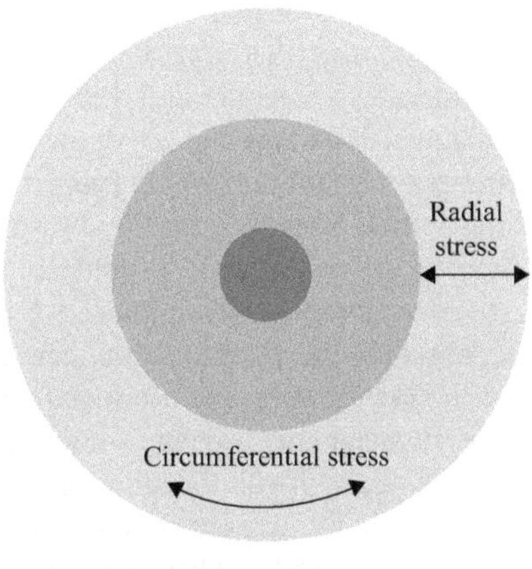


*Figure 3.6 Radial and circumferential stresses in a hollow cylinder flywheel rotor*

materials for the rotor needs to satisfy the requirements for the minimal stresses caused by the centrifugal force. Table 3.2 compares the physical properties of different flywheel materials (Bolund *et al.*, 2007). The material 'Carbon T1000' outperforms other materials significantly. There are mainly two stresses: radial stress and circumferential stress. To exemplify this, a design is based on a hollow cylinder as shown in Figure 3.6. For an isotropic material, the radial stress and the circumferential stress can be represented as follows (Bolund *et al.*, 2007):

$$\sigma_r = \frac{\upsilon + 3}{8}\rho\omega^2\left(r_i^2 + r_o^2 - r^2 - \frac{r_i^2 r_o^2}{r^2}\right) \tag{3.1}$$

$$\sigma_c = \frac{\upsilon + 3}{8}\rho\omega^2\left(r_i^2 + r_o^2 + \frac{r_i^2 r_o^2}{r^2} - \frac{3\upsilon + 1}{\upsilon + 3}r^2\right) \tag{3.2}$$

where $\upsilon$ is the Poisson ratio, $\rho$ is the mass density, $\omega$ is the machine speed, $r_i$ is the inner radius of the rotor, $r_o$ is the outer radius of the rotor and $r$ represents the radius of a point within the rotor.

In the case of a catastrophic failure, the flywheel is more likely to develop circumferential cracks and less likely to produce free-flying projectile fragments.

Therefore, the flywheel should be made by fibre-reinforced composite rotors with circumferential orientation. In most circumstances, a rotational speed can be dropped up to 50% of the full speed, therefore the mechanical energy is dropped to 25% of the stored energy and the depth of discharge is 75%. In summary, the geometry of the flywheel and the operating speed determines the energy storage capability, whilst the electric machine and the power converter determine the power capability.

### 3.2.2 Bearing

High friction and the associated energy losses are the result of ultrahigh-speed operation of mechanical bearings, which require lubrication and periodic maintenance. In order to reduce losses, magnetic bearings are proposed for totally or partially replacing the mechanical bearings in FESSs. A magnetic bearing utilizes magnetic forces to levitate the shaft without direct contact, therefore there is no wear and tear problem. Also, no lubrication or maintenance is required. There are two types of magnetic bearings, namely, passive magnetic bearings and active magnetic bearings.

Passive magnetic bearings consist of permanent magnets that support the weight of the flywheel by repelling forces. However, this kind of magnetic bearing should be used together with other bearing types due to its inherent unstable nature. Active magnetic bearings consist of electromagnets produced by coils that can adjust the electromagnetic forces based on the shaft position, hence achieving stability by using a complicated feedback system. Magnetic bearings based on high-temperature superconductors utilize the diamagnetic property of superconductors when the superconducting temperature is reached with the aid of a cryogenic cooling system. The rotor can be self-levitated without the control system for positioning. However, it still needs auxiliary mechanical bearings when the magnetic bearings encounter a failure.

### 3.2.3 Motor/generator

An electric machine is a key component of an FESS. It performs electromechanical interconversion in the course of charge and discharge. When the FESS operates in charge mode, it works as a motor to accelerate the flywheel for converting electricity into mechanical energy. When the FESS operates in discharge mode, it works as a generator to decelerate the flywheel for converting mechanical energy into electricity. Since the speed of the flywheel is variable in the entire process, variable speed AC machines are considered suitable candidates for FESS application.

Electric machines possessing high power density, high torque density, high efficiency and high reliability features are much preferred. However, most common types of electric machines, such as induction machines, synchronous machines and reluctance machines, have various limitations, such as high losses, low ruggedness and so on. In order to improve the performance for ultrahigh-speed operation, the machine design and control strategy should be specially designed and implemented with high-strength and low-loss materials and advanced control schemes, such as feedforward (FF) and disturbance decoupling (DD) control. Detailed design and control issues are presented in the following sections.

### 3.2.4 Power converter

A power converter is also a must in FESSs. During the charge mode, the power converter acts as a motor driver to realize the power flow from the power source to the flywheel. During the discharge mode, the power converter works together with the generator to serve as a power source for providing electricity to the traction loads.

A bidirectional power converter, generally a voltage source inverter (VSI), is adopted for coupling the FESS to a DC-link (Pena-Alzola *et al.*, 2011). The appropriate control of the power switches relies on the voltage and current capabilities and the switching frequency. The VSI is controlled to coordinate the machine to operate as a motor or generator according to the requirement.

Generally, electrical machine control for FESSs consists of two closed loops: (1) the inner loop is to control the stator current to make a fast response and (2) the outer loop is to the control the torque and hence the power, which is slow compared to the current control loop. Usually, FF current control is used to provide a more accurate model to avoid instability. The aim of current control is to produce the required torque for maximizing the energy efficiency.

In order to stabilize the DC-link voltage, the torque of the electric machine needs to be controlled based on the load variation. When the DC-link voltage increases, the electric machine is controlled as a motor for accelerating the flywheel to charge the FESS. When the DC-link voltage decreases, the electric machine is controlled as a generator for decelerating the flywheel to discharge the FESS. The VSI controls the DC-link voltage and allows the commanded torque reference for speed control.

### 3.2.5 Enclosure

Air friction torque is proportional to the flywheel speed and also to the density and pressure of the gas surrounding the flywheel. In order to reduce windage losses, the flywheel should be placed in a container with partial vacuum or at least with a gas that is less dense than air. Furthermore, for completely eliminating the windage losses, the flywheel can be implanted into a container with absolute vacuum. However, the absolute vacuum environment may also lead to a complicated structure and a high cost. In addition, it may be problematic for cooling the electrical machine and lubricating the mechanical bearings. To solve this problem, the flywheel and the electric machine can be implanted in different ways as shown in Figure 3.2.

For the sake of other safety concerns, the enclosure must have the capability of preventing the catastrophic problem caused by the flywheel flying out or disintegrating. Flywheels made of composite material fibres may disintegrate into rotating fragments, which can easily be contained in the enclosure. In the case of flywheels made of steel, a fierce explosion may occur in the form of a few translational fragments, and this is much more difficult to contain. Therefore, for ultrahigh-speed operation of FESSs, the enclosure must weigh half of the flywheel weight, and more than twice the flywheel weight for low-speed operation by storing the same energy (Pena-Alzola *et al.*, 2011).

## 3.3 Electric machines

As discussed in the previous section, the stored energy in FESSs is proportional to the square of rotating speed of the flywheel. The power density can therefore be significantly improved by increasing the rotating speed rather than increasing the rotor mass. Thus, electric machines with high rotating speed are preferred for FESSs. In this chapter, the electric machines discussed are ultrahigh-speed machines that generally operate beyond 18,000 rpm (Kalpakjian and Schmid, 2014). However, ultrahigh-speed machines generally encounter many more difficulties in electrical, structural and thermal aspects than conventional-speed machines: high core losses and the associated power electronic losses, mechanical losses caused by mechanical bearing friction, a huge centrifugal force and thermal concerns due to the vacuum enclosure. The high losses may decrease the overall efficiency, the centrifugal force may affect the rotor mechanical integrity and the thermal issues may reduce the reliability. To solve these problems, electrical, magnetic, mechanical and metallurgical schemes have been proposed and implemented, such as the use of low iron loss electrical steel, which not only improves the efficiency but also avoids the thermal problems, high-strength steel or casing, mechanical bearings with appropriate lubrication, or adoption of magnetic bearings. In this section, induction machines, permanent magnet brushless (PMBL) machines, synchronous reluctance (SynR) machines, switched reluctance machines (SRMs) and homopolar machines for ultrahigh-speed design are mainly discussed and presented.

### *3.3.1 Induction machine*

Since it is relatively mature, the manufacturing cost is comparatively low and it can be readily operated without speed or position feedbacks, an induction machine is very attractive for operating in the ultrahigh-speed regime. In contrast with a wound-rotor induction machine, its cage-rotor counterpart is more suitable for high-speed operation, due to its compact structure, as shown in Figure 3.7.

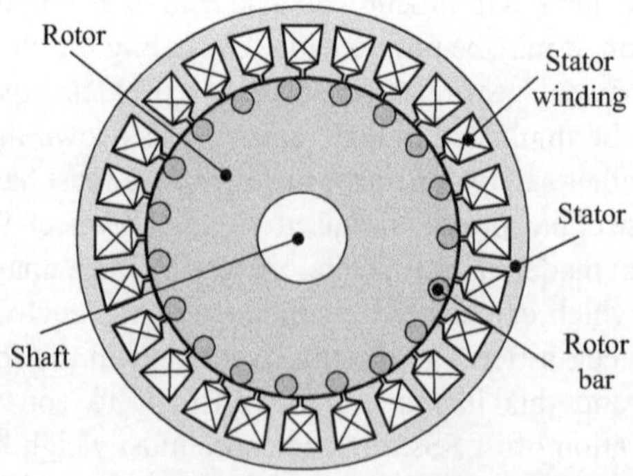


*Figure 3.7 Cage-rotor induction machine*

However, there is a distinct conflict between selecting a laminated rotor, which can improve the efficiency, or a solid rotor, which can enhance the rotor's ruggedness. In order to minimize the core loss whilst retaining the mechanical integrity, some special design techniques are proposed.

The rotor should adopt high-silicon steel laminations to avoid high core losses and improve strength in the high-speed regime. Advanced high-strength high-conductivity conductor material for the rotor cage and end rings should be adopted. Moreover, the stray load losses can be minimized by reducing the stator slot opening.

For the rotor design, the selection of rotor bars is also important. The rotor bars not only carry the induced current but also keep the rotor lamination together. AL-15 from Glidcop was selected for the rotor bars (Soong *et al.*, 2000). This material is created by copper powder mixed with aluminium oxide. The mixture is poured into normal copper tubes, heated and drawn down to consolidation. Thus it has both high conductivity and high strength. Moreover, the rotor slots should be closed to avoid excessive centrifugal forces and reduce windage losses.

For the lamination design, 0.356 mm high-silicon (3.25%) non-oriented silicon steel lamination material was adopted (Soong *et al.*, 2000). To further reduce high-frequency core losses, a special heat treatment was developed to produce an optimum grain size. The laminations were heat treated to produce large grains in order to minimize hysteresis loss.

For the stator design, the stray load losses were minimized by reducing the stator slot opening. For the heat treatment, stator for lowest losses, rotor for best strength and ductility with adequate electromagnetic properties should be adopted.

A double-fed induction machine is referred to as an adjustable-speed rotational condenser capable of both active power control and reactive power control. The rotor windings are excited by three-phase AC currents, which are supplied via slip rings by either a cycloconverter or a voltage-fed pulsewidth modulation (PWM) rectifier-inverter (Akagi and Sato, 2002). The AC excitation on account of a position feedback control to attain stable variable-speed operation. The electrical power flow can be controlled by adjusting the rotor speed. However, since the rotating part utilizes the slip rings for excitation, it is only applicable for low-speed FESSs and not suitable for ultrahigh-speed operations.

### *3.3.2 PMBL machines*

In contrast to induction machines, PMBL machines offer high torque density, high power density and high efficiency, and are suitable candidates for FESS applications. Designing an ultrahigh-speed PMBL machine is always problematic due to the existence of PMs. Besides the electrical, mechanical and thermal problems discussed before, there are three major concerns over PMBL machines: the PM demagnetization problem due to overheating, the PM retention problem and the mechanical strength of PM materials due to the large centrifugal force.

In order to reduce the core loss, stator slot number and PM pole number usually adopt a combination with a small number of slots and poles, such as 3-slot/2-pole (Noguchi *et al.*, 2007). For surface-mounted PM (SPM) machines, parallel-magnetized solid cylinder PMs are used and held in a hollow section retention sleeve, which is generally made of a high-strength material such as titanium, Inconel, carbon fibre or glass fibre, for holding PMs in position. For interior PM machines, the magnet retention sleeve usually adopts silicon-iron alloy (SiFe) (Gerada *et al.*, 2014).

Two radial-flux SPM machines, namely outer-rotor and inner-rotor, are presented as follows. In the outer-rotor machines, PMs are surface-mounted on the outer rotor, which results in a simple magnet retention configuration, as shown in Figure 3.8. However, the shape factor of a hollow cylinder is poor. Since the armature winding in the inner stator should axially pass through the rotor bearing, the bearing bore needs to be large and the outer brace needs to be rotating. The large diameter bearings for ultrahigh-speed operation cause severe bearing losses. Therefore, magnetic bearings are more suitable for this topology (Ooshima *et al.*, 2010).

As shown in Figure 3.9, for inner-rotor radial-flux PMBL machines, PMs are surface-mounted on the inner rotor, and need to be held in position by a retaining sleeve. This sleeve should be made of high-strength nonmagnetic steel, or a layer of

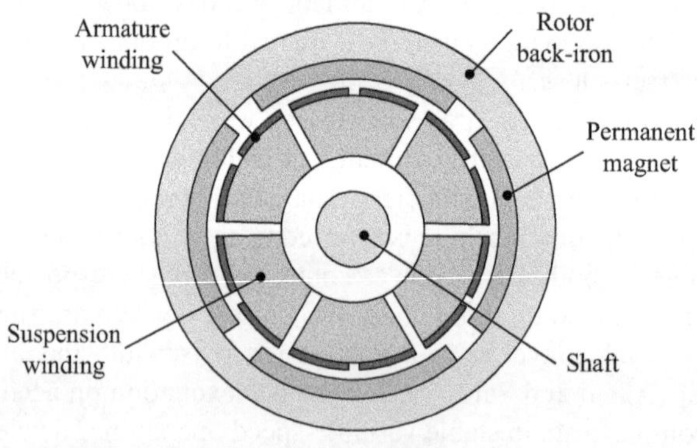


*Figure 3.8 An outer-rotor radial-field PMBL machine using magnetic bearing*

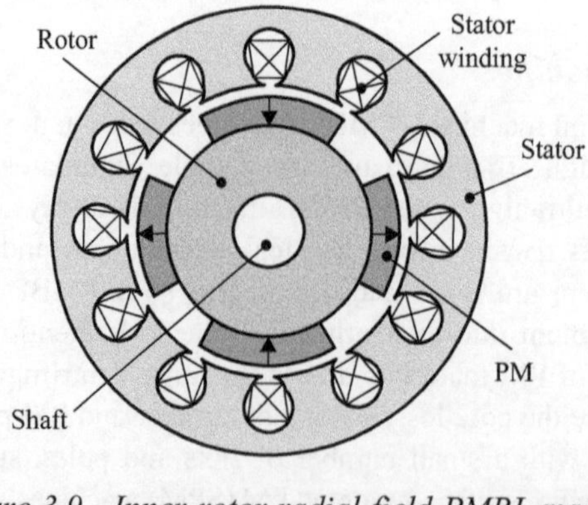


*Figure 3.9 Inner-rotor radial-field PMBL machine*

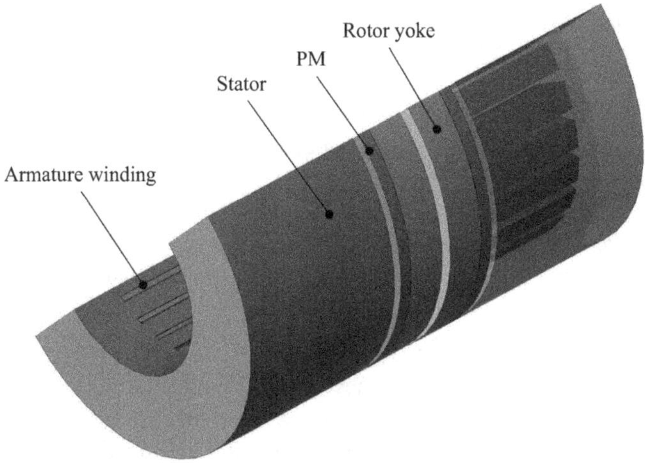


*Figure 3.10 Double-stator axial-field PMBL machine*

filament-wound carbon fibre (Acarnley *et al.*, 1996). The carbon fibre sleeve has a greater centrifugal radial expansion than the steel rotor. Therefore, it can avoid PMs separating from the rotor at high speed.

Compared to radial-field machines, axial-field machines have a higher torque-to-size ratio, easier implementation for cooling, and lower noise and vibration (Chan, 1987). They can operate at a significantly greater speed of 100,000 rpm when the ratio of axial length to outside diameter is appropriately set. Moreover, the resultant axial forces exerted on the rotor can be completely eliminated when dual-rotor or dual-stator configurations are adopted, namely, either a single stator, sandwiched between two rotors, or a single rotor, sandwiched between two stators (Nguyen *et al.*, 2011) as shown in Figure 3.10.

The main drawback of the PMBL motor is its limited constant power range due to the restricted field-weakening capability. The field provided by PMs can only be weakened through the production of a stator armature field component that decreases the PM field.

### *3.3.3 Switched reluctance machine*

A switched reluctance machine (SRM) is inherently a cost-effective machine. Its rotor only consists of iron core laminations, which results in a very rugged and low-cost mechanical structure. Since no windings or PMs are located in the rotor, the SRM rotor is capable of operating at very high speed without retention problems. The armature windings usually adopt a concentrated manner and are wound in the stator salient poles, which have shorter end windings and a lower assembly cost for fabrication.

An SRM is also capable of operating in very harsh environments. Because of its magnetless topology, the maximum operating temperature of an SRM only depends on the machine insulation. An SRM for operation at ambient temperatures up to 400 °C was designed and implemented (MacMinn and Jones, 1989).

In addition, an SRM is very suitable for high fault-tolerant operation. The machine phase windings are both physically and electromagnetically decoupled. The phase-to-phase faults can be minimized and faulty phases can be isolated from the other, healthy phases. Furthermore, since there is no excitation source on the rotor, a faulty phase is less likely to generate a braking torque when the machine is in motoring operation.

The SRM drive topology has the phase winding in series with the two power switching devices in each phase leg. This topology has no shoot-through fault for direct short circuit of the DC-link which is found in conventional AC drive converters.

### 3.3.4 Synchronous reluctance machine

Synchronous reluctance (SynR) machines have many advantages as a motor/generator in FESSs, such as low cost, ease of flux weakening control, large overload capability and zero losses for idle operation due to the single excitation. However, some significant disadvantages need to be solved.

- The saliency ratio is generally low due to a solid rotor, thereby resulting in a low power factor and a low specific torque density.
- The harmonic fields may cause serious eddy current losses in the rotor.
- The basic rotor configuration causes a decreased energy storage density.
- The rotor may result in a higher windage loss due to the salient poles.

A rugged rotor with high saliency ratio is essential for ultrahigh-speed FESSs. Several approaches have been proposed for constructing rotors for SynR machines. A rotor with salient poles is a conventional configuration. For this configuration, the reluctance in the quadrature axis is much larger than the reluctance in the direct axis. This results in a small saliency ratio in the range of 2–3, which largely leads to a circulating flux in the rotor pole faces. The use of laminations with flux barriers is another approach for constructing rotors. However, the flux barriers reduce the ruggedness of the rotor structure, which is not suitable for the ultrahigh-speed design. Lamination for the rotor in the axial direction is also proposed. However, this is also not suitable for the ultrahigh-speed design. Another approach is that the rotor consists of ferromagnetic steel and nonmagnetic steel interleaving each other in a zebra manner, as shown in Figure 3.11. The thicknesses of the steels should be appropriately designed. The width of the ferromagnetic steel segments should be the same as the slot pitch of the stator. Therefore, the ferromagnetic steel segments always overlap with a stator tooth pitch regardless of the position of the rotor, which may minimize flux variations and rotor iron losses.

To fabricate a rotor, a joining technique known as explosion bonding is proposed (Hofmann and Sanders, 1996). Explosion bonding adopts explosive energy to merge metal sheets at high pressures.

### 3.3.5 Homopolar machine

A homopolar machine utilizes a field winding, PMs or hybrid excitation provided by both PMs and windings. Since the magnetic flux in this type of machine flows along

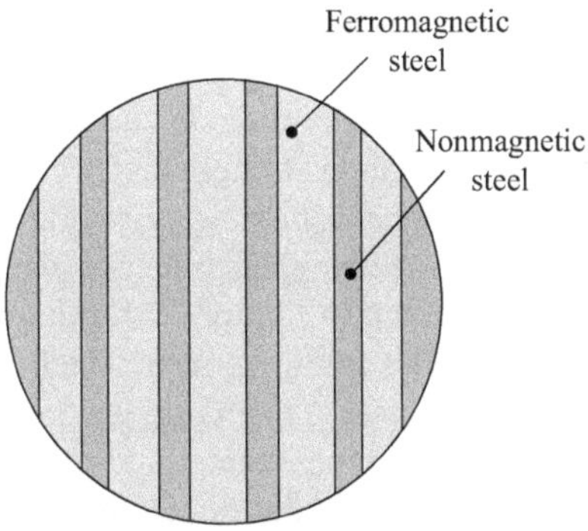


*Figure 3.11 Rotor design for SynR machine*

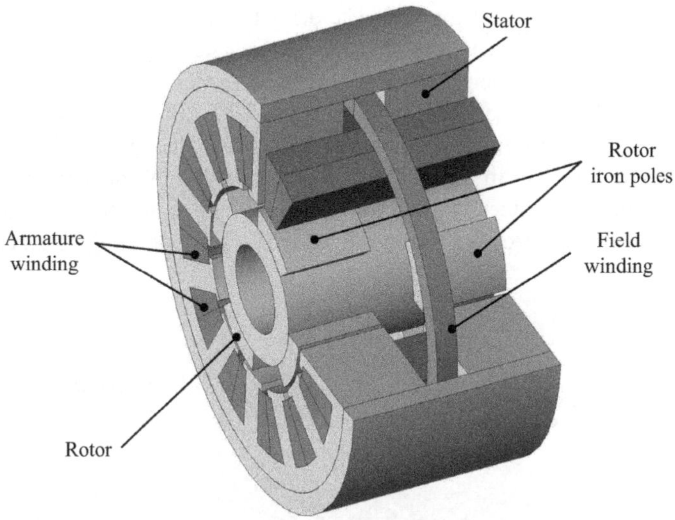


*Figure 3.12 A homopolar machine excited by a field winding*

the homopolar axis, it is called the homopolar machine, synchronous homopolar machine or homopolar inductor machine. As shown in Figure 3.12, the field winding is located in the stator and encircles the rotor. It should be noted that the rotor pole faces on the upper part of the rotor are offset from the pole faces on the lower part. Although the structure is rather different from the conventional synchronous machine, its operating principle is the same. There are several merits to implement the field winding in the stator, such as the elimination of slip rings, ease in cooling the field winding, the increased space available for this winding and greatly simplified rotor construction. The large space for the field winding allows high flux levels to be achieved efficiently. Thus, the stator design can adopt slotless topology.

Furthermore, it is practical to construct the rotor from a single piece of high-strength steel which makes homopolar motors very attractive for high-speed operation. Additional homopolar rotor design issues include laminations, PMs and other non-magnetic structural elements to increase strength and reduce windage losses.

As described previously, magnetic bearings can dramatically reduce friction losses and improve efficiency. A homopolar machine with magnetic bearings is proposed (Li *et al.*, 2015). As shown in Figure 3.13, this machine consists of two pairs of the stator segment and the rotor segment which artfully share the same stator yoke and rotor yoke for integrating them into one machine. The iron poles of the two segments are deployed in an interleaved manner and are a half pole pitch away from each other in the circumferential direction.

The stator houses the field winding, the armature winding and the suspension winding. Figure 3.14 shows the winding configuration of the armature winding and the suspension winding, which share the same slot space. The 3-phase 8-pole armature winding adopts a concentrated manner and the 2-phase 2-pole suspension winding adopts a distributed manner. The armature winding spans the total stack length and the suspension winding only spans one stator segment length. The field winding is deployed between the two segments. The field winding is fed with DC currents for the field excitation. By adjusting the DC current flowing in the field winding, the magnetic field can be controlled readily. The amplitude and polarity of magnetization in the air gap can be easily adjusted. With these merits, flux weakening can be realized for motoring operation, especially for high-speed operation, and flux control for voltage regulation when operating as a generator.

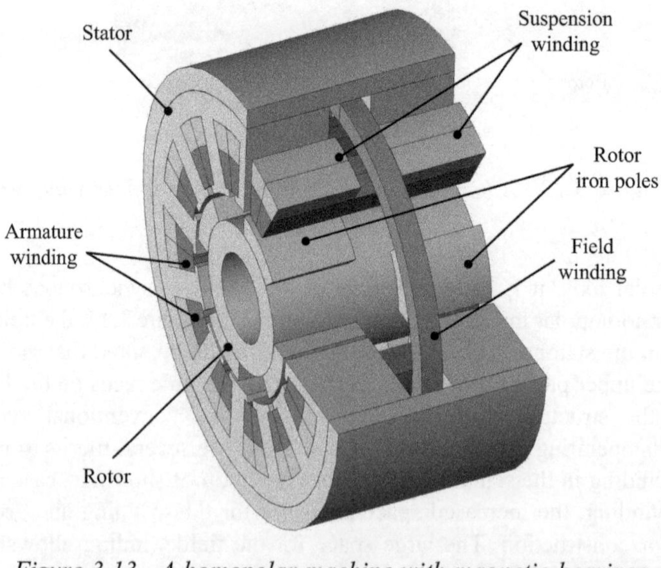


*Figure 3.13 A homopolar machine with magnetic bearings*

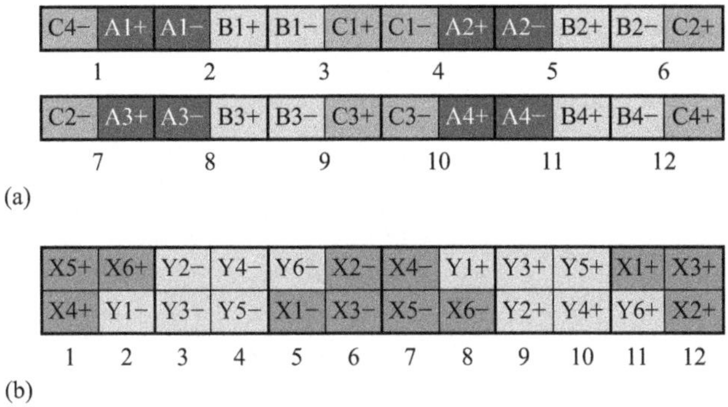


*Figure 3.14 Winding arrangement: (a) armature winding and (b) suspension winding*

Because of the three groups of windings in the proposed machine, the air gap flux is more complicated than the conventional synchronous machines. In this section, the air gap flux density is calculated in order to analyse the operating principle of the proposed machine, especially the suspension force generation by the 2-phase 2-pole suspension winding.

By ignoring the magnetic saturation effect, the flux fringing effect and the magnetic reluctance of the iron core, the air-gap permeance can be expressed as:

$$P_{ag}(\theta_m) = \begin{cases} \dfrac{2\mu_0 \pi l r_0}{k_0 g_0}, & \text{for facing iron poles} \\ 0, & \text{otherwise} \end{cases} \tag{3.3}$$

where $\mu_0$ is the permeability of the free space, $l$ is the length of one stator segment, $r_0$ is the mean radius of the air gap under the salient poles, $k_0$ is Carter's coefficient and $g_0$ is the air-gap length.

Therefore, the resultant air gap flux density can be expressed as:

$$B_{ag}(\theta_m) = \frac{(F_f + F_a + F_s)P_{ag}(\theta_m)}{2\pi l r_0} \tag{3.4}$$

where $F_f$, $F_a$ and $F_s$ are the magnetomotive forces (MMFs) due to the field winding, the armature winding and the suspension winding, respectively.

With the expression of the air-gap flux density, the machine's performances, such as the induced voltage and the developed torque, can be deduced, which is similar to the conventional machine calculations. In this chapter, only the suspension force calculation is elaborated. Figure 3.15 depicts the suspension force generation principle, where the solid line with arrow represents the magnetic flux induced by the field winding and the armature winding, and the dashed line with arrow represents the magnetic flux induced by the suspension winding. As shown in Figure 3.15(a), the flux produced by the phase X of the suspension winding strengthens the flux of the armature winding on the right hand side and weakens it on the left hand side. Therefore, a suspension force towards the right can be

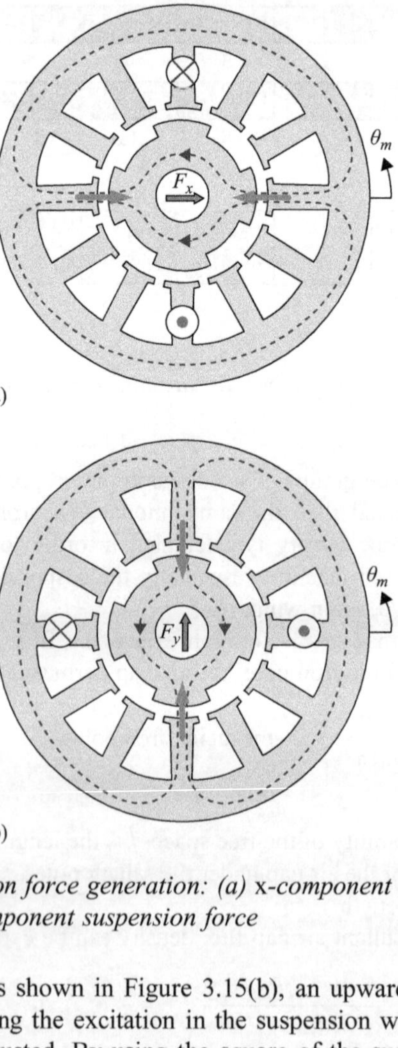


*Figure 3.15 Suspension force generation: (a) x-component suspension force and (b) y-component suspension force*

generated. Similarly, as shown in Figure 3.15(b), an upward suspension force is generated. By controlling the excitation in the suspension winding, the two force components can be adjusted. By using the square of the sum of the air-gap flux density, the two suspension force components can be expressed as:

$$\begin{cases} F_x = \dfrac{lr_0}{2\mu_0}\displaystyle\int_0^{2\pi} B_g(\theta_m)^2 \cos\theta_m \, d\theta_m \\ F_y = \dfrac{lr_0}{2\mu_0}\displaystyle\int_0^{2\pi} B_g(\theta_m)^2 \sin\theta_m \, d\theta_m \end{cases} \tag{3.5}$$

When the pole number is greater than 8, a constant suspension force can be generated by injecting DC currents into the suspension winding without position sensors.

## 3.4 Control strategies

The function of FESSs for EVs mainly lies in the power augmentation and regenerative energy recuperation. Therefore, the control strategy must provide a fast response for the large amount of power requirement and regenerative power absorption. Meanwhile, the depleted energy can be recovered in the normal vehicle operation condition. In this section, the motor/generator control, the FESS control, and the charge and discharge control schemes are presented. In addition, the condition for operation mode transition is also discussed.

### *3.4.1 Motor/generator control*

Since the motor/generator is responsible for energy conversion between electrical energy and mechanical energy, control of the motor/generator realizes the control of the power flow between the power source and the FESS. Torque control is usually implemented, which regulates the motoring during the charge operation and the generating during the discharge operation. During the charge mode, the flywheel can be charged at a constant power rate with the appropriate torque command, while during the discharge mode, the flywheel supplies the necessary power to regulate the DC-link voltage to a reference value and supply the loads.

To realize the torque control, rotor position information is required for PMBL machines, SRMs, SynR machines and homopolar machines, while for induction machines that information can be omitted. In order to improve the reliability and reduce the cost, sensorless control techniques are adopted for FESS control. The importance of sensorless control is greatly relies on the estimation for the position information. For different machine topologies, the estimation approaches are different. For PMBL machines, the estimation approaches are relatively mature, and can be estimated by using the back-electromotive force method, disturbance observers, sliding-mode observers and extended Kalman filters (Zhao *et al.*, 2013). Here, only estimation strategies for SRMs, SynR machines and homopolar machines are discussed.

#### 3.4.1.1 Sensorless technologies for SRMs

Many methods have been proposed for estimating the rotor position for SRM drives. Signal injection approaches measure the current of the inactive phase by applying a voltage with small amplitude into the inactive phases (Acarnley *et al.*, 1985). The rotor position can be determined via the rise and fall times of the measured current. In addition, some modulation techniques, such as frequency modulation (Ehsani *et al.*, 1992; Ehsani and Fahimi, 2002) and amplitude modulation (Suresh *et al.*, 1998), are added to these approaches to obtain the inductance variation information, which is position-dependent. However, they are not suitable for high-speed operation, due to the short conduction time of each phase. The direct measurement approach determines the rotor position direction from the voltage and current of the active phases (Lyons and MacMinn, 1992). Moreover, with the current and voltage, the phase flux linkage can also be calculated easily. Other computational methods, such as the state observer techniques (Lumsdaine and Lang, 1990), fuzzy logic (Cheok and Ertugrul, 2000) and intelligent method (Mese and Torrey, 2002), are deemed unsuitable for ultrahigh-speed operation since the processing time is limited and the complicated computation is time-consuming. The current gradient sensorless (CGS) scheme is very promising (Gallegos-Lopez *et al.*, 1998). It determines the rotor position via the phase current gradient. When the phase inductance increases, the current gradient decreases. The phase current gradient equation for SRM is as follows:

$$\frac{di_a}{dt} = \frac{V_a - i_a R_a - \omega_m(\partial\psi_a/\partial\theta_m)}{\partial\psi_a/\partial i_a} \tag{3.6}$$

where $V_a$, $i_a$, $R_a$ and $\psi_a$ are phase voltage, phase current, phase resistance and phase flux linkage, respectively, $\theta_m$ is the rotor position and $\omega_m$ is the rotor angular speed.

By observing the point at which the current gradient begins to decrease, the rotor position will be obtained. This approach only needs the phase current information and no other parameters are required, thus it is more attractive for ultrahigh-speed FESS applications.

#### 3.4.1.2 Sensorless technologies for SynR machines

Current, voltage and high-frequency signals are used for rotor position estimation (Consoli *et al.*, 1999; Ha *et al.*, 1999; Matsuo and Lipo, 1995). Since the estimation accuracy is independent of the rotor speed, these methods are well suited for low-speed operation. The fundamental component of voltage or current is also used for rotor position estimation (Capecchi *et al.*, 2001; Shinnaka, 2002). Since the amplitude of voltage is larger in the high-speed range, this method is suitable for high-speed operation. Due to the variation in the stator inductance caused by the magnetic saturation, the accuracy of position determination in the high-speed range is compromised. An online parameter identification method taking into consideration magnetic saturation, the extended electromotive force (EEMF) model, was proposed to solve the aforementioned problem (Ichikawa *et al.*, 2006). Two EEMF models have been established using dynamic inductance and static inductance, respectively. Then, the rotor position information can be obtained from the difference between the two EEMF models.

The dynamic inductances are given by:

$$L_d^\Delta = i_d\frac{dL_d}{di_d} + L_d \tag{3.7}$$

$$L_q^\Delta = i_q\frac{dL_q}{di_q} + L_q \tag{3.8}$$

where $L_d^\Delta$, $L_q^\Delta$, $L_d$, $L_q$, $i_d$ and $i_q$ are $d$-axis dynamic inductance, $q$-axis dynamic inductance, $d$-axis static inductance, $q$-axis static inductance, $d$-axis stator current and $q$-axis stator current, respectively.

The EEMF model with saturation can be expressed in the following equation:

$$\begin{bmatrix} v_d \\ v_q \end{bmatrix} = \begin{bmatrix} R + pL_d^{\Delta} & -\omega_{re}L_q \\ \omega_{re}L_q & R + pL_d^{\Delta} \end{bmatrix} \begin{bmatrix} i_d \\ i_q \end{bmatrix} + \left\{ (L_d - L_q)\omega_{re}i_d - \left( L_d^{\Delta} - L_q^{\Delta} \right) i_q \right\} \begin{bmatrix} 0 \\ 1 \end{bmatrix} \tag{3.9}$$

where $v_d$, $v_q$, $R$, $p$, and $\omega_{re}$ are $d$-axis voltage, $q$-axis voltage, stator resistance, differential operator and electrical angular velocity, respectively.

The second term in (3.9) is called the EEMF term, which only needs two inductances to make the estimation. With this estimated EEMF term, the position can be accurately determined regardless of the saturation effect.

#### 3.4.1.3 Sensorless technologies for homopolar machines

Due to the decoupled nature of the homopolar machine, two independent loops for sensorless control have been designed as shown in Figure 3.16 (Tsao *et al.*, 2002). One closed loop is a proportional–integral (PI) module for controlling the quadrant-axis current $i_q$ for providing the desired response for tracking the commanded active current, which finally feeds into the armature winding. The other closed loop is an integral module for controlling the direct-axis current $i_d$, which finally feeds into the field winding to ensure a unity power factor.

Unlike the field-oriented control schemes, an observer is not required to resolve the reference frame. The inverter voltage can determine the reference frame angle $\theta$. By activating the control loop from the inverter switching, the angular orientation of the reference frame can be obtained. Therefore, rotor position information is not required and sensorless control is achieved.

### *3.4.2 FESS control*

As shown in Figure 3.17, the main power source, the FESS and the propulsion motors in an EV are coupled together by a DC-link via power electronic circuits. The main power source provides the necessary power for the FESS and the vehicle

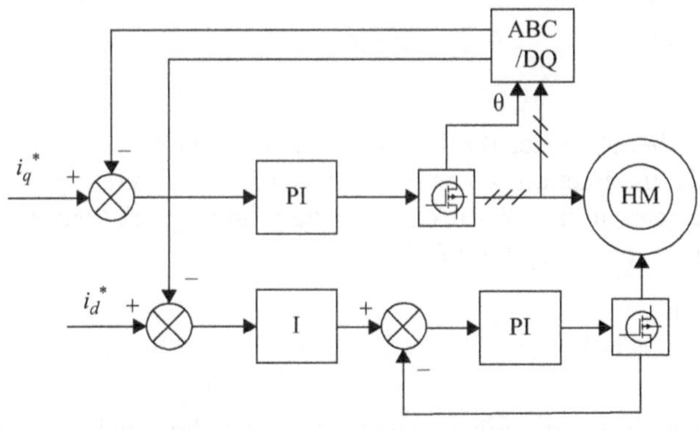


*Figure 3.16 Sensorless control of homopolar machine for FESS*

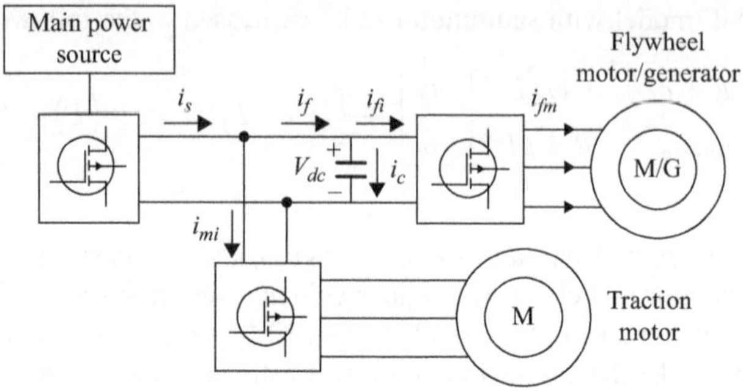


*Figure 3.17 Block diagram model for EV energy system with FESS*

propulsion. The FESS assists the main power source for providing power augmentation and recuperates the regenerative power from the traction motor during the braking. Therefore, the FESS mainly have three operation modes: charge, partial charge and discharge (Kenny *et al.*, 2005). In the charge mode, the main power source produces enough current to both charge the flywheel at its set point and provides the required current for driving the traction motor. The power converter of the main power source regulates the DC-link voltage in the charge operation. Therefore, the currents in the DC-link satisfy the following relationship and the main power source regulates the DC-link voltage:

$$i_s = i_f + i_{mi} \tag{3.10}$$

$$i_f = i_c^* \tag{3.11}$$

where $i_s$ is the current of the power converter connected to the main power source, $i_f$ is the DC-link current, $i_{mi}$ is the current of the power inverter connected to the traction motor and $i_c^*$ is the charge current reference.

In the partial charge mode, the recovery power returns to the FESS from regenerative braking. In this circumstance, the currents in the DC-link satisfy the following relationship and the DC-link voltage is regulated by the FESS:

$$i_{mi} = i_f \tag{3.12}$$

$$i_f < i_c^* \tag{3.13}$$

In the discharge mode, the FESS provides power to assist the main power source to drive the traction motor for starting-up or accelerating. In this circumstance, the currents in the DC-link satisfy the following relationship and the DC-link voltage is regulated by the FESS:

$$i_{mi} = i_s + i_f \tag{3.14}$$

$$i_f < 0 \tag{3.15}$$

To provide a steady power supply, the DC-link voltage must be kept unchanged. The FESS has the capability of regulating the DC-link voltage during both the charge and discharge modes. The relationship between the power inverter current $i_{fi}$ and the motor/generator current $i_{fm}$ can be obtained based on the condition of power balance where the power flows into the power inverter should be equal to the power flows into the motor (Kenny *et al.*, 2005). By neglecting the power inverter losses, the relationship of the power flow is governed by the following equation:

$$V_{dc} i_{fi} = m v_{fm} i_{fm} \tag{3.16}$$

where $V_{dc}$ is the DC-link voltage, $m$ is the number of phases and $v_{fm}$ is the motor/generator voltage.

The motor/generator voltage $v_{fm}$ can be expressed as:

$$v_{fm} = R_{fm} i_{fm} + L_{fm} \frac{di_{fm}}{dt} + \omega_m \lambda_f \tag{3.17}$$

where $R_{fm}$ and $L_{fm}$ are the motor/generator internal resistance and the inductance, respectively, $\omega_m$ is the motor/generator speed and $\lambda_f$ is the no-load flux linkage.

When the steady-state conditions are assumed, the derivative term in (3.17) can be omitted. By combining (3.16) and (3.17), the following equation can be obtained:

$$V_{dc} i_{fi} = m v_{fm} i_{fm} = m R_{fm} i_{fm}^2 + m \omega_m \lambda_f i_{fm} \tag{3.18}$$

The first term in (3.17) is very small compared to the back-EMF voltage, especially for ultrahigh-speed operation. By neglecting this term and rearranging (3.18) between the motor/generator current $i_{fm}$ and the average power inverter current $i_{fi}$, the following equation can be obtained:

$$i_{fm} \approx \frac{V_{dc} i_{fi}}{m \omega_m \lambda_f} \tag{3.19}$$

This relationship between the motor/generator current $i_{fm}$ and the average power inverter current $i_{fi}$ has become the basis for controlling the motor/generator in all modes of operation. In the charge mode, it is used as an FF term in the controller, and in the partial charge mode and the discharge mode it is used as a DD term and the FESS regulates the DC-link voltage.

### *3.4.3 Charge and discharge control*

In the charge mode, the flywheel is charged at a constant power via the constant DC current from the main power source. The charge control scheme regulates the acceleration of the flywheel motor so that the DC current is maintained at a reference value. The block diagram of the charge control is shown in Figure 3.18.

As shown in Figure 3.18, two modules are included in the charge controller, namely the PI module and the FF module (Kenny *et al.*, 2005). The FF module uses the DC charging current reference $i_c^*$ as the input and converts it into a motor current reference based on (3.19), with the measured DC-link voltage and the estimated

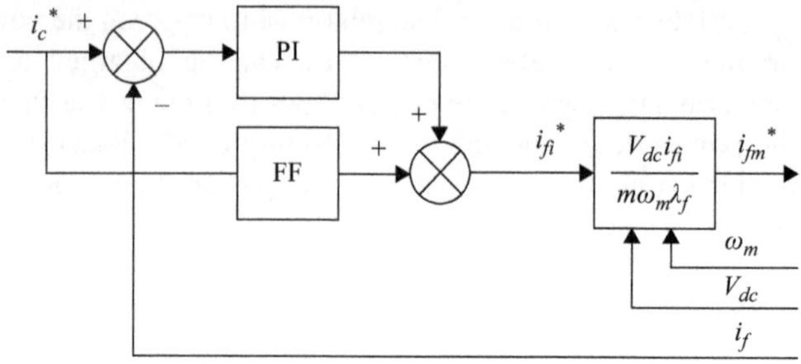


*Figure 3.18 Control block diagram for the charge mode*

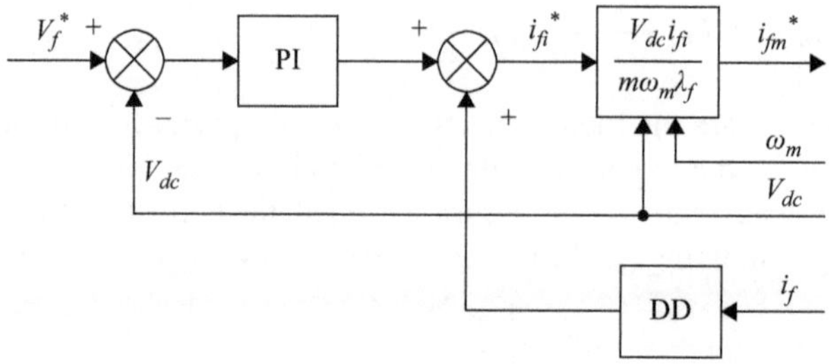


*Figure 3.19 Control block diagram for the partial charge and discharge mode*

motor speed. The PI module aims to compensate for any inaccuracies and ensures zero steady-state error. Therefore, fast and accurate performance can be achieved.

In the partial charge and discharge modes, the flywheel motor/generator must decelerate at an appropriate rate to maintain a constant DC-link voltage. For the discharge mode, it must also supply the necessary current to the traction motor. The corresponding control block diagram is shown in Figure 3.19.

There are two modules in the controller, namely the PI module and a DD module (Kenny *et al.*, 2005). In the DD module, the DC flywheel current $i_f$ is measured and adopted as an early indicator to the controller to check for any increase or decrease that has happened in the traction motor. If there is an abrupt increase of current in the traction motor, such as vehicle acceleration or climbing a hill, the capacitor will initially maintain the DC-link voltage and there will be an increase in the DC current to supply the new load. This increase is measured and used to calculate the corresponding flywheel motor current $i_{fm}^*$. The flywheel motor responds by decelerating more quickly based on the requirement, avoiding a drop in the DC-link voltage to trigger the PI module.

In the PI module, the measured DC-link voltage $V_{dc}$ is compared to the DC-link voltage reference $V_f^*$ and the relevant error signal is processed. The average power inverter current needs to maintain the DC-link voltage by adjusting the current to an appropriate value. The PI module aims to compensate for any inaccuracies in the DD module and maintains the DC-link voltage.

As previously mentioned, the control strategy for the charge mode is a current regulation scheme. The main power source provides sufficient current to meet the requirements for both vehicle propulsion and FESS charging. The control strategy for the partial charge and discharge mode is a voltage regulation scheme. The FESS regulates the DC-link voltage for stabilization.

The control schemes shift between the current regulation and the voltage regulation based on different operation modes (Kenny *et al.*, 2005). When the current $i_s$ is sufficient to provide both the load and the charging current, the main power source will regulate the DC-link voltage to a set value and slightly higher than the flywheel regulation set value. When the EV needs to accelerate or climb a hill, the DC-link voltage decreases due to the large current drawn by the traction motor. This condition is detected in the controller by comparing the difference between the actual DC-link voltage and the FESS set value. The FESS begins to operate in the discharge mode and regulates the DC-link voltage. The other condition is that when the EV decelerates and the regenerative braking is activated, the current $i_{mi}$ reverses its direction and the DC-link voltage increases. The FESS regulates the DC-link voltage and meanwhile absorbs the power from the regenerative braking.

Similarly, as the EV resumes cruising, the main power source supplies sufficient current. When the current exceeds the demand from the traction motor, the integrator in the current regulator module is reset and the FESS shifts back to the charge mode where the FESS regulates the current into the flywheel motor and the main power source regulates the DC-link voltage.

## 3.5 Summary

In this chapter, ultrahigh-speed flywheel energy storage technology for EVs is presented. First, the configuration of FESSs coupled into the HEVs and BEVs is introduced. Second, the key components of the FESS, namely the flywheel, the motor/generator, the bearings, the power converter and the enclosure are briefly addressed. Third, the motor/generator for FESSs is thoroughly discussed, especially in relation to some practical issues, namely magnetic, electrical, mechanical and thermal, for ultrahigh-speed machine design. Fourth, the control strategies for FESSs are presented, including the motor/generator control, the flywheel control and the charge and discharge control. The transition conditions of FESS control for different operating modes are also discussed.

Since FESSs exhibit very attractive advantages, such as high specific energy and power, long cycle life, high energy efficiency, reduced maintenance and so on, they are very promising for application in EVs in the near future.

## Acknowledgements

This work was supported in part by two grants from The University of Hong Kong, Hong Kong, China (Project Nos. 201309176197 and 201309176048).

## References

Acarnley, P. P., Hill, R. J., and Hooper, C. W. (1985) Detection of rotor position in stepping and switched motors by monitoring of current waveforms. *IEEE Transactions on Industrial Electronics*, 32, 215–222.

Acarnley, P. P., Mecrow, B. C., Burdess, J. S., Fawcett, J. N., Kelly, J. G., and Dickinson, P. G. (1996) Design principles for a flywheel energy store for road vehicles. *IEEE Transactions on Industry Applications*, 32, 1402–1408.

Akagi, H., and Sato, H. (2002) Control and performance of a doubly-fed induction machine intended for a flywheel energy storage system. *IEEE Transactions on Power Electronics*, 17, 109–116.

Bolund, B., Bernhoff, H., and Leijon, M. (2007) Flywheel energy and power storage systems. *Renewable & Sustainable Energy Reviews*, 11, 235–258.

Capecchi, E., Guglielmi, P., Pastorelli, M., and Vagati, A. (2001) Position-sensorless control of the transverse-laminated synchronous reluctance motor. *IEEE Transactions on Industry Applications*, 37, 1768–1776.

Chan, C. C. (1987) Axial-field electrical machines – design and applications. *IEEE Transactions on Energy Conversion*, 2, 294–300.

Chen, H. S., Cong, T. N., Yang, W., Tan, C. Q., Li, Y. L., and Ding, Y. L. (2009) Progress in electrical energy storage system: a critical review. *Progress in Natural Science*, 19, 291–312.

Cheok, A. D., and Ertugrul, N. (2000) High robustness and reliability of fuzzy logic based position estimation for sensorless switched reluctance motor drives. *IEEE Transactions on Power Electronics*, 15, 319–334.

Consoli, A., Russo, F., Scarcella, G., and Testa, A. (1999) Low- and zero-speed sensorless control of synchronous reluctance motors. *IEEE Transactions on Industry Applications*, 35, 1050–1057.

Cosgriff, L. M., and Baaklini, G. Y. (2003) Ultrasonic spectroscopy of composite rings for space station flywheel rotors. *Materials Evaluation*, 61, 1013–1019.

Dhand, A., and Pullen, K. (2015) Review of battery electric vehicle propulsion systems incorporating flywheel energy storage. *International Journal of Automotive Technology*, 16, 487–500.

Diaz-Gonzalez, F., Bianchi, F. D., Sumper, A., and Gomis-Bellmunt, O. (2014) Control of a flywheel energy storage system for power smoothing in wind power plants. *IEEE Transactions on Energy Conversion*, 29, 204–214.

Ehsani, M., and Fahimi, B. (2002) Elimination of position sensors in switched reluctance motor drives: state of the art and future trends. *IEEE Transactions on Industrial Electronics*, 49, 40–47.

Ehsani, M., Husain, I., and Kulkarni, A. B. (1992) Elimination of discrete position sensor and current sensor in switched reluctance motor-drives. *IEEE Transactions on Industry Applications*, 28, 128–135.

Gallegos-Lopez, G., Kjaer, P. C., and Miller, T. J. E. (1998) A new sensorless method for switched reluctance motor drives. *IEEE Transactions on Industry Applications*, 34, 832–840.

Gerada, D., Mebarki, A., Brown, N. L., Gerada, C., Cavagnino, A., and Boglietti, A. (2014) High-speed electrical machines: technologies, trends, and developments. *IEEE Transactions on Industrial Electronics*, 61, 2946–2959.

Ha, J. I., Kang, S. J., and Sul, S. K. (1999) Position-controlled synchronous reluctance motor without rotational transducer. *IEEE Transactions on Industry Applications*, 35, 1393–1398.

Hebner, R., Beno, J., and Walls, A. (2002) Flywheel batteries come around again. *IEEE Spectrum*, 39, 46–51.

Hofmann, H., and Sanders, S. R. (1996) Synchronous reluctance motor/alternator for flywheel energy storage systems. *Proceeding of the IEEE Power Electronics in Transportation,* Dearborn, Michigan, USA, pp. 199–206.

Ichikawa, S., Tomita, M., Doki, S., and Okuma, S. (2006) Sensorless control of synchronous reluctance motors based on extended emf models considering magnetic saturation with online parameter identification. *IEEE Transactions on Industry Applications*, 42, 1264–1274.

Kalpakjian, S., and Schmid, S. R. (2014) *Manufacturing Engineering and Technology*. Upper Saddle River, NJ, Pearson.

Kenny, B. H., Kascak, P. E., Jansen, R., Dever, T., and Santiago, W. (2005) Control of a high-speed flywheel system for energy storage in space applications. *IEEE Transactions on Industry Applications*, 41, 1029–1038.

Lawrence, R. G., Craven, K. L., and Nichols, G. D. (2003) Flywheel ups. *IEEE Industry Applications Magazine*, 9, 44–50.

Lee, H., Jung, S., Cho, Y., Yoon, D., and Jang, G. (2013) Peak power reduction and energy efficiency improvement with the superconducting flywheel energy storage in electric railway system. *Physica C – Superconductivity and Its Applications*, 494, 246–249.

Li, W. L., Chau, K. T., Ching, T. W., Wang, Y. B., and Chen, M. (2015) Design of a high-speed superconducting bearingless machine for flywheel energy storage systems. *IEEE Transactions on Applied Superconductivity*, 25, 5700204:1–4.

Lumsdaine, A., and Lang, J. H. (1990) State observers for variable-reluctance motors. *IEEE Transactions on Industrial Electronics*, 37, 133–142.

Lyons, J. P., and MacMinn, S. R. (1992). Lock detector for switched reluctance machine rotor position estimator, *U.S. Patent* No.5, 140,244.

MacMinn, S. R., and Jones, W. D. (1989) A very high speed switched-reluctance starter-generator for aircraft engine applications. *Proceedings of IEEE National Aerospace and Electronics Conference*, Dayton, Ohio, USA, pp. 1758–1764.

Matsuo, T., and Lipo, T. A. (1995) Rotor position detection scheme for synchronous reluctance motor based on current measurements. *IEEE Transactions on Industry Applications*, 31, 860–868.

Mese, E., and Torrey, D. A. (2002) An approach for sensorless position estimation for switched reluctance motors using artificial neural networks. *IEEE Transactions on Power Electronics*, 17, 66–75.

Miller, J. M. (2004) *Propulsion Systems for Hybrid Vehicles*. Stevenage, UK, Institution of Electrical Engineers.

Nguyen, T. D., Tseng, K. J., Zhang, S., and Nguyen, H. T. (2011) A novel axial flux permanent-magnet machine for flywheel energy storage system: Design and analysis. *IEEE Transactions on Industrial Electronics*, 58, 3784–3794.

Noguchi, T., Takata, Y., Pyamashita, Y., Komatsu, Y., and Ibaraki, S. (2007) 220,000-r/min, 2-kw pm motor drive for turbocharger. *Electrical Engineering in Japan*, 161, 31–40.

Ooshima, M., Kobayashi, S., and Tanaka, H. (2010) Magnetic suspension performance of a bearingless motor/generator for flywheel energy storage systems. *Proceedings of IEEE Power and Energy Society General Meeting*, Minneapolis, Minnesota, USA, pp. 1–4.

Pena-Alzola, R., Sebastian, R., Quesada, J., and Colmenar, A. (2011) Review of flywheel based energy storage systems. *Proceedings of International Conference on Power Engineering, Energy and Electrical Drives*, Malaga, Spain, pp. 1–6.

Read, M. G., Smith, R. A., and Pullen, K. R. (2015) Optimisation of flywheel energy storage systems with geared transmission for hybrid vehicles. *Mechanism and Machine Theory*, 87, 191–209.

Shinnaka, S. (2002) Mirror-phase characteristics of synchronous reluctance motor and salient-pole orientation methods for sensorless vector controls. *Electrical Engineering in Japan*, 139, 61–71.

Soong, W. L., Kliman, G. B., Johnson, R. N., White, R. A., and Miller, J. E. (2000) Novel high-speed induction motor for a commercial centrifugal compressor. *IEEE Transactions on Industry Applications*, 36, 706–713.

Suresh, G., Fahimi, B., and Ehsani, M. (1998) Improvement of the accuracy and speed range in sensorless control of switched reluctance motors. *Proceedings of Thirteenth Annual Applied Power Electronics Conference and Exposition*, Anaheim, California, USA, pp. 771–777.

Tsao, P., Senesky, M., and Sanders, S. (2002) A synchronous homopolar machine for high-speed applications. *Proceedings of 37th IAS Annual Meeting*, Pittsburgh, Pennsylvania, USA, pp. 406–416.

Vazquez, S., Lukic, S. M., Galvan, E., Franquelo, L. G., and Carrasco, J. M. (2010) Energy storage systems for transport and grid applications. *IEEE Transactions on Industrial Electronics*, 57, 3881–3895.

Zhao, Y., Wei, C., Zhang, Z., and Qiao, W. (2013) A review on position/speed sensorless control for permanent-magnet synchronous machine-based wind energy conversion systems. *IEEE Journal of Emerging and Selected Topics in Power Electronics*, 1, 203–216.

# Chapter 4
# Hybridization of energy sources for electric and hybrid vehicles

*Y.S. Wong*[1]

The internal combustion engine vehicle (ICEV) loses a large amount of energy in heat in the engine and idling. The electric vehicle (EV) and hybrid EV (HEV) are more efficient vehicles in the market. The EV has the highest energy efficiency among the EV, the HEV and the ICE. The EV allows diversification of energy resources; enables load equalization of power systems; delivers zero local and minimal global exhaust emissions and operates quietly. However, the commercialization of EVs is hindered by short driving range and high upfront cost. These barriers cannot be easily solved by the available EV energy source technologies, including batteries, fuel cells, capacitors and flywheels. Hybridization of the high-energy energy source and the high-power energy source in the electrical drivetrain of EVs is a viable approach to prolong driving range of EVs. On the other hand, the HEV is designed to hybridize the ICE drivetrain and electrical drivetrain to boost up efficiency of the ICE system. Hybridization of energy sources in EVs and hybridization of powertrains in HEVs can significantly boost system efficiency and driving ranges of the EV and HEV.

The onboard energy source is the most important part in drivetrain hybridization in the HEV and energy source hybridization in the EV. This work discusses the energy sources for the EV and HEV applications. The drivetrain topologies of HEVs will be elaborated with the emphasis on functionality of the electrical powertrain. The topologies of hybrid energy systems (HESs) in battery EVs (BEVs) and fuel cell EVs (FCEVs) will also be discussed.

## 4.1 Introduction

The commercialization of EVs is hindered by the lack of technically and commercially viable energy sources. The hybridization of engine and electric motor drivetrains in an HEV is an interim solution for improvement of vehicular efficiency. The hybridization of energy sources in the electrical drivetrain of EVs is also

[1]Automotive Parts and Accessory Systems (APAS) R&D Centre, Hong Kong Productivity Council, Hong Kong, China

a potential approach to prolong driving range of EVs and lower its upfront and operating costs.

The HEV greatly extends the limited EV driving range by two to four times and offers rapid refuelling by liquid gasoline or diesel. The HEV requires no change in the existing energy supply infrastructure. The HEV offers more efficient and less emission than the ICEV. With the assist of electric motors, the engine in an HEV can operate in its most efficient mode, yielding low emissions and low fuel consumption. Some HEVs are designed to operate in the EV mode for a short period of time. The HEV is not only an interim solution of zero-emission vehicles, but also a practical solution for commercialization of super-ultra-low-emission vehicles. The basic principle of HEV design is the integration of the EVs electric propulsion system into the ICEVs mechanical propulsion system (Chan *et al.*, 2009).

There is no single energy source that can fulfil the performance and cost requirements of EVs. Some energy sources can deliver high energy, while some energy sources can deliver high power, but none of them can deliver both high power and high energy. Instead of using a single energy source, EVs can use multiple energy sources, so-called hybridization of energy sources. The control and packaging complexities of energy source hybridization are complicated if too many energy sources are involved. The viable approach is the hybridization of two energy sources, in which, one energy source provides high energy for long driving range while the other one provides high power for acceleration and efficiency improvement. In a typical battery–battery hybrid energy source (HES) system as shown in Figure 4.1, the high-energy energy source supplies the primary power to the electric motor in normal driving condition. During acceleration, both energy sources supply the energy to drive the electric motor simultaneously. During braking, the high-power energy source absorbs the braking energy by regenerative braking. With the integration of the high-power energy source and high-energy energy source, the HES system can deliver high power and store more energy onboard for long driving range.

## 4.2 Characteristics of engine and electrical powertrains

The ICE is intrinsically inefficient at converting the fuel's chemical energy to mechanical energy, losing energy to engine friction, pumping air into and out of the engine, and wasted heat. In EVs, the electric motor does not consume energy at standstill and it has higher peak and average efficiency than the ICEs. The electric motor also has maximum torque at standstill and low speeds and it allows bidirectional energy flow such that mechanical energy can be converted into electrical energy and stored in the traction batteries during regenerative braking.

Comparing vehicular efficiencies of the ICEVs and EVs, the ICEV has low thermal and mechanical efficiencies such that most of the fuel is dissipated as heat in the engine. As illustrated in Figure 4.2, the vehicular efficiency of a typical ICEV is about 15%. Around 62% of the fuel's energy is lost in the engine and 17% and 6% of the energy are lost in idling and in the driveline, such as acceleration and braking (USDOE, 2011a; Taylor, 1998). On the other hand, the vehicular efficiency

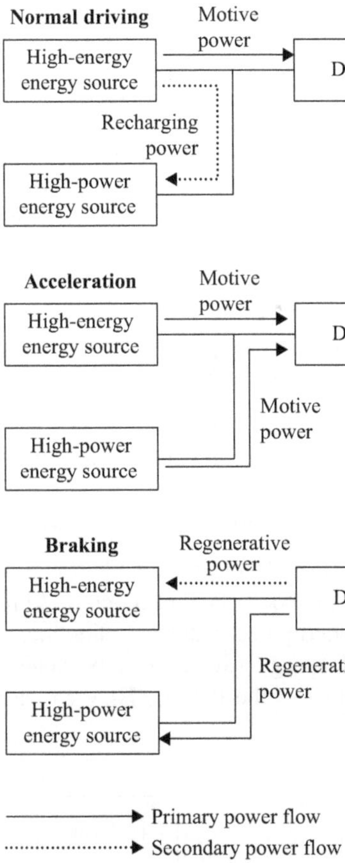


*Figure 4.1 Principle of energy hybridization in the EV*

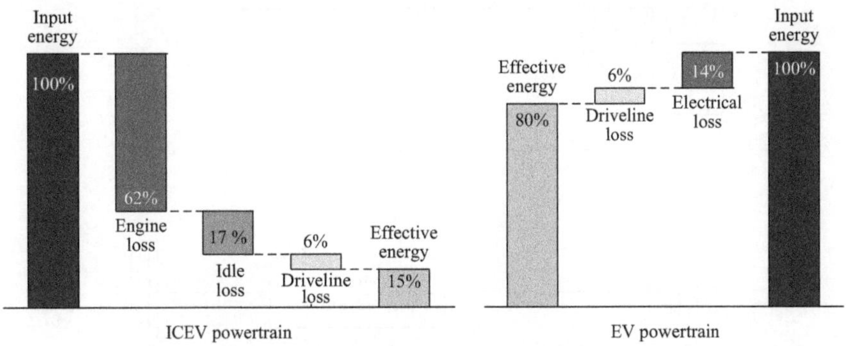


*Figure 4.2 Vehicular efficiencies of ICEVs and EVs*

of a BEV is around 80%. The vehicular efficiency of the HEV is between the ICEVs and the EVs, depending on the drivetrain topology and control strategies.

### 4.2.1 Energy efficiency improvement in HEVs

In an HEV, the motor can supplement motive power to the ICE or be the sole source of the traction power. The engine can charge up the battery in light load. The kinetic energy during braking and down-slope driving can be captured by the motor and stored in the vehicle batteries or ultracapacitors in regenerative braking. In the power boost mode, the electric motor provides supplementary power to the engine such that the engine can operate in the most efficient point or region. Figure 4.3 shows the system architecture of a series–parallel HEV.

The electrical drivetrain boosts an HEVs fuel economy in the following ways:

- Assist vehicle launch and acceleration
- Assist load enhancement for the ICE for faster warm-up and better efficiency
- Undertake regenerative braking
- Deliver pure electric driving for certain period of time

The energy saving potential of HEVs achieved by the electrical powertrain is shown in Figure 4.4. The HEV can boost the fuel economy from its ICE counterpart by about 6% by reducing the energy losses during idling and stop-and-start in city driving. An additional 6% of energy savings can be achieved by regenerative braking. By optimizing efficiency of the ICE, an HEV can achieve an aggregated

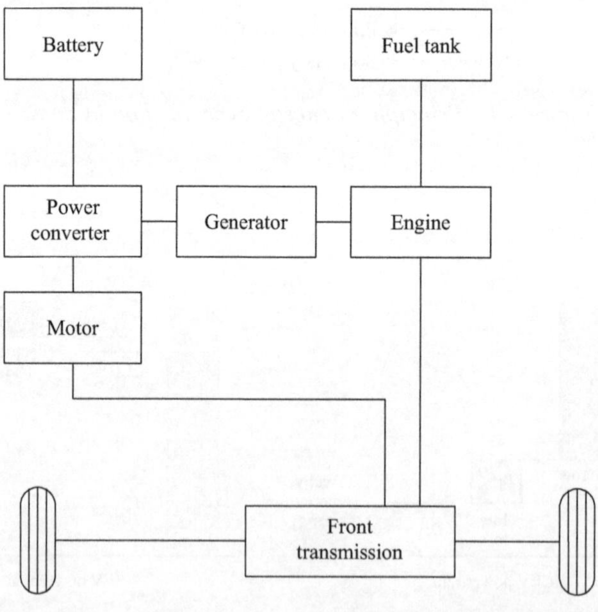


*Figure 4.3 System architecture of a series–parallel HEV*

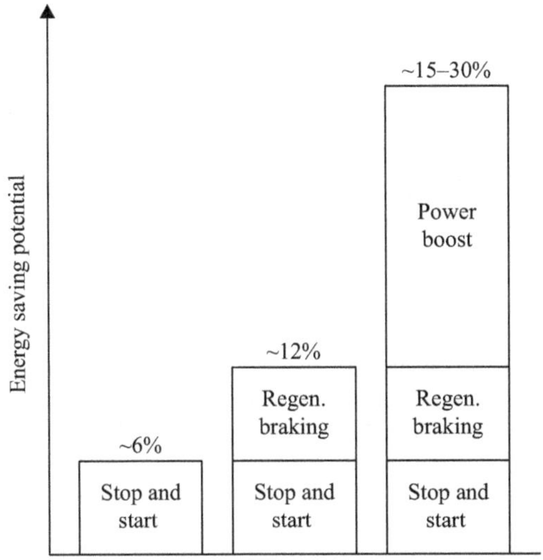


*Figure 4.4 Energy savings potential in HEVs*

savings potential of 15–30% in fuel consumption relative to the ICE vehicle, subject to the vehicle design and energy management strategies. The HEVs can achieve a reduction of 3.9–20.3% in fuel consumption in the highway driving cycle and a higher reduction of 8.4–46.7% in the urban driving cycle (Wong *et al.*, 2014).

## *4.2.2 Drivetrain design of BEVs and HEVs*

The key considerations in designing the HEV drivetrain are summarized below:

- Optimal engine operating point – the optimal operating point on the torque–speed plane of the engine can be based on the maximization of fuel economy, the minimization of emissions, or even a compromise between fuel economy and emissions.
- Optimal engine operating line – in case the engine needs to deliver different power demands, the corresponding optimal operating points constitute an optimal operating line.
- Optimal engine operating region – the engine has a preferred operating region on the torque–speed plane, in which the fuel efficiency remains optimum.
- Minimum engine dynamics – the engine operating speed needs to be regulated in such a way that any fast fluctuations are avoided, hence minimizing the engine dynamics.
- Minimum engine speed – when the engine operates at low speeds, the fuel efficiency is very low. The engine should be cut off when its speed is below a threshold value.
- Minimum engine turn-on time – the engine should not be turned on and off frequently; otherwise, it results in additional fuel consumption and emissions. A minimum turn-on time should be set to avoid such drawbacks.

- Proper battery capacity – the battery capacity needs to be kept at a proper level so that it can provide sufficient power for acceleration and can accept regenerative power during braking or going downhill. When the battery capacity is too high, the engine should be turned off or operated idly. When this capacity is too low, the engine should increase its output to charge the battery.
- Safe battery voltage – the battery voltage may be significantly altered during discharging, generator charging or regenerative charging. This battery voltage should not be too high or too low; otherwise, the battery may be permanently damaged.
- Relative distribution – the distribution of power demand between the engine and battery should be proportionally divided during the driving cycle.
- Geographical policy – in certain cities or areas, the HEV needs to be operated in the pure electric mode. The changeover should be controlled manually or automatically.

The energy source has to be intrinsically tolerant of abusive conditions such as overcharge, short circuit, crush, fire exposure, mechanical shock and vibration. Energy sources are connected in series and/or parallel in a vehicle's energy system. The key requirements for onboard energy source are high specific energy, high specific power, long cycle life, high efficiency, wide operating temperature and low cost for commercialization.

## 4.3 Energy sources for EV and HEV applications

Among the energy storage technologies, the viable energy sources for EV and HEV applications are rechargeable electrochemical batteries (loosely called batteries), fuel cells, ultrahigh-capacitance capacitors (loosely called ultracapacitors) and ultrahigh-speed flywheels, as shown in Figure 4.5. Among them, the batteries, capacitors and flywheels are energy storage systems in which electrical energy is stored by electrical charging, whereas the fuel cell and engine are energy generation systems in which electricity or motive energy is generated by chemical reaction. The hydrogen/methanol and gasoline undergo irreversible chemical or electrochemical reactions to drive the vehicle such that they cannot absorb any braking energy. The ultracapacitor, ultrahigh-speed flywheels, high power and high energy lithium-ion (Li-Ion) batteries, nickel-metal (Ni-Metal) battery, and valve-regulated lead-acid (VRLA) battery are high-power sources, while the metal/air battery, gasoline and fuel cell systems are high-energy sources (Ghoniem, 2011).

The battery has been identified as the major energy source for EV and HEV applications, because of its technological maturity and reasonable cost. The fuel cells, ultracapacitors and ultrahigh-speed flywheels are potential energy sources for EV and HEV applications. The introduction of high performance Li-Ion batteries into EVs and HEVs increased BEVs driving range from 160 km to 250–300 km. The energy density of Li-Ion batteries is close to twice that of the nickel-metal hydride (Ni-MH) battery. However, it is still much less than the corresponding values of combustion engines using gasoline or diesel. There is no single energy

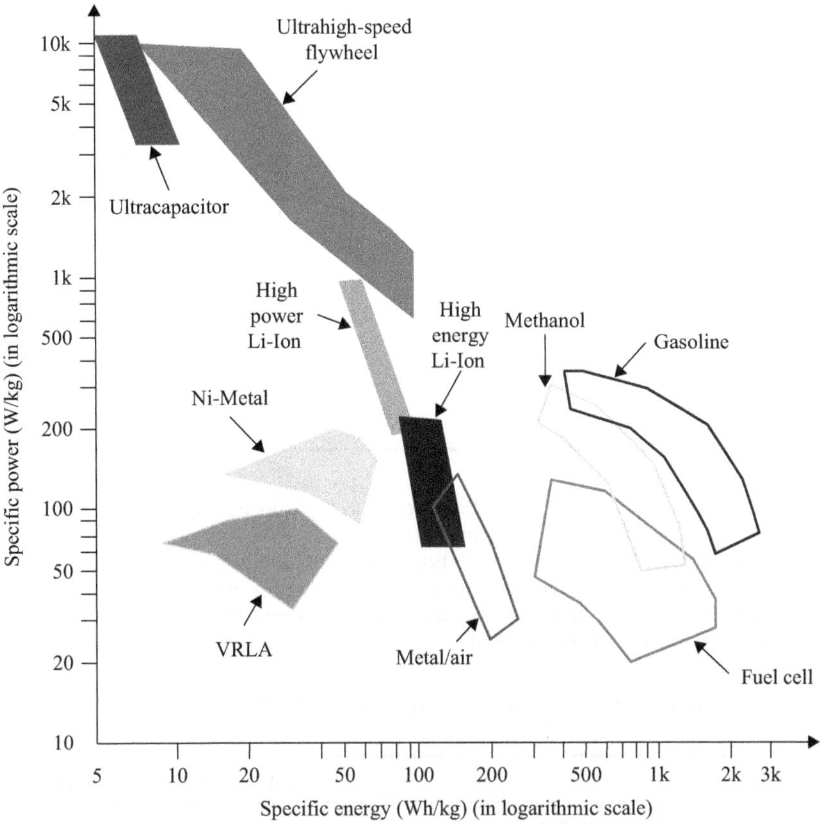


*Figure 4.5 Specific energy and specific power of energy sources for vehicle application*

source that can fulfil the performance requirements of the EV. Some energy sources can deliver high energy, while some energy sources can deliver high power. The ultracapacitor and ultrahigh-speed flywheel have much higher charging and discharging power density compatible with regenerative braking at high speeds, and power surge during fast acceleration.

## *4.3.1 Batteries*

The battery refers to the rechargeable electrochemical battery. The battery has to be intrinsically tolerant of abuse conditions such as overcharge, short circuit, crush, fire exposure, mechanical shock and vibration. Battery cells are connected in series and/or parallel in a vehicle battery system, thus, cells' state of charges (SoCs) have to be balanced to prevent undercharge and overcharge. The operating temperature range of the battery is wide such that active thermal management systems are needed (Chan and Chau, 2001). The EV needs batteries with high specific power for quick charge and with high specific energy for long driving range. Table 4.1 shows some

*Table 4.1 Typical USABC goals for batteries in EV applications*

| Parameters | Unit | System level |
|---|---|---|
| Discharge specific power at 80% DoD* for 30 s | W/kg | 470 |
| Regenerative specific power at 20% DoD† for 10 s | W/kg | 200 |
| Power density | W/L | 1000 |
| Onboard energy capacity | kWh | 45 |
| Specific energy at C/3 discharge rate | Wh/kg | 235 |
| Energy density at C/3 discharge rate | Wh/L | 500 |
| Calendar life | year | 15 |
| Cycle life to 80% DoD‡ | cycle | 1000 |
| Operating temperature | °C | −30 to +52 |
| Selling price | USD/kWh | <125 |
| Normal recharge time | H | 7 |
| High recharge rate | min | 15 (80% ΔSoC) |

Notes:
*The higher the degree of discharge (DoD), the lower the battery's discharging power. The discharging power at 80% DoD can effectively indicate the battery's characteristics during discharging.
†The lower the DoD, the lower the battery's recharging power. The recharging power at 20% DoD can effectively indicate the battery's recharging power during regenerative braking.
‡The higher the DoD in cyclic applications, the lower the battery's cycle life. The cycle life in 80% DoD can effectively indicate the battery's cycle life.

*Table 4.2 Typical USABC goals for batteries in PHEV applications*

| Parameters | Unit | PHEV-20 | PHEV-40 |
|---|---|---|---|
| Reference equivalent electric range | mile | 20 | 40 |
| Peak pulse discharge power for 10 s | kW | 37 | 38 |
| Peak pulse discharge power for 2 s | kW | 45 | 46 |
| Peak regenerative power for 10 s | kW | 25 | 25 |
| Available energy for charge depleting mode at 10 kW discharge rate | kWh | 5.8 | 11.6 |
| Available energy for charge sustaining mode | kWh | 0.3 | 0.3 |
| Cold cranking power at −30 °C | kW | 7 | 7 |
| Calendar life at 30 °C | year | 15 | 15 |
| Maximum system weight | kg | 70 | 120 |
| Maximum system volume | L | 47 | 80 |
| Maximum operating voltage | V | 420 | 420 |
| Operating temperature | °C | −30 to +52 | −30 to +52 |

United States Advanced Battery Consortium (USABC) key goals for batteries in EV applications (USCAR, 2015).

The plug-in HEV (PHEV) needs batteries with high specific power but the amount of onboard energy source varies with the targeted pure electric driving range. Table 4.2 shows some USABC key goals for batteries in PHEV applications with the pure electric driving range at 20–40 miles (USCAR, 2015).

The viable batteries for vehicle applications includes the VRLA, nickel-zinc (Ni-Zn), Ni-MH, zinc/air (Zn/Air), aluminium/air (Al/Air), sodium/sulphur (Na/S),

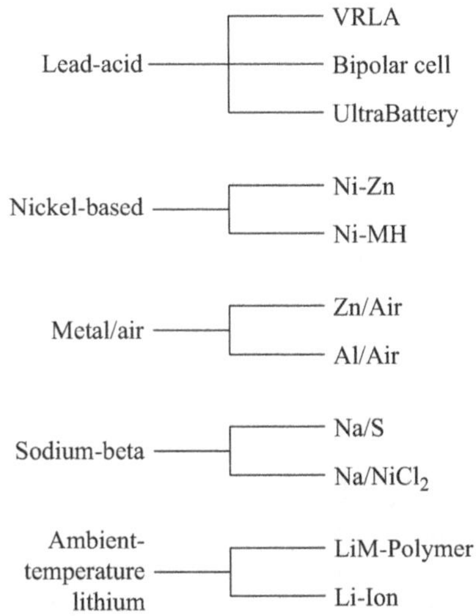


*Figure 4.6 Classification of vehicle batteries*

sodium/nickel chloride (Na/$NiCl_2$), lithium metal-polymer (LiM-Polymer) and Li-Ion batteries. The specific energy and specific power of these batteries are shown in Figure 4.5. These batteries are classified into lead-acid (Pb-Acid), nickel-based, metal/air, sodium-beta and ambient-temperature lithium batteries, as shown in Figure 4.6 (Köhler, 2009; Gutmann, 2009).

#### 4.3.1.1 Pb-Acid batteries

The Pb-Acid battery has been a successful commercial product for over a century. The Pb-Acid battery is mature and low cost. It has a nominal cell voltage of 2 V, specific energy of 35 Wh/kg, energy density of 90 Wh/L and specific density of 200 W/kg. In the sealed Pb-Acid battery, a special porous separator is employed in the cell such that the evolved oxygen is transferred from the negative electrode to the positive electrode and then combines with hydrogen to form water. Thus, it provides a definite advantage of maintenance-free operation. Moreover, the immobilization of the gelled (Gel) electrolyte or absorbed electrolyte with absorptive glass mat separators allows the battery to operate in different orientations without spillage. The sealed Pb-Acid battery is so-called VRLA battery (Rand and Moseley, 2009).

The VRLA battery has maintained its prime position for more than a century, there are a number of advantages contributing to this outstanding position: proven technology and mature manufacturing, low cost, high cell voltage, good high-rate performance that are suitable for vehicle applications, good low temperature and high temperature performances, high energy efficiency (75–80%) and available in a variety of sizes and designs. The VRLA battery's specific energy and energy density are relatively low, typically, 35 Wh/kg and 70 Wh/L. Its self-discharge rate is relatively high at about 1% per day at 25 °C.

Advanced Pb-Acid batteries with improved performance are being developed for vehicle applications. Improvements of the VRLA battery in specific energy over 40 Wh/kg and energy density over 80 Wh/L with the possibility of rapid recharge have been attained. The bipolar VRLA battery and UltraBattery™ are promising Pb-Acid batteries for vehicle applications (Wong and Chan, 2013).

#### 4.3.1.2 Nickel-based batteries

There are many kinds of electrochemical batteries using nickel oxyhydroxide as the active material for the positive electrode, including the nickel-cadmium (Ni-Cd), Ni-Zn and Ni-MH. Among them, the Ni-MH battery has been well accepted for EV and HEV applications because of its proven technology and good performance.

The Ni-Zn battery is still under development. The Ni-Zn battery has high specific energy and low material cost; however, it has not achieved any commercial importance because of the short life in the zinc electrode (Cairns, 2009). The Ni-Zn battery nominally operates at 1.6 V and has energy and power densities of 60 Wh/kg, 120 Wh/L and 300 W/kg. It uses zinc as the negative electrode and nickel oxyhydroxide as the positive electrode. The electrolyte is an alkaline potassium hydroxide solution. The Ni-Zn battery has the advantages of higher specific energy and power than the Ni-Cd battery, high cell voltage (the highest of the nickel-based family), non-toxicity (more environmental friendliness than the Ni-Cd), tolerance of overcharge and overdischarge, capable of high discharge and recharge rates, and wide operating temperature (from −20 °C to 60 °C). However, the major and serious drawback of the Ni-Zn battery is its short cycle life (about 300 cycles). It is mainly due to the partial solubility of zinc species in the electrolyte.

The Ni-MH battery has been on the market since 1992. Its characteristics are similar to the Ni-Zn battery. The principal difference between them is the use of hydrogen, absorbed in a metal hydride, for the active negative electrode material in the Ni-MH battery (Hariprakash *et al.*, 2009). The Ni-MH battery has a nominal voltage of 1.32 V and attains specific energy of 65–110 Wh/kg for EV applications and 45–60 Wh/kg for HEV applications. It operates in a temperature from −20 °C to +45 °C. A number of battery manufacturers, such as GM Ovonic, GP, GS, Panasonic, SAFT, VARTA and YUASA, have actively engaged in the development of this battery for HEVs.

#### 4.3.1.3 Metal/air batteries

The rechargeable metal/air batteries include the electrically or mechanically rechargeable Zn/Air battery and the mechanically rechargeable Al/Air battery. These metal/air batteries have very high specific energy and energy density (as high as 600 Wh/kg and 400 Wh/L for Al/Air) and they are low cost, environmental friendly. In addition, those mechanically rechargeable batteries have two distinct advantages which are very essential for EV applications, namely, fast and convenient refuelling (comparable to petrol refuelling with a few minutes) and centralized recharging and recycling (most efficient and environmentally sound use of electricity). The disadvantages associated with rechargeable metal/air batteries are low specific power (at most 105 W/kg for Zn/Air), narrow operating temperature window, carbonation of alkaline electrolyte due to carbon dioxide in air and evolution of hydrogen gas from corrosion in electrolyte.

The Zn/Air battery has been developed as an electrically and mechanically rechargeable battery. Although both of them have been applied to EV applications, the mechanically rechargeable battery is more favourable (Haas and Van Wesemael, 2009). The electrically rechargeable Zn/Air battery nominally operates at 1.2 V and has the specific energy of 180 Wh/kg, energy density of 160 Wh/L and specific power of 95 W/kg. The mechanically rechargeable Zn/Air battery avoids the use of bidirectional air electrode and the shape change problem. Hence, it can offer a higher specific energy of 230 Wh/kg and a higher specific power of 105 W/kg. The depleted zinc negative electrode cassettes can be replaced robotically by a mechanically refuelling system at a fleet servicing point or at a public service station. The discharged fuel is then electrochemically recharged at central facilities. A mechanically rechargeable Zn/Air battery was developed for field test. A 160-kWh Zn/Air battery was installed and tested in a Mercedes-Benz 180E van in 1994. The driving range at a constant speed of 64 km/h was 689 km.

The Al/Air battery has a nominal voltage of 1.4 V. The Al/Air battery with a saline electrolyte is attractive only for low power applications. On the other hand, the alkaline Al/Air battery offers high specific energy and energy density of 250 Wh/kg and 200 Wh/L, respectively, and is suitable for high power applications. Nevertheless, the corresponding specific power is low. Because of its exceptionally low specific power, the Al/Air battery is seldom used as the sole energy source for EVs and it is commonly used in conjunction with other batteries in a HES system.

The battery developer Phinergy developed a silver-based catalyst that only allows oxygen from the ambient air into the positive cathode. The oxygen then combines with the liquid electrolyte, releasing the latent electrical energy stored in the aluminium anode. In one Citroën C1 prototype EV, the Phinergy demonstrated that 25 kg of aluminium cells can deliver a total energy of 100 kWh, which gives a 600-plus mile range.

#### 4.3.1.4 Sodium-beta batteries

The sodium-beta battery refers to the Na/S and Na/$NiCl_2$ batteries, which have liquid sodium as one reactant and beta-alumina ceramic as the electrolyte. The Na/S battery operates at 300–350 °C with a nominal cell voltage of 2 V, specific energy of 170 Wh/kg, energy density of 250 Wh/L and specific power of 390 W/kg.

In Na/$NiCl_2$ battery, the active materials are molten sodium for the negative electrode and solid nickel chloride for the positive electrode. In addition to the beta-alumina ceramic electrolyte as used in the Na/S, there is a secondary electrolyte, namely, sodium-aluminium chloride, in the positive electrode chamber. The secondary electrolyte conducts sodium ions from the primary beta-alumina electrolyte to the solid nickel chloride positive electrode (Sudworth and Galloway, 2009).

The Na/$NiCl_2$ battery operates at 155–350 °C with a nominal cell voltage of 2.58 V. Based on the battery configuration, the battery pack performance parameters are the specific energy of 86–120 Wh/kg and specific power of 150–300 W/kg. Comparing with the Na/S battery, the Na/$NiCl_2$ battery has higher open circuit cell voltage, wider operating temperature, safer products of reaction (less corrosive than molten $Na_2S_x$) and better freeze–thaw durability (smaller temperature difference).

The AEG Zero Emission Battery Research Activity (ZEBRA) has been the major developer of the Na/$NiCl_2$ battery. The ZEBRA battery, namely, Z12, offered a specific energy of 103 Wh/kg and a specific power of 180 W/kg.

#### 4.3.1.5 Ambient-temperature lithium batteries

There are a number of approaches being taken in the design of rechargeable ambient-temperature lithium batteries. One approach is to use metallic lithium for the negative electrode and a solid inorganic intercalation material for the positive electrode. The electrolyte can be a solid polymer, leading to name as the LiM-Polymer battery. Another approach is the use of a lithiated carbon material as the negative electrode such that lithium ions move forth and back between the positive and negative electrodes during cycling. The 'rocking-chair' movements of lithium ions lead to name as the Li-Ion battery (Yamaki, 2009).

##### *4.3.1.5.1 LiM-Polymer batteries*

The LiM-Polymer battery uses lithium metal and a transition metal intercalation oxide ($M_yO_z$) for the negative and positive electrodes, respectively. A thin solid polymer electrolyte (SPE) is used, which offers the merits of improved safety and flexibility in design (Kobayashi *et al.*, 2009). By using a lithium foil as a negative electrode and vanadium oxide ($V_6O_{13}$) as a positive electrode, the Li/SPE/$V_6O_{13}$ cell is a typical LiM-Polymer battery. It operates at a nominal voltage of 3 V and has the specific energy of 155 Wh/kg, energy density of 220 Wh/L and specific power of 315 W/kg. The advantages are high cell voltage (3 V), very high specific energy and energy density (155 Wh/kg and 220 Wh/L, respectively), very low self-discharge rate (about 0.5% per month) and capability of fabrication in a variety of shapes and sizes. However, its low temperature performance is weak.

##### *4.3.1.5.2 Li-Ion batteries*

Since the commercialization of the Li-Ion battery by Sony Energytec in 1990, the Li-Ion battery has been considered to be the most promising rechargeable battery of the future. The Li-Ion battery has already gained acceptance for HEV applications. The specific energy of Li-Ion battery was 98 Wh/kg in 1990 and increased to 195 Wh/kg in 2008 (Yamaki, 2009).

The Li-Ion battery consists of two electrodes, namely, a porous separator impregnated with electrolyte and two current collectors. Lithium cobalt oxide ($LiCoO_2$) typically serves as an active electrode material for the positive electrode. The negative electrode is usually made of lithiated carbon or graphite ($LiC_6$). Electrodes are electrically isolated by the separator, where the space between them is filled by electrolyte. Copper foil is used for the negative current collector and aluminium for the positive current collector.

##### *4.3.1.5.3 Vehicle applications*

The Li-Ion battery can be made from different advanced positive electrode and negative electrode materials. The mature positive electrode materials are $LiCoO_2$, $LiMn_2O_4$, $LiFePO_4$, lithium nickel manganese cobalt (NMC) oxide ($LiNiMnCoO_2$) and lithium nickel cobalt aluminium (NCA) oxide ($LiNiCoAlO_2$). The mature

*Table 4.3 Advanced Li-Ion batteries*

| Positive electrode | Negative electrode | Manufacturers | Key feature |
|---|---|---|---|
| $LiCoO_2$ | Graphite | Sony | Mature |
| $LiMn_2O_2$ | Graphite | NEC, GS, Yuasa, LG | High power |
| NCA/NMC | Graphite | SAFT, Samsung, Sanyo, Evonik | High energy |
| $LiFePO_4$ | Graphite | A123, Valence Tech, BYD | Highly stable |
| $LiMn_2O_4$ | Titanate | Toshiba, Enerdel | High discharge rate |

*Table 4.4 Key features of promising batteries*

| Type | Pb-Acid | Nickel-based | Lithium | |
|---|---|---|---|---|
| Feature | VRLA | Ni-MH | Li-Ion | Li-Titanate |
| Specific energy (Wh/kg) | 30–40 | 60–70 | 160 | 70–90 |
| Cycle life at 100% DoD (cycle) | 50–80 | 300–500 | 500–750 | 25,000 |
| Safety | Fire hazard | Fire hazard | Fire hazard | Safest |
| Charge time (0–90% SoC) (h) | 8 | 2 | 2 | 0.1 |
| Operating temperature (°C) | −10 to 60 | 0–40 | 0–40 | −40 to 70 |
| Environmental impact | Toxic | Low | Minimal | Minimal |
| Memory effect | Very low | Moderate | None | None |
| Power delivery | Good | Moderate | Moderate | High |
| Manufacturability | Easy | Adequate | Easy | Easy |
| Maintenance | Moderate | Moderate | Moderate | Moderate |
| Market position | High volume | Modest | Good | Rising |
| Cost | Low | Tied to Ni | Moderate | Moderate |

negative electrode materials are graphite and titanate. Table 4.3 shows the potential Li-Ion batteries for vehicle applications.

#### 4.3.1.6 Batteries for EV and HEV applications

The battery is the most significant factor of commercialization of EVs and HEVs. Developments of vehicle batteries are continued and accelerated to meet requirements of EVs and HEVs, namely, safety, high specific energy, high specific power, short recharge time, long life cycle and low cost. The mature and promising batteries for EVs and HEVs are VRLA, Ni-MH and Li-Ion batteries. The specific power and specific energy of the batteries are shown in Figure 4.5 and their features are compared in Table 4.4. The VRLA battery is still popular for mild HEVs and low cost EVs due to its maturity and cost effectiveness. The Ni-MH battery is mainly used in HEVs. The Li-Ion batteries are the promising battery in the future. The advanced Li-Ions batteries have demonstrated the potentials of improvements in specific power, specific energy, charge rate and safety.

### *4.3.2 Fuel cells*

The fuel cell generates electrical energy rather than stores it. The fuel cell EV can achieve a long driving range comparable to an ICEV. The hydrogen for fuel cells can be refilled in a much shorter time than battery recharging (except for those mechanically rechargeable ones). Fuel cell's lifetime is generally much longer than that of batteries, and they generally require less maintenance than batteries.

Fuel cells are generally classified by the type of electrolyte, namely, acid, alkaline, molten carbonate, solid oxide and solid polymer. Instead of using hydrogen as the fuel, carbon monoxide and methanol have also been adopted by some fuel cells. However, the by-product of these fuel cells becomes carbon dioxide, rather than plain water. The molten carbonate and solid oxide fuel cells suffer from very high-temperature operation, respectively, over 600 °C and 900 °C, making them practically difficult to be applied to EVs. The acid fuel cells, alkaline fuel cells (AFCs) and solid polymer fuel cells (SPFCs) are technically possible for EV applications.

#### 4.3.2.1 Acid fuel cells

The acid fuel cell is generally characterized by the ionic conduction of hydrogen ions, and platinum or platinum alloys as electrocatalysts for both the anode and cathode. In the early development of acid fuel cells, many different acids were investigated to be the electrolyte, such as the sulphuric acid, hydrofluoric acid and phosphoric acid. Finally, the phosphoric acid won this competition because of its attractive features of stable operation at temperatures up to at least 225 °C, reasonably high conductivity at temperatures above 150 °C, as well as efficient rejection of product water and waste heat at the operating temperature. The phosphoric acid fuel cell (PAFC) is the only acid fuel cell ready for applications. The PAFC generally operates at 150–210 °C and at atmospheric or slightly higher pressures, offering the power density of 0.2–0.25 W/cm$^2$. Its projected life can be over 40 thousand hours. The major disadvantage of the PAFC is its dependence on noble metal electrocatalysts.

#### 4.3.2.2 AFCs

The AFC can adopt low-cost non-noble metal or oxide electrocatalysts, such as nickel for the anode and lithiated nickel oxide for the cathode, to provide reasonable performances. The lower working temperature further enhances the AFC to be more attractive than the PAFC for EV applications. However, there are two major challenges for widespread applications of the AFC. First, since it generally operates at less than 100 °C, the water rejection and heat removal must be design well to maintain its efficiency and reliability. Second, carbon dioxide must be completely removed from the inlet hydrogen and air before their entry into the cell. Even a small amount of carbon dioxide is sufficient to carbonate the electrolyte and form solid deposits in the porous electrode.

#### 4.3.2.3 Proton exchange membrane fuel cells

The proton exchange membrane fuel cell (PEMFC), also named as the SPFC, uses a solid polymer membrane as the electrolyte. This membrane is sandwiched between two platinum-electrocatalysed porous electrodes, namely, the anode and cathode.

The PEMFC works at a lower temperature than the PAFC, it operates at 50–100 °C and at atmospheric or slightly higher pressures. With low platinum loading electrodes, it can offer the power density of 0.35–0.6 W/cm$^2$. The projected life is over 40 thousand hours. The PEMFC has five definite advantages in EV applications. First, the power density of the PEMFC is the highest among all available types of fuel cells. Second, its low-temperature operation and rapid start-up are desirable for EV application. Third, its electrolyte, being solid, does not change, move about or vaporize from the cell. Fourth, since the only liquid in the cell is water, the possibility of any corrosion is essentially eliminated. Fifth, the PEMFC is insensitive to carbon dioxide in the inlet hydrogen and air. However, similar to the PAFC, the major drawback is its dependence on noble metal electrocatalysts.

#### 4.3.2.4 Direct methanol fuel cells

Instead of using hydrogen, methanol can be directly used as the fuel for a fuel cell, so-called the direct methanol fuel cell (DMFC). There are some definite motivations of using the DMFC. First, methanol is the simplest organic fuel which can economically and efficiently be produced on a large scale from the relatively abundant fossil fuels, namely, coal and natural gas. Second, it is the most electro-active organic fuel. Third, methanol is a liquid fuel which can be easily stored, distributed and marketed for EV applications, whereas hydrogen always suffers from the difficulty of storage and distribution.

The DMFC is relatively immature among the aforementioned fuel cells. At the present status of DMFC technology, it generally operates at 50–100 °C and at atmospheric pressure, offering the power density of 0.04–0.23 W/cm$^2$. Its projected life is over 10 thousand hours.

#### 4.3.2.5 Fuel cells in EV applications

Among those EV fuel cells, the PAFC is less attractive than the AFC because of its relatively higher working temperature and higher projected cost. The AFC has ever been implemented in practical EVs. With the advancement of PEMFC technology, the PEMFC takes advantages over the AFC for EV applications. The major reasons are due to its higher power density and longer projected life while maintaining the low working temperature and economical projected cost. Thus, recent research and development on fuel cells for EVs have been focused on the PEMFC technology. Starting from 1993, a Canadian firm, namely Ballard Power Systems, has pioneered the development of compressed hydrogen gas (CHG)-fuelled PEMFC technology for EVs. Including the container and auxiliaries, its CHG-fuelled PEMFC system has achieved the specific energy of about 500 Wh/kg, which is much higher than the maximum value offered by any batteries. However, the corresponding system specific power has been only of 50–180 W/kg, limiting its sole application to those EVs desiring high acceleration rate and hill-climbing capability. Nevertheless, Ballard Power Systems and Daimler-Benz jointly produced a CHG-fuelled PEMFC powered bus in 1997, namely, the NEBUS. This bus has installed 21-kg CHG in seven fibreglass-reinforced aluminium tanks and employed five 25-kW PEMFC stacks, hence delivering the power of 250 kW and guaranteeing the range of 250 km. As of fall 2015, commercially available fuel cell vehicles for sale or lease in the United States are Honda FCX Clarity, Hyundai

Tucson Fuel Cell and Toyota Mirai. All these three fuel cell vehicles are equipped with the PEMFC.

### 4.3.3 Ultracapacitors

The ultracapacitor (UC) or supercapacitor (SC), formerly electric double-layer capacitor (EDLC), is a high-capacity electrochemical capacitor with capacitance values greater than 1,000 F at 1.2 V that bridge the gap between electrolytic capacitors and rechargeable batteries. They typically store 10–100 times more energy per unit volume or mass than electrolytic capacitors. The UC can accept and deliver charge much faster than batteries, and tolerate many more charge and discharge cycles than rechargeable batteries.

#### 4.3.3.1 UC in EV and HEV applications

The UC can deliver high power but short-duration power. Its specific energy is extremely low such that it cannot be used as a sole energy source for EVs. The promising application is the integration of UC with a battery in a HES in the EV drivetrain. The UC can provide much faster and more efficient energy recovery during regenerative braking of EVs. Maxwell Technologies produces UC modules for vehicle and industrial applications. The 48 V UC module can offers a high capacitance of 165 F at 48 V in mass production. As listed in Table 4.5, its specific energy and specific power can achieve 3.9 Wh/kg and 3,300 W/kg, respectively.

### 4.3.4 Ultrahigh-speed flywheels

The use of flywheels for storing energy in mechanical formats is not a new concept. More than 25 years ago, the Oerlikon Engineering Company in Switzerland made the first passenger bus solely powered by a massive flywheel. This flywheel, weighed 1,500 kg and operated at 3,000 rpm, was recharged by electricity at each bus stop. The traditional flywheel is a massive steel rotor with hundreds of kg that spins at the order of ten hundreds of rpm. On the contrary, the advanced flywheel is a lightweight composite rotor with tens of kg and rotates at the order of ten thousands of rpm, so-called the ultrahigh-speed flywheel.

*Table 4.5 Specifications of a 48 V ultracapacitor module*

| | 48 V module |
|---|---|
| Rated capacitance (F) | 165 |
| Nominal voltage (V) | 48 |
| Absolute maximum current (A) | 1,900 |
| Dimensions (mm) | *L*: 418, *W*: 194, *H*: 179 |
| Weight (kg) | 13.5 |
| Specific energy (Wh/kg) | 3.9 |
| Specific power (W/kg) | 3,300 |
| Operating temperature (°C) | −40 to 65 |
| Projected cycle life at 25 °C (cycles) | 1,000,000 |

The concept of ultrahigh-speed flywheels appears to be a feasible means for fulfilling the stringent energy storage requirements for EV applications, namely, high specific energy, high specific power, long cycle life, high energy efficiency, quick recharge, maintenance free and cost effective. When the flywheel is conceived as a high power energy source in a HES in an EV, it stores energy in mechanical form during periods of cruising or regenerative braking while generates electrical energy to meet the peak power demands during periods of starting, acceleration or hill climbing. The potential benefits of using the ultrahigh-speed flywheel as the high-power source in a hybrid configuration with the battery for EVs are similar to those of using the ultracapacitor.

#### 4.3.4.1 Ultrahigh-speed flywheel in EV applications

The Lawrence Livermore National Laboratory (LLNL) in the United States has actively engaged on the development of ultrahigh-speed flywheels. The LLNL has developed a modular flywheel storage system based on the use of some advanced technologies, including passive magnetic bearings and a novel type of electrostatic generator that is very light in weight. It has near 100% efficiency. Computer-based simulations predicted that the use of this flywheel module may allow a doubling of an EV equipped with the same battery pack weight of the flywheel system.

## 4.4 Hybridization of energy sources in EVs and HEVs

The basic principle of energy hybridization in EVs and HEVs is the integration of both high specific energy and high specific power sources in the drivetrain, as illustrated in Figure 4.1. First, during the normal driving condition, the high-energy energy source supplies the propulsion energy to the driveline. To enable the system ready for sudden power demand, this source can charge the high-power power source in the light-load period. Second, during the acceleration or hill-climbing condition, both energy sources supply the propulsion energy simultaneously. Third, during the braking or downhill condition, the braking energy can be used to recharge the high-power energy source with regenerative braking. When the high-power source is fully charged, braking energy can be diverted to recharge the high-energy energy source provided that it is energy receptive.

The principle of HEV design is the integration of the EVs electric propulsion system into the ICEVs mechanical propulsion system. It is the hybridization of drivetrains in HEVs. In EVs, the whole vehicle is driven by electric motor(s) but the electric motor is powered by dual energy sources in the HES. It is the hybridization of energy sources in EVs.

### *4.4.1 Hybridization of drivetrains in HEVs*

The HEV can be divided into series, parallel, series–parallel and complex hybrids with reference to their drivetrain topologies. Figure 4.7 shows the system architecture, in which the electrical link is bidirectional; the hydraulic link is unidirectional and the mechanical link (including brakes, clutches and gears) is also

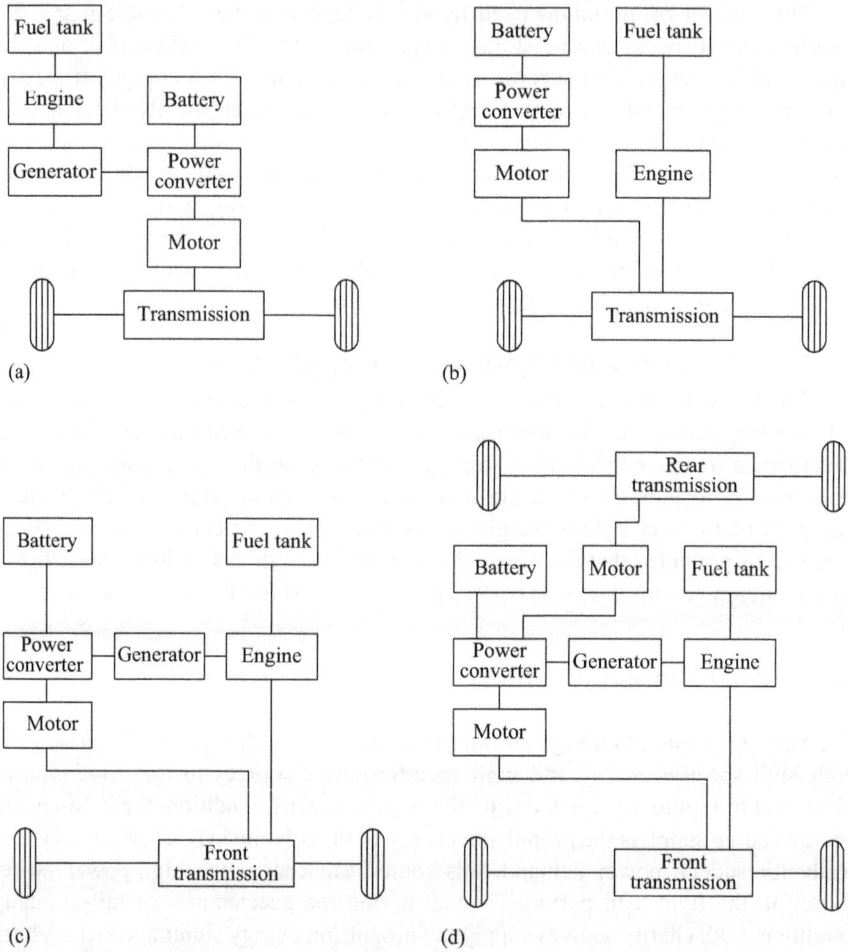


*Figure 4.7 System architecture of HEVs: (a) series hybrid, (b) parallel hybrid, (c) series-parallel hybrid, and (d) complex hybrid*

bidirectional. The series hybrid couples the engine with the generator to produce electricity for pure electric propulsion, whereas the parallel hybrid couples both the engine and electric motor with the transmission via the same drive shaft to propel the wheels. The series–parallel hybrid is a direct combination of both the series and parallel hybrids. On top of the series–parallel hybrid operation, the complex hybrid can offer additional and versatile operating modes by adding one more electric motor to deliver four-wheel drive (4WD) operations.

The series hybrid is the simplest kind of HEV. The engine's mechanical output is first converted into electricity using a generator. The converted electricity either charges the battery or bypasses the battery to propel the wheels via the same electric motor and mechanical transmission. Conceptually, it is an engine-assisted EV,

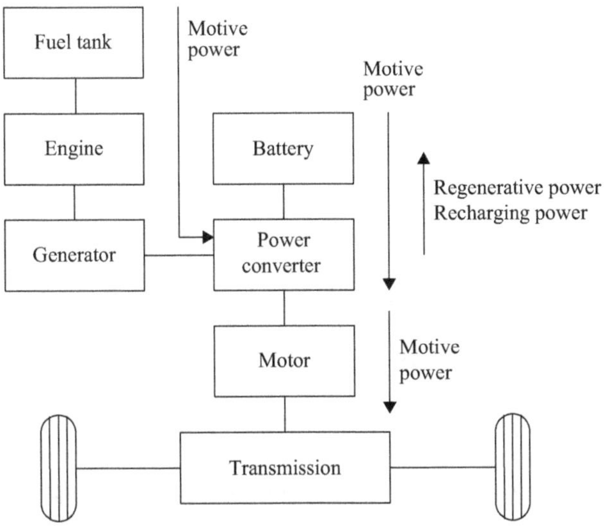


*Figure 4.8 Operating models of series hybrids*

aiming to extend the driving range comparable with that of the ICEV. Due to the absence of clutches throughout the mechanical link, it has the definite advantage of flexibility for locating the engine-generator set. Although it has an added advantage of simplicity of its drivetrain, it needs three propulsion devices – the engine, the generator and the electric motor. Another disadvantage is that all these propulsion devices must be sized for maximum sustained power if the series HEV is designed to climb a long grade. The Chevrolet Volt has a series hybrid drivetrain. The Chevrolet Volt is propelled by a 111 kW electric motor and 16 kWh Li-Ion batteries. The driving range is extended by a 1.4 L ICE and a 53 kW generator.

In the series hybrid system, the energy flow can be illustrated in Figure 4.8. During start-up, normal driving or acceleration of the series HEV, both the engine (via the generator) and battery deliver electrical energy to the power converter which then drives the electric motor and then the wheels via the transmission. At light loads, the engine output is greater than the output required to drive the vehicle so that the generated electrical energy is also used to charge the battery until the battery capacity reaches a proper level. During braking or deceleration, the electric motor acts as a generator which transforms the kinetic energy of the wheels into electricity, hence charging the battery via the power converter. The battery can also be charged by the engine via the generator and power converter, even when the vehicle is standstill.

The parallel HEV allows both the engine and electric motor to deliver power in parallel to drive the vehicle. Both the engine and electric motor are generally coupled to the drive shaft of the wheels via two clutches such that the propulsion power may be supplied by the engine, the electric motor or both. Conceptually, it is an electrically assisted ICEV for achieving lower emissions and fuel consumption.

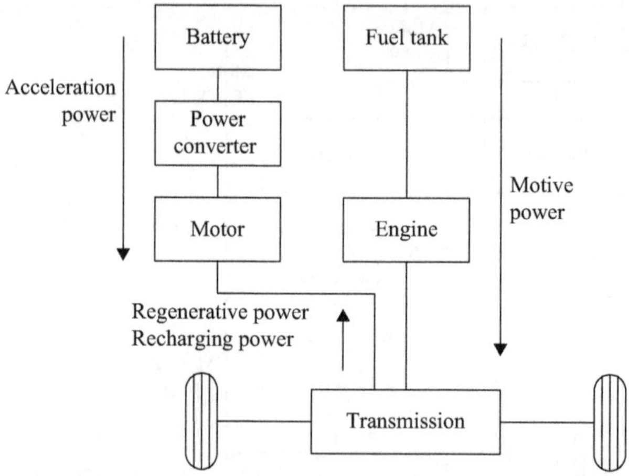


*Figure 4.9 Operating modes of parallel hybrids*

The electric motor can be used as a generator to charge the battery by regenerative braking or absorbing power from the engine when its output is greater than that vehicle demand. The parallel hybrid needs only two propulsion devices – the engine and the electric motor. Another advantage over the series case is that a smaller engine and a smaller electric motor can be used to get the same performance until the battery is depleted. The Honda Civic Hybrid is a parallel HEV, which is propelled by a 1.3 L ICE and a 12 kW electric motor integrated into the powertrain to boost efficiency of the engine.

The energy flow of a parallel HEV is illustrated in Figure 4.9. During start-up or full-throttle acceleration, both the engine and electric motor proportionally share the required power to propel the vehicle. Typically, the power distribution between the engine and electric motor is 80–20%. During normal driving, the engine solely supplies the necessary power to propel the vehicle while the electric motor remains in the off mode. During braking or deceleration, the electric motor acts as a generator to charge the battery via the power converter. Also, since both the engine and electric motor are coupled to the same drive shaft, the battery can be charged by the engine via the electric motor when the vehicle is at light load.

In the series–parallel hybrid, the configuration incorporates the features of both the series and parallel hybrids. It has advantageous features of both the series and parallel hybrids, but it is relatively more complicated and costly. The Toyota Prius is the first series–parallel HEV on the market. The series–parallel hybrid has more freedom to boost the system efficiency of the vehicle. Series–parallel hybrids can be further divided into engine-heavy and electric-heavy hybrids. Engine-heavy denotes that the engine is more active than the electric motor, whereas electric-heavy indicates that the electric motor is more active. Figure 4.10 shows an engine-heavy series–parallel hybrid system and six operating modes. At start-up, the battery solely provides the necessary power to propel the vehicle while the engine is in the off mode. During full-throttle acceleration, both the engine and electric motor

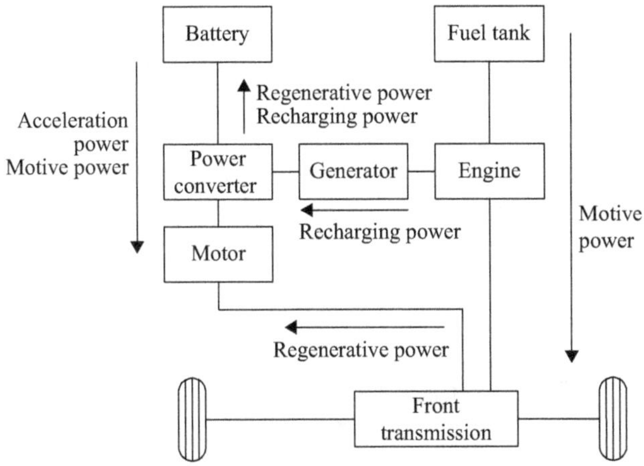


*Figure 4.10 Operating modes of engine-heavy series–parallel hybrids*

proportionally share the required power to propel the vehicle. During normal driving, the engine solely provides the necessary power to propel the vehicle while the electric motor remains in the off mode. During braking or deceleration, the electric motor acts as a generator to charge the battery via the power converter. For battery charging during driving, the engine drives not only the vehicle but also the generator to charge the battery via the power converter. When the vehicle is at a standstill, the engine can drive the generator to charge the battery.

The complex hybrid seems to be similar to the series–parallel hybrid; however, the key difference is the bidirectional power flow of the electric motor in the complex hybrid, comparing to the unidirectional power flow of the generator in the series–parallel hybrid. This bidirectional power flow can allow for versatile operating modes, especially the three propulsion power (due to the engine and two electric motors) operating mode which cannot be offered by the series–parallel hybrid. Similar to the series–parallel HEV, the complex hybrid suffers from higher complexity and costliness. The Lexus RX 400 h was built with this topology. The front wheels of the RX 400 h are propelled by both the 3.3 L ICE and the 123 kW electric motor in the series–parallel hybrid mode while the rear wheels are propelled by a 50 kW electric motor. There is no direct mechanical coupling between the ICE and the 50 kW rear motor, but they are electrically connected by the 82 kW generator.

The energy management system of the complex hybrid is focused on the dual-axle propulsion system. In this system, the front-wheel axle and rear-wheel axle are separately driven. There is no propeller shaft to connect the front and rear wheels, so it enables a more lightweight propulsion system and increases the vehicle packaging flexibility. Moreover, regenerative braking on all four wheels can significantly improve the vehicle fuel efficiency. Figure 4.11 shows a dual-axle complex hybrid system, where the front-wheel axle is propelled by a hybrid drivetrain and the rear-wheel axle is driven by an electric motor. During start-up,

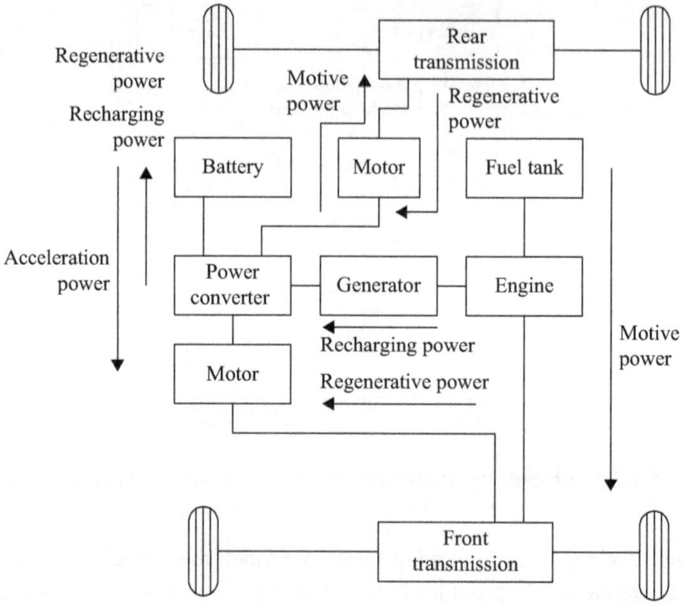


*Figure 4.11 Operating modes of dual-axle (front-hybrid rear-electric) complex hybrids*

the battery delivers electrical energy to drive both the front and rear electric motors to individually power the front and rear axles of the vehicle while the engine is off. For full-throttle acceleration, both the engine and front electric motor power the front axle while the rear electric motor drives the rear axle. Notice that this operating mode involves three propulsion devices (one engine and two electric motors) to simultaneously propel the vehicle. During normal driving and/or battery charging, the engine output is split to power the front axle and to drive the electric motor, which works as a generator to charge the battery. The engine, front electric motor and front axle can be mechanically coupled by planetary gear sets. When driving at light load, the battery delivers electrical energy to the front electric motor only to drive the front axle whereas both the engine and rear electric motor are off. During braking or deceleration, both the front and rear electric motors act as generators to simultaneously charge the battery. A unique feature of this dual-axle system is the capability of axle balancing. In case the front wheels slip, the front electric motor works as a generator to absorb the change of engine output power. Through the battery, this power difference is then used to drive the rear wheels to achieve axle balancing.

According to the level of electric power contribution and functionalities of the electrical powertrain, HEVs can be classified into micro hybrid vehicle (MHV), mild HEV (MHEV), full HEV (FHEV) and plug-in HEV (PHEV). On the other hand, EVs are classified into BEV and FCEV.

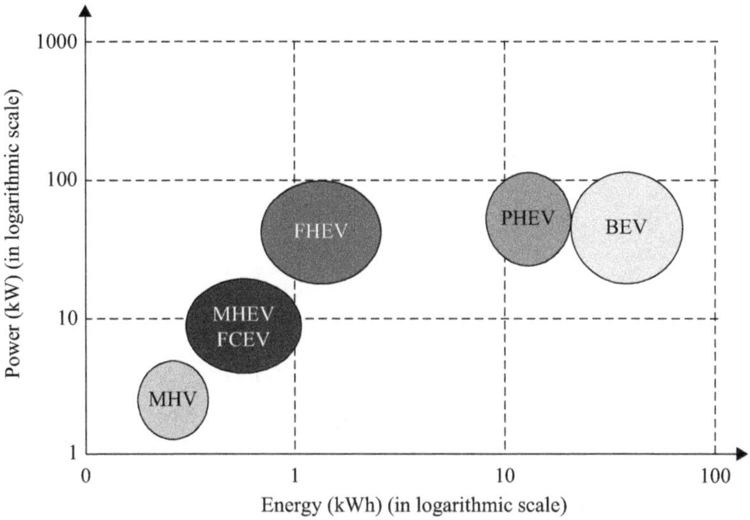


*Figure 4.12 Power and energy requirements of batteries for various EVs and HEVs*

The battery is still the most important component of these vehicles but the requirements on power, energy, cycle life and system voltage are different. Key requirements for vehicle batteries are high specific energy, high specific power, long cycle life, high efficiency, wide operating temperature and low cost for commercialization. Figure 4.12 shows the power and energy requirements of battery for various EVs and HEVs. Functionality of the electrical powertrain and the favourable battery voltages in these vehicles are shown in Figure 4.13.

The United States Council for Automotive Research LLC (USCAR) and the USABC have set technical targets for vehicle batteries. The HEV needs batteries with long cycle life, high specific power for power boost and regenerative braking. Table 4.6 shows some USABC key goals for batteries in HEV and FCEV applications (Wong and Chan, 2013).

#### 4.4.1.1 MHV

The MHV has an electric motor with peak power of about 2.5 kW. The electrical powertrain is driven by a battery system at 12–42 V. The motor is small and simple in structure and serves a function similar to the starter and alternator in an ICE vehicle. The electrical and engine powertrains in an MHV are governed by an automatic stop–start mechanism, in which, the engine shuts down under vehicle braking and rest. The MHV is favourable for city driving, where there are frequent stops and starts. An MHV's fuel economy can be 5–10% higher than that of an ICE vehicle in city driving. The Citroen C3 is an MHV using the Valeo motor system. The battery discharges frequently in cranking the engine in MHVs. Thus, there is a demand for high cycle life for batteries in MHVs. Table 4.7 lists some key technical data of batteries for MHVs.

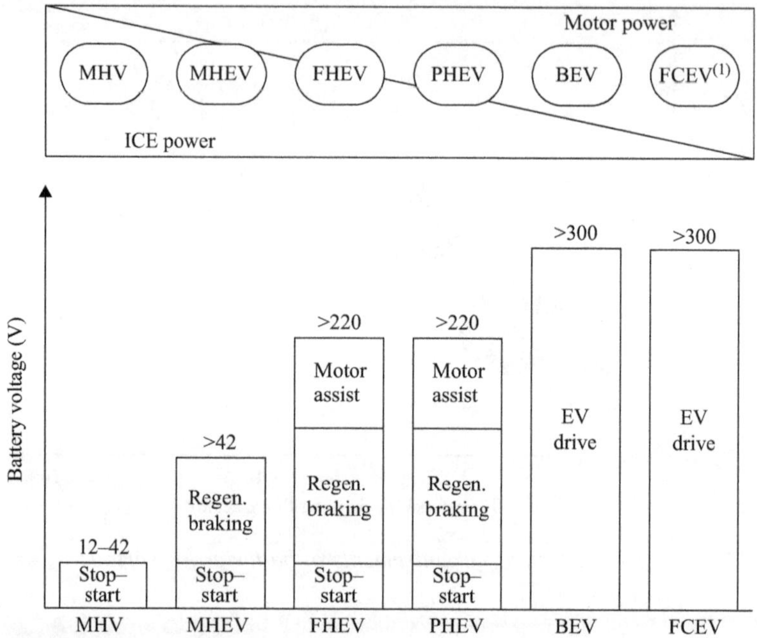


Remarks:
(1) : Assume the FCEV adopts a fuel cell and battery hybrid system

*Figure 4.13 Battery operating voltages in EVs and HEVs*

*Table 4.6 Typical USABC goals for batteries in HEV and FCEV applications*

| **Parameters** | **Unit** | **MHV** | **MHEV** | **FHEV** | **FCEV** |
|---|---|---|---|---|---|
| Discharge pulse power | kW | 6 | 13 to 25 | 40 | 20 |
| Regenerative pulse power | kW | NA | 8 to 20 | 35 | 25 |
| Energy capacity | kWh | 0.25 | 0.30 | 0.50 | 0.25 |
| Calendar life | year | 15 | 15 | 15 | 15 |
| Cycle life | cycle | 150,000 | 300,000 | 300,000 | NA |
| Maximum operating voltage | V | 48 | 400 | 400 | 440 |
| Operating temperature | °C | −30 to +52 | −30 to +52 | −30 to +52 | −30 to +52 |

NA: not applicable

#### 4.4.1.2 Mild HEV

The MHEV has a more powerful electrical powertrain than an MHV. The typical electric motor power of a sedan MHEV is about 10–20 kW at 100–200 V. The motor is directly coupled with the engine. The motor has a large inertia such that it can replace the original flywheel of the engine. The motor and the engine are generally coupled in parallel. The electrical powertrain is designed to crank the engine and

*Table 4.7 Technical data of batteries for MHVs*

| Parameters | Unit | MHV |
|---|---|---|
| Voltage | V | 12–42 |
| Discharge power | kW | 4.2–6 |
| Low temperature (−28 °C) discharge power | kW | >3 |
| Energy capacity | kWh | 0.2–1 |
| Operating temperature | °C | −30 to +52 |
| Calendar life | year | >3 |

*Table 4.8 Technical data of batteries for MHEVs*

| Parameters | Unit | MHEV |
|---|---|---|
| Voltage | V | 42–200 |
| Discharge power | kW | >15 |
| Low temperature (−28 °C) discharge power | kW | >4 |
| Recharge power | kW | >15 |
| SoC Window | % | 40–70 |
| Recharge pulse power | kW | >20 |
| Energy capacity | kWh | 0.8–1 |
| Operating temperature | °C | −30 to +52 |
| Calendar life | Year | >10 |

offer regenerative braking during braking. There are demands of high specific power and long service life for batteries in MHEVs. Table 4.8 shows some technical data of batteries for MHEVs. The battery's charge and discharge power depend on its SoC. The battery's discharge power decreases with its SoC. The minimum operating SoC is around 40–50% to uphold sufficient power for launch and acceleration support. On the other hand, the battery's recharging power drops when the SoC is high; thus, the maximum operating SoC is regulated at around 70–80% to maintain sufficient recharge power for regenerative braking. Typically, the batteries operate in a SoC window between 40% and 70%. Comparing with an ICE vehicle, the MHEV can boost the fuel economy by 20–30% in city driving. Examples of MHEVs are Honda Insight Hybrid, Honda Civic Hybrid and Ford Escape Hybrid.

#### 4.4.1.3 Full HEV

The FHEV has a high power electrical powertrain to drive the vehicle purely by electricity in a short driving range. The typical electric motor power for sedan FHEV is about 50 kW at 200–350 V. Generally, the motor, generator and engine are coupled in series–parallel configuration. With the aid of power split devices, which are mainly built by planetary gear sets and clutches, the energy management system of the engine, motor and generator is designed to maximize energy efficiency and minimize emissions. Table 4.9 shows the technical data of batteries for FHEVs.

*Table 4.9 Technical data of batteries for FHEVs*

| Parameters | Unit | FHEV |
|---|---|---|
| Voltage | V | 220–350 |
| Discharge power | kW | >35 |
| Low temperature (−28 °C) discharge power | kW | >4 |
| Recharge power | kW | >30 |
| SoC Window | % | 40–80 |
| Recharge pulse power | kW | >40 |
| Energy capacity | kWh | 1–2 |
| Operating temperature | °C | −30 to +52 |
| Calendar life | year | >10 |

The FHEV can be driven in pure EV mode and hybrid mode. The electrical powertrain assists the engine, not only at the starting, but also during acceleration in the hybrid mode, which is also called charge sustaining mode. In the charge sustaining mode, the battery is recharged not only by regenerative braking but also by the engine to maintain the SoC in a high and narrow window. The FHEV can achieve higher fuel economy than that of the ICE vehicle by 30–50% in city driving. Examples of FHEVs are the Toyota Prius, Toyota Highlander and Lexus RX 400 h.

#### 4.4.1.4 Plug-in HEV

The PHEV is similar to that of an FHEV. The key differences are the additional battery pack and the functionality of grid recharging. In addition to the charge sustaining mode, the PHEV can operate in the charge depletion mode, in which the PHEV operates in pure EV mode. Thus, the battery SoC drops in the charge depletion mode.

The electrical drivetrain of a PHEV works in a high voltage at 220–350 V. The battery energy capacity in PHEVs is the largest among all HEVs and it is determined by the designated pure electric driving range. The PHEV operates in the charge depletion mode first and then the charge sustaining mode. In the charge depletion mode, the battery SoC decreases from 100% to a threshold SoC (typically 20–30%), which triggers the operation mode change. In the charge sustaining mode, the battery SoC oscillates around the threshold SoC. The battery is recharged from the grid at the end of the trip. Similar to the EV, the PHEV suffers from complexity and costliness. However, the PHEV delivers longer driving range than the EVs that is comparable to conventional ICEVs. Table 4.10 shows technical data of batteries for PHEVs.

The BYD F3DM is the world's first mass production PHEV, which went on sale to the government agencies and corporations in China in December, 2008. Toyota also worked on a plug-in version of the Prius. The plug-in Prius was converted from the Prius by adding additional 1.3 kWh battery pack into the car and a charging unit. A PHEV can also be implemented in a series hybrid topology. The GM Chevrolet Volt is a series hybrid PHEV, which is also called extended-range EV (EREV). The EREV is driven by one sole electrical powertrain, powered by the battery and a small engine.

*Table 4.10 Technical data of batteries for PHEVs*

| **Parameters** | **Unit** | **PHEV** |
|---|---|---|
| Voltage | V | 220–350 |
| Discharge power | kW | >50 |
| Low temperature (−28 °C) discharge power | kW | >6 |
| Recharge power | kW | >30 |
| SoC Window | % | 20–100 |
| Recharge pulse power | kW | >20 |
| Energy capacity | kWh | 5–20 |
| Charge time | h | <5 |
| Operating temperature | °C | −30 to +52 |
| Calendar life | year | >10 |

*Table 4.11 Technical data of batteries for BEVs*

| **Parameters** | **Units** | **BEV** |
|---|---|---|
| Voltage | V | >300 |
| Discharge power | kW | >50 |
| Low temperature (−28 °C) discharge power | kW | >40 |
| Recharge power | kW | >50 |
| SoC Window | % | 20–100 |
| Energy capacity of the HES | kWh | 200–300 |
| Charge time | h | <8 |
| Operating temperature | °C | −30 to +52 |
| Calendar life | year | >10 |

### *4.4.2 Hybridization of energy sources in EVs*

Based on the available technology of various energy sources, there are several viable hybridization schemes for EVs in the near term: battery–battery hybrids, battery–ultracapacitor hybrids, and fuel cell–battery hybrids.

#### 4.4.2.1 Battery-battery and battery-ultracapacitor HES in BEVs

Battery is the sole energy source for the electrical powertrain and accessory systems in a BEV. The typical electric motor power for sedan BEV is about 50–80 kW and the battery operates in a high voltage, over 300 V. Table 4.11 lists some key data of batteries for BEVs. In city driving, the EVs fuel economy can be double that of an ICE vehicle (Gutmann, 2009).

In the battery–battery hybrids, one battery provides high specific energy while another battery offers high specific power. Taking into account the maturity and cost, the Zn/Air–Li-Ion hybrid seems to be a natural choice. It combines the merits of the 230 Wh/kg of Zn/Air for long driving range and the 300 W/kg of Li-Ion for acceleration and hill climbing. This choice also overcomes the incapability of the mechanically rechargeable Zn/Air which cannot accept the regenerative energy during braking or downhill. In the battery–battery HES system, the energy flow is illustrated in Figure 4.14, where Zn/Air is the high-energy energy source, while

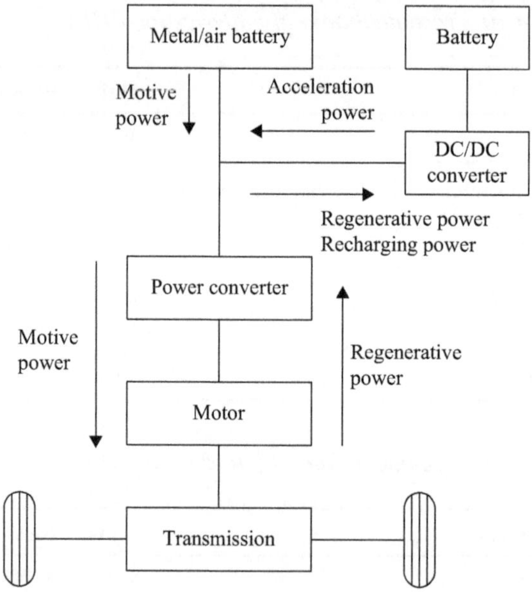


*Figure 4.14 Operating modes of battery–battery HES in BEVs*

Li-Ion is the high-power energy source. During start-up, normal driving or acceleration, both the high-energy energy source and the high-power energy source discharge to drive the electric motor and then the wheels via the transmission. At light loads, the high-energy energy source drives the vehicle and charge the high-power energy source. During braking or deceleration, the electric motor acts as a generator which transforms the kinetic energy of the wheels into electricity, hence charging the high-power energy source via the power converter.

The ultracapacitor can offer high charging and discharging power, but its energy density is extremely low. The ultracapacitor has to work together with other EV batteries in a battery–ultracapacitor HES system. As shown in Figure 4.15, an additional two-quadrant DC–DC converter should be placed between the battery source and the ultracapacitor source, because the working voltage of the ultracapacitor source is quite low (generally less than 100 V even many ultracapacitors have already been internally stacked up). In a Li-Ion–UC HES, the Li-Ion is the primary energy source while the peak power is delivered by the ultracapacitor source. The ultracapacitor is recharged during regenerative braking or from the Li-Ion at light load.

#### 4.4.2.2 Fuel cell–battery HES in FCEVs

Although the fuel cell can offer outstanding specific energy, it suffers from low specific power and incapability of accepting regenerative energy. A FCEV is generally propelled by a fuel cell–battery HES system, in which, the fuel cell stack and the battery are working at a high voltage for efficient regenerative braking and higher acceleration power. The battery operates in a SoC window between 40% and 70%. Figure 4.16 shows the energy flow in a typical FCEV and Table 4.12 shows technical data of batteries for FCEVs.

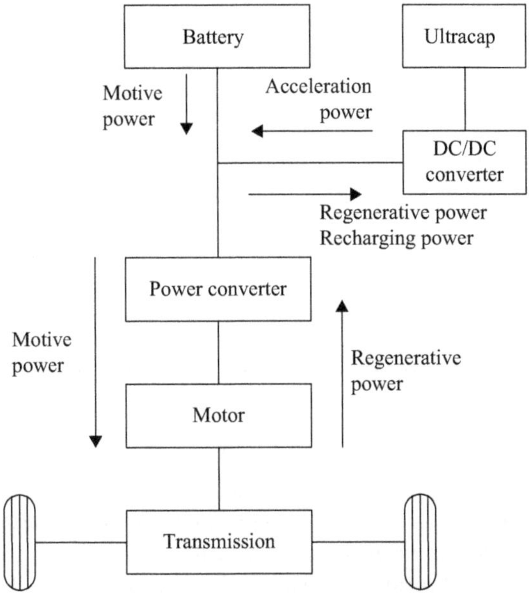


*Figure 4.15 Operating modes of battery–ultracapacitor HES in BEVs*

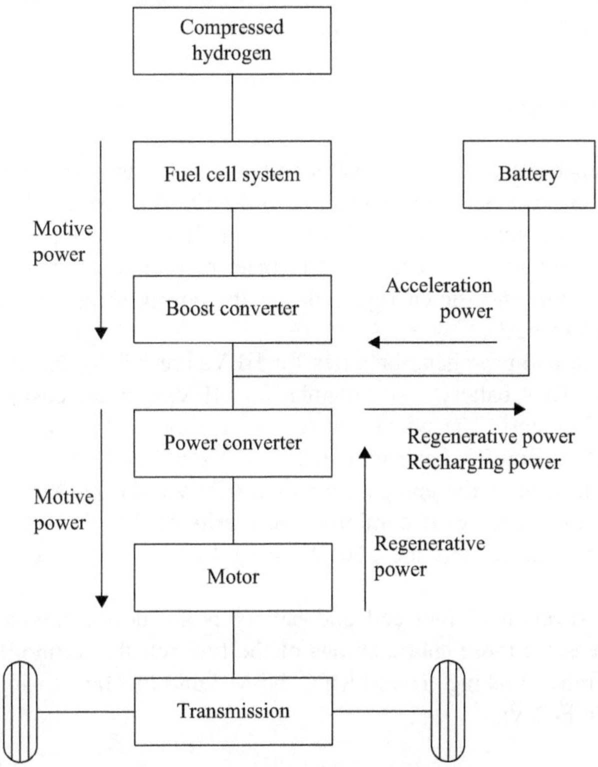


*Figure 4.16 Operating modes of fuel cell–battery HES in FCEVs*

*Table 4.12 Technical data of batteries for FCEVs*

| Parameters | Unit | FCEV |
|---|---|---|
| Voltage | V | 200–440 |
| Discharge power | kW | >15 |
| Low temperature (−28 °C) discharge power | kW | >4 |
| Recharge power | kW | >15 |
| SoC window | % | 40–70 |
| Recharge pulse power | kW | >20 |
| Energy capacity | kWh | 0.6–1.5 |
| Operating temperature | °C | −30 to +52 |
| Calendar life | year | >10 |

The fuel cell–battery hybrid is a good collaboration in which the battery can be purposely selected to compensate those shortcomings of the fuel cell. The PEMFC–Ni-MH and PEMFC–Li-Ion hybrids are typical choices since these batteries are of high specific power and rapid recharge capability.

In long term, the ultracapacitor should be improved to such a level that its specific energy is sufficiently high to provide all necessary instantaneous energy for acceleration and hill climbing, as well as to accept all regenerative energy during braking and downhill. Similarly, the ultrahigh-speed flywheel should also be able to attain such specific energy in the long term.

## 4.5 Conclusions

The onboard energy source is the most important part for drivetrain hybridization in the HEV and energy source hybridization in the EV. The battery is the most significant factor of commercialization of EVs and HEVs. Developments of vehicle batteries are continued and accelerated to meet requirements of EVs and HEVs, namely, safety, high specific energy, high specific power, short recharge time, long life cycle and low cost.

The mature and promising batteries for HEVs are VRLA, Ni-MH and Li-Ion batteries. The VRLA battery is still popular for MHVs and low-cost EVs due to its maturity and low cost. The Ni-MH battery is mainly used in HEVs. The Li-Ion batteries are the promising battery in HEVs, BEVs and FCEVs.

In hybridization of the energy sources for BEVs, the Zn/Air–Li-Ion battery hybrid is a natural choice. It combines the merits of the 230 Wh/kg of Zn/Air for long driving range and the 300 W/kg of Li-Ion for acceleration and hill climbing.

The hybridization of fuel cell and battery is an ideal combination that the battery compensates those shortcomings of the fuel cell that cannot be recharged during regenerative braking. The PEMFC–Ni-MH and PEMFC–Li-Ion hybrids are typical HES in FCEVs.

## References

Arai H, and Hayashi M (2009) Positive electrode: lithium cobalt oxide. *Encyclopedia of Electrochemical Power Sources*, Amsterdam, Elsevier, pp. 258–263.

Cairns EJ (2009) Nickel–zinc. *Encyclopedia of Electrochemical Power Sources*, Amsterdam, Elsevier, pp. 528–533.

Chan CC (2007) The state of the art of electric, hybrid, and fuel cell vehicles. *Proceedings of the IEEE,* 95 (4), 704–718.

Chan CC, and Chau KT (2001) *Modern Electric Vehicle Technology.* Oxford: Oxford University Press.

Chan CC, Wong YS, Bouscayrol A, and Chen K (2009) Powering sustainable mobility: roadmaps of electric, hybrid, and fuel cell vehicles. *Proceedings of the IEEE*, 97 (4), 603–607.

Chau KT, and Wong YS (2002) Overview of power management in hybrid electric vehicles. *Energy Conversion and Management*, 43, 1953–1968.

Ghoniem AF (2011) Needs, resources and climate change: clean and efficient conversion technologies, *Progress in Energy and Combustion Science*, 37 (1), 15–51.

Goodenough JB (2009) Positive electrode: layered metal oxides. *Encyclopedia of Electrochemical Power Sources*, Amsterdam, Elsevier, pp. 243–248.

Gutmann G (2009) Electric vehicle: batteries. *Encyclopedia of Electrochemical Power Sources*, Amsterdam, Elsevier, pp. 219–235.

Haas O, and Van Wesemael J (2009) Zinc–air: electrical recharge. *Encyclopedia of Electrochemical Power Sources,* Amsterdam, Elsevier, pp. 384–392.

Hariprakash B, Shukla AK, and Venugoplan S (2009) Nickel–metal hydride: overview. *Encyclopedia of Electrochemical Power Sources*, Amsterdam, Elsevier, pp. 494–501.

Holze R (2009) Sodium–sulfur. *Encyclopedia of Electrochemical Power Sources*, Amsterdam, Elsevier, pp. 302–311.

Kabza H (2009) Hybrid electric vehicles: overview. In J Garche (Ed.), *Encyclopedia of Electrochemical Power Sources* (pp. 249–268). Amsterdam: Elsevier.

Kanno R (2009) Positive electrode: lithium nickel oxide. *Encyclopedia of Electrochemical Power Sources*, Amsterdam, Elsevier, pp. 297–306.

Kobayashi Y, Seki S, and Terada N (2009) Lithium-ion polymer batteries. *Encyclopedia of Electrochemical Power Sources*, Amsterdam, Elsevier, pp. 375–382.

Köhler U (2009) Hybrid electric vehicles: batteries. *Encyclopedia of Electrochemical Power Sources*, Amsterdam, Elsevier, pp. 269–285.

Kramm F, and Niepraschk H (2009) Valve-regulated batteries: gel. *Encyclopedia of Electrochemical Power Sources*, Amsterdam, Elsevier, pp. 727–734.

Lam LT, and Furukawa J (2009) Supercap hybrid (UltraBattery$^{TM}$). *Encyclopedia of Electrochemical Power Sources*, Amsterdam, Elsevier, pp. 755–763.

Liu DW, Cao GZ, and Wang Y (2009) Positive electrode: nanostructured transition metal oxides. *Encyclopedia of Electrochemical Power Sources*, Amsterdam, Elsevier, pp. 337–355.

Loyns AC (2009) Bipolar batteries. *Encyclopedia of Electrochemical Power Sources*, Amsterdam, Elsevier, pp. 750–754.

Rand DAJ, and Moseley PT (2009) Lead-acid system overview. *Encyclopedia of Electrochemical Power Sources*, Amsterdam, Elsevier, pp. 550–575.

Sudworth JL, and Galloway RC (2009) Sodium–nickel chloride. *Encyclopedia of Electrochemical Power Sources*, Amsterdam, Elsevier, pp. 312–323.

Taylor C (1998) Automobile engine tribology—design considerations for efficiency and durability. *Wear*, 221 (1), 1–8.

USCAR (2015) *Energy Storage System Goals: USABC Goals for Advanced Batteries for EVs.* USCAR, http://www.uscar.org [Accessed 2 Dec 2015].

USDOE (2011a) *Fuel Economy: Where the Energy Goes*. http://www.fueleconomy.gov/feg/atv.shtml [Accessed 21 Aug 2011]

USDOE (2011b) *Find a Car*. http://www.fueleconomy.gov/feg/findacar.htm [Accessed 21 Aug 2011]

Weighall MJ (2009) Valve-regulated batteries: absorptive glass mat. *Encyclopedia of Electrochemical Power Sources*, Amsterdam, Elsevier, pp. 715–726.

Wohlfahrt-Mehrens M (2009) Positive electrode: manganese spinel oxides. *Encyclopedia of Electrochemical Power Sources*, Amsterdam, Elsevier, pp. 318–327.

Wong YS, and Chan CC (2013) Vehicle energy storage: batteries. In M Ehsani *et al.* (eds.), *Transportation Technologies for Sustainability*, New York: Springer, pp. 1082–1102.

Wong YS, Chan CC, and Nazir S (2014) Basic consideration. In D. Crolla *et al.* (eds.) *Encyclopedia of Automotive Engineering*. Chichester, West Sussex: John Wiley & Sons, pp. 1023–1038.

Yamaki J (2009) Lithium rechargeable systems – lithium-ion overview. *Encyclopedia of Electrochemical Power Sources*, Amsterdam, Elsevier, pp. 183–191.

Zaghib K, Mauger A, Gendron F, Julien CM, and Goodenough JB (2009) Positive electrode: lithium iron phosphate. *Encyclopedia of Electrochemical Power Sources*, Amsterdam, Elsevier, pp. 264–296.

*Chapter 5*

# Solar energy harvesting for electric vehicles

*King Hang Lam*[1]

The harvesting of solar energy has gained much impetus in recent years as part of the solutions to tackle the ever increasing global energy demand amid the increasing threats from climate change and pollution issues. Its large-scale deployment, similar to the situation of electric vehicles (EVs), has many advantages but also is facing many challenges at the same time. Nevertheless, promoting the application of both solar energy and EVs is probably one of the ways to resolve some problems that we are facing worldwide.

In this chapter, the basic ideas about the means to harvest solar energy, and the ways to harvest it specifically for EVs will be described. The appropriate type of solar energy – photovoltaic (PV) will be introduced and the respective electrical characteristics and performance will be explained by referring to a mathematical model developed. Finally, two case studies will be presented for illustrating the practical application of harvesting solar energy for vehicles.

## 5.1 How to harvest solar energy?

The human race has a long history of harvesting solar energy. We had been using the thermal energy from the sun to dry our clothes and food since early civilisation. We all enjoy the warmth under the sun in the cold weather, it is therefore natural for the human race to utilise the solar energy in the form of thermal energy. The first type of widely applied solar energy system was understandably the solar thermal system. Generally, it absorbs the heat from the sun to heat up a media of thermal storage for further use (e.g. heat up water for domestic application). Other more sophisticated applications include concentrating the sunlight to a solar tower to collect the heat for generation of electricity. These complex systems are still developing and not widely applied up till now. Nevertheless, the domestic use of solar thermal systems is already a commonplace application nowadays. Figure 5.1 shows a domestic solar thermal system for preheating the water to supplement hot water supply in a residential building.

[1]Department of Electrical and Electronic Engineering, The University of Hong Kong, Hong Kong, China

*Figure 5.1 A domestic solar thermal system installed on the roof of a house in Hong Kong (Photo courtesy: Mr. Jackson FUNG)*

Although harvesting the thermal energy from the sun is a convenient and economical way, it would require another step to convert it to the common type of energy carrier that modern people are enjoying – electricity. Therefore, the most widely applied type of solar thermal system is to utilise the thermal energy for heating up the domestic water. On the other hand, electricity is also the specific energy carrier that could link up solar energy and EV without the need of further energy conversion. Hence, this chapter will be devoted to the technology of harvesting solar energy directly by transforming sunlight into electricity. This type of technology is called PV, which will be explained in more details in the following section, followed by two case studies illustrating how this technology could be applied to vehicles.

### 5.1.1 Brief history and types of PV technology

PV technology – the direct conversion of electricity from sunlight – was rooted from the discovery of photogalvanic effect by Becquerel in 1839 (Messenger and Venture, 2000). The first PV cell, however, was only successfully materialised in 1954 with the advancement of material science and manufacturing of solid state devices. It is now widely accepted as a benign technology to generate electricity from sunlight with minimal pollution (Martin, 1997), and as one of the possible alternatives to fossil fuel energy sources. Kurokawa (2001) showed that PV had the merits of simplicity as modular energy source which can be easily deployed and installed in the urban environment.

There are many ways to apply PV for harvesting the solar energy. The application with the longest history was to power the energy requirements of satellites at such a location where no other means of reliable and maintenance-free power source could be possible. With the advancement in manufacturing of semi-conducting devices which mainly consist of silicon, the development of PV cells from the by-products of semi-conductor industry could lower the cost and made civil use of PV feasible (Hegedus and Luque, 2003).

Throughout the history of development of the PV technology, its dominating applications have long been the stand-alone applications. In a stand-alone PV system, PV is the only source of energy powering all the electrical loads by itself. To cater for the demand when there is no sunlight, it requires some kind of energy storage: by means of battery, water tanks, fuel cells, etc. These types of systems are widely used for the rural and remote applications that required modest power or where the supply from utility was not easily available. An EV running on the road, of course, can be a good candidate of getting at least part of the electricity supply if PV modules are mounted on its body. Only towards the end of last century (International Energy Agency PVPS T1-23, 2013), these stand-alone PV systems fell below 50% of total worldwide PV installations. The percentage of stand-alone systems went down rapidly to below 20% by the end of 2004 (ditto). Nowadays, the dominating type of PV application is the grid-connected PV system producing clean power for contributing to part of the electricity grid.

The operational principle of grid-connecting PV system is to convert the direct current (DC) electricity from the PV modules to alternating current (AC) with sophisticated power electronic devices and then supply the power to the electrical loads in conjunction with the utility grid. It serves as a supplementary power source to main-stream generation with fossil fuel, nuclear or other conventional means from the utility. Effectively, the grid-connecting PV systems use the grid power as the storage and also the complementary source. It would supply the loads together with the grid when there is not enough power from the sun, and would generate excess power into the grid when the PV system could provide more than the loads is needed. This is the most fast-increasing type of PV application and became the dominant type by the end of 1999. Luther *et al.* (2003) argued that this type of PV system will be contributing to the main-stream power production for the reduction of $CO_2$ emission in industrialised countries. Hence, the course of change in PV application indicates a development of its role in supplying power to remote areas where no other means of electricity could be easily available and then to supporting the modern society in reducing the adverse effect of burning fossil fuels. Table 5.1 summarises the history of development of PV (Hegedus and Luque, 2003). It demonstrates that alongside the beginning of the new century, PV industry is entering the era of grid-connected applications.

Although grid-connecting PV systems are now the dominating type, they can only help in powering the EVs when the EV charging station has such systems connected to it. The more direct way to utilise solar energy is to install PV cells onto the body of the EV for providing all or part of the energy needed. This will be explained more in the next section.

### *5.1.2 Harvesting solar energy for EVs*

When an EV is running on the road during daytime, it is constantly receiving radiant energy from the sun unless there is thick clouds covering the sky. Under

*Table 5.1 Development and forecast of PV technology*

| | |
|---|---|
| 1839 | Discovery of photogalvanic effect by Becquerel |
| 1954 | First silicon PV cell by Bell Laboratory |
| 1959 | First satellite utilising 20 W PV cells |
| 1973 | Worldwide oil crisis stimulated nations to adopt renewables |
| 1974 | First Government programme: Project Sunshine initiated in Japan |
| 1985 | Crystalline silicon PV panels sold at US$10/W |
| 1986 | First commercial amorphous silicon (a-Si) PV panel available |
| 1990 | German 1,000 roof programme |
| 2000 | Grid-connected PV constituted over 50% of the market |
| 2000 | World installed PV capacity over 1,000 MW |
| *2020* | *Generation cost of electricity by PV matches that from fossil fuel* |
| *2030* | *10% of global electricity from PV* |

normal situation, part of that radiant energy would be converted into heat that would raise the internal temperature of the vehicle (including the temperature of the electronic circuits and the battery bank), hence decreasing the overall efficiency of the vehicle. There were different trials of mounting PV cells onto the body of the vehicle to provide part of the electricity required by the EV. To promote the research and development (R&D) activities for such applications, a team of researchers is organising a semi-annual event called the World Solar Challenge (http://www.worldsolarchallenge.org/). The race requires the EVs to run 3,000 km all the way from Darwin to Adelaide, Australia; and the only source of energy is from the sun. The vehicles taking part in the competition demonstrate the extreme of the possibility of harvesting solar energy for EVs. Every vehicle would need high efficiency PV cells, sophisticated power management system (PMS) and effective battery bank to enable it to run over 3,000 km. Nevertheless, these vehicles are setting a high and yet reachable goals for our future research into this particular area. In the official website of World Solar Challenge, they describe the participating EVs being 'arguably the most efficient electric vehicles'. Figure 5.2 shows the champion car in the World Solar Challenge 2013.

In order to appreciate the challenges we are facing if we intend to power the entire EV by PV cells, we have to understand the operating principles, conversion efficiencies and limitations of the PV cells when they operate in the actual outdoor environment. The following section will then outline these background information.

## 5.2 PV cell technologies

As mentioned in section 5.1.1, the materialisation of PV technology was promoted by the growth in manufacturing of solid state devices. This was naturally dominated by the use of silicon since it is the most abundant naturally occurring semi-conductor on earth. Up till recent years, more than 90% of the market share was attributed to the silicon-based technology (Fraunhofer Institute for Solar Energy Systems, 2015).

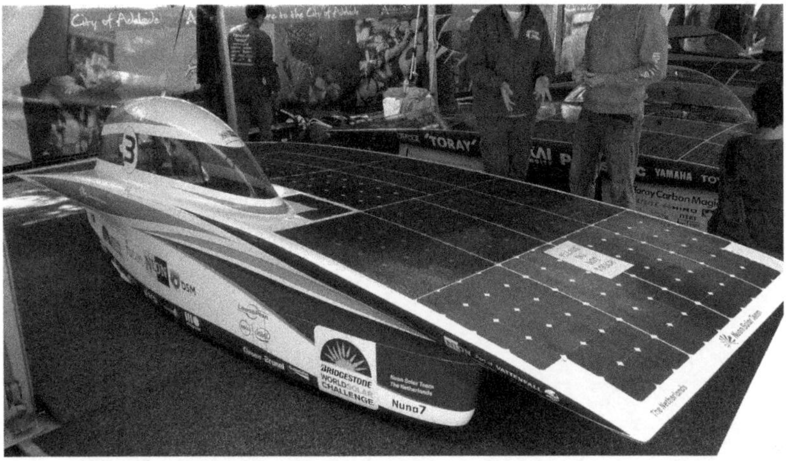

*Figure 5.2 The champion car in the World Solar Challenge 2013 (source of image: https://commons.wikimedia.org/wiki/File:Nuna_7.jpg)*

With the emergence of new material science technologies and manufacturing processes, PV cells that made of different materials are now becoming more readily available in the market. This section will describe the basic characteristics of different PV technologies and the influence of these characteristics on their performance in generating electricity.

### *5.2.1 Crystalline silicon*

Crystalline silicon is the most mature type of PV industry technology and has been developed since the 1950s. It has the relatively high conversion efficiency of about 13–22% at Standard Test Conditions (STC). STC is the convention adapted by the PV industry in benchmarking the performance of a PV cell under particular testing circumstances. It is defined to be the tests being conducted when the PV cell is operating at 25 °C, receiving a 1,000 W/m$^2$ solar irradiance, at which the spectrum of light rays shall be as it passes through a relative air mass of 1.5. Relative air mass is defined to be 1.0 when the sun is directly overhead. This way of characterisation of PV modules has its limitations (Lam *et al.*, 2012) and will be discussed in more details in section 5.3. Conversion efficiency, on the other hand, is defined as the percentage ratio of output power from a PV cell to that of incoming solar radiant power. The amount of solar radiant power that we could receive on a horizontal plane on the surface of the earth varies according to the weather condition, in particular the cloud coverage. The amount of peak solar power reaching the earth right outside the atmosphere, however, is relatively constant and is found to be 1,362 W/m$^2$ (denoted as the solar constant). Hence in many locations on earth, at around solar noon, the peak solar power we receive from the sun on earth's surface is around 1,000 W/m$^2$. This peak power of solar resources is then taken as one of the industry standard in defining the conversion efficiency. To illustrate the idea of conversion efficiency, we may consider this: an array of PV cells occupying 1 m$^2$, receiving 1,000 W/m$^2$ from the sun would imply that the PV cells are receiving 1,000 W. Hence if it is giving out 190 W at that moment, then the conversion efficiency of the cells is 19%.

Due to the relative abundance of silicon and development of technology in the manufacturing process (Goetzberger and Hoffmann, 2005), crystalline silicon is still leading the PV market and is expected to be so for the near future. Furthermore, silicon in its crystal structure resembles the diamond lattice structure as shown in Figure 5.3. This structure is very stable and less susceptible to degradation due to

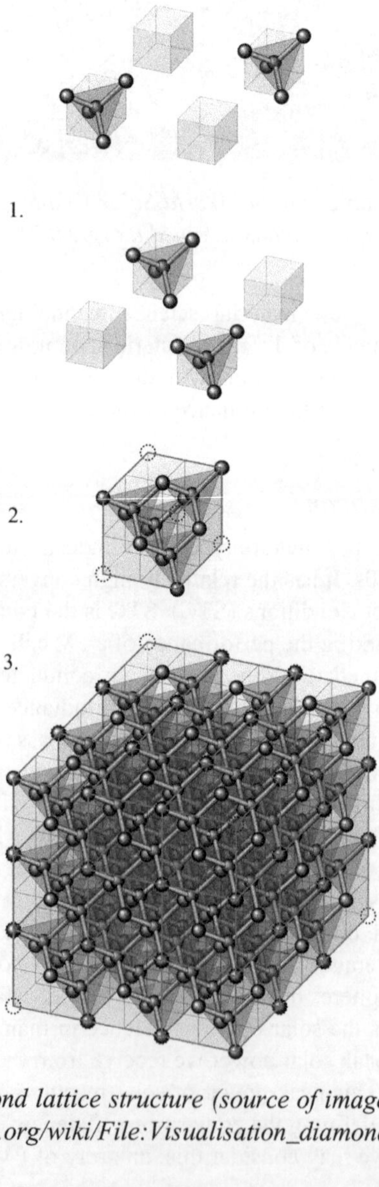


*Figure 5.3 The diamond lattice structure (source of image: https://commons.wikimedia.org/wiki/File:Visualisation_diamond_cubic.svg)*

light and heat. This specific property enables crystalline silicon PV to last for over 20 years and remain functional (Quintana *et al.*, 2000). Many of the PV module manufactures are offering a warranty of 25 years for their modules made of crystalline silicon cells. This is another reason for their dominating the market.

Crystalline silicon is further subdivided into mono-crystalline silicon (m-Si) and poly-crystalline silicon (p-Si). m-Si is the earliest commercialised technology for terrestrial applications. It requires high energy input to produce the raw materials (in the form of silicon ingot or ribbon) for m-Si cells by single crystal growth. A less energy demanding casting procedure of silicon crystal was therefore developed to lower the production cost. This is the manufacturing of p-Si by the melting and then solidification processes. When the PV cells consist of many regions of individual crystalline, this would inevitably increase the recombination of electron–hole pairs at the grain boundary and thus lowers the conversion efficiency (Masters, 2004). However, due to its lower manufacturing costs, p-Si is becoming increasingly popular. These two manufacturing technologies together dominate the PV market, and in year 2014 they accounted for over 90% market share (Fraunhofer Institute for Solar Energy Systems, 2015).

The thickness of crystalline silicon PV cells is in the range of 0.13–0.3 mm and needs proper protection for mechanical reason (Hegedus and Luque, 2003). To provide that mechanical protection, the common form of PV module for crystalline silicon is the cells embedded in glass layers or glass-PV-substrate arrangement. This is by far the most widely adapted arrangement of PV modules, which may not always be suitable for EV application. Another way is to encapsulate the PV cells under a layer of protection by epoxy or other similar synthetic materials. This layer of protection, however, may deteriorate in terms of transparency after years of exposing to sunlight if the type chosen is not appropriate. Hence, it is necessary to choose a proper type of encapsulation and maintenance would be needed to enhance its life span.

### *5.2.2 a-Si*

Amorphous is the term describing the characteristics of silicon atoms grouped together forming the PV cell. The word amorphous literally means formless. This structure contrasts with its crystalline counterpart, which is much more stable and of higher efficiency. One of the main reasons for developing a-Si PV technology is the lower production cost. This lower production cost is brought about by the lower energy input for the deposition of silicon layer onto the substrate instead of high temperature crystal formation (Luther *et al.*, 2003); and the significantly less amount of silicon required for the cell production (Deng and Schiff, 2003). The thickness of silicon coating on the substrate is usually in the range of about 0.001 mm, i.e. less than one-hundredth of that of crystalline silicon PV cells. This significant reduction in raw material used enables a-Si modules to be sold at a lower price per unit area than that of crystalline silicon modules, especially at time of ever increasing price of silicon due to the competition between manufacturing of computer chips and PV cells.

In its amorphous state, the photo-electric property of silicon is subjected to light-induced degradation. This effect was reported by Staebler and Wronski – therefore known as the Staebler-Wronski effect (Goetzberger and Hoffmann, 2005). After an initial degradation of about 10–20% decrease in conversion efficiency, the electrical properties of the a-Si cells will be stabilised. It is now a common practice for the manufacturers to publish the stabilised efficiency. The stabilised STC efficiencies of commercial a-Si PV modules are around 6–13%. In order to improve the conversion efficiency of the a-Si modules, some manufacturers stack layers of silicon depositions to capture more photons. Subsequently, tandem junction a-Si, triple junction a-Si and multi-junction a-Si are commercially manufactured, and they all can achieve a higher efficiency than traditional a-Si. Although with lower efficiency, a-Si PV cells are widely applied in indoor applications for small electronic appliances (e.g. calculators and toys) attributing to its better response to weak light. It is the third most widespread type of PV technology after crystalline silicon.

Since the thin layer of a-Si can be deposited onto various types of substrate, this enables the form of PV modules made of a-Si having a much greater variety. Furthermore, silicon in its amorphous form does not have the limitations on thickness or mechanical strength as in crystalline form. These special properties facilitate its application in a flexible format. This is very suitable for producing PV modules for building applications. Together with its performance being less affected by the heat built up in the module, it can be a good candidate for applications in hot conditions. Figure 5.4 shows a flexible PV module made of a-Si coated on polymer substrate that acts as roofing material. This is an important

*Figure 5.4 Flexible a-Si PV modules (source of image: https://commons.wikimedia.org/wiki/File:Thin_Film_Flexible_Solar_PV_Installation_2.JPG)*

advantage if we put it to use in EVs. Nevertheless, since the EVs are requiring a large power with very limited space available for mounting of PV cells, we have to plan carefully for the type of PV cells to be used.

### *5.2.3 Other thin-film PV cells*

Besides using silicon as the materials, there are many other types of semi-conductors available for manufacturing PV cells. Thin-film is the general term for PV cells produced other than those thick (130–300 μm) layers of crystalline silicon. a-Si is the earliest commercialise thin-film technology (Table 5.1). Its development was mainly from the drive to lower production cost by using significantly less raw material and energy input during the manufacturing process. Other new comers comprise copper indium diselenide ($CuInSe_2$ or simply CIS) and cadmium telluride (CdTe) etc. This section shall discuss these thin-film technologies.

In 1975, a CIS cell of 12% efficiency was successfully demonstrated at the USA Bell Laboratory and work at the National Renewable Energy Laboratory (NREL) developed the technology under the US Department of Energy thin-film program during the 1980s, consolidating the technology's promise (Zweibel, 1990). Due to the collapse of research funding in the United States, it was almost another 10 years before Siemens Solar Industries (SSI) produced the first commercially CIS modules in late 1990s. A whole family of similar compounds has also been developed: $CuInSe_2$, $CuGaSe_2$, $CuInS_2$, $Cu(InGa)(SSe)_2$, etc. The coating of thin-film CIS onto substrate was found to be more flexible and easily manipulated (Goetzberger and Hoffmann, 2005). Therefore, CIS is the key material used within this family of compounds.

Similar to a-Si, CIS can be coated onto many types of substrate and thus the choices of CIS modules are wider. For example, flexible CIS modules and modules on polymer backsheets can be found on the market. However, due to the rapidly evolving technology, the manufacturing process and even the compound used for CIS modules are ever changing. Second generation CIS modules were reported by Palm *et al.* (2004) and more development happened within the last decade. With the flexibility in manufacturing the modules, CIS can be one of the candidates for applying onto EVs.

Cadmium telluride (CdTe) has a similar long history in the whole group of thin-film PV modules. Due to its tolerance to defects and grain boundaries, simple and easy to handle deposition process is possible for this material (Luther *et al.*, 2003). Technology-wise this type of material should have been booming. The major hurdle now is the market acceptance since both cadmium and telluride are toxic materials although CdTe itself is stable and harmless to the environment. Furthermore, different countries have different regulations regarding cadmium-containing materials (Deb, 2002). This uncertainty might increase the risk factor in investing production plant for CdTe. For instance, BP solar was reported to have closed their CdTe plant (Luther *et al.*, 2003).

Synergising with the technology advancement in device handling and manufacturing process in semi-conductor industry, there are emerging technologies for thin-film PV cells (Sopori, 2003). These include crystalline silicon thin-film (m-SiTF), micro-crystalline silicon (μm-Si), etc. However, since they are either newly or even not yet commercialised, their applications in EVs are not expected to gain much impetus in the near future. There is, however, one particular type of thin-film PV that may have better potential in EV application. It is the gallium arsenide (GaAs) PV cells, which has high conversion efficiency (over 28%). The problem with GaAs nowadays is the high cost and limited supply. More research is needed to bring the cost down before this type of technology to become popular in EV application. Nonetheless, given the very limited mounting space on the body of EVs, this type of high efficiency technology can be promising since we need to produce as much power as we could under the existing constraints.

Besides those PV cells made of inorganic semi-conductors, there are also organic materials that can produce electricity under the sunlight. Their common drawbacks are their degradation issues when they are exposed to sunlight. The author would therefore not expect any substantial application of these organic materials in EVs before these issues can be overcome.

## 5.3 Electrical characteristics and performance of PV cells

PV cells can be mounted onto the body of an EV to generate electricity and reduce heat gain at the same time. But since the primary function of the PV cells is the generation of electricity, hence the electrical characteristics are the prime concern. As mentioned in section 5.2.1, STC is the PV industry convention for benchmarking the conversion efficiency of a PV cell under a set of specific testing conditions. Although STC is widely accepted as the industry standard, there are certain researchers asserted that these measuring conditions do not correspond to the real operating conditions (Bucher *et al.*, 1998; Lam *et al.*, 2004). The following section shall then discuss various factors that would affect the electrical characteristics and performance of PV cells to help the readers in better planning for harvesting solar energy for EVs.

### *5.3.1 Does PV technology matter?*

Different types of PV technology were briefly introduced in section 5.2. Due to the longer history of development, crystalline silicon PV modules are still more easily available as proprietary products and thus widely being applied. But due to their fragility, they have to be well protected by layers of front glass and backing or substrate. This intrinsic nature of crystalline silicon determines the formation and may limit its applications on EVs. Thin-film technology, on the other hand, can be applied on more options of substrate. For thin-film PV, flexible modules are also available besides glass-glass modules and glass-film modules that are common in crystalline silicon technology. The size, form and material used are all factors to be considered during the design process. Hence, a wider range of choices of materials for embedding the PV cells is an advantage of thin-film modules over crystalline silicon. Nevertheless, the cost and ease in fixing the PV cells onto the body of the vehicle is another important factor to consider.

Furthermore, the electrical properties also vary with different technology other than the structural formation. The ability of capturing the light from the sun and convert it into electricity differs from technology to technology. Even with the same technology, the power it delivers is not a constant when the environmental conditions are changing. Durisch *et al.* (2001) proposed that different technologies of PV could be characterised in a general formula relating the cell efficiency $\eta$ of the PV cell as a function of global irradiance $G_n$ (impinging vertically onto the module), cell temperature $\vartheta$ and relative air mass (AM). When represented in a mathematical manner, it is:

$$\eta = f\{G_n, \vartheta, AM\} \tag{5.1}$$

Instantaneous power would be calculated by multiplying the cell efficiency to the solar resources available, which is the product of global irradiance per unit area and total cell area:

$$P_{PV} = \eta * G_n * A_{PV} \tag{5.2}$$

In subsequent research, Durisch *et al.* (2002) demonstrated the same functional dependence of $\eta$ as in (5.1) could be applied to different types of PV technologies, and generalised the model as:

$$\eta = p\left[q\frac{G_n}{G_{n_0}} + \left(\frac{G_n}{G_{n_0}}\right)^m\right]\left[1 + r\frac{\vartheta}{\vartheta_0} + S\frac{AM}{AM_0}\right] \tag{5.3}$$

The parameters $p$, $q$, $r$, $m$ and $s$ are to be determined for each type of technology individually. Whereas $G_{n_0}$, $\vartheta_0$ and $AM_0$ are the reference values of global irradiance, cell temperature and relative air mass at STC, respectively. To illustrate the importance of this dependence, a two-dimensional graph of PV cell efficiency against cell temperature is shown in Figure 5.5. From the graph, it is clear that the ability of the PV cell to convert the sunlight into electricity decreases linearly with the rise in cell temperature. This decrease is rather significant, and is denoted by the term 'Temperature Coefficient'. For example, for a first generation CIS module tested, it represents a relative decrease of 20% in conversion efficiency if we compare the PV cell operating at 70 °C with its STC efficiency. In the design of PV system, the ventilation of the PV modules would therefore be very critical. To conclude, the choice of technology would have an impact on the real performance of the PV system.

From (5.3), we can see that besides the cell temperature, other important factors affecting the power output from the PV cells are the global solar irradiance $G_n$ and the relative air mass. Amongst which, the solar irradiance is the most important factor. First, the output is directly related to the amount of solar resources available – the solar irradiance. Second, Durisch *et al.* (2002) demonstrated that the conversion efficiency of a PV cell is a function of the solar irradiance as well.

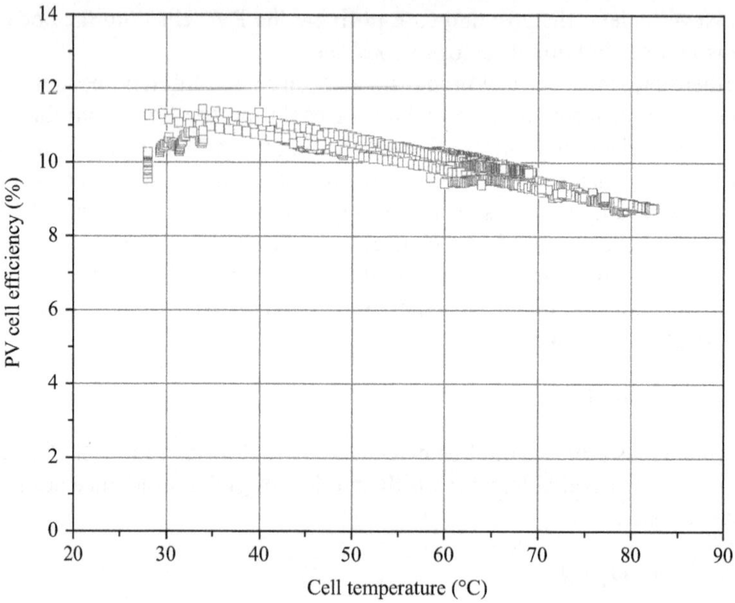


*Figure 5.5 A first generation CIS PV cell efficiency varies with PV cell temperature.*

By substituting (5.3) into (5.2), the power output from the PV cells with area $A_{PV}$ would then be:

$$P_{PV} = p\left[q\frac{G_n^2}{G_{n_0}} + \left(\frac{G_n^{m+1}}{G_{n_0}^m}\right)\right]\left[1 + r\frac{\vartheta_c}{\vartheta_{c_0}} + S\frac{AM}{AM_0}\right]A_{PV} \tag{5.4}$$

For more detailed analysis on the responses of different PV technologies to changes in the environmental conditions, the readers may refer to Durisch *et al.* (2007). The important point to note is that we cannot simply take the STC efficiency in estimating the power output from the PV cells mounted onto the EVs since the actual operating conditions would affect significantly the effective energy yield. Hence, a careful design process involves a better understanding on the specific responses. One of the ways is to apply the above empirical models developed by Durisch.

### 5.3.2 Energy yield calculations

Once the parameters *p*, *q*, *r*, *m* and *s* in (5.4) of a particular type of PV cell were made available, then a detailed energy yield calculation under different weather conditions can be performed. This calculation would allow the designer to have a better understanding on the possibility of the EV under investigation could be powered entirely by PV or not. The designer could then also estimate when and by how much extra energy needed to charge the battery in case the energy from the PV alone could not support the prescribed mileage. The general formula in estimating the total output from the PV cells during a specific period could be found by:

$$E = \sum_{k=0}^{n} P_{PV} * \Delta t_k \tag{5.5}$$

In the above equation, the parameter $\Delta t_k$ is the incremental duration of time during which the independent variables in (5.4) is constant enough for an accurate calculation of energy yield from PV during that period of time. Whereas $n$ represents the total number of incremental time period for the duration of evaluation. For example, if environmental data for one whole year is input to the equation, and the $\Delta t_k$ is 1 h, then, $n = 24 \times 365 = 8{,}760$. Although it is widely accepted that hourly environmental data are used for PV systems energy yield calculation, Gansler *et al.* (1995) asserted that hourly data are only suitable for systems responding slowly or linearly to changes in solar radiation. This is unquestionably not the case for PV systems which are responding almost instantaneously to changes in solar irradiance. Also, Durisch *et al.* (2002) successfully demonstrated that the change in PV conversion efficiency with respect to solar irradiance is non-linear. Gansler *et al.* (1995) showed that the error in estimation of PV energy yield by using hourly solar radiation data as compared with that of 1-min data could be as high as 35%. Hence, the author would recommend adopting $\Delta t_k$ to be 1 min whenever possible for a more accurate energy yield estimation. Furthermore, if the analysis requires estimation of instantaneous power output, the use of high resolution data at 1-min interval would be more appropriate. The application of (5.4) and (5.5) in designing PV systems for EVs shall be illustrated with two case studies mentioned in section 5.4.

### *5.3.3 Power management for EVs*

One particular characteristic about PV technology is the photo-electric effect of the semi-conductor. The photo-current is converted from the energy contained in the photons falling onto the surface of the semiconductor at specific physical conditions. Castañer and Silvestre (2002) summarised the electrical characteristics of a PV cell as follows:

$$I_m = I_{SC} - I_0\left(e^{(V_m + I_m R_S)/V_T} - 1\right) \tag{5.6}$$

$$V_m = V_T \ln\left\{1 + \frac{I_{SC} - I_m}{I_{SC}}\left(e^{(V_{OC}/V_T) - 1}\right)\right\} - I_m R_S \tag{5.7}$$

In the above equations, $I_m$ and $V_m$ are maximum current and maximum voltage values at maximum power point (MPP), respectively; whereas $V_{OC}$ and $I_{SC}$ are the open circuit voltage and short circuit current, respectively. These are the important parameters for analysis on the electrical power output of the PV cells. It should be noted that the electrical characteristics of a PV cell is non-linear. The typical current–voltage characteristics when plotted together is called the *IV* curve and is shown in Figure 5.6.

Hence the maximum power could only occur at $I_m$ and $V_m$ which are constantly shifting according to different environmental conditions. Continuous adjustment is

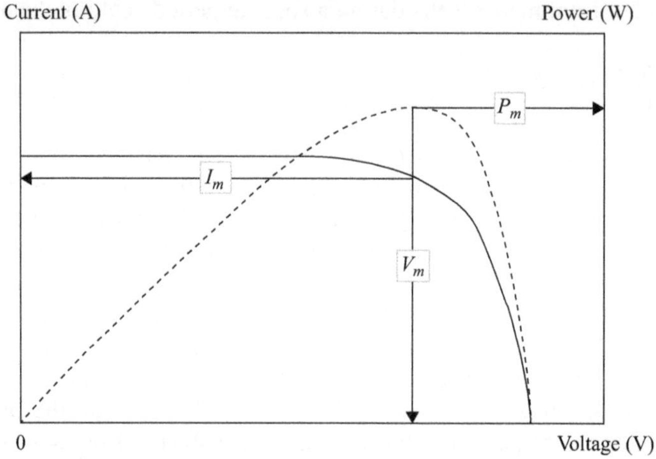


*Figure 5.6 A typical IV curve of a PV cell*

therefore needed to capture the maximum output from the PV Cell, and the process is called maximum power point tracking (MPPT). After regulating the output current and voltage of the PV cell, the power available would then be fixed according to the solar resources and other physical conditions. In the case of ordinary grid-connecting systems, the DC power will be converted to AC and then fed into the electricity grid. But for the case of EVs, since the system would be operating as DC, it will remain to be so for the rest of the conversion process. A few more steps are involved in the process of power management.

For application in EVs, the next step after MPPT would be voltage regulation. Since the MPP voltage $V_m$ may not be the optimal voltage for battery charging or driving the motor in case of the EVs, a second step of DC–DC voltage conversion is therefore needed. Besides the MPPT mode, there could exist other modes of operation during the time when the battery in the EV is not delivering power to the motor, dependent on the state of charge of the battery. The three basic modes of battery charging can be summarised as:

- Constant current mode: when the battery voltage falls below the cut-off voltage, the battery is charged using constant current which is the minimum of the inductor peak current and the PV peak current
- MPPT mode: between cut-off voltage and float voltage, the battery is charged in the MPPT mode. The MPPT controller will continuously track the peak power point of the PC cells. The MPPT calibration of the PV cell is done periodically as well as can be triggered by sudden changes in PV output voltage. The current through the battery at any time during the charging is given by power from the PV cells divided by the battery voltage.
- Trickle charge mode: when battery state of charge (SoC) indication is full with voltage (equals to the float voltage), the battery is trickle charged with minimal duty cycle applying to the synchronous buck converter.

These three different modes can be related to the battery voltage as shown in Figure 5.7.

The above charging mechanism is only detailing the relationship between PV and the battery. The modes of operation would be much more complex when the other sources of power together with the regenerative load – the motor is taken into consideration (it could be even more so with hybrid EVs). One of the possible ways to handle this complexity is to separate the charge bus and discharge bus, although the discharge bus would accept charging current from the motor during regenerative braking. To conclude, the mode of operation of the PV cells is determined by a number of factors, and their operation at MPP may not always be possible. A simplified schematic diagram showing one of the possible PMSs for EVs incorporating PV is shown in Figure 5.8.

From the above simplified diagram, it can be observed that the energy loss in different stages could vary a lot in different operation mode. In case the EV is

| **Constant current mode** | **MPPT charge mode** | **Trickle charge mode** |
|---|---|---|
| Battery charging is limited to a programmable maximum current | The PV operates at the maximum power point | Battery draws power only to compensate loss due to leakage |

Cut-off voltage Float voltage

**Battery voltage (V)** →

*Figure 5.7 Three charging modes of the MPPT controller for PV cells*

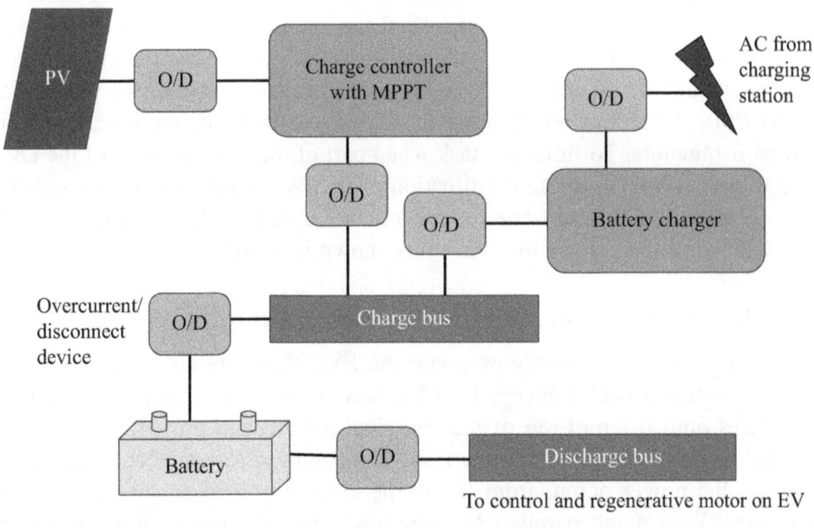


*Figure 5.8 The simplified schematic diagram of the possible PMSs for PV on EV*

running on the road under the sunlight, the electricity generated from the PV cells can be utilised more directly without going through the processes of charging and discharging of battery. During regenerative braking, however, the power from PV would be used to charge up the battery for future use. Furthermore, the efficiency of charging of battery depends on the mode of charging as shown in Figure 5.7. Hence, the overall efficiency of power conversion from PV to the final step of driving the EV may not be easily estimated. Despite these limitations, applying (5.5) is still useful in evaluating the maximum energy available from PV cells on the EV over a specific period of time. This valuable information can be used for further improving the overall efficiency of the EVs. The idea of incorporating the modelling of PV for aiding the PMS on-board will be discussed in the next section.

### *5.3.4 Incorporating solar energy into PMS*

The power output from the PV cells is not a constant but depends on the environmental factors, the PV characteristics and the operating conditions. One of the most important nature of a PV cell is its passiveness in output power and the intermittence. This intermittence can be attributed to the environmental factors influencing the PV cell, and also to the environment that it situated (e.g. when it is mounted directly onto the body of EV, its operating temperature will be affected seriously). This intermittent property of PV requires some sort of energy storage and power management. An accurate forecast on the PV power production would also be a useful tool in ensuring the system stability and reliability when the supply and load can be predicted with higher resolution. Furthermore, the PMS is playing an important role to regulate the power flow for enhancing energy efficiency and system stability. If the precise knowledge of the efficiency characteristics of different PV technologies is available, then the modelling of instantaneous power output from the PV cells can be realised as part of the PMS in providing necessary control strategy for the power flow. This can be achieved by incorporating a mathematical model of different PV technologies that can predict the instantaneous output given suitable input parameters. Also, the proper control of the power flow with the PMS dependent on accurate monitoring of both the environmental and electrical parameters. To fulfil this task when part of the power on-board the EV is provided by the PV, an accurate estimation of the PV power output is essential in providing information necessary for the control strategy. A possible schematic diagram showing the control mechanism is shown in Figure 5.9.

### *5.3.5 Harvesting solar energy for charging station*

Besides harvesting solar energy on-board the EVs, there are also other beneficial techniques to harvest solar energy for EVs. One of them is by means of installing PV modules onto the roof-top of the charging station, and providing part of the electricity necessary for charging the EVs. The common way to accomplish this is to connect the power output from PV to the electricity grid on the AC side and charge the EVs via the parallel AC supplies. The advantages of adopting this approach include the availability of proprietary grid-connect inverters for PV, ease

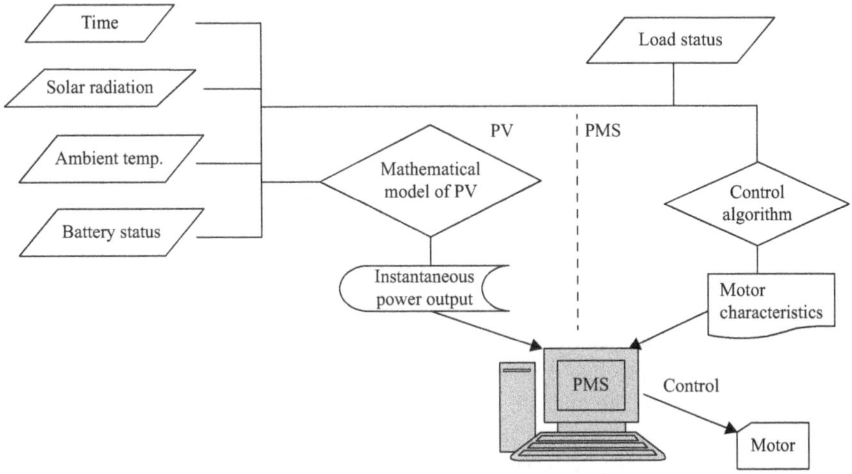


*Figure 5.9 Data-flow diagram of the PMSs incorporating PV on EV*

in operation and maintenance of the systems, etc. The power loss, however, through this common scheme would be a bit higher than coupling the power output from PV on the DC side. A proper DC charging network coupled with renewable energy sources with DC in nature could further enhance the overall system efficiency. This is because one less step of conversion would be needed in such cases, where the DC output (e.g. from PV) can be utilised directly in the charging of EVs, without the need to convert to AC, then back to DC for charging. In either case of DC or AC coupling, the utilisation of the solar energy will no longer be limited to charging of the battery in one EV. There could be the possibility of charging multiple EVs within the same station and hence the energy loss in the forced mode of trickle charge would be minimised. Additionally, during the period when there is no EV within the charging station, the surplus electrical power can be delivered to the grid via the grid-connecting mechanism. The generation profile of PV usually peaks around noon time and therefore coincides with the peak loads of the local electricity network if the charging station situated in cities. This characteristic enabling the charging station with PV installed poses a very good potential in supplying part of the peak electricity loads. If smart grid control is in place, then the flexibility in managing the electrical power flow will be higher since there could be moments when the charging station is taking energy from the battery in the EVs and supply back to the grid during some peak load situations. In such cases, the developed mathematical model of different PV technologies would be very useful in providing valuable information in forecasting the instantaneous output of the PV under different operating conditions. This accurate prediction of the performance of the PV system is crucial to the control strategy in a smart grid implementation.

One of the obvious applications in smart grid control that can be related to PV power generation would be the load-shifting control. It is the reduction of electricity consumption during peak periods, by means of re-scheduling of some of the controllable loads (Paracha and Doulai, 1998). The control process involves sophisticated load profile monitoring, prediction, and then finally decision-making by means of artificial intelligence. The contribution from PV systems would then be critical if certain part of the electrical load is going to be supplied by the sun. In performing the function of load-shifting, one of the major pre-requisites is the ability of the smart grid control to predict the shape of the load profile in the day, so that re-scheduling of loads would then be possible. When the smart grid has better knowledge on both the load profile (the demand) and the grid in co-generation with the PV system, its ability in deciding on the load shifting strategy would be more accurate. The resulting load profile will be taken as the total power requirement of the entire local power network obtained by subtracting the contribution of the PV system from the actual load before any smart grid control strategy. The battery in the EVs connected to the charging station then would be valuable in its capability and flexibility in providing buffer for getting surplus power from grid supplemented with PV; and also possibly delivering power into the grid when needed. For further details about the PV modelling in smart grid applications, the reader may refer to Lam *et al.* (2011).

## 5.4 Case studies

There has already been quite a number of trials to install PV cells onto EVs to provide at least part of the energy consumption on-board. The impressive ones include the World Solar Challenge (see Figure 5.2) in which the high performance EVs are powered solely by PV. Other applications for EVs are also possible given proper planning ahead. This section will illustrate the real applications of harvesting solar energy on vehicles, supplemented with an example of the design process in estimating the possible peak power and daily energy yield, respectively.

### *5.4.1 PV module as roof for electrical cart*

With its special usage of short daily range, one of the main types of EV application is the form of electrical golf-cart. Since the majority time of use being in the day when there is sunlight, putting PV onto it would then be a logical choice to extend further the possible range of the cart. The PV can also be serving as the roof to the cart and provide shading to the passengers underneath. A photo showing the set-up can be found in Figure 5.10.

Since the dimensions of some of the proprietary PV modules (around 2 m × 1 m) suit that of a golf-cart, they can easily be put as the roof without any modification. Furthermore, the battery bank of the cart can readily accept the DC output from the charge controller, this type of application requires the least modification in the cart construction. The photo shows a case in which the cart was installed with a 300 W (measured at STC) PV module and adapted a PMS as shown in Figure 5.8. Its simplicity has the advantages of low cost in deployment and maintenance. More sophisticated modes of operation, for example supporting the smart grid control,

*Figure 5.10 Putting a PV module as the roof of a golf-cart. (Photo courtesy: Mr. Jackson FUNG)*

may not be possible with such simple configurations. Nevertheless, the expected available power from PV can be calculated using (5.4) and a daily output profile can be produced in advance for a better understanding on the modes of operation during the day. All these useful information would help the designer and client to have a better understanding on the performance of the cart with PV on it.

To estimate the power available from the PV mounted onto EV, a computer program was developed by the author. It can generate a chronological series of data of PV system power output according to a set of formatted meteorological data input followed by the other details such as the environmental factors and design parameters. All the characteristics of each type of PV technology and the installation details are to be input separately. The parameters $p$, $q$, $r$, $m$ and $s$ of different PV technologies for (5.4) are resolved by performing non-linear regression on adequate data sets collected in advance. The details of the process of resolving these parameters were illustrated with figures and graphs in Lam *et al.* (2004). Besides aiding the design process, another key feature of the computer program developed is its ability in performing accurate prediction on PV power output from the real-time input data. This is a prerequisite for the PMS to achieve proper control mechanism.

To resolve the daily power output profile of specific PV, the author fed into the computer program with the weather data formatted as typical meteorological year for the target site together with a suitable range of system parameters on the environment and system design for a detail energy yield calculation. In the process, if the data interval is frequent enough, a good estimation of the instantaneous power output form the PV module year round can be produced.

In this case study, the PV module is to be analysed with formatted weather data for Kunming, China. The meteorological data set containing the records of the solar and wind resources is made available by the project Solar and Wind Energy Resource Assessment (SWERA) under the UN Environment Programme (Schillings *et al.*, 2002). The SWERA project is providing data in hourly interval, and hence the computer program is estimating the PV output at 1-h interval. Although this is not the optimal resolution for PV system simulation as asserted by Craggs *et al.* (2000), this is the best resolution of data available at the time of research.

To demonstrate the capability of the mathematical model in assisting design of PV system, a set of meteorological data from the SWERA project is taken on for finding the most suitable technology to be applied in Kunming. The programme initially simulates a total area of 100 m$^2$ (cell area) for PV flat roof application. This is an arbitrary roof area for a building application for future consideration. The expected energy available from the PV module on the cart will then be calculated by proportionally multiplying the PV cell area. The results displayed in the graphical interface are shown in Figure 5.11.

After simulation, the computer program calculates various parameters for four types of PV technologies and the results are summarised in Table 5.2.

The PV technology chosen by the client for the golf-cart is p-Si, with the STC power output being 300 W. Hence the yearly energy yield of this p-Si PV in Kunming would be 371 kWh, and the average daily energy yield from PV would be 1.02 kWh. Dependent on the range need to be covered by the cart, the energy from

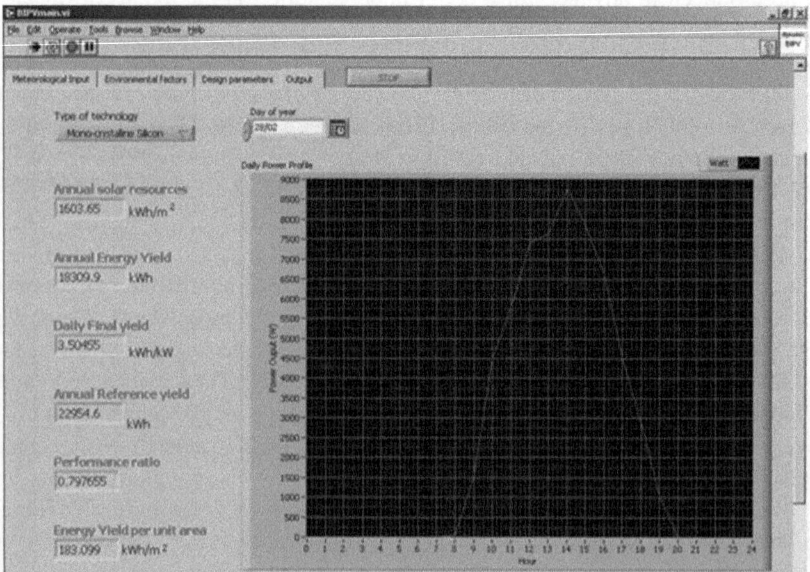


*Figure 5.11 Graphical interface (computer program developed by the author) displaying the simulation results for Kunming, China*

*Table 5.2 Table summarising the simulation results for Kunming, China*

| | Annual energy yield (kWh) | Daily final yield (kWh/kW) | Performance ratio |
|---|---|---|---|
| m-Si | 18,310 | 3.50 | 0.80 |
| p-Si | 16,243 | 3.39 | 0.75 |
| CIS | 12,524 | 3.26 | 0.74 |
| a-Si | 9,829 | 3.19 | 0.73 |

the PV may not be sufficient for providing all the energy requirement for running the cart for a day. But certainly with solar energy harvested, we can extend the interval of time between charges required. The main advantages of this case study are its low cost in implementation and operation, with satisfactory performance.

The above case study demonstrates that the mathematical model can serve as a useful tool in supporting the PV system design. It estimates various performance indicators and also displays the instantaneous power output from the PV system in graphical formats. These indicators can help the designer to select the appropriate PV technology according to different design intent. All these indicators besides simply the STC efficiency are valuable information a designer would need to know before going into the detailed design of a PV system to be adapted at a particular location. Furthermore, to fully utilise the mathematical model during the design stage, a more detailed analysis on the power output from the PV system in relation to the total loads on-board the EV can be performed to evaluate the capability of the PV systems in supporting the load at different time of the day. In case the high resolution weather data are available, a detailed study on the actual possible power output of the PV cells mounted onto the EV can be performed by checking the daily output profile. The peak power output, the time at which this happen and the expected rate of changes in power can all be checked accordingly. These information can be useful in simulation of actual performance of the EVs incorporating PV.

## *5.4.2 PV modules mounted on roof of ICEV*

The above case study demonstrated the possibility of powering part of the energy needed in an EV by solar. There is another real application of PV on vehicle to illustrate further other possibilities as well. This case, however, is solar energy deployed on an internal combustion engine vehicle (ICEV). Although this case is not on EVs, the practical results can still be made reference in case similar PV systems are to be applied on EVs.

The ICEV is a single-deck coach specially converted into a mobile laboratory for the students to conduct biological experiments inside. The program to run this coach is called the Sik Sik Yuen (SSY) Biotechnology Mobile Laboratory, where SSY is a non-profit organisation in Hong Kong. The readers may refer to the website: http://mobilelab.hoyu.edu.hk/ for more details. This coach, although runs on internal combustion engine, has numerous electrical equipment on-board for conducting experiments. There are also computers and monitors on the coach for the teachers to conduct lectures for the students. Hence the electrical loads on-board are quite demanding. A generator set is installed on the coach to cope with the electrical energy requirements. The main problem here is, there is noise generated once the generator set is in operation. Hence solar energy would be the preferred source as long as there are students on the coach. The system designed is to provide all electrical loads on-board by solar while some energy has been stored in the battery bank inside the coach. The generator set will only come to operation when the battery level drops below certain level. This would ensure a quiet environment for the students to learn on the coach. Figure 5.12 shows the roof and outlook of the mobile laboratory.

Due to the very limited space on the roof of the coach, high efficiency m-Si PV was chosen. There are totally ten PV modules on the roof of the coach, each with peak power output 100 W at STC, giving the peak power of the PV system to be

*Figure 5.12 The roof of the mobile laboratory with PV (above) and the outlook of the mobile laboratory (below). (Photo courtesy: Dr. William MAK)*

1,000 W. There are altogether eight pieces of 12 V 100 Ah battery for electrical energy storage. PV charge controller with MPPT function is installed. Sophisticated hybrid bi-directional inverter that can take AC input from the generator set to charge battery and supply AC power to the equipment, or take DC from the battery to supply the AC loads is installed as the main power management equipment. Since this coach is running in the Hong Kong territory, the weather data of Hong Kong were applied to the computer program for simulation study. The 1 kW system was estimated to be able to generate 1,069.5 kWh/year or on average 2.92 kWh/day in Hong Kong. Given sufficient electrical energy was stored in the battery bank, and the coach is out for service on a sunny day, usually the students can conduct the experiments in a quiet environment on that day. This application is rather different from EVs but we can still can appreciate the idea how much solar energy could provide to the vehicle under different scenarios.

## 5.5 Conclusions

EV is a rapidly developing technology that may hold one of the solutions to the very challenging situation we are facing: climate change and pollution problems. Harvesting of solar energy, on the other hand, shares many advantages that EV has. It would be beneficial if we could examine the possibility of harvesting solar energy for EVs. This chapter investigated a few important elements in doing so. First, different PV technologies were presented, and their respective characteristics and limitations were discussed, in particular the important point we need to take note of when we intend to put them onto EVs. Second, a mathematical model that could predict the actual performance of PV under real operating conditions was introduced and illustrated with two case studies. It is the author's personal vision to see more real application of PV on EVs in the near future.

## Acknowledgements

The author would like to express thank to Dr William MAK of the SSY Biotechnology Mobile Laboratory and Mr Jackson FUNG of Renewable Energy Group Limited for their kind permission to use the photos showing solar energy systems and PV mounted on their vehicles.

## References

Bucher, K., Kleiss, G., and Batzner, D. (1998) Photovoltaic modules in buildings: performance and safety. *Renewable Energy*, 15, pp. 545–551.

Castañer, L., and Silvestre, S. (2002) *Modelling of Photovoltaic Systems using PSpice.* Chichester: John Wiley & Sons, pp. 44–70.

Craggs, C., Conway, E.M., and Pearsall, N.M. (2000) Statistical investigation of the optimal averaging time for solar irradiance on horizontal and vertical surfaces in the UK. *Solar Energy*, 68, pp. 179–187.

Deb, S.K. (2002) Current progress and future opportunities for thin film solar cells. *Proceedings of World Renewable Energy Congress VII*, Cologne, Germany, pp. 1–6.

Deng, X., and Schiff, E.A. (2003) Amorphous silicon-based solar cells. In *Handbook of Photovoltaic Science and Engineering*, eds. A. Luque and S.S. Hegedus, Chichester: John Wiley & Sons.

Durisch, W., Bitnar, B., Mayor, J.-C., Kiess, H., Lam, K.H., and Close, J. (2007) Efficiency model for photovoltaic modules and demonstration of its application to energy yield estimation. *Solar Energy Materials & Solar Cells*, 91, pp. 79–84.

Durisch, W., Lam, K.H., and Close, J. (2002) Behaviour of a copper indium gallium diselenide module under real operating conditions. *Proceedings of World Renewable Energy Congress VII*, Cologne, Germany, pp. 1–6.

Durisch, W., Mayor, J.-C., and Gray, A. (2001) Application of a generalized model for solar cells to outdoor measurements on a commercial module. *Proceedings of Sharjah Solar Energy Conference*, Sharjah, UAE, pp. 1–5.

Fraunhofer Institute for Solar Energy Systems ISE (2015) Photovoltaics report. Available from: https://www.ise.fraunhofer.de/en/downloads-englisch/pdf-files-englisch/photovoltaics-report-slides.pdf [Accessed 25 Aug 2015]

Gansler, R.A., Klein, S.A., and Beckman, W.A. (1995) Investigation of minute solar radiation data. *Solar Energy*, 55, pp. 21–27.

Goetzberger, A., and Hoffmann, V.U. (2005) *Photovoltaic Solar Energy*. Berlin: Springer, p. 65.

Hegedus, S.S. and Luque, A. (2003) Status, trends, challenges and the bright future of solar electricity from photovoltaics. In *Handbook of Photovoltaic Science and Engineering*, eds. A. Luque and S.S. Hegedus, Chichester: John Wiley & Sons.

International Energy Agency (2013) *Trends in Photovoltaic Applications: Survey Report of Selected IEA Countries Between 1992 and 2012, IEA-PVPS T1–23:2013*.

Kurokawa, K. (2001) PV system in urban environment. *Solar Energy Materials & Solar Cells*, 67, pp. 469–479.

Lam, K.H., Close, J., and Durisch, W. (2004) Modelling and degradation study on a copper indium diselenide module. *Solar Energy*, 77, pp. 121–127.

Lam, K.H., Lai, T.M., Lo, E.W.C., and To, W.M. (2012) The application of dynamic modelling techniques to the grid-connected PV (photovoltaic) systems. *Energy*, 46, pp. 264–274.

Lam, K.H., Lo, E.W.C., and Lai, T.M. (2011) Intermittent nature of PV system – the applications of dynamic modelling in smart grid. *Proceedings of International Conference on Electrical Engineering*, Hong Kong, pp. 1–6.

Luther, J., Preiser, K., and Willeke, G. (2003) Solar modules and photovoltaic systems. In *Photovoltaics Guidebook for Decision Makers: Technology Status and Potential Role in Energy Economy*, eds. A. Bubenzer, and J. Luther, Heidelberg: Springer, pp. 41–103.

Martin, J.A. (1997) A total fuel cycle approach to reducing greenhouse gas emission: solar generation technologies as greenhouse gas offsets in U.S. utilities systems. *Solar Energy*, 59, pp. 195–203.

Masters, G.M. (2004) *Renewable Energy and Efficient Electric Power Systems.* Hoboken: John Wiley & Sons, p. 491.

Messenger, R., and Venture, J. (2000) *Photovoltaic Systems Engineering.* Florida: CRC Press, p. 15.

Palm, J., Probst, V., and Karg, F.H. (2004) Second generation CIS solar modules. *Solar Energy*, 77, pp. 757–765.

Paracha, Z.J. and Doulai, P. (1998) Load management: techniques and methods in electric power system. *Proceedings of International Conference on Energy Management and Power Delivery,* Singapore, pp. 213–217.

Quintana, M.A., King, D.L., Hosking, F.M., *et al.* (2000) Diagnostic analysis of silicon photovoltaic modules after 20-year field exposure. *Proceedings of IEEE 28th Photovoltaic Specialists Conference*, Anchorage, Alaska, USA, pp. 1420–1423.

Schillings, C., Pereira, E.B., Perez, R., Meyer, R., Trieb, F., and Renn, D. (2002) High resolution solar energy resource assessment within the UNEP – Project SWERA. *Proceedings of World Renewable Energy Congress VII*, Cologne, Germany, pp. 1–6.

Sopori, B. (2003) Thin-film silicon solar cells. In *Handbook of Photovoltaic Science and Engineering*, eds. A. Luque and S.S. Hegedus, Chichester: John Wiley & Sons, pp. 307–357.

Zweibel, K. (1990) *Harnessing Solar Power – The Photovoltaic Challenge.* New York: Plenum Press, pp. 161–180.

# Chapter 6
# On-board electromagnetic energy regeneration for electric vehicles

*T.W. Ching*[1] *and Wenlong Li*[2]

Compared with conventional vehicles, electric vehicles (EVs) have a unique feature, that they can recover the kinetic energy during braking for battery recharging. Meanwhile, the vibrational energy of all kind of vehicles during normal operation is also recoverable, and can readily be used for battery charging in EVs.

In this chapter, on-board electromagnetic energy regeneration for EVs is presented, with emphasis on energy recovery from braking and energy recuperation from suspension system. Both of their system configurations and control strategies are discussed in detail.

## 6.1 Introduction

EVs, including the pure EV (PEV), hybrid EV (HEV) and fuel-cell EV (FEV) (Chau, 2014), have been identified to be the most viable green transportation form the foreseeable future. HEVs have also been labelled as Super Ultra Low Emission Vehicles (Chan, 2007). They have their individual merits and demerits, leading them to have their unique roles in modern society.

### *6.1.1 Vehicle energy*

According to the US Department of Energy (DOE) website (http://www.fueleconomy.gov/feg/atv.shtml) regarding fuel economy in a gasoline-dominated market, only 14–30% of the chemical energy contained in the fuel is put to work moving an internal combustion engine vehicle (ICEV) and powering the accessories, depending on the drive cycle. An electric powertrain can be four times more efficient than its mechanical counterpart. EVs can be three times more efficient than hydrogen-fuelled vehicles, and, more importantly, this can already be achieved using existing technology and power distribution infrastructures.

In a conventional ICEV, only 18–23% of the energy stored in the fuel is converted into motion, whereas EVs utilize up to 75% of the electricity taken from

[1]Department of Electromechanical Engineering, University of Macau, Macao, China
[2]Department of Electrical and Electronic Engineering, The University of Hong Kong, Hong Kong, China

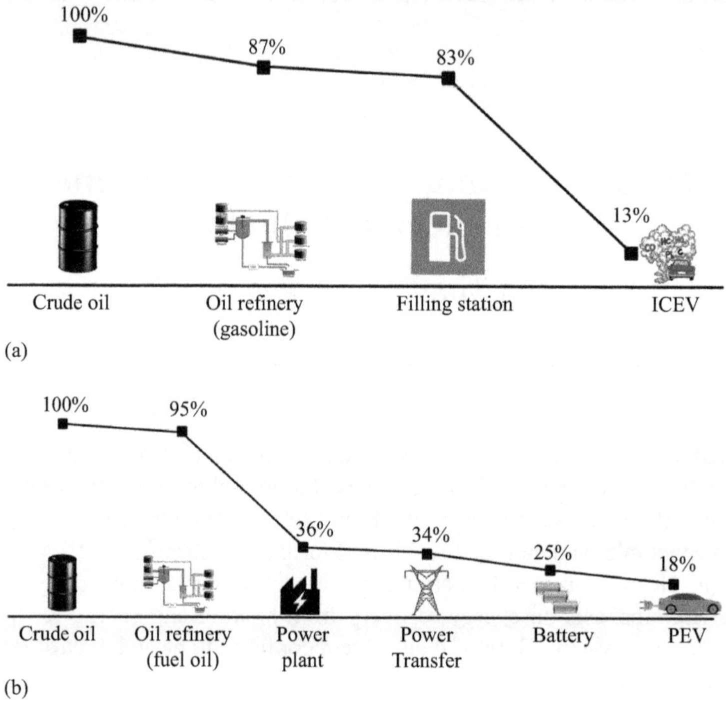


*Figure 6.1 Comparison of vehicle energy efficiencies: (a) ICEV and (b) PEV*

the grid (Kendall, 2008). In order to compare the overall energy efficiency of EVs with that of ICEVs, their energy conversion processes from crude oil to road load are illustrated in Figure 6.1, indicating that EVs are more energy-efficient than ICEVs (Chau, 2015). However, EVs can recover kinetic energy during braking and utilize it for battery recharging, whereas ICEVs wastefully dissipate this kinetic energy as heat in the brake discs. With this regenerative braking technology, the energy efficiency of EVs can be further increased by up to 10% (Chau, 2015). Another major source of loss is vibration energy dissipated by shock absorbers in a vehicle suspension system under excitations from road irregularity, vehicle acceleration and deceleration (Zuo *et al.*, 2010). With a suitable suspension regeneration system, about 5–10% of vehicle power consumption might be restored (David and Bobrovsky, 2011).

Figure 6.2 shows a typical energy flow of an ICEV in which about 40% is lost in the form of waste heat and exhaust gas, whereas only about 25% of the fuel combustion can be utilized for vehicle operation (Stabler, 2002). The braking energy dissipates as heat by using a conventional mechanical braking system. Vibration suppressed by the suspension system also dissipates energy as heat. Therefore, if the energy losses due to mechanical braking and suspension systems can be recuperated, this would greatly extend the driving range, and hence, improve the energy efficiency of vehicles.

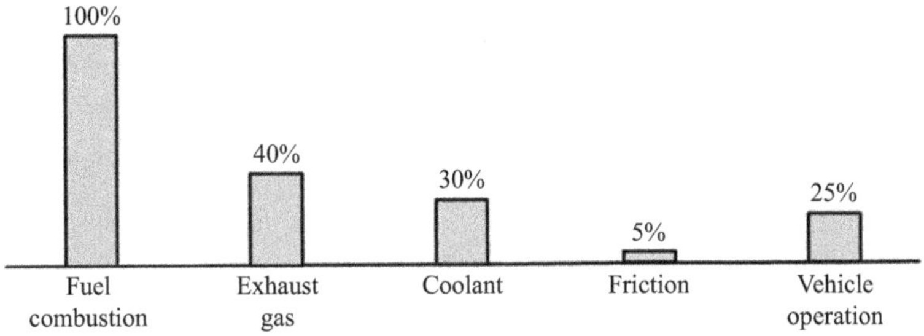


*Figure 6.2 Energy consumption of fuel combustion*

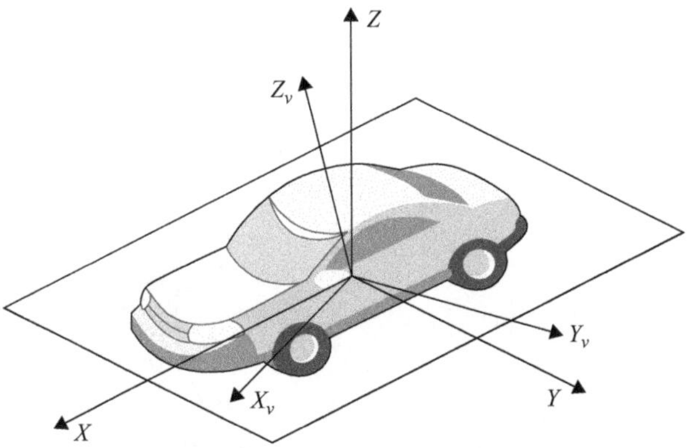


*Figure 6.3 DOF of a vehicle*

### 6.1.2 Vehicle dynamics

In order to understand various energy losses associated with mechanical braking and suspension systems, a vehicle's dynamic will be reviewed in this section. The driving dynamics of a vehicle are presented in Figure 6.3. They can be divided into three categories according to three degrees of freedom (DOF) of the movements (Spichartz and Sourkounis, 2013) as follows:

- Longitudinal dynamics describes driving, acceleration and braking processes in the longitudinal direction of a vehicle and allows for calculations of power and energy demands.
- Lateral dynamics is related to stability, especially when a vehicle is making a turn. It is very important for the implementation of stability controls or electric differentials (Han *et al.*, 2011).
- Vertical dynamics characterizes the vibration behaviour of a vehicle in the vertical direction and has much influence on driving comfort and safety.

Figure 6.4 illustrates the forces acting on a vehicle (Tur *et al.*, 2007). There are four kinds of forces, namely, rolling resistance force, $F_w$; aerodynamic drag force, $F_a$; slope frictional force, $F_s$; and braking force, $F_{br}$.

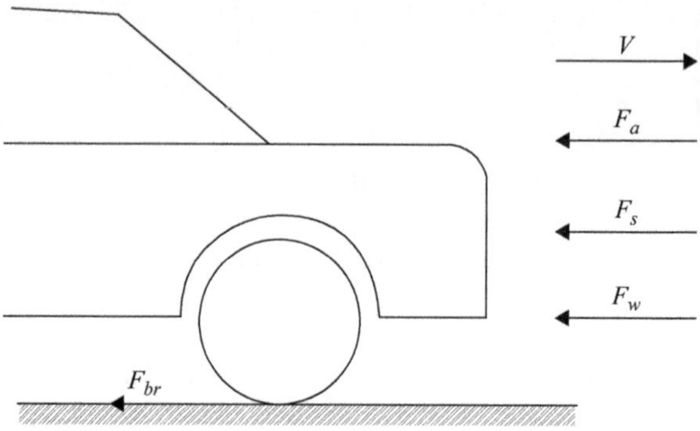


*Figure 6.4 Forces acting on a vehicle*

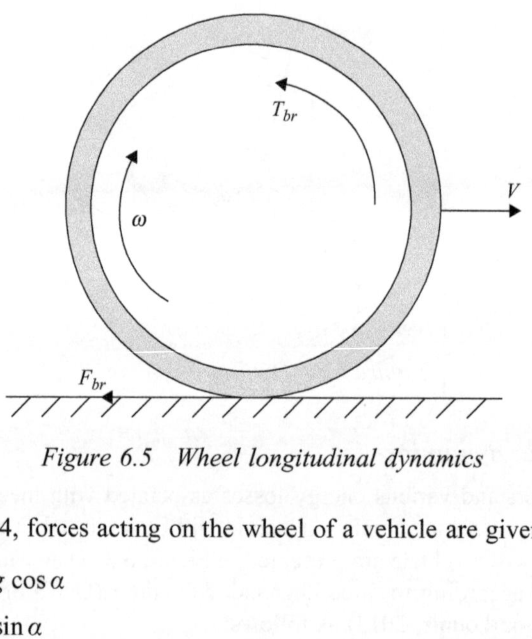


*Figure 6.5 Wheel longitudinal dynamics*

In Figure 6.4, forces acting on the wheel of a vehicle are given by:

$$F_w = c_t mg \cos \alpha \tag{6.1}$$

$$F_s = mg \sin \alpha \tag{6.2}$$

$$F_a = 0.5 c_r \delta A_f V^2 \tag{6.3}$$

where $c_t$ is the wheel rolling resistance coefficient, $m$ is the mass of the quarter vehicle, $\alpha$ is the slope angle, $c_r$ is the aerodynamic coefficient, $\delta$ is the air density, $A_f$ is the vehicle frontal area and $V$ is the velocity of the vehicle.

As shown in Figure 6.5, the longitudinal dynamics of a quarter vehicle during braking is:

$$-F_{br} - F_w - F_s - F_a = m\frac{dV}{dt} \tag{6.4}$$

$$F_{br}R - T_{br} = J\frac{d\omega}{dt} \tag{6.5}$$

where $R$, $T_{br}$, $J$ and $\omega$ are wheel radius, braking torque, wheel inertia and wheel angular speed, respectively. The purpose is to optimize the braking effectiveness and maintain vehicle stability under various road conditions. It can be achieved through controlling the maximum braking force to be applied on the wheels.

## 6.2 Electromagnetic energy regeneration from braking

Regenerative braking can reduce vehicle emissions significantly, and hence, improve energy efficiency (Clarke *et al.*, 2010). Such a system permits a vehicle's kinetic energy, which is conventionally dissipated by the brake pads during braking, to be recaptured and stored for later use in acceleration; it can be incorporated into both HEVs and EVs. In this section, the system configuration, mathematical models and control strategies will be presented.

Substantial efficiency gains can be realized through regenerative braking technology, whereby motive energy is recovered to recharge the battery as the vehicle brakes, instead of simply being lost to the brake discs as waste heat (Kendall, 2008). Regenerative braking is very essential for EVs and can extend the vehicle driving range by up to 12% (Rajashekara, 1994; Ching, 2005). Battery-powered EVs desire frequent regenerative braking. During braking, the motor operates as a generator to convert the kinetic energy into electrical energy while the converter must allow for reverse power flow to restore the electrical energy to the battery system (Chau *et al.*, 1999; Ching and Chau, 2001). This energy-recovery feature is particularly attractive to battery-powered EVs (Ching, 2006). Toyota reported that the greatest factor for the improved fuel efficiency of HEVs is regenerative braking, which accounts for about 35% of the total energy efficiency improvement, as evident in the Toyota Prius. Studies showed that HEVs have remarkably improved their fuel efficiencies by 30–40% through regenerative braking (Ko *et al.*, 2015). In the power-assist HEV, the recuperation of vehicle braking energy via regeneration through the electric drive subsystem to the battery partially offsets the operating cost of electricity generation (Miller, 2004). A recently launched PEV, the LEAF, is equipped with an 'eco mode' that functions to conserve energy. When selected, acceleration power is saved, the regenerative braking system becomes more aggressive and power is conserved in the auto climate control system. By activating the eco mode, the driving range can be extended by approximately 10% in city driving (Tokuoka, 2010).

### *6.2.1 Electric machines and power electronic drives*

Motor drives are the core technology for EVs and convert the on-board electrical energy to the desired mechanical motion. Meanwhile, electric machines are the key element of motor drive technology. Figure 6.6 shows the classification of electric machines for EVs (Chan and Chau, 2001) in which the bold types are those that have been applied to EVs, including the series DC, shunt DC, separately excited DC, permanent magnet (PM) DC, cage-rotor induction, PM brushless AC (BLAC), PM brushless DC (BLDC) and switched reluctance machines (SRMs) (Chau, 2015).

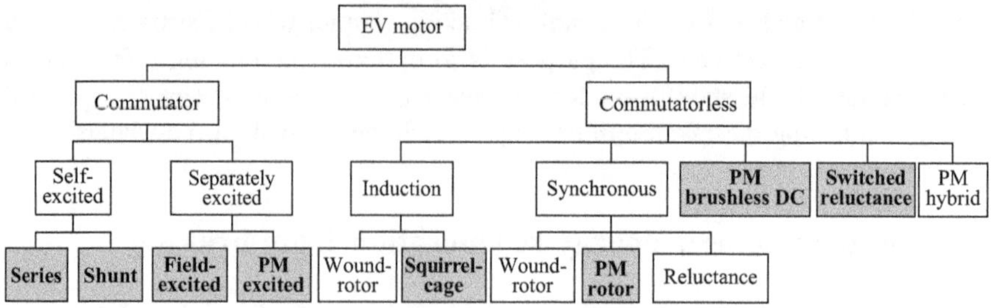


*Figure 6.6 Classification of EV motors*

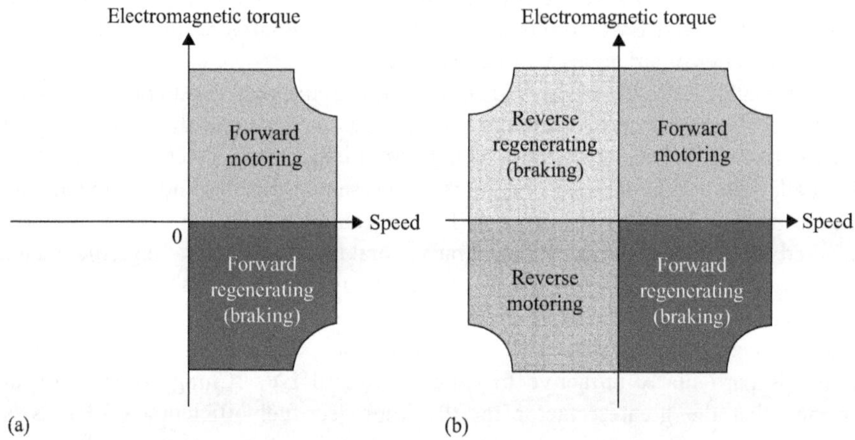


*Figure 6.7 DC motor operations: (a) 2Q and (b) 4Q*

Power converters play a vital role in the success of braking energy recovery. Electric propulsion systems require that the power electronic converter performs multi-quadrant operations and allows reverse power flow to restore electrical energy to the battery system (Ching, 2009a), as shown in Figure 6.7.

In a DC motor drive, a two-quadrant (2Q) chopper is preferred as it converts battery DC voltage to variable DC voltage during the motoring mode and reverts the power flow during regenerative braking (Chau *et al.*, 1999). A four-quadrant (4Q) DC chopper is employed for reversible and regenerative speed control of DC motors (Ching, 2005). The 2Q and 4Q DC choppers, in which regenerative currents could return to the battery system via the anti-parallel diodes of power switches, are shown in Figure 6.8.

Current-fed inverters are seldom used for EV propulsion. In fact, voltage-fed inverters are almost exclusively used because they are simple and can have power flow in either directions (Ching, 2007). A typical three-phase full-bridge voltage-fed inverter for induction and brushless motor drives is shown in Figure 6.9. The output waveform of an inverter may be rectangular or with pulse width modulation (PWM),

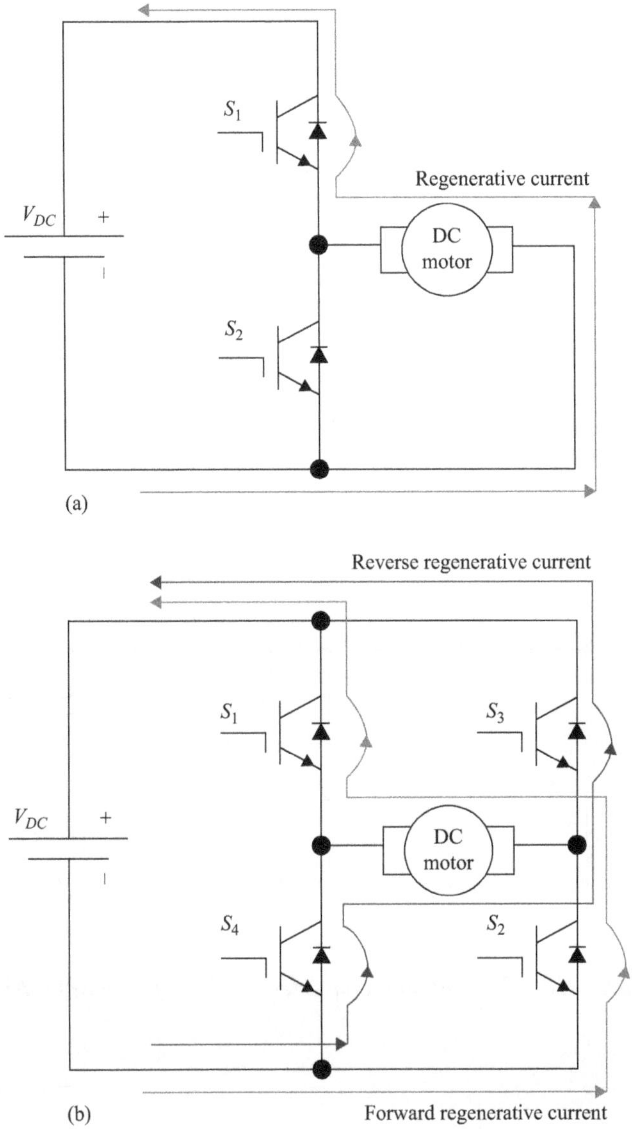


*Figure 6.8 DC choppers: (a) 2Q and (b) 4Q*

depending on the switching strategy for different applications. For example, a PWM output waveform is for an induction motor and PM BLAC motor while a rectangular output waveform is produced for a PM BLDC motor. The regenerative currents could return to the battery system via the anti-parallel diodes of power switches.

A conventional converter for three-phase SRM drives is shown in Figure 6.10, and the corresponding conduction modes are shown in Figure 6.11. It can be found

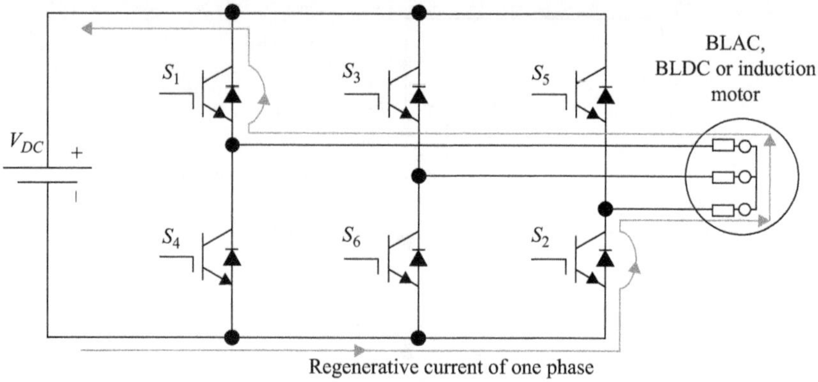


*Figure 6.9 Three-phase full-bridge voltage-fed inverter*

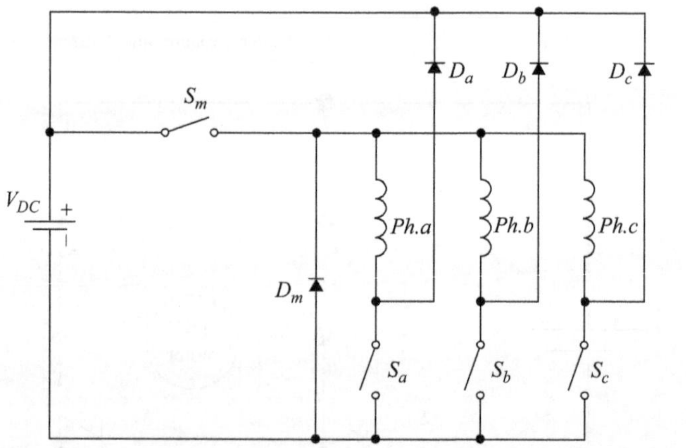


*Figure 6.10 Conventional converter for three-phase SRM drives*

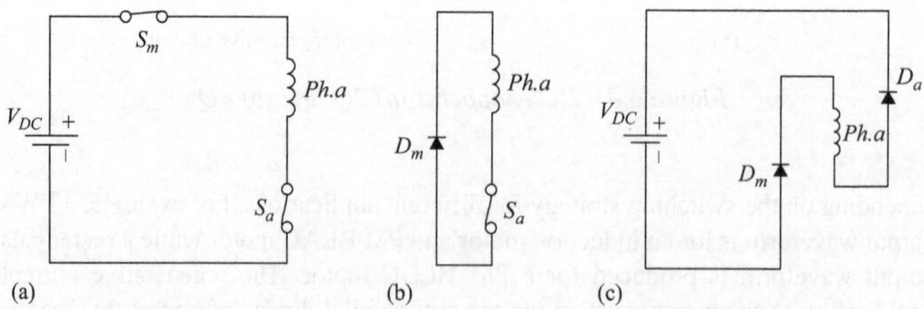


*Figure 6.11 SRM drive conduction modes for one phase: (a) powering, (b) freewheeling and (c) regenerating*

that, whenever $S_a$ is turned off, the current is returned to the battery system through $D_m$ and $D_a$. Thus, the SRM drive can readily offer regenerative braking, which is a key demand for EV application (Ching, 2009b).

### 6.2.2 System configuration for braking energy recovery

Quantitative analysis in this section gives an overview of energy recovery during the general braking process, and the analysis can be simplified and limited to longitudinal movements on a straight test track.

Figure 6.12 shows the power flow of regenerative braking energy in which the energy is recovered from the wheels to the battery and back to the wheels again (Spichartz *et al.*, 2014a):

$$E_{rec} = \int P_b \eta_M \eta_G \eta_R \eta_{Bi} \, dt \tag{6.6}$$

where $E_{rec}$ $P_b$, $\eta_M$, $\eta_G$, $\eta_R$ and $\eta_{Bi}$ are the recuperation energy, the braking power, and the efficiencies of mechanical part, generator, rectifier and battery charging, respectively.

Similarly, the fraction of the brake power, which is usable for a subsequent acceleration, $P_{acc}$, is calculated as follows:

$$P_{acc} = P_b \eta_M \eta_G \eta_R \eta_{Bi} \eta_{Bo} \eta_I \eta_{Mo} \eta_{Me} \tag{6.7}$$

where $\eta_{Bo}$, $\eta_I$, $\eta_{Mo}$ and $\eta_{Me}$ are the efficiencies of battery discharging, inverter, motor and mechanical transmission, respectively.

With regard to efficiency, several solutions are proposed to avoid unnecessary losses and undesired recuperations. When a driver really wants to brake, the driver must use the brake pedal; as long as no brake pedal is activated, the EV sails. Undesired recuperation is prevented as the driver has to use the brake pedal to activate regenerative braking.

A small part of the power is dissipated through the air and through rolling resistances; however, most of the power has to be absorbed by the brakes. Assuming that all brake power is provided by the drivetrain with the objective of total recuperation, all main drivetrain parts in a commercially available EV would be overloaded. These components are not dimensioned for these kinds of accelerations and power values as the EV would become too heavy and, above all, too expensive.

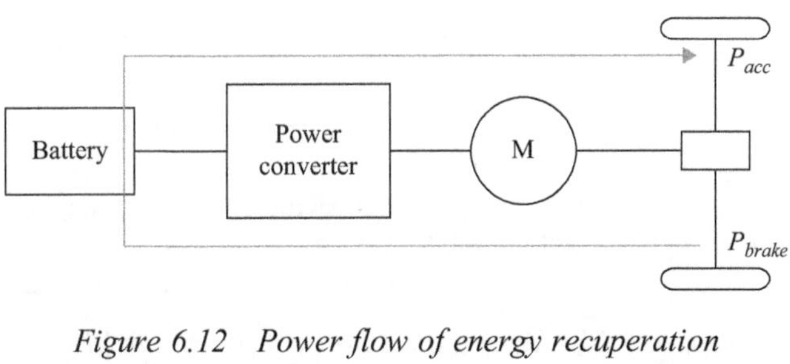


*Figure 6.12 Power flow of energy recuperation*

Batteries have a comparatively low power density. Therefore, batteries may not be able to deliver or store high power peaks, and beyond this, current peaks would reduce the lifetime of batteries. It was proposed to use ultracapacitors (ultracaps), which have a lower energy density but a higher power density in comparison to batteries, in parallel to the battery (Rotenberg *et al.*, 2011). The structure is shown in Figure 6.13.

In order to optimize power conversion between the electric machine and the ultracap, a DC/DC converter is required to accommodate the changeability of the voltage of both the ultracap and the electric machine. A digital machine controller is usually used to control the power of the motor during acceleration, as well as to control regenerative braking power to ensure smooth braking as shown in Figure 6.14 (Clarke *et al.*, 2010). A bidirectional power electronic converter

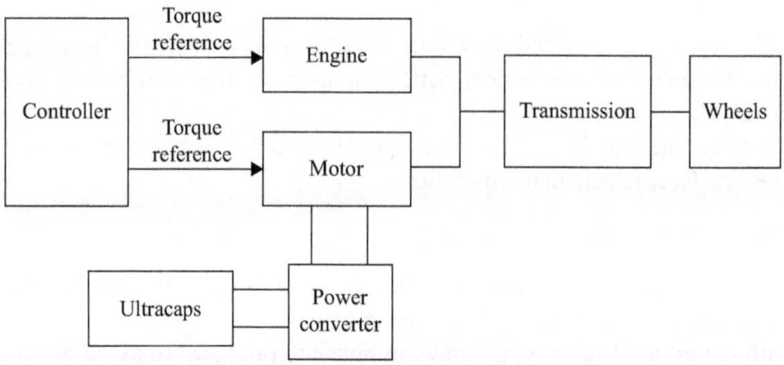


*Figure 6.13 Structure of parallel hybrid with ultracap energy storage*

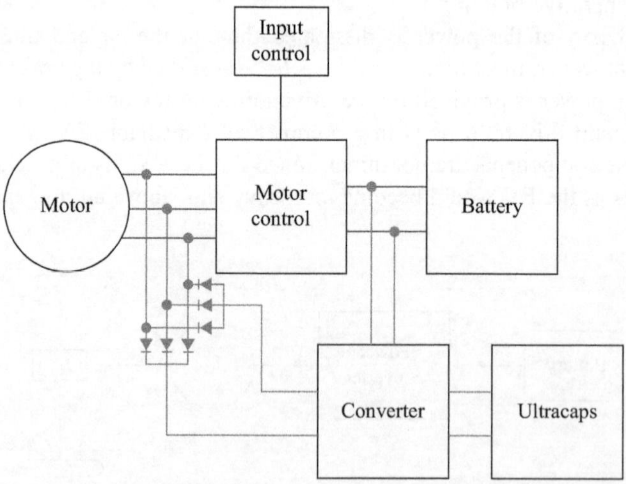


*Figure 6.14 Regenerative braking system for EVs*

(Chau *et al.*, 1998, 1999; Ching and Chau, 2001) is essential for the charging and discharging of ultracaps in this system.

### *6.2.3 Modelling of braking energy recovery*

Experimental studies have been conducted recently to measure the recuperable energies from an EV (Spichartz *et al.*, 2014b). While driving, a collective tractional resistance force, $F_{res}$ works against the moving direction. The EV drivetrain has to provide a driving force, $F_d$ at the wheels with the same absolute value in the moving direction to hold the desired driving condition, as shown in Figure 6.15 (Spichartz and Sourkounis, 2013):

$$F_{res} = -F_d \tag{6.8}$$

While the energy transfer to the road is fulfilled by the four wheel contact patches, in a steady state with constant velocity, three types of tractional resistance exist, namely air resistance, $F_{air}$, climbing resistance, $F_{cl}$ and the resistance of the four wheels, $F_{wh}$. During a straight forward movement on a dry road surface, the latter is approximately equal to the rolling resistance since the others in $F_{wh}$, including resistances like water displacement or bearing friction, have no or only a comparatively small influence.

During acceleration processes, the force of inertia, $F_a$, has to additionally be considered. Consequently, the absolute values of tractional resistance and driving force are described by the following equation:

$$F_{drive} = F_{es} = F_{air} + F_{wh} + F_{cl} + F_a \tag{6.9}$$

During a journey with a distance, $S_x$, the total energy, $E_{tot}$, is transformed at the wheels to overcome the tractional resistance:

$$E_{tot} = \int E_{drive}\, ds \tag{6.10}$$

The energy for overcoming air resistance, $F_{air}$, and the resistance of the four wheels, $F_{wh}$, are converted into thermal energy and are transferred to the air and to

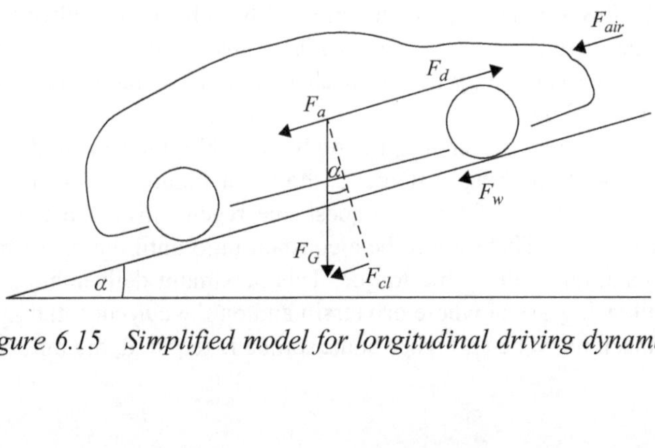


*Figure 6.15 Simplified model for longitudinal driving dynamics*

the road surface, respectively. In contrast, the energy for overcoming the two other forces is stored in the vehicle in its state and driving condition. While moving uphill and overcoming climbing resistance, $F_{cl}$, potential energy, $E_{pot}$, is stored in the vehicle and can be recovered during a downhill journey.

$$E_{pot} = mgh \tag{6.11}$$

During acceleration, the velocity of the vehicle, $v$, and the kinetic energy, $E_{kin}$, will increase. In an HEV or PEV, this energy can also be recuperated.

$$E_{kin} = \frac{1}{2}mv^2 \tag{6.12}$$

Consequently, the total energy can be divided into two parts: dissipating and recuperable energy (Spichartz *et al.*, 2014a):

$$E_{tot} = E_{rec} + E_{dis} \tag{6.13}$$

$$E_{dis} = \int_0^{S_x} \left[F_{air}(s) + F_{wh}(s)\right] ds \tag{6.14}$$

$$E_{rec} = \int_0^{S_x} \left[F_{cl}(s) + F_a(s)\right] ds = E_{kin} + E_{pot} \tag{6.15}$$

The distribution between these two parts depends on the journey travelled. During trips on highways with constant high velocities, air resistance, $F_{air}$, is the major tractional resistance force. While, for urban traffic, air resistance plays a minor part, and the force of inertia, $F_a$, has more influence on the total tractional resistance because of many brakings and accelerations (Spichartz and Sourkounis, 2013). Therefore, recuperation can help to reduce energy consumption and extend the driving range, especially in the city.

### 6.2.4 Control strategies for regenerative braking

As most of the currently available EVs are driven by one electric machine at the front or at the rear axle, a single motor-drive system is considered in this section.

In consideration of all the limitations in a drivetrain, blending methods are usually used in many EVs, i.e. the torque of the electric machine and the mechanical brakes is combined as a function of the velocity and other conditions as shown in Figure 6.16 (Spichartz *et al.*, 2014a). Thus, the braking power is partially converted into thermal energy by the mechanical brakes and partially recuperated by regenerative braking.

There are two major blending methods available. One method is called the parallel regenerative braking system, as shown in Figure 6.17 (Spichartz *et al.*, 2014a). For every value of torque request, the regenerative and the mechanical brakes work together. There might be a constant ratio until the regenerative brake reaches its maximum deliverable torque. This maximum deliverable torque relies on the overall sizing of the whole drivetrain and on the current rotational speed of the electric machine. When a higher brake torque is required, the ratio shifts to the

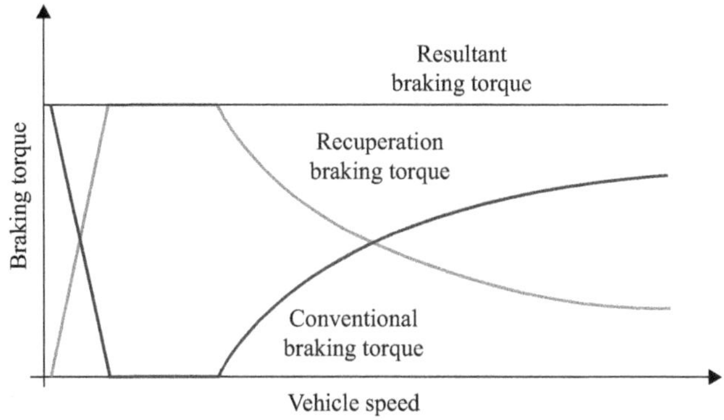


*Figure 6.16 Combination of mechanical braking and electrical braking*

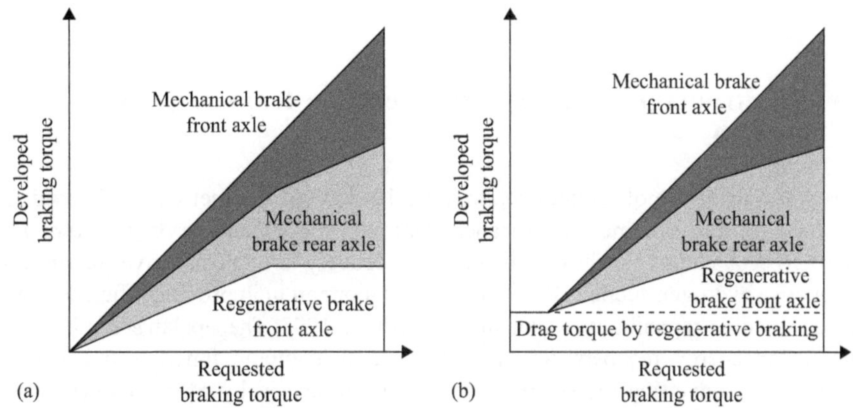


*Figure 6.17 Parallel regenerative braking system: (a) without drag torque and (b) with drag torque*

mechanical brakes. It has to be considered that the rear axle must not lock before the front axle during a brake application in order to stabilize the EV. Consequently, the brake torque for the front axle is normally set higher than the brake torque for the rear axle and is usually kept at a constant ratio.

Another major method is the series regenerative braking system, as shown in Figure 6.18 (Spichartz *et al.*, 2014a). As long as the requested torque is lower than the available brake torque from the electric machine, the machine delivers it. The mechanical brakes are not used until the requested torque exceeds the current maximum value of the machine, provided that the other parts of the drivetrain can handle the power.

It is obvious that the series braking system is more reasonable regarding energy recovery in most circumstances. The difficulties lie in the higher complexity of implementation and compliance with safety regulations. Therefore, most of the available EVs use a parallel braking system.

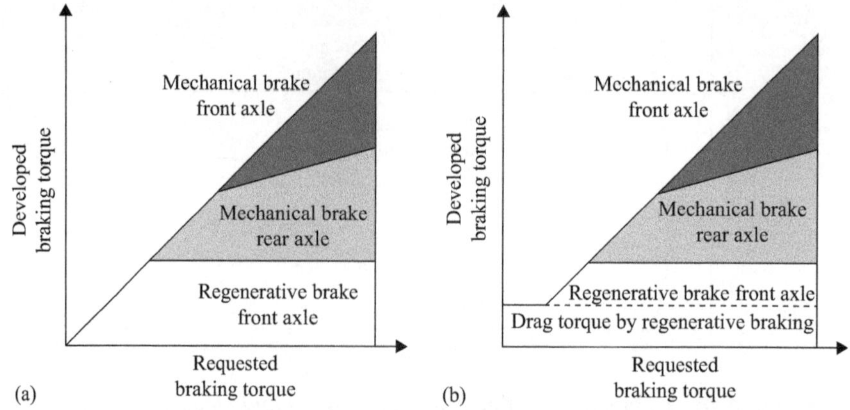


*Figure 6.18 Series regenerative braking system: (a) without drag torque and (b) with drag torque*

## 6.3 Electromagnetic energy regeneration from suspension system

During the daily use of an automobile, only 10–16% of the fuel energy is used to move the car to overcome the resistance from road friction and air drag (Zuo *et al.*, 2010). Recent studies have shown that energy recovery through an active suspension system in a hydrogen-fuelled HEV may play a major part in improving efficiency and range extending capabilities (Rajamani and Hedrick, 1995; Thanapalan *et al.*, 2012b). Suspension energy recuperation can achieve both damping function and energy recovery by converting the suspension vibration produced by road roughness into useful electrical energy, which can stored in the battery system to be reused later.

### *6.3.1 Suspension systems of vehicles*

When vehicles travel on a road, they are always subjected to excitation from road irregularities, acceleration forces, braking forces and inertial forces on a curved track, which leads to discomfort for the driver, and hence, influences overall manoeuvrability. Viscous shock absorbers, connected in parallel with suspension springs, have been widely used to suppress vibrations by dissipating the vibration energy into waste heat. To achieve better ride comfort and road handling, active suspensions have been explored by researchers (Sharp and Crolla, 1987; Okada and Harada, 1996; Nakano *et al.*, 2003; Gysen *et al.*, 2011; Gohrle *et al.*, 2015). However, active suspension requires a significant amount of energy, which limits its wide implementation. Hence, regenerative suspensions have been proposed, which have the potential to harvest vibration energy from suspension systems while reducing the vibration (Zhang *et al.*, 2009; Zuo *et al.*, 2010). Theoretical results revealed that a maximum of 10% fuel efficiency can be recovered from vehicle suspension system by realizing regenerative shock absorbers (Zhang, 2010).

Vehicle suspension is used to attenuate unwanted vibrations from various road conditions, such as road roughness. This system not only provides ride comfort but also provides safety by maintaining good road contact. Three types of suspension systems have been proposed and successfully realized: passive, active and semi-active (Rajamani and Hedrick, 1995; David and Bobrovsky, 2011; Ebrahimi *et al.*, 2011; Li *et al.*, 2012; Jayachandran and Krishnapillai, 2013; Ben *et al.*, 2014).

Passive suspension is established by an oil damper that provides a simple design and cost-effectiveness. Performance limitations are inevitable due to the damping force not being controllable (Seong *et al.*, 2011). A conventional suspension system of this kind usually consists of a spring and a viscous shock absorber. The shock absorber dissipates vibration energy into wasted heat.

Semi-active suspension relies on an actuator's ability to control the damping characteristics according to dynamic conditions. It can offer enhanced performance, generally in the active mode, without requiring a large power source or expensive component. Such a scheme is also energy absorbing. The damping profile is 'active' when both masses move in the same direction and 'passive' for the rest of the vibration cycle. The most common actuator developed recently is based on the flow of a magneto-rheological fluid having variable viscosity that can be controlled electromagnetically (Seong *et al.*, 2011; Poussot-Vassal *et al.*, 2012).

Fully active suspension is based on a variable damping principle similar to the semi-active schemes, and control can be applied over the damping characteristic throughout the entire vibration cycle. Hence, within some parts of the cycle, the element is energy consuming rather than energy absorbing. This can be provided by an external power source and continuous switching between consuming and absorbing operation modes. This system provides a near ideal and completely isolating damping profile (David and Bobrovsky, 2011).

### *6.3.2 System configuration of shock absorbers*

An electromagnetic regenerative suspension system transforms shock energy into electrical energy, which is more convenient to store and reuse, and has better performance, increased efficiency and less space requirements.

According to its working principle, a regenerative suspension system can be classified into two categories: mechanical and electromagnetic regenerative suspensions. In this section, only the electromagnetic regenerative suspension will be discussed. According to the electric machine utilized for energy conversion, it can further be categorized into rotational and linear generators.

Energy recovery suspension can achieve both the damping function and energy recovery by converting the suspension input vibration produced by road irregularities into reusable electrical energy, thus recovering energy and achieving damping through the same system. The configuration of a suspension system is shown in Figure 6.19 (Thanapalan *et al.*, 2012a). The control unit will detect the harvestable energy supply from the suspension subsystem through sensor circuits. The recovered energy will then be stored in the EV's main electrical energy storage (EES) system, such as a battery, to be utilized subsequently for range extension.

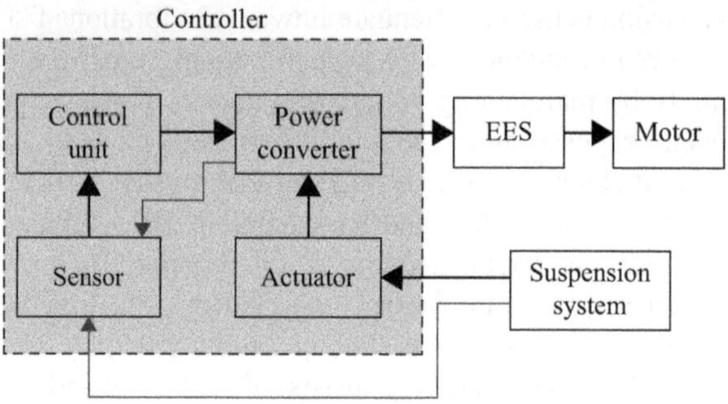


*Figure 6.19 Schematic of vehicle suspension system*

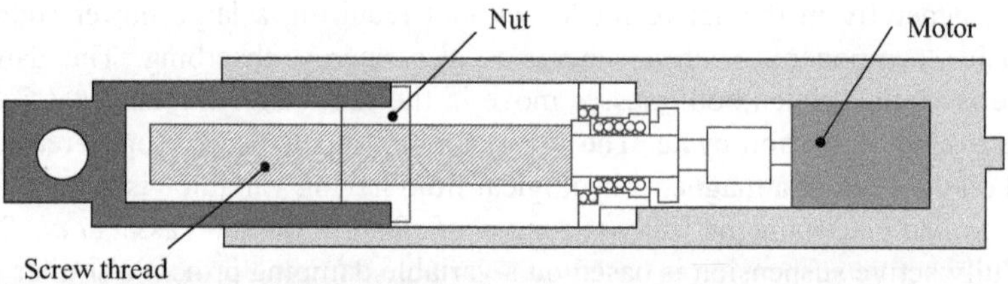


*Figure 6.20 Ball screw based energy harvester*

### 6.3.3 Energy harvester based on rotational electric machine

A PM machine is preferable in electromagnetic suspension (EMS), which provides an active force when operating in the actuator mode or a damping force when operating in the generator mode. The damping force can be changed simply by the tuning of shunt resistances. There are several types of electromagnetic regenerative suspensions classified by their structure and configuration, which are summarized in this section.

A ball screw is a traditional transmission device that converts linear motion into rotation. An electric shock absorber with a ball screw was developed to harvest vibration energy (Arsem, 1971). A ball screw electric damper was also proposed, and its damping force can be altered by controlling the shunt resistance (Murty, 1989). Figure 6.20 shows a ball screw–based harvester, and its dynamic and regenerative characteristics were also analysed (Suda *et al.*, 2000).

A rack-pinion can also convert linear motion into rotation. A rack-pinion damper incorporates a bevel gear to change the motor axis in parallel with the linear motion, as shown in Figure 6.21 (Zhang, 2010). A regenerative active suspension system combining a rack-pinion and rotary motor was also proposed (Suda and Shiiba, 1996). An electronically controlled active suspension system is a system, which utilizes the rack-pinion configuration. Experimental results revealed that the speed limit and manoeuvrability of vehicle were enhanced significantly with such a suspension system (Weeks *et al.*, 2000; Beno *et al.*, 2002).

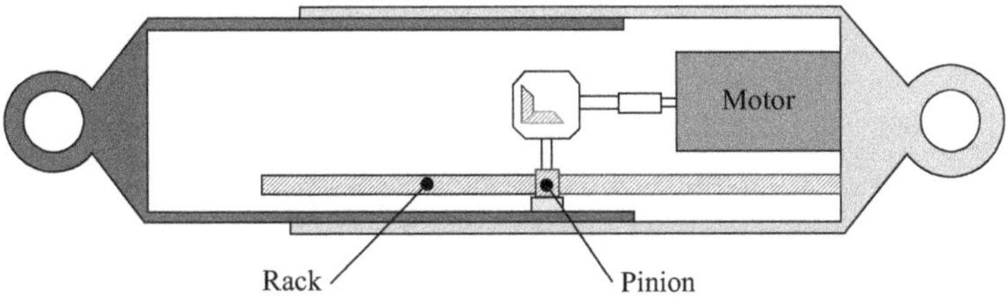


*Figure 6.21 Rack-pinion-based energy harvester*

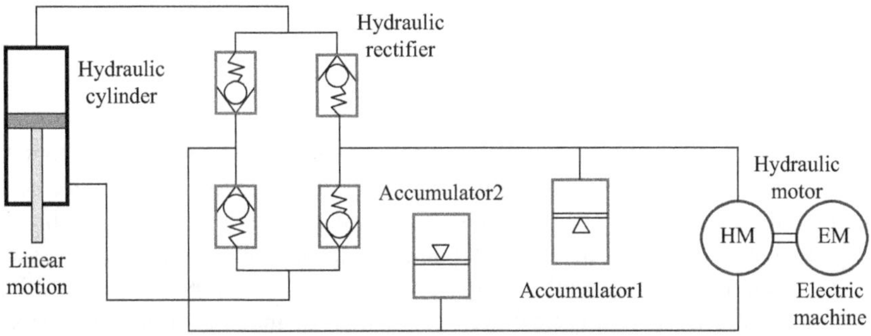


*Figure 6.22 Hydraulic-based energy harvester*

A hydraulic electromagnetic shock absorber (HESA) was recently developed (Fang *et al.*, 2013), which can not only isolate vibration but also recover part of the energy originally dissipated by heat. The HESA shown in Figure 6.22 comprises a hydraulic cylinder, hydraulic rectifier, hydraulic motor, accumulators and a generator connected with pipelines. The operating principle of a hydraulic rectifier is analogous to a Wheatstone bridge. The piston inside the hydraulic cylinder is driven to reciprocate under external stimulus, and the high-pressure oil therefore flows into the hydraulic rectifier. At the same time, oil flows out of the hydraulic rectifier from the specified export in the compression stroke or extension stroke and then flows through the accumulator for weakening the fluctuation, which drives the hydraulic motor to generate electricity.

### *6.3.4 Energy harvester based on linear electric machine*

A traditional shock absorber is usually replaced by a linear PM machine in direct-drive EMS systems. A linear machine converts mechanical energy (the relative motion between the vehicle chassis and wheels) into electrical energy without any transmission devices. An active and regenerative vibration control suspension using a linear actuator was proposed (Okada *et al.*, 2003), which can realize vibration isolation and energy regeneration simultaneously.

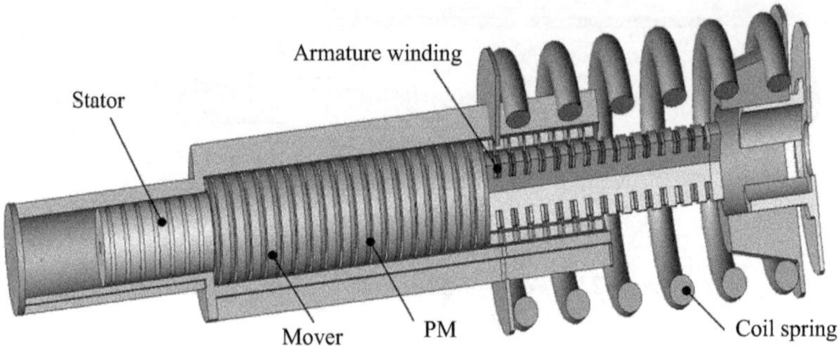


*Figure 6.23 Linear machine–based energy harvester assisted by a passive coil spring*

A self-powered active vibration control system with two linear motors for truck cabins was also proposed (Nakano and Suda, 2004). In this system, an electric generator installed in the suspension system of the chassis regenerates vibration energy and stores it in a condenser. An actuator in the cabin suspension provides active vibration control using the energy previously stored in the condenser. Since the mass of the chassis of a typical heavy-duty truck is much greater than the mass of the cabin, vibration energy in the suspension of the chassis is anticipated to be greater than that in the cab suspension. So, this system can be self-powered.

Furthermore, a self-powered active vibration control system utilizing a single electric actuator was proposed (Gysen *et al.*, 2011). The proposed system shown in Figure 6.23 consists of a passive coil spring to support the sprung mass and a direct-drive brushless tubular PM actuator to deliver active forces. For safety reasons, the suspension system should provide damping even when there is a power failure; hence, a passive damper should also be incorporated into the active suspension system. This condition can be achieved through an oil-filled damper connected in parallel. Since an electromagnetic actuator is considered, it is obtained from eddy currents that are induced in the conducting solid stator.

Generally, lower passive damping will increase the level of regenerated energy since the actuator has to provide the damping function; however, safety is compromised because less passive damping is present when there is a power failure. Therefore, during a design process, the amount of passive damping will be considered as a variable initially, and the influence on performance and power consumption will then be analysed.

A regenerative shock absorber integrated with a linear generator (Zuo *et al.*, 2010) is shown in Figure 6.24. The mechanism converts the kinetic energy from vibration between the wheel and a sprung mass into useful electrical energy. The shock absorber consists of an integrated magnet and a coil assembly. The integrated magnet is made with ring-shaped PMs and ring-shaped high permeability spacers stacked on a rod of high reluctance material. The magnets are inserted with like-poles of adjacent magnets facing each other to redirect the magnetic flux to a radial

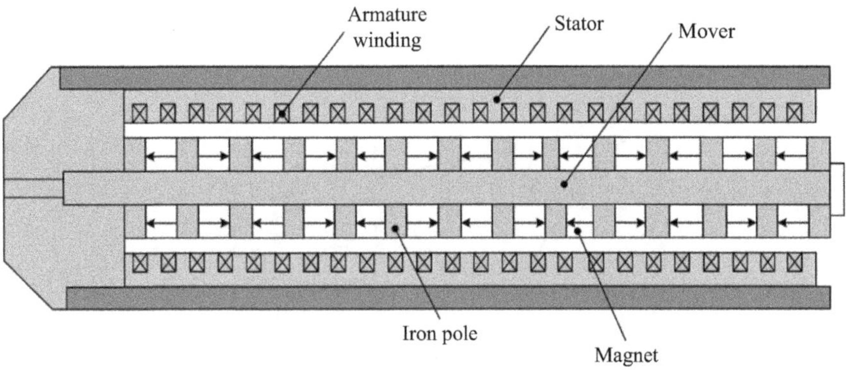


*Figure 6.24 Linear machine–based energy harvester*

direction. A concentric outer cylinder made of a high permeability material is used to reduce the reluctance of the magnetic circuits, and hence, further enhances the magnetic flux density in the coils. The coil assembly is made of copper coils wound on a Delrin tube. As the copper coils traverse inside the magnetic field, a voltage will be generated and the coils will be connected to rectifier circuits.

### *6.3.5 Modelling of suspension systems*

The vehicle suspension system can be established in three models, namely, the full vehicle model, half-vehicle model and quarter-vehicle model, as shown in Figure 6.25 (Chi *et al.*, 2008; Montazeri-Gh and Soleymani, 2010; Marzbanrad *et al.*, 2013). The full vehicle model is most comprehensive with seven DOF. These comprise three DOF for the vehicle body (pitch, bounce and roll) and a further vertical DOF at each of the four unsprung masses. The half-vehicle model has four DOF where the roll information is not accounted for. The quarter-vehicle model has two DOF where the pitch information is lost (Happian-Smith, 2002).

In the designing of shock absorbers and the establishing of suspension system models, full-vehicle and half-vehicle models are more comprehensive than the quarter-vehicle model. The quarter-vehicle model is acceptable and with less complexity in the mathematical formula. The suspension system is considered in this section because it is easier to regenerate energy. An active suspension control system based on the quarter-vehicle model is shown in Figure 6.26 (Okada and Harada, 1996).

Equation of motion of the suspension system is given by:

$$m\ddot{x}(t) + k\left[x(t) - x_0(t)\right] = f(t) \tag{6.16}$$

where $m$ is the sprung mass, $k$ is the physical spring constant, $x$ and $x_0$ are the displacement of sprung mass and unsprung mass, respectively, and $f$ is the actuator force.

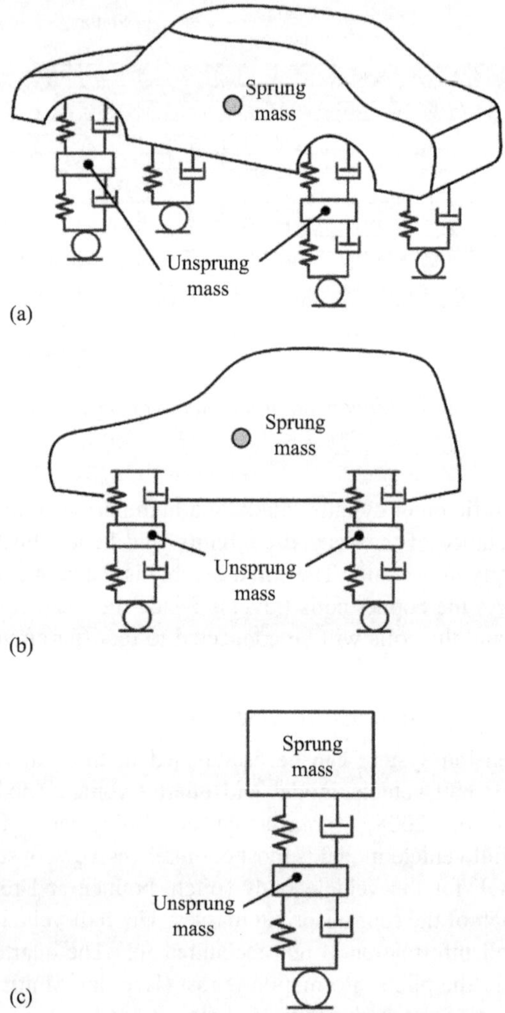


*Figure 6.25 Modelling of vehicle suspension systems: (a) full-vehicle model, (b) half-vehicle model and (c) quarter-vehicle model*

Similar to the other damping systems, the energy can be regenerated only with high speed motion. For the low speed motion, active control will be applied (Okada and Harada, 1996). The actuator force is calculated by:

$$f(t) = \begin{cases} -\dfrac{\psi^2}{R}(v - v_0) + \dfrac{\psi}{R}e_b, & \dfrac{e_b}{\psi} < (v - v_0) \\ -K_p x - K_v v, & -\dfrac{e_b}{\psi} < (v - v_0) < \dfrac{e_b}{\psi} \\ -\dfrac{\psi^2}{R}(v - v_0) + \dfrac{\psi}{R}e_b, & (v - v_0) < -\dfrac{e_b}{\psi} \end{cases} \tag{6.17}$$

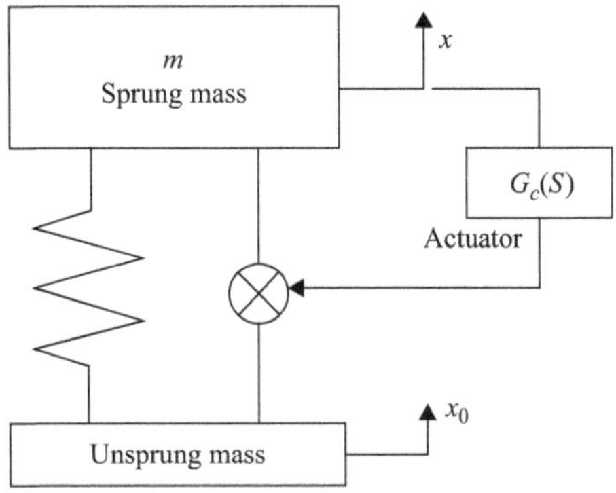


*Figure 6.26 Active suspension control system*

where $\psi$ is the actuator constant, $R$ is the load resistance, $v$ is the actuator velocity, $v_0$ is the boundary velocity for energy regeneration, $e_b$ is the battery voltage, $K_p$ is the displacement gain and $K_v$ is the velocity gain.

The regenerative energy is calculated using the following equations (Okada and Harada, 1996):

$$W = \begin{cases} \frac{\psi}{R}\left[\psi(v - v_0) - e_b\right](v - v_0) - Ri^2, & \frac{e_b}{\psi} < (v - v_0) \\ Ri^2 - (v - v_0)(K_p x + K_v v), & -\frac{e_b}{\psi} < (v - v_0) < \frac{e_b}{\psi} \\ \frac{\psi}{R}\left[\psi(v - v_0) + e_b\right](v - v_0) - Ri^2, & (v - v_0) < -\frac{e_b}{\psi} \end{cases} \tag{6.18}$$

A quarter-vehicle model is used as shown in Figure 6.27 (Huang *et al.*, 2011), with the assumption that all springs are linear within the maximum available working space and tire damping is negligible. The output force of the actuator comprises the electromagnetic vertical force, $F_i$, and the damping force, $F_c$. The latter comes from mechanical friction and is assumed to be linear and also proportional to the stroke speed of the actuator.

The electromagnetic vertical force, $F_i$, and the damping force, $F_c$, are expressed by:

$$F_i = k_i i = \frac{2\Phi \cot \varphi}{r} i \tag{6.19}$$

$$F_c = C_s v = C_s(\dot{x}_w - \dot{x}_b) \tag{6.20}$$

where $\Phi$ is the stator flux linkage, $k_i$ is the vertical force coefficient, $r$ is the effective radius for force conversion, $\varphi$ is the screw lead angle, $C_s$ is the damping coefficient, $x_w$ is the unsprung mass displacement and $x_b$ is the sprung mass displacement.

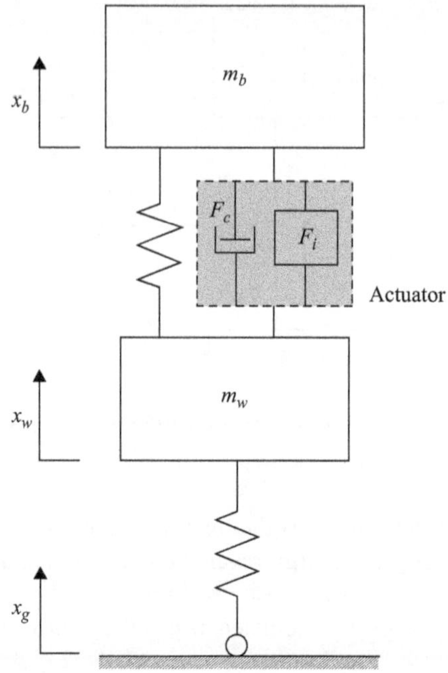


*Figure 6.27 Quarter-vehicle model*

Thus, the model equations of motion can be written as follows:

$$m_b\ddot{x}_b = k_s(x_w - x_b) + C_s(\dot{x}_w - \dot{x}_b) + k_i i \tag{6.21}$$

$$m_w\ddot{x}_w = -k_s(x_w - x_b) + k_t(x_g - x_w) - C_s(\dot{x}_w - \dot{x}_b) - k_i i \tag{6.22}$$

where $m_b$ is the sprung mass, $m_w$ is the unsprung mass, $k_s$ is the spring stiffness and $k_t$ is the stiffness of a tire.

In conventional passive suspension systems, a significant amount of kinetic energy will be dissipated as heat, especially when excited by high-frequency input attenuation. Another important fact is that undesired high-frequency vibrations are usually orthogonal to the direction of a vehicle's movement. Therefore, the mentioned attenuation does not produce additional load to the drivetrain. The operating principle of a regenerative shock absorber is a magnetic element moving inside a coil that induces an electromotive force in the coil. The relative movement of the sprung and unsprung masses due to road roughness or inertial forces is the mechanical power input for the regeneration process. Moreover, the current of the coil induces a motion opposing mechanical force, producing a controlled variable attenuation of the undesired movements. The configuration not only attenuates vibrations but also transforms the kinetic energy into useful electrical energy, which is stored in the battery system for range extension.

Power regeneration is coupled to the damping level, based on the following model (David and Bobrovsky, 2011). The induced electromotive force, $E_{Ind}$, is proportional to the relative velocity, $v_{Rel}$:

$$E_{Ind} = Kv_{Rel} \tag{6.23}$$

where $K$ is the electromagnetic force constant.

The input power, whether electrical or mechanical is given by

$$P_{In\_Mech} = Fv_{Rel} = E_{Ind}I = P_{In\_Ele} \tag{6.24}$$

where $F$ is the applied mechanical force and $I$ is the current in the coils.

Substituting (6.23) into (6.24), the motion opposing force becomes

$$F_{Em\_Opp} = \frac{E_{Ind}I}{v_{Rel}} = KI \tag{6.25}$$

The generator output voltage, $V_{Out}$, is

$$V_{Out} = E_{Ind} - R_{Loss}I \tag{6.26}$$

where $R_{Loss}$ is the resistance of the generator coils.

And finally the output power for charging the battery is obtained as

$$P_{Out\_Eff} = V_{Out}I = (E_{Ind} - R_{Loss}I)I = P_{In\_Ele} - I^2R_{Loss} \tag{6.27}$$

The trade-off between the amount of power regenerated and vibration damping is clearly indicated in the mathematical model. Intensive regeneration stiffens the suspension and prevents effective vibration attenuation, since precise damping force control is paramount, while the energy regeneration intensity is determined by the desired dynamic mechanical behaviour (David and Bobrovsky, 2011).

### 6.3.6 Control strategies for regenerative suspension

The system architecture of an active regenerative suspension system is depicted in Figure 6.28 (David and Bobrovsky, 2011). The control system is implemented with a dual loop scheme: an outer low-bandwidth inertial rate loop and an inner

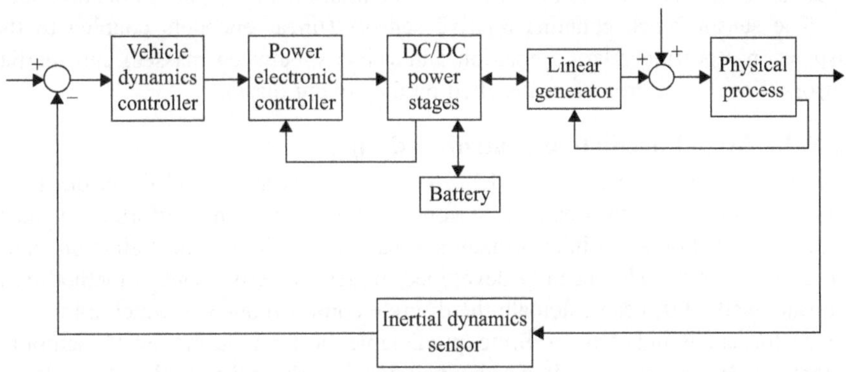


*Figure 6.28 Control block diagram of active regenerative suspension system*

high-bandwidth current loop. Both loops are based on state feedback. The basic external configuration for the control system is more desirable. The core of the control scheme is formed by three major components: a vehicle dynamics controller, power electronic controller and DC/DC power stages.

For the vehicle dynamics controller, a system states feedback error is input to the block as well as to the desired operation mode. The sprung mass inertial vertical velocity is the measured state feedback. This feedback is related to the zero inertial command in static states and forms the external inertial loop error. Additionally, four unsprung masses and four relative vertical velocities between sprung masses are measured through linear encoders for the skyhook controller. This block simultaneously serves the entire vehicle (commands four actuators in parallel) as it emulates different kinds of shock absorbing elements (passive damper, active damper, etc.) using electrical means. The desired four actuator forces (one for each wheel) are the output of this block and are sent to the power electronic controller.

Inputs to the power electronic controller are four force commands, as well as the operation mode and the power electronic feedback measurements. The input and output voltage and the current of the DC/DC power stage are measured feedback. This feedback is related to the force command, which forms the force error. The output of this block is the control signal over time sent to the power stage. Moreover, this block defines a two switched bridges topology. The input/output ports bridge defines the energetic mode, the consumption from the battery or from charging it, and the rectification bridge is correlated to the sign of the input voltage (transforming the voltage from bipolar to unipolar).

For the DC/DC power stage, the same control signal is the input to the block, and it is affected by the peripheral electronic topology (switched bridges). This block is based on the gyrator model (Singer, 1988), and it softly connects the battery and the generator, allowing energy regeneration and force control simultaneously. This block realizes the hybrid system concept, and its activity results in two electrical consequences: battery input/output power control and generator/motor current control.

The system plant includes the linear generators and the vehicle's body and responds to the sum of the control forces and additional road pattern disturbances.

The sensor block contains relative sensors (linear encoders coupled to the suspension reporting relative position and velocity between masses) and inertial sensors (linear accelerometers installed on the sprung mass).

#### 6.3.6.1 Model predictive controller design

Due to the inherent constraints of EMS systems, the conventional linear quadratic optimal control algorithm cannot be applied to minimize the performance index since it is based on an infinite horizon and cannot handle the constraints on states and process inputs. The recently developed model predictive control method (Qin and Badgwell, 2003) can calculate the discrete control sequence, which minimizes the performance index on a finite predictable horizon during every sampling instant. Process constraints can also be included directly in the problem formulation so that constraint violation can be prevented and predicted.

#### 6.3.6.2 Skyhook control

Skyhook control is a simple and very effective control algorithm. Ease in implementing the logic of the skyhook controller is also well known for real applications. The principle of skyhook control is to design an active or semi-active suspension control so that the sprung mass is linked to the sky in order to suppress the vertical oscillations of the sprung mass. Figure 6.29 shows the conceptual scheme of skyhook control for a vehicle suspension system.

A regenerative suspension system using skyhook control is shown in Figure 6.30 (Hsieh *et al.*, 2014). It consists of a sprung mass, a DC machine (DCM), a planetary gear, a physical spring, an algebraic screw and a load resistor.

The damping force for a single DOF system with a skyhook damper can be written as:

$$F_{sky} = C_{sky}\dot{x} \tag{6.28}$$

where $C_{sky}$ is the skyhook damping ratio.

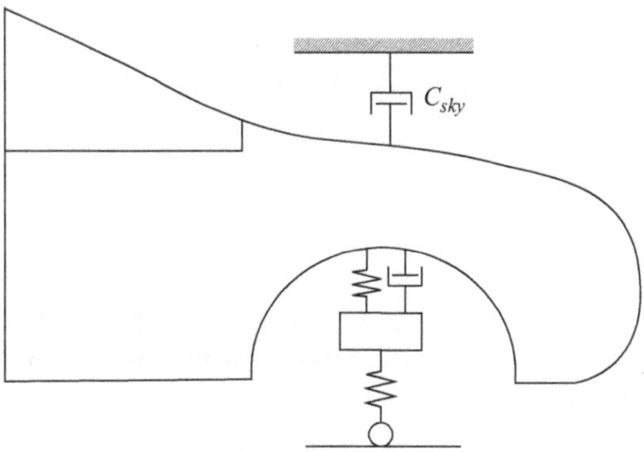


*Figure 6.29 Skyhook control for regenerative suspension system*

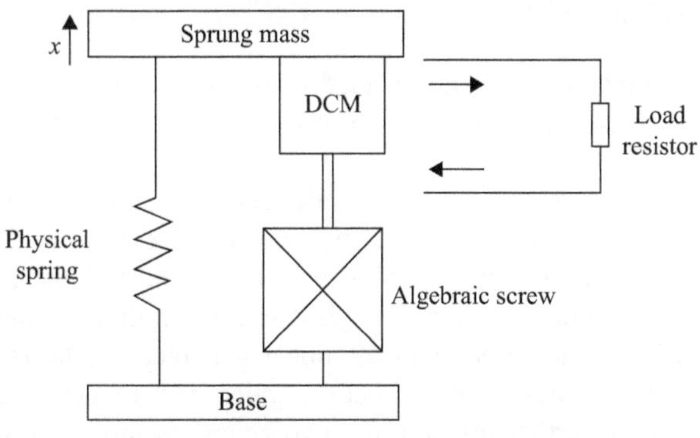


*Figure 6.30 Regenerative suspension system using skyhook control*

Using a semi-active damper, the skyhook damping force must be realized in terms of the relative velocity $\dot{z}$. Therefore, the conventional semi-active skyhook control algorithm is given by:

$$F_{sky} = \begin{cases} C_{sky}\dot{z}, & \dot{x}\dot{z} \geq 0 \\ 0, & \dot{x}\dot{z} < 0 \end{cases} \tag{6.29}$$

According to the limitation of a semi-active damper, when $\dot{x}$ and $\dot{z}$ have opposite signs, the damper cannot provide a force opposite to $\dot{z}$. As a result, it is better to supply no force in that situation.

As for the active skyhook control strategy, the damper can supply a negative force from a negative damping ratio. So, the active skyhook control algorithm can be written as:

$$F_{sky} = \begin{cases} C_{sky}\dot{z}, & \dot{x}\dot{z} \geq 0 \\ -C_{sky}\dot{z}, & \dot{x}\dot{z} < 0 \end{cases} \tag{6.30}$$

The desired positive and negative damping ratios can be realized by a switch-mode power electronic converter by synthesizing an adjustable load resistance, $R_{load}$. Specifically, when a positive damping ratio is desired, the load resistance can be calculated as:

$$R_{load}^{+} = \frac{k_e k_t k_g^2}{(C_{sky} - C_f)d^2} - R_{\text{int}} \tag{6.31}$$

where $k_e$, $k_t$ and $k_g$ are the motor torque constant, angular velocity constant and gear ratio, respectively, $C_f$ is the physical friction, $d$ is the rotor diameter and $R_{int}$ is the motor internal resistance.

Similarly, when a negative damping ratio is desired, the load resistance can be calculated as follows:

$$R_{load}^{-} = -\frac{k_e k_t k_g^2}{(C_{sky} + C_f)d^2} - R_{\text{int}} \tag{6.32}$$

By actively adjusting the value of load resistance, both positive and negative damping ratios are realizable, and a corresponding continuous damping force can be generated.

According to the current flow in the linear generator, the control current range can be divided into several operating zones, which correspond to different motor operating and energy flow states, as shown in Figure 6.31 (Huang *et al.*, 2011). Based on linear generator operating modes, it can be divided into two zones, namely, the linear actuator mode and the linear generator mode. Based on linear generator operating modes and the battery operating modes, the Zone II in Figure 6.31(a) can be further divided into two zones, as shown in Figure 6.31(b), namely, the linear generator mode with the battery discharging mode (Zone III) and

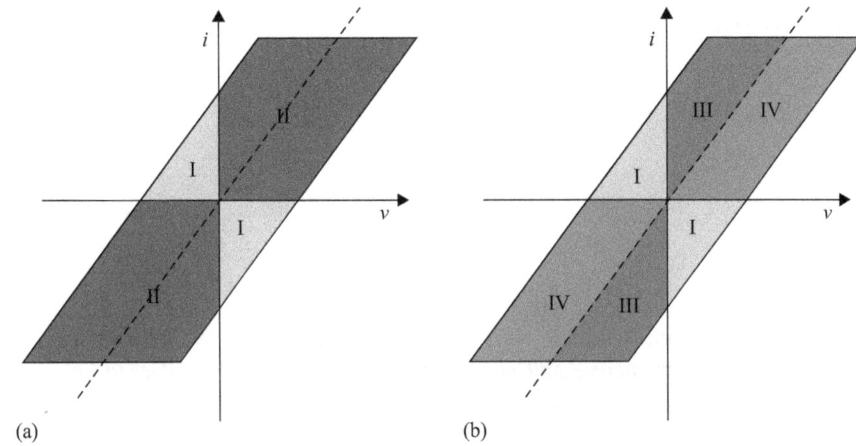


*Figure 6.31 Energy flow for different operating modes: (a) two zones and (b) four zones*

the linear generator mode with battery charging mode (Zone IV). In Zone I, the linear generator operates as a linear actuator. The current flows from the battery into the positive terminal of a motor, and energy is consumed from the battery. In Zone III, the linear generator operates as a generator, but the reclaimed energy is insufficient to suppress the sprung mass vibration caused by a rough road; thus, extra energy is required from the battery, and hence current flows from the battery to the negative terminal of the motor. Only in Zone IV does the current flow into the positive terminal of the battery, and the motor charges the battery as a generator with reclaimed energy from vibration.

## 6.4 Summary

Improvement of the driving range is one of the main research challenges for all kinds of EVs. An optimized braking recuperation strategy can help to extend the range without increasing the capacity of a battery system. Improved braking energy recovery can also extend the driving range and, therefore, can raise the acceptance and suitability of EVs for daily use.

The suspension system in a vehicle is used to suppress undesirable vibration from road roughness. The system provides not only ride comfort, but also provides safety by maintaining good road contact. A conventional suspension system consists of a spring and viscous shock absorber. The shock absorber dissipates vibration energy into wasted heat. A fully active suspension system provides better isolation and improves control performance. However, it has a main disadvantage of high power consumption. In addition to passive and fully active suspension systems, semi-active suspension systems have been proposed to provide a compromise between isolation control and energy savings and are deemed suitable for range extension for EVs.

## Acknowledgements

This work was supported in part by grants MYRG2015-00218-FST and MYRG067 (Y1-L2)-FST13-CTW from Research Council of the University of Macau.

## References

Arsem, H. B. (1971) Electric shock absorber, Google Patents.

Ben, L. Z., Hasbullah, F., and Faris, F. W. (2014) A comparative ride performance of passive, semi-active and active suspension systems for off-road vehicles using half car model. *Int. J. Heavy Veh. Syst.*, 21, 26–41.

Beno, J. H., Weeks, D. A., Bresie, D. A., Guenin, A. M., Wisecup, J., and Bylsma, W. (2002) Experimental comparison of losses for conventional passive and energy efficient active suspension systems, SAE Technical Paper 2002-01-028.

Chan, C. C. (2007) The state of the art of electric, hybrid, and fuel cell vehicles, *Proc. IEEE*, 95 (4). 704–718.

Chan, C. C. and Chau, K.T. (2001) *Modern Electric Vehicle Technology*, Oxford: Oxford University Press. Chau, K.T. (2014) Pure electric vehicles. In *Alternative Fuels and Advanced Vehicle Technologies for Improved Environmental Performance*, ed. R. Folkson, Amsterdam: Woodhead Publishing.

Chau, K.T. (2015) *Electric Vehicle Machines and Drives: Design, Analysis and Application*, Singapore: Wiley-IEEE Press.

Chau, K. T., Ching, T. W., and Chan, C. C. (1998) Bidirectional soft-switching converter-fed DC motor drives. *Proc. IEEE Power Electronics Specialists Conference*, Fukuoka, Japan, pp. 416–422.

Chau, K.T., Ching, T.W., and Chan, C.C. (1999) A new two-quadrant zero-voltage transition converter for DC motor drives. *Int. J. Electron.*, 86, 217–231.

Chi, Z. Z., He, Y. P., and Naterer, G. F. (2008) Design optimization of vehicle suspensions with a quarter-vehicle model. *Trans. Can. Soc. Mech. Eng.*, 32, 297–312.

Ching, T. W. (2005) Four-quadrant zero-voltage-transition converter-fed DC motor drives for electric propulsion. *J. Asian Electric Veh.*, 3, 651–656.

Ching, T. W. (2006) Four-quadrant zero-current-transition converter-fed DC motor drives for electric propulsion. *J. Asian Electric Veh.*, 4, 911–917.

Ching, T. W. (2007) Soft-switching converters for electric vehicle propulsion. *J. Asian Electric Veh.*, 5, 1019–1026.

Ching, T. W. (2009a) Review of soft-switching technologies for high-frequency switched-mode power conversion. *Int. J. Electron. Eng. Educ.*, 46, 104–119.

Ching, T. W. (2009b) Analysis of soft-switching converters for switched reluctance motor drives for electric vehicles. *J. Asian Electric Veh.*, 7, 893–898.

Ching, T. W., and Chau, K. T. (2001) A new two-quadrant zero-current transition converter for DC motor drives. *Int. J. Electron.*, 88, 719–735.

Clarke, P., Muneer, T., and Cullinane, K. (2010) Cutting vehicle emissions with regenerative braking. *Transp. Res. Part D: Transp. Environ.*, 15, pp. 160–167.

David, S. B., and Bobrovsky, B. Z. (2011) Actively controlled vehicle suspension with energy regeneration capabilities. *Veh. Syst. Dyn.*, 49, 833–854.

Ebrahimi, B., Bolandhemmat, H., Khamesee, M. B., and Golnaraghi, F. (2011) A hybrid electromagnetic shock absorber for active vehicle suspension systems. *Veh. Syst. Dyn.*, 49, 311–332.

Fang, Z., Guo, X., Xu, L., and Zhang, H. (2013) Experimental study of damping and energy regeneration characteristics of a hydraulic electromagnetic shock absorber. *Adv. Mech. Eng.*, 2013, pp. 1–9.

Gohrle, C., Schindler, A., Wagner, A., and Sawodny, O. (2015) Road profile estimation and preview control for low-bandwidth active suspension systems. *IEEE-ASME Trans. Mechatron.*, 20, 2299–2310.

Gysen, B. L. J., van der Sande, T. P. J., Paulides, J. J. H., and Lomonova, E. A. (2011) Efficiency of a regenerative direct-drive electromagnetic active suspension. *IEEE Trans. Veh. Technol.*, 60, 1384–1393.

Han, J., Park, Y., and Park, Y. S. (2011) Adaptive regenerative braking control in severe cornering for guaranteeing the vehicle stability of fuel cell hybrid electric vehicle. *Proc. IEEE Vehicle Power and Propulsion Conference*, Chicago, Illinois, USA, pp. 1–5.

Happian-Smith, J. (2002) *An Introduction to Modern Vehicle Design*, Warrendale: SAE International.

Hsieh, C. Y., Huang, B., Golnaraghi, F., and Moallem, M. (2014) Sky-hook control for a regenerative suspension system. *Proc. 2014 IEEE 80th Vehicular Technology Conference (Vtc Fall)*, Vancouver, British Columbia, Canada, pp. 1–6.

Huang, K., Yu, F., and Zhang, Y. (2011) Active controller design for an electromagnetic energy-regenerative suspension. *Int. J. Autom. Technol.*, 12, 877–885.

Jayachandran, R., and Krishnapillai, S. (2013) Modeling and optimization of passive and semi-active suspension systems for passenger cars to improve ride comfort and isolate engine vibration. *J. Vib. Control*, 19, 1471–1479.

Kendall, G. (2008) *Plugged In – The End of the Oil Age*, Brussels: WWF European Policy Office.

Ko, J., Ko, S., Son, H., Yoo, B., Cheon, J., and Kim, H. (2015) Development of brake system and regenerative braking cooperative control algorithm for automatic-transmission-based hybrid electric vehicles, *IEEE Trans. Veh. Technol.*, 64, 431–440.

Li, C., Liang, M., Wang, Y. X., and Dong, Y. T. (2012) Vibration suppression using two-terminal flywheel. Part II: application to vehicle passive suspension. *J. Vib. Control*, 18, 1353–1365.

Marzbanrad, J., Poozesh, P., and Damroodi, M. (2013) Improving vehicle ride comfort using an active and semi-active controller in a half-car model. *J. Vib. Control*, 19, 1357–1377.

Miller, J. M. (2004), *Propulsion Systems for Hybrid Vehicles*, Stevenage: The Institution of Engineering and Technology.

Montazeri-Gh, M., and Soleymani, M. (2010) Investigation of the energy regeneration of active suspension system in hybrid electric vehicles. *IEEE Trans. Ind. Electron.*, 57, 918–925.

Murty, B. V. (1989) Electric, variable damping vehicle suspension, U.S. Patent No. 4,815,575.

Nakano, K., and Suda, Y. (2004) Combined type self-powered active vibration control of truck cabins, *Veh. Syst. Dyn.*, 41, 449–473.

Nakano, K., Suda, Y., and Nakadai, S. (2003) Self-powered active vibration control using a single electric actuator. *J. Sound Vib.*, 260, 213–235.

Okada, Y., and Harada, H. (1996) Regenerative control of active vibration damper and suspension systems. *Proc. 35th IEEE Conference on Decision and Control*, Kobe, Japan, pp. 4715–4720.

Okada, Y., Kim, S., and Ozawa, K. (2003) Energy regenerative and active control suspension. *Proc. ASME Design, Engineering Technical Conference*, Chicago, Illinois, USA, pp. 2135–2142.

Poussot-Vassal, C., Spelta, C., Sename, O., Savaresi, S. M., and Dugard, L. (2012) Survey and performance evaluation on some automotive semi-active suspension control methods: a comparative study on a single-corner model. *Ann. Rev. Control*, 36, 148–160.

Qin, S.J., and Badgwell, T. A. (2003) A survey of industrial model predictive control technology. *Control Engineering Practice*., 11, 733–764.

Rajamani, R., Hedrick, J. K. (1995) Adaptive observers for active automotive suspensions: theory and experiment. *IEEE Trans. Control Syst. Technol.*, 3, 86–93.

Rajashekara, K. (1994) History of electric vehicles in General Motors. *IEEE Trans. Ind. Appl.*, 30, 897–904.

Rotenberg, D., Vahidi, A., and Kolmanovsky, I. (2011) Ultracapacitor assisted powertrains: modelling, control, sizing, and the impact on fuel economy. *IEEE Trans. Control Syst. Technol.*, 19, 576–589.

Seong, M.-S., Sung, K.-G., and Choi, S.-B. (2011) Control strategies for vehicle suspension system featuring magnetorheological (MR) damper, In F. Beltran-Carbajal (Ed.), *Vibration Analysis and Control – New Trends and Developments*. Rijeka, Croatia: INTECH Open Access Publisher.

Sharp, R., and Crolla, D. (1987) Road vehicle suspension system design—a review. *Veh. Syst. Dyn.*, 16, 167–192.

Singer, S. (1988) Loss-free gyrator realization. *IEEE Trans. Circ. Syst.*, 35, 26–34.

Spichartz, P., Dost, P., and Sourkounis, C. (2014a) Recuperation strategies for electric vehicles. *Proc. Ninth International Conference on Ecological Vehicles and Renewable* Energies, Monte-Carlo, Monaco, pp. 1–7.

Spichartz, P., Dost, P., and Sourkounis, C. (2014b) On examination and measurement for recuperation modes targeting range extension in electric vehicles. *Proc. 40th Annual Conference of the IEEE Industrial Electronics Society*, Dallas, Texas, USA, pp. 3036–3042.

Spichartz, P., and Sourkounis, C. (2013) Measurement of braking energy recuperation in electric vehicles. *Proc. 15th European Conference on Power Electronics and Applications*, Lille, France, pp. 1–9.

Stabler, F. (2002) Automotive applications of high efficiency thermoelectrics. *DARPA/ONR Program Review and DOE High Efficiency Thermoelectric Workshop*, San Diego, California, USA, pp. 1–26.

Suda, Y., and Shiiba, T. (1996) A new hybrid suspension system with active control and energy regeneration, *J. Veh. Syst. Dyn.*, 25, 641–654.

Suda, Y., Suematsu, K., Nakano, K., and Shiiba, T. (2000) Study on electromagnetic suspension for automobiles—simulation and experiments of performance. *Proc. 5th International Symposium on Advanced Vehicle Control*, Ann Arbor, Michigan, USA, pp. 699–704.

Thanapalan, K., Zhang, F., Premier, G., Guwy, A., and Maddy, J. (2012a) Renewable hydrogen hybrid electric vehicles and optimal energy recovery systems. *Proc. UKACC International Conference on Control*, Cardiff, Wales, UK, pp. 935–940.

Thanapalan, K., Zhang, F., Premier., G., Maddy, J. and Guwy, A. (2012b) Onboard renewable hydrogen production system for hydrogen hybrid vehicles. *Proc. World Renewable Energy Forum*, Denver, Colorado, USA, pp. 1–8.

Tokuoka, S. (2010) Nissan LEAF product outline. *Proc. 25th World Battery, Hybrid and Fuel Cell Electric Vehicle Symposium & Exhibition*, Shenzhen, China, pp. 1–4.

Tur, O., Ustun, O., and Tuncay, R. N. (2007) An introduction to regenerative braking of electric vehicles as anti-lock braking system. *Proc. IEEE Intelligent Vehicles Symposium*, Istanbul, Turkey, pp. 944–948.

Weeks, D. A., Beno, J., Guenin, A., and Bresie, D. (2000) Electromechanical active suspension demonstration for off-road vehicles. SAE Technical Paper 2000-01-0102.

Zhang, J., Peng, Z., Zhang, L., and Zhang, Y. (2013) A review on energy-regenerative suspension systems for vehicles, *Proc. World Congress on Engineering*, London, UK, pp. 1–4.

Zhang, P. S. (2010) Design of electromagnetic shock absorber for energy harvesting from vehicle suspensions, MS thesis, Advisor L. Zuo, Stony Brook University, Stony Brook, NY.

Zhang, Y., Yu, F., and Huang, K. (2009) A state of art review on regenerative vehicle active suspension. *Proc. 3rd International Conference on Mechanical Engineering and Mechanics*, Beijing, China, pp. 1689–1695.

Zuo, L., Scully, B., Shestani, J., and Zhou, Y. (2010) Design and characterization of an electromagnetic energy harvester for vehicle suspensions. *Smart Mater. Struct.*, 19, 1–10.

Zuo, L., and Zhang, P. (2013) Energy harvesting, ride comfort, and road handling of regenerative vehicle suspensions. *J. Vib. Acoust.*, 135, 1–8.

*Chapter 7*

# On-board thermoelectric energy recovery for hybrid electric vehicles

*Shuangxia Niu*[1] *and Chuang Yu*[1]

With growing concern on environment pollution and energy conservation, the development of energy-efficient technologies for automotive industry has taken on an accelerated pace (Chau and Chan, 2007). In an internal combustion engine, a portion of energy is inevitably wasted in the form of waste gas. If this waste energy can be collected and recovered, the overall fuel efficiency will be significantly improved (Saidur *et al.*, 2012). On the other hand, during the past decade, thermoelectric generation (TEG) technology has been developed and widely studied as a potential green energy source, which is based on the thermoelectric effect to directly convert the temperature difference to electricity (Elsheikh *et al.*, 2014). Since the internal combustion engine is an indispensable part in the hybrid electric vehicle (HEV), this chapter is to integrate the TEG into the HEV to recover the waste heat energy, hence offering a compact on-board auxiliary energy system.

## 7.1 TEG

Thermoelectric effect describes the conversion between thermal energy and electrical energy. In 1821, Thomas Johann Seebeck first discovered that a voltage existed between ends of a metal bar when a temperature gradient existed in the bar, which is named as the Seebeck effect to describe the conversion of temperature differences directly into electricity. In contrast, in 1834, Jean Peltier found the reverse of the Seebeck effect, namely the Peltier effect. Obviously, the TEG system based on the Seebeck effect takes the advantages of no moving parts, silent operation and very reliable.

There is an important feature to describe the high energy conversion efficiency of thermoelectric material, which is defined as the figure of merit (LaGrandeur *et al.*, 2006):

$$ZT = \frac{S^2 T}{\kappa\rho} \tag{7.1}$$

[1]Department of Electrical Engineering, The Hong Kong Polytechnic University, Hong Kong, China

where $S$ is the thermoelectric power, $T$ is the absolute temperature, $\kappa$ is the total thermal conductivity, and $\rho$ is the electrical resistance. Before 1990s, the highest *ZT* values of all thermoelectric materials remained below 1. With the development of material technology in the past decades, the characteristics of thermoelectric materials have been significantly improved, not only in terms of the highest *ZT* value up to about 3.5, but also in terms of the temperature range covering 200–900 K.

Figure 7.1 shows the *ZT* characteristics of typical thermoelectric materials (Wang *et al.*, 2013). It can be seen that both the N-type and P-type $Bi_2Te_3$ offer the optimal *ZT* values at the lower temperature range (<200 °C) of exhaust gas; the N-type PbTe and the P-type Te-Ag-Ge-Sb (TAGS) alloy offer the optimal *ZT* values at the medium temperature range (200–500 °C); and the skutterudites,

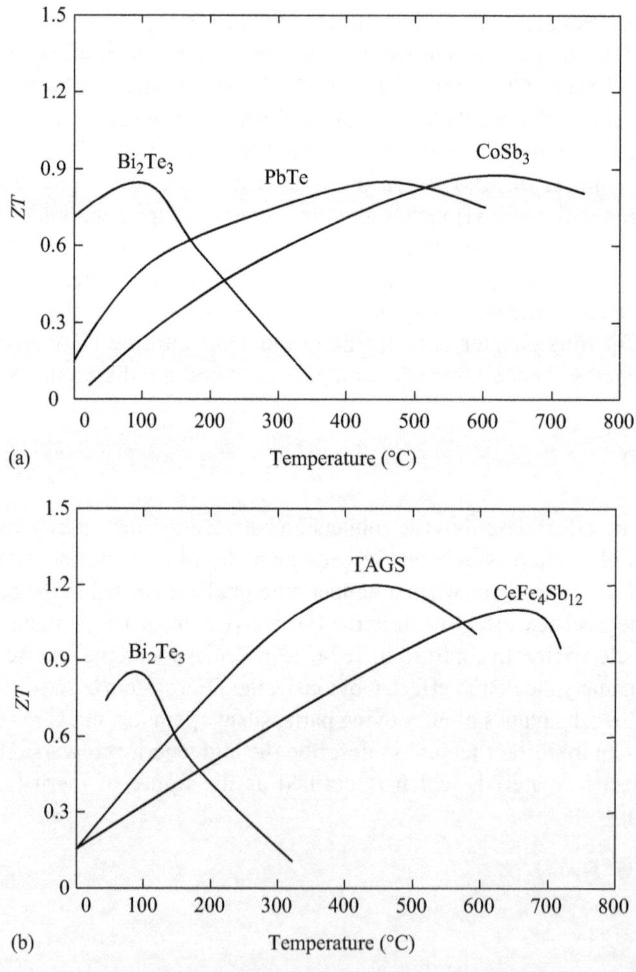


*Figure 7.1 ZT values of typical thermoelectric materials: (a) N-type and (b) P-type*

namely, N-type $CoSb_3$ and P-type $CeFe_4Sb_{12}$, offer the optimal *ZT* values at the higher temperature range (500–800 °C).

Figure 7.2(a) depicts the configuration of a typical single TEG couple in which an N-type thermoelectric semiconductor and a P-type thermoelectric semiconductor are connected thermally in parallel and electrically in series. When the heat passes from hot sides to cold sides, the accumulation of charge carriers in both semiconductors generates an electrical potential between the ends of this TEG couple. Typically, in a commercial TEG device as shown in Figure 7.2(b), more than 100 such TEG couples are connected in series to bring the total output voltage up to a useful level (He *et al.*, 2015).

In the market, a practical TEG system usually contains dozens of TEG devices. They are connected in series, and then sandwiched tightly between a heater and a radiator. The heat source always comes from hot water, hot oil, or hot gas, while the radiator mainly adopts air-cooling method or water-cooling method.

In order to figure out the electrical characteristics of a single TEG device, an experimental prototype is built as shown in Figure 7.3, where a water-cooled aluminum radiator serves as the cold side of the TEG device, while an electric heater works as its hot side. The TEG device is based on $Bi_2Te_3$ semiconductor material. When heating the hot-side temperature of TEG device to different values of 100, 150, 200, and 250 °C, and keeping the cold-side temperature at 50 °C. The corresponding open-circuit

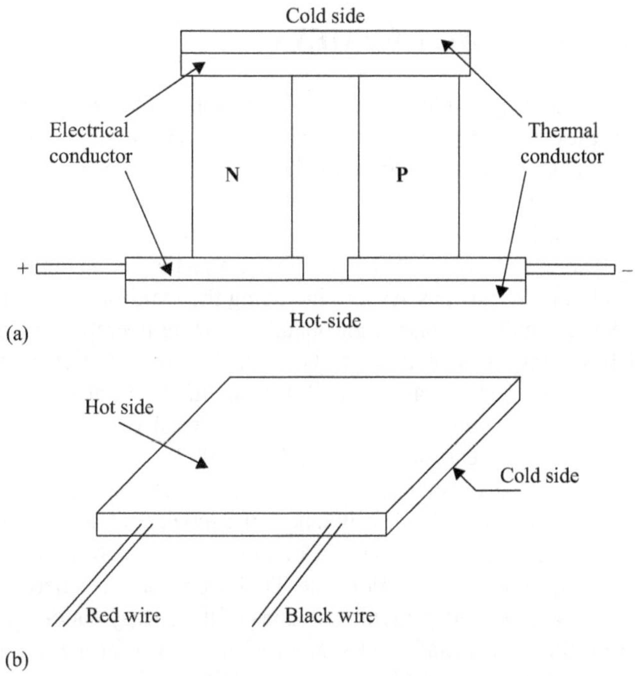


*Figure 7.2 Configuration of TEG: (a) single TEG couple and (b) TEG device*

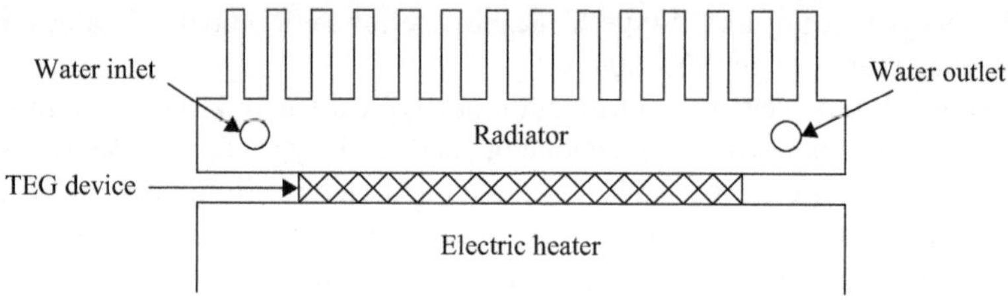


*Figure 7.3 Configuration of an experimental TEG system*

output voltages are measured as 2.02 V, 3.52 V, 4.56 V, and 5.25 V, respectively. Then, by varying the load resistance from 2 Ω to 10 Ω, the output voltage, output current, and output power of the device versus the load resistance under different hot-side temperatures are obtained as shown in Figure 7.4(a)–(c), respectively.

Further, if ten such TEG devices are connected in series, the output voltage and the corresponding internal resistance of the whole TEG system versus temperature difference are obtained as shown in Figure 7.5. It can be observed that both voltage and resistance increase with the temperature difference. In order to enable the maximum power transfer from the source to the load, the internal resistance needs to be equal to the load resistance. Thus, the maximum power transfer can be achieved when the load resistance is controlled in accordance with temperature difference.

## 7.2 Waste heat recovery for HEVs

Although HEVs have been identified as the most feasible energy-efficient road transportation, they are not of absolutely zero emission due to the use of internal combustion engines which are inefficient. Figure 7.6 shows a typical energy flow path of an internal combustion engine, in which only about 25% of the fuel combustion energy is utilized for vehicle operation, whereas about 40% is lost in the form of waste heat of exhaust gas (Stabler, 2002). As a result, the temperature of all components in HEV exhaust gas system, including the manifold, pipe, muffler, and catalytic converter, will increase significantly. The temperature distribution at normal vehicle operation is plotted as shown in Figure 7.7 (LaGrandeur *et al.*, 2005). A recent study has revealed that if 10% of the thermal energy carried by exhaust gas can be captured and converted to electrical energy, the overall fuel economy will directly increase up to 20% (Yang, 2005).

By utilizing the TEG technology to recover such heat losses and generate electricity without adding engine load, a practical thermoelectric waste heat recovery system for HEVs is proposed as shown in Figure 7.8, in which the heat exchanger serves as the energy collection section, the TEG works as the energy conversion section, and the power converter circuit functions as the energy conditioning section.

Considering the limit installation space for such waste heat recovery system, the heat exchanger always combines with the TEG devices to form a compact structure (Bari and Hossain, 2013). On the other hand, the heat exchanger should

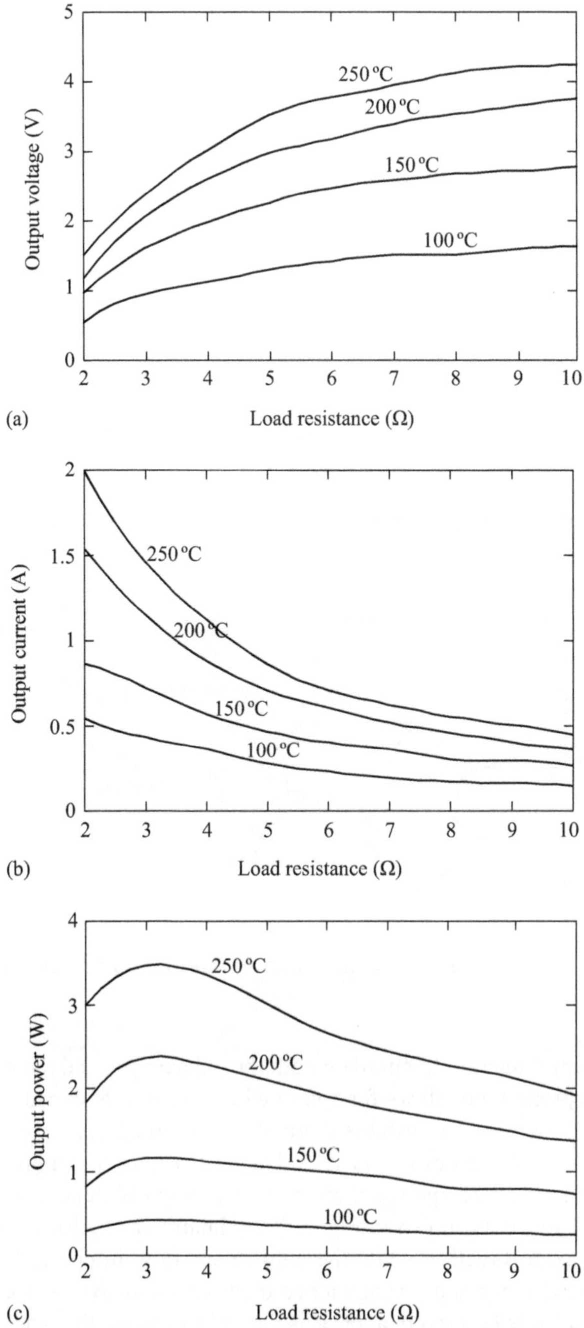


*Figure 7.4 Output characteristics of a single TEG device: (a) voltage, (b) current and (c) power*

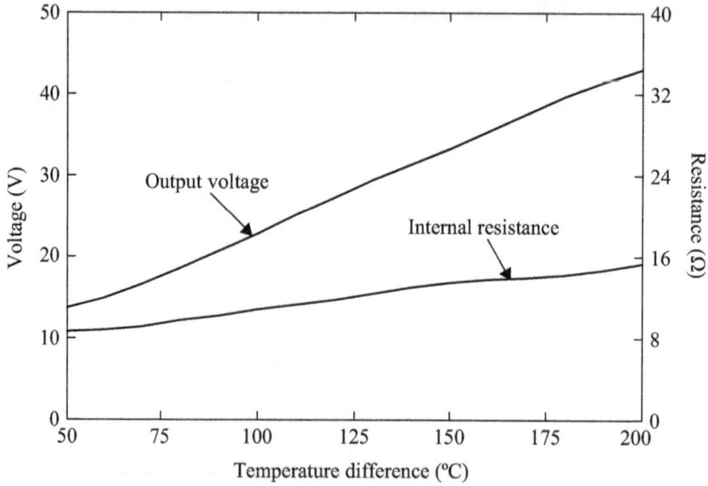


*Figure 7.5 Output voltage and internal resistance characteristics of TEG system*

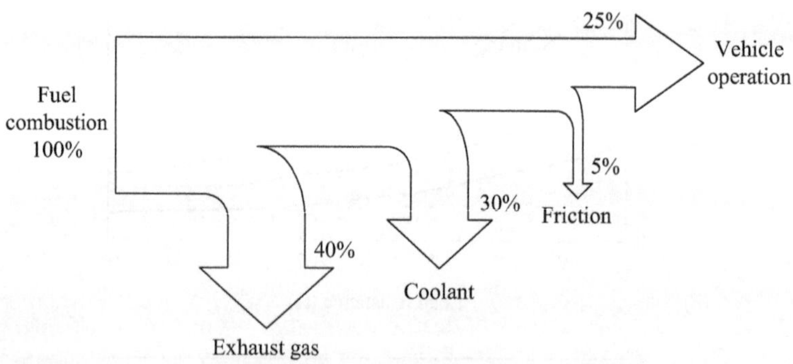


*Figure 7.6 Typical energy flow path in internal combustion engine*

integrate the pump function to circulate both the exhaust gas and the coolant liquid. The use of this pump loop allows for a very wide variation of exhaust gas flow and coolant liquid flow, hence adjustable thermal energy to be converted. It should be pointed out that the design of this heat exchanger involves a compromise between the gas or liquid flow rate and the thermal exchanging efficiency. A reported heat exchanger is designed as shown in Figure 7.9 (Hatami *et al.*, 2014), where the hot fluxes flow in axial direction while the cold fluxes flow in circumferential direction, and the TEG devices are sandwiched between them. Another concern is the location to install this heat exchanger. It is natural to choose the place with highest temperature in Figure 7.7, namely, the exhaust manifold. However, taking into account the allowable working temperature of the selected TEG device and the mounting convenience, another compromise should be made.

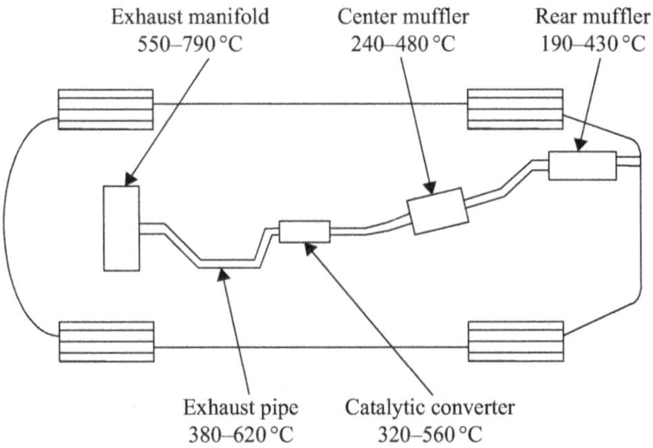


*Figure 7.7 Typical temperature distribution of exhaust gas system in HEV*

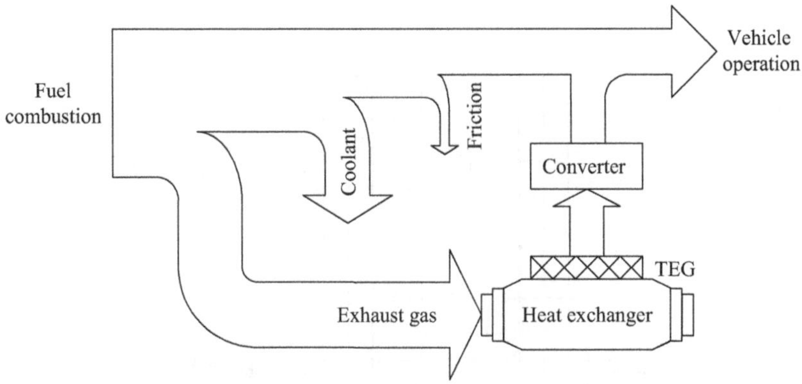


*Figure 7.8 Proposed thermoelectric waste heat energy recovery system for HEV*

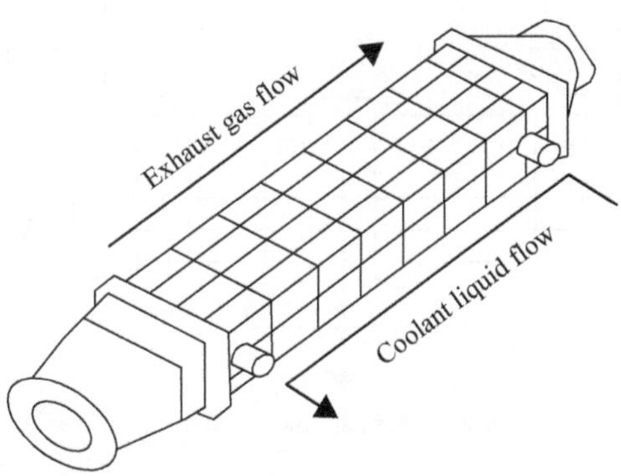


*Figure 7.9 Heat exchanger*

In order to achieve the highest average *ZT* value over the whole temperature range of exhaust gas during HEV operation, the TEG should employ different thermoelectric materials, and hence different thicknesses of each thermoelectric segment or layer to manage the thermoelectric compatibility both within each element and between individual elements. A traditional TEG configuration is shown in Figure 7.10(a). As can be seen, the thermoelectric elements with different thicknesses and different thermal expansion coefficients will cause an unstable structure during the practical operation. Another alternative configuration is proposed in recent year as shown in Figure 7.10(b), which can accommodate thermoelectric elements of different thicknesses and thermal expansion coefficients and maintain a very stable structure (Bell, 2004).

The electricity generated by TEG should be stored in the battery or directly supplied to the vehicle power network. Rather than simply connecting the TEG output to the battery pack, a power conditioning system needs to work with the TEG in such a way that the maximum power transfer to the battery can be ensured for different

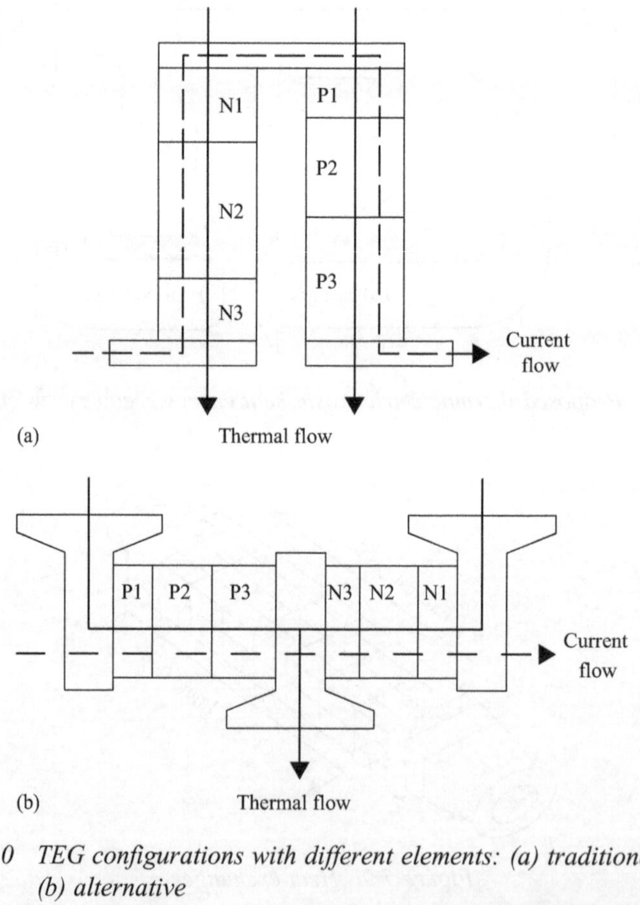


*Figure 7.10 TEG configurations with different elements: (a) traditional and (b) alternative*

temperatures. The two key issues are the design of a proper power converter and the devise of an effective control algorithm, which will be discussed in later sections.

## 7.3 Thermoelectric energy system

According to the energy management system of electric vehicles, the TEG system can be arranged in the series connection or parallel connection. In the series connection system, the DC–DC converter will process all the power generated by the TEG, whereas in the parallel connection system, the bidirectional DC–DC converter balances the power between the TEG and the load (Gao *et al.*, 2012).

### *7.3.1 System configuration with series connection*

The configuration of thermoelectric energy system with series connection is depicted as shown in Figure 7.11, which is composed of a TEG, a unidirectional DC–DC converter, a controller, a driver circuit, a battery, and a load. The unidirectional converter handles all the output power from the TEG with a simple system structure and control method. The floating battery stables the DC bus in a relatively small range. The controller measures the output voltage and current of the TEG, namely, the input voltage and current of the DC–DC converter, and the output voltage and temperature of the battery via analog/digital (A/D) conversion channels.

According to the maximum power point tracking (MPPT) algorithm, the pulse-width modulation (PWM) switching signal is generated from the controller and then amplified by the driver circuit before sending to the converter. The measured battery terminal voltage and body temperature data help the controller to protect the battery from the overvoltage and overheat problems. Once the overvoltage or overheat fault is found, the controller will switch off the converter to protect the battery.

This auxiliary energy system can be connected with the low voltage battery or DC bus of the multiple voltage level energy system of HEVs. Within this connection topology, the MPPT can be realized by sharing the hardware of the

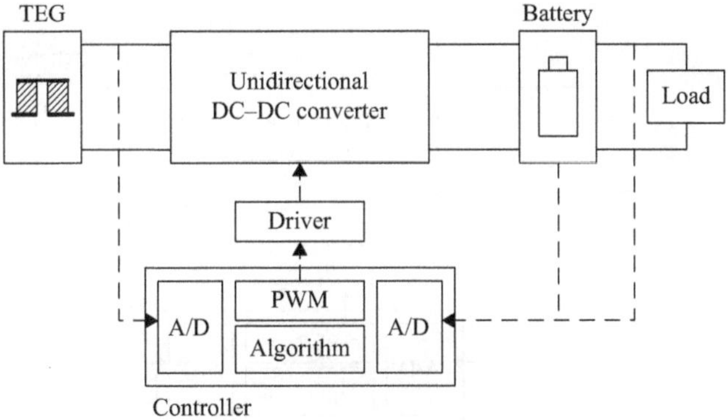


*Figure 7.11 Thermoelectric energy system with series connection*

controller in the energy management system. Then this auxiliary energy system can be easily integrated into the whole energy system of HEVs. When the output voltage of the TEG is higher than the preset minimum value $V_{TE,\min}$ and the battery voltage is lower than a preset minimum value $V_{bat,\min}$, the charging process is started with MPPT control. When the maximum battery voltage $V_{bat,\max}$ is reached, the battery is regarded as fully charged and the charging process is terminated.

There are three possible operation modes in this system with series connection. First, if the power provided by the TEG is greater than the power drawn by the load, the battery will be charged by the balanced power; second, if the power provided by the TEG is less than the power drawn by the load, the battery will discharge to the load to provide the balanced power; third, if the external load is too low and the battery is fully charged, the DC–DC converter will be switched off by the controller to protect the system from overvoltage or overheat.

### *7.3.2 System configuration with parallel connection*

The configuration of thermoelectric energy system with parallel connection is depicted as shown in Figure 7.12, which is composed of a TEG, a bidirectional DC–DC converter, a controller, two driver circuits, a battery, and a load. The dash lines and solid lines in the figure represent the signal and power connections, respectively. Two DC voltage buses are provided at both sides of the bidirectional converter. The controller measures the output voltage and current of the TEG, and the input voltage and temperature of the battery via A/D conversion channels.

Although the structure and control scheme of this system are more complex, the bidirectional DC–DC converter technology has been widely applied in the electrical distribution system of HEVs and the control scheme can be integrated into the energy management controller of the power network of HEVs. So this configuration is one preferable option of the auxiliary energy system for HEVs, and its overall system efficiency is higher than that of the system with series connection.

This thermoelectric energy system will operate in pure battery discharging mode, when the power provided by TEG is very small or even zero, and the battery

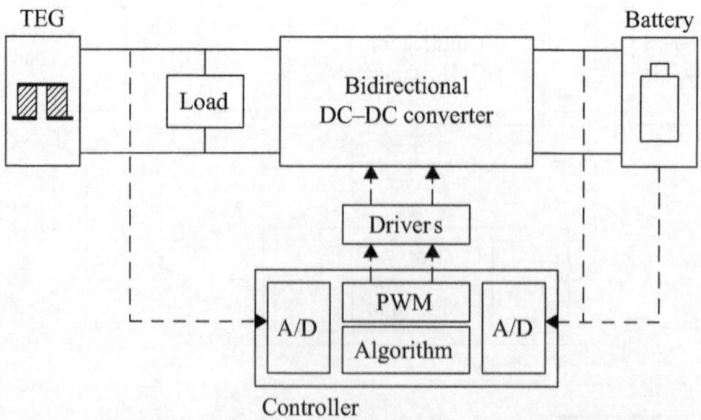


*Figure 7.12 Thermoelectric energy system with parallel connection*

has stored enough energy for discharging. In other words, at this moment, the output voltage of the TEG is smaller than the preset minimum value $V_{TE,\min}$, or the output power of the TEG at maximum power point (MPP) is smaller than the preset minimum value $P_{TE,\min}$, and the battery voltage is higher than the preset minimum value $V_{bat,\min}$. In this mode, only the battery discharges to drive the load at a specific voltage level. Except for this mode, the MPPT control will be activated in the system.

When the TEG is used to supply the load directly without the interface of the power conditioning circuit (PCC), the power extracted from the TEG will vary dramatically with the changing of the external load. As illustrated in Figure 7.13, the operating power point of the TEG varies significantly under different external loads without the PCC and MPPT. According to the power demand by the load, there are three modes, namely large load mode, optimal load mode, and small load mode, which correspond to $R_L < k_1 R_{TE}$, $k_1 R_{TE} < R_L < k_2 R_{TE}$, and $R_L > k_2 R_{TE}$, respectively, where $R_L$ is the load resistance and $R_{TE}$ is the TEG internal resistance.

The pure battery discharging mode is governed by the voltage control loop, while the other three modes are governed by the MPPT control. In the large load mode, both the TEG and battery provide power to the load. In the small load mode, the extra power is used to charge the battery to balance the power difference between the TEG output and the load demand. In the optimal load mode, the maximum power is drawn from the TEG and transferred to the load totally, and the battery does not draw power from the TEG or deliver power to the load. It can be found that, when the load is not connected into the system, namely $R_L/R_{TE} = \infty$, the power provided by TEG totally charges the battery, which also occurs in the series-connected thermoelectric energy system without the external load. Consequently, the PCC and MPPT of the series-connected thermoelectric energy system can be regarded as one special case of the parallel-connected thermoelectric energy system.

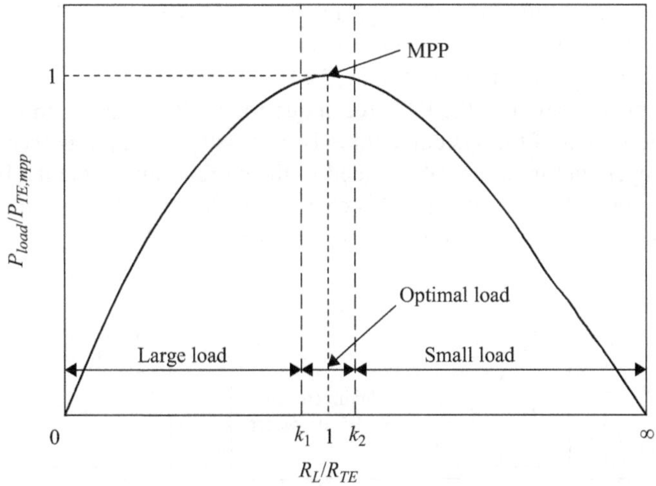


*Figure 7.13 Extracted power with respect to external load in parallel-connected thermoelectric energy system*

## 7.4 MPPT

MPPT control has been widely implemented in photovoltaic power systems, and its effectiveness is well proven (Zhang and Chau, 2011). There are many MPPT control techniques that have been proposed, among which, the perturb-and-observe (PAO) technique, incremental-conductance technique, parasitic-capacitance technique, and constant-voltage technique are considered to be viable for practical implementation (Lyden and Haque, 2015; Liu *et al.*, 2015).

The PAO technique is an iterative approach that perturbs the operating point of the photovoltaic power system and then observes the output power so as to determine the direction of change for maximizing the output power. Thus, the maximum power can be eventually achieved by forcing the derivative of output power to be equal to zero under power feedback control. It takes the definite advantages that there is no need to have an accurate model of the photovoltaic power system, and the online searching process can automatically handle minor parameter variations. However, this PAO technique is unsuitable for those systems under rapidly changing atmospheric condition.

As revealed by Figure 7.4, the TEG power system has similar electrical characteristics as the photovoltaic power system, and, on the other hand, the operating temperature of the exhaust gas system does not have a rapid change, so the PAO technique can be readily employed in the thermoelectric energy system, and its shortcoming is avoided.

### *7.4.1 MPPT for thermoelectric energy system with series connection*

Since the temperature difference and the load resistance may vary in a large range, the MPP of the thermoelectric energy system may also dynamically change. The PCC can track the MPP of the TEG by tuning its duty cycle of the PWM switching signal to enable the equivalent input resistance, $R_{in,s} = V_{i,s}/I_{i,s}$, equal to the TEG internal resistor $R_{TE}$, as shown in Figure 7.14.

The implementation of the PAO technique for MPPT is shown in Figure 7.15. The operation point of the system is initially at $A$ with the output power $P_0$, then a small duty cycle perturbation $\Delta\delta$ is added to the current duty cycle $d_0$. In the next moment, the output power at point $B$ is changed to $P_1$. Since $P_0$ is smaller than $P_1$, it means that the operating point is approaching to the MPP and the searching

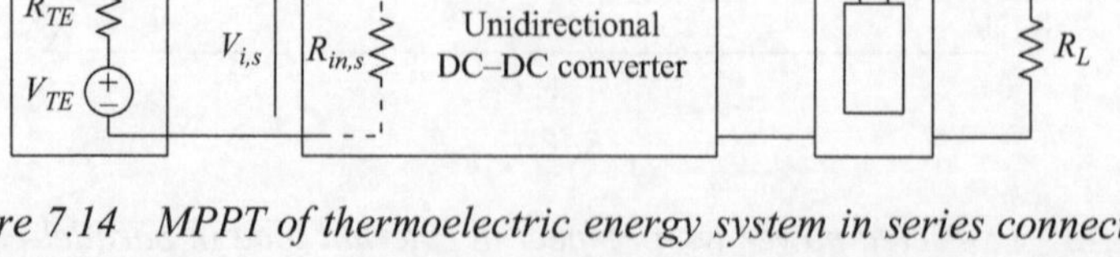


*Figure 7.14 MPPT of thermoelectric energy system in series connection*

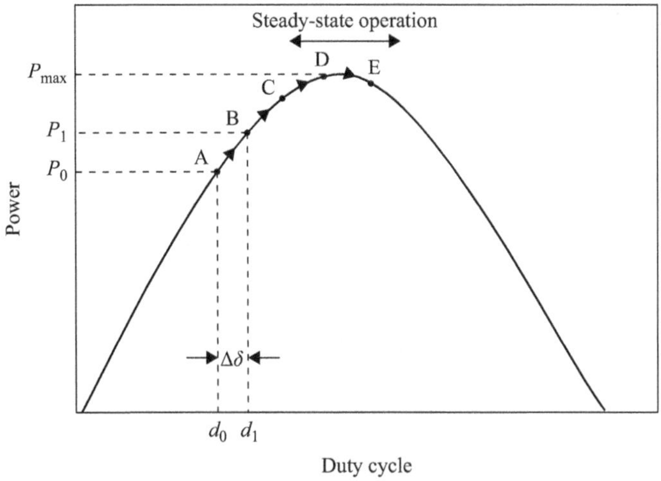


*Figure 7.15 PAO method in MPPT control*

direction is right. Consequently, small perturbations will be added to the duty cycle continuously from point $B$ to point $C$ and to point $D$, until the operating point exceeding the MPP to point $E$. Then since the output power at point $E$ is smaller than the power at point $D$, it reveals that the operating point is departing from MPP. Then, a small perturbation should be subtracted from the current duty cycle, which will change the searching direction back to point $D$ and to point $C$. Under such searching logic, the steady state of the PAO method is dynamically oscillating around the MPP between points $C$, $D$, and $E$. When $\Delta\delta$ is small enough, the MPPT can be achieved. As can be seen, rather than estimating the exact system MPP model, the controller only acquires the information of output voltage and current. Basically, the PAO technique can be described as:

$$\alpha = \mathrm{sgn}(P_{kT} - P_{(k-1)T}) \tag{7.2}$$

$$\Delta D_{kT} = \alpha \Delta D_{(k-1)T} \tag{7.3}$$

$$D_{(k+1)T} = D_{kT} + \Delta D_{kT} \tag{7.4}$$

where $D_{iT}$ and $P_{iT}$ are the duty cycle of switching signal and the output power of DC–DC converter at the sampling instant $iT$ ($i = 1, 2, \ldots, k, k+1, \ldots$), respectively, and $\Delta D_{iT}$ is initialized as 0.5%.

### *7.4.2 MPPT for thermoelectric energy system with parallel connection*

The PCC of thermoelectric energy system with parallel connection can track the MPP of the TEG by tuning its duty cycle of the PWM switching signal to enable the equivalent load resistance, $R_{in,p}//R_L$, equal to the TEG internal resistor $R_{TE}$, as

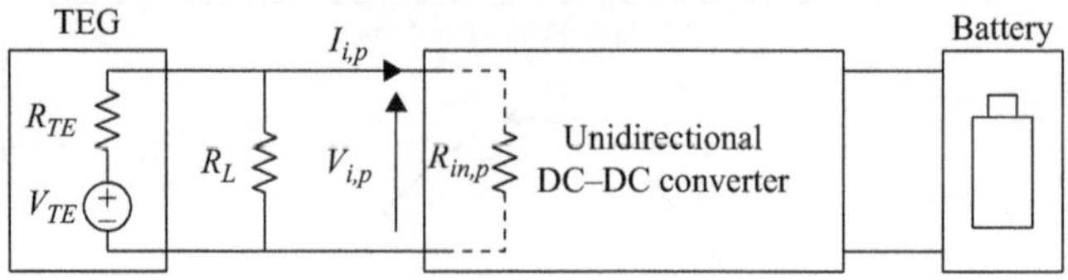


*Figure 7.16 MPPT of thermoelectric energy system with parallel connection*

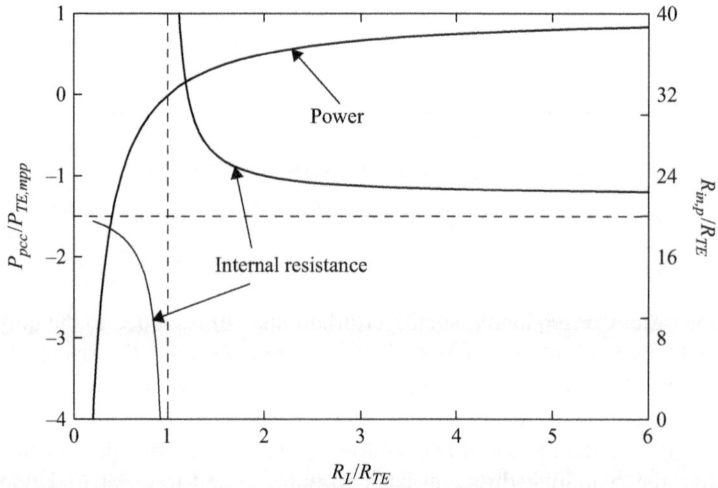


*Figure 7.17 Normalized power and input resistance of bidirectional DC–DC converter*

shown in Figure 7.16. Due to the bidirectional DC–DC converter in this system, its input resistance, $R_{in,p} = V_{i,p}/I_{i,p}$, may have a positive or negative value.

By defining the ratio between $R_L$ and $R_{TE}$ as $k = R_L/R_{TE}$, and the maximum power being extracted from TEG as $P_{TE,mpp} = V_{TE}^2/4R_{TE}$, the input resistance and the power handling by the bidirectional DC–DC converter at MPP can be derived according to the changing load:

$$R_{in,p} = \frac{k}{k-1} R_{TE} \tag{7.5}$$

$$P_{pcc} = \frac{V_{TE}^2}{4R_{TE}} \left(1 - \frac{1}{k}\right) \tag{7.6}$$

and the normalized converter power and input resistance are plotted as shown in Figure 7.17.

The MPPT flowchart for the thermoelectric energy system with parallel connection is depicted as shown in Figure 7.18, where $D_1$ and $D_2$ are the duty cycles of

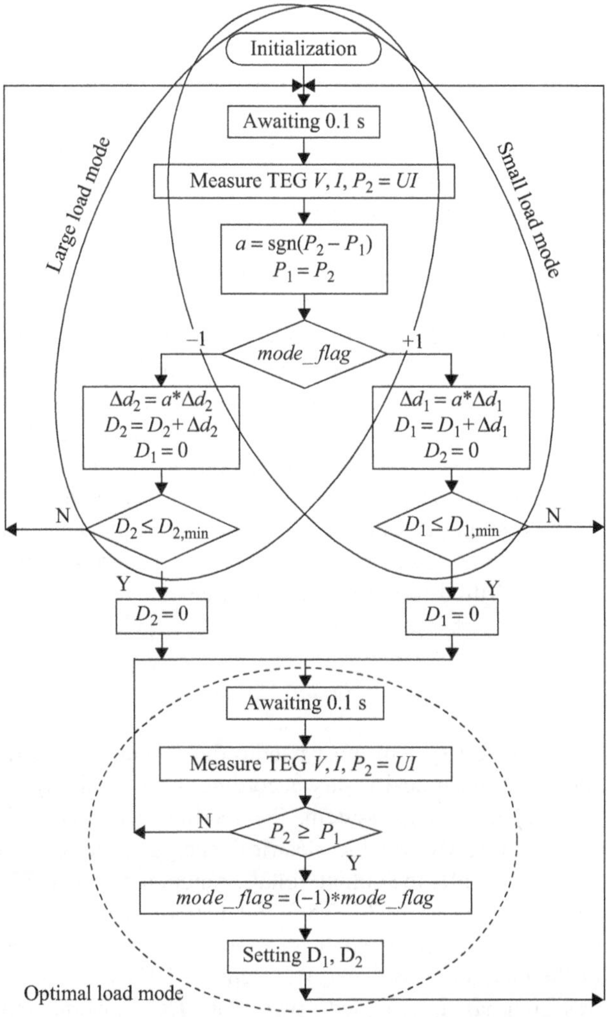


*Figure 7.18 MPPT flowchart of thermoelectric energy system with parallel connection*

switching signals for battery charging and discharging, respectively. Three ellipses denote three different control algorithms in the large load mode, small load mode, and optimal load mode.

In the large load mode, both the TEG and the battery provide power to the load. So the switching duty cycle for battery charging $D_1$ is set to be zero. The flag value for this mode is $mode\_flag = -1$. The switching duty cycle $D_2$ is determined by the PAO method according to (7.2)–(7.4). As the load resistance becoming larger and the external load becoming smaller, less power is required from the battery with smaller $D_2$. When $D_2$ is smaller than $D_{2,\min}$, it means that $R_L$ is larger than $k_1R_{TE}$ and the system will drift from the large load mode to the optimal load mode.

In the small load mode, the extra power is used to charge the battery to balance the output power of the TEG and required power of the load. So the switching duty cycle for battery discharging $D_2$ is set to be zero. The flag value for this mode is *mode_flag* = +1. The switching duty cycle $D_1$ is determined by the PAO method according to (7.2)–(7.4). As the load resistance becoming smaller and the external load becoming larger, less power is charged to the battery with smaller $D_1$. When $D_1$ is smaller than $D_{1,\min}$, it means that $R_L$ is smaller than $k_2R_{TE}$ and the system will drift from the small load mode to the optimal load mode.

In the optimal load mode, the maximum power is drawn from the TEG and transferred totally to the load. The converter is switched off with both $D_1$ and $D_2$ equal to zero. Taking the optimal load mode being drifted from the large load mode as example, the value of the TEG output power at last sampling instant in large load mode is stored as $P_1$. The system is still in the optimal load mode if the measured output power of the TEG is larger than the $P_1$. Otherwise, the mode of the system will drift out of the optimal load mode. Then, the flag of the small load mode and the duty cycle of $D_1$ will be reset again. The similar procedure will also occur if the optimal load mode is drifted from the small load mode.

## 7.5 PCC

The thermoelectric energy system must handle the output voltage from the TEG which may widely and dynamically vary according to external physical factors. In addition, as an auxiliary energy system, the thermoelectric energy system will charge the battery or directly provide electrical energy to the DC bus of vehicle power network. For example, the output voltage range of a typical TEG is 0–25 V, which has to be converted to 12.3–16.5 V for battery charging. So the DC–DC converter with step-up and step-down characteristics is required to serve for power conditioning of the thermoelectric energy system for HEVs.

Previous research has investigated several DC–DC converter topologies with step-up and step-down characteristics: Buck-Boost, Boost-Buck-cascaded, Buck-Boost-cascaded, Zeta, Flyback, Canonical switching cell, SEPIC, Ćuk, etc. It has pointed out that SEPIC and Ćuk topologies are the most favourable for the renewable energy application considering properties like input ripple, device stress, size of magnetic components, and input–output isolation. Compared with those of the SEPIC, both of the input current and output current of the Ćuk converter are nonpulsating. By properly winding the input and output inductors on the same core, the ripple amplitude can be reduced to zero. This merit can significantly minimize the disturbance on the TEG operating points, and potentially enhance the battery life (Yu and Chau, 2009).

A basic Ćuk converter is shown in Figure 7.19. When the switching frequency $f_s$ is high enough or the capacitance $C_1$ is large enough, the converter usually operates with a continuous capacitor voltage (CCV), namely the CCV mode. In this

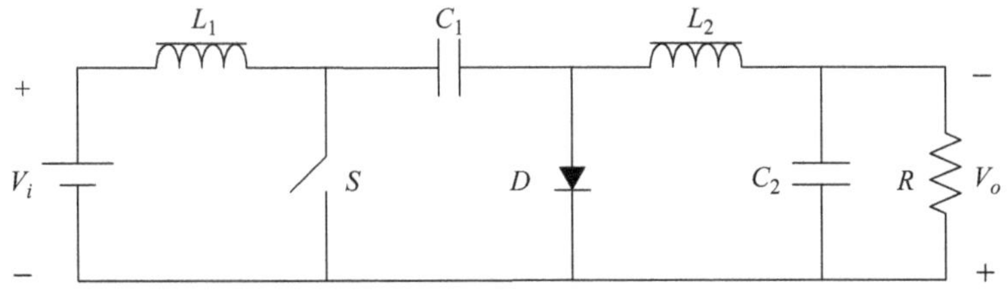


*Figure 7.19 Basic Ćuk converter*

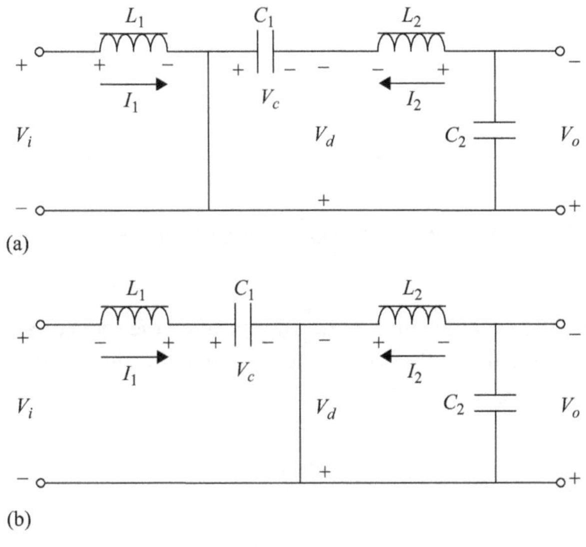


*Figure 7.20 Topological stages of Ćuk converter in CCV mode: (a) turn-on period and (b) turn-off period*

mode of operation, there are two topological stages as shown in Figure 7.20, in which the switch $S$ is turned on within the period of $\delta T_s$ and turned off within the period of $(1 - \delta)T_s$, where $\delta$ is the duty cycle and $T_s$ is the switching period. The corresponding theoretical waveforms are further shown in Figure 7.21.

From its principle of operation, the voltage conversion gain, namely, the ratio of output voltage $V_o$ to input voltage $V_i$, can be readily deduced as:

$$\frac{V_o}{V_i} = -\frac{\delta}{1 - \delta} \tag{7.7}$$

which indicates that $V_o$ can be larger or smaller than $V_i$, depending on the value of $\delta$, while the minus sign denotes that the terminals of $V_o$ and $V_i$ are of opposite

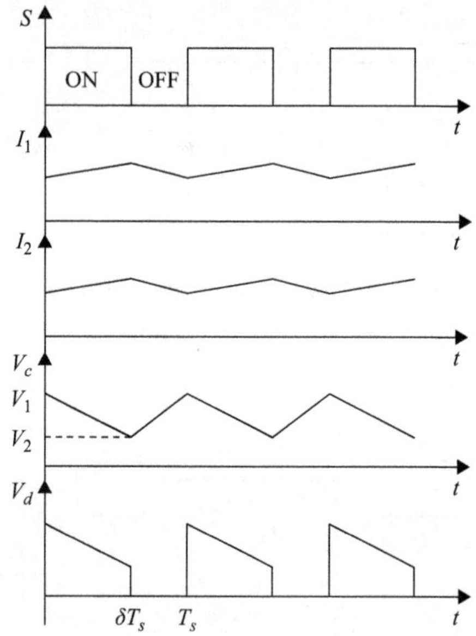


*Figure 7.21 Theoretical waveforms of Ćuk converter in CCV mode*

polarity. Taking the average currents flowing through the inductors $L_1$ and $L_2$ be $I_1$ and $I_2$, respectively, the capacitor voltage $V_c$ and the diode voltage $V_d$ can be expressed as:

$$V_c = \begin{cases} V_2 + \dfrac{I_2}{C_1}(-t + \delta T_s) & \text{for } 0 < t < \delta T_s \\ V_2 + \dfrac{I_1}{C_1}(t - \delta T_s) & \text{for } \delta T_s < t < T_s \end{cases} \tag{7.8}$$

$$V_d = \begin{cases} V_c & 0 < t < \delta T_s \\ 0 & \delta T_s < t < T_s \end{cases} \tag{7.9}$$

where $V_2$ is the minimum value of $V_c$. At the steady state, the average voltage across $L_1$ and $L_2$ are both zero. Hence, $V_i$ is equal to the average value of $(V_c - V_d)$, which can be deduced as:

$$V_i = \frac{1}{T_s}\int_0^{T_S} (V_c - V_d)\, dt = V_2(1-\delta) + \frac{I_1 T_s}{2C_1}(1-\delta)^2 \tag{7.10}$$

Consequently, the input resistance of the converter can be obtained as:

$$R_i = \frac{V_i}{I_1} = \frac{V_2(1-\delta)}{I_1} + \frac{(1-\delta)^2}{2f_s C_1} \tag{7.11}$$

from which, it can be found that $R_i$ changes with $\delta$, namely $R_i$ increases with the decrease of $\delta$ and vice versa when $\delta$ is within 0–1.

## 7.6 Experimental implementation

A simple TEG with power conditioning system is implemented in laboratory to emulate the proposed on-board waste heat energy recovery system for HEVs. The experimental system consists of a TEG, a Ćuk converter, an MPPT controller, and a battery pack. The TEG system is configured by connected six pieces of BiTe thermoelectric devices electrically in series and three branches in parallel. The MPPT controller is digitally implemented by a low-cost microcontroller which possesses an ADC module to capture the feedback signals and a direct PWM module to drive the power switches of the Ćuk converter. The temperature difference between the hot-side and cold-side is adjusted by an electric heater and the water velocity of a cooling system. In order to ensure the temperature uniformity, a copper plate with 5 cm thickness is inserted between the iron plate and the thermoelectric devices. The thermal insulation glue is also applied around the thermoelectric devices to improving the efficiency. The temperatures of both sides of thermoelectric devices are online recorded by a data logger.

It is known that the maximum power transfer occurs when internal resistance of the TEG equals the input resistance of the Ćuk converter. Meanwhile, (7.11) shows that the Ćuk converter input resistance is a function of duty cycle. Therefore, by tuning the duty cycle properly, the input resistance can match with the internal resistance, hence achieving maximum power transfer at different temperatures. This power conditioning system offers two important features.

First, both the battery voltage and battery current are online measured and used to calculate the output power. Thus, the variation of battery voltage during charging can be taken into account. It should be noted that this voltage variation may be over 20% for the commonly used battery. Second, rather than maximizing the TEG output power, the MPPT controller maximizes the converter output power which is the actual power that can be stored in the battery. It should be noted that the maximum battery power may not be coincident with the maximum TEG output power, since the converter power loss may not be constant. So, the use of load power for MPPT can virtually maximize the power output of the whole power conditioning system.

By connecting the Ćuk converter with the TEG in series configuration, the thermoelectric energy system with series connection is tested. For verification, by tuning the duty cycle of the Ćuk converter, the input power of the converter under different temperatures can be obtained as shown in Figure 7.22. As expected, the duty cycle of the Ćuk converter to achieve the MPP decreases with the increasing of temperature, which can be explained that the higher the temperature difference, the higher the internal resistance of the TEG, the smaller the duty cycle is required to produce a matching input resistance of the Ćuk converter.

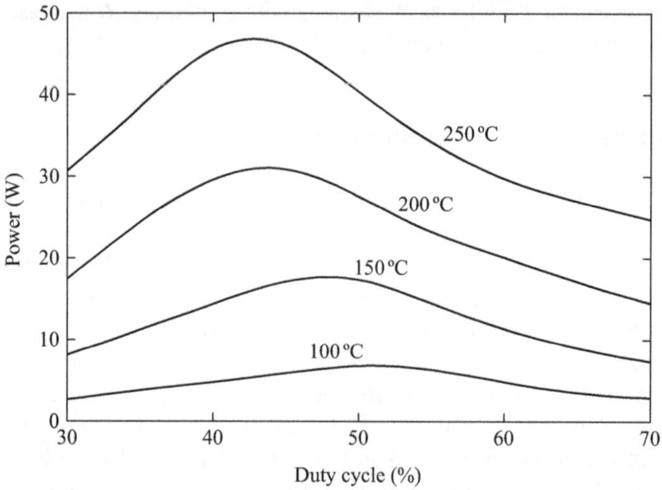


*Figure 7.22 Input power of Ćuk converter versus duty cycle*

By connecting the Ćuk converter with the TEG in parallel configuration, the thermoelectric energy system with parallel connection is tested under the maximum output power of the thermoelectric energy system with hot side temperature of 250 °C. During this experiment, the lower boundary of the handling power of the Ćuk converter is set to be 10% of maximum output power of the TEG. The minimum values of battery charging duty cycle and battery discharging duty cycle are set as $D_{1,min} = 0.3$ and $D_{2,min} = 0.35$, respectively.

The output power of the TEG is measured to show the successful MPPT implementation when the load power is changing, as illustrated in Figure 7.23. Meanwhile, the handling power of the Ćuk converter and the transferred power to or from the battery are also shown in Figure 7.24, in which the positive value and negative value represent the battery charging and discharging processes, respectively. The tested results reveal that the MPP can be tracked with various load powers.

The efficiency comparison is made among the thermoelectric energy system with series connection under MPPT control, the thermoelectric energy system with parallel connection in pure battery discharging mode, and the thermoelectric energy system with parallel connection under MPPT control, namely, the small load mode, the large load mode, and the optimal load mode, as shown in Figure 7.25. Obviously, the overall efficiency of the system with parallel connection is better than that of the system with series connection, which is because of first, only partial power being handled in the large load mode and small load mode; second, almost 100% efficiency being obtained in the optimal load mode. Moreover, the synchronous rectifier technology can be easily applied to the bidirectional converter to further improve the overall efficiency of the parallel-connected thermoelectric energy system.

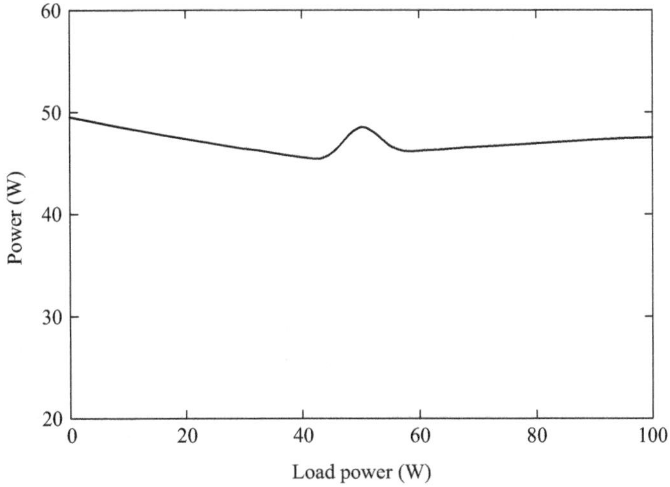


*Figure 7.23 Output power of the TEG with changing load power*

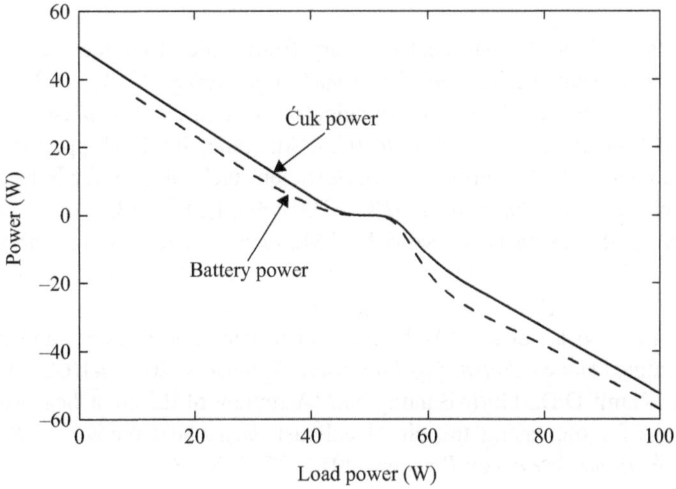


*Figure 7.24 Handling powers of the Ćuk converter and battery with changing load power*

It should be point out that the efficiency of the system in pure battery discharging mode is very low. One possible solution is that when the output power of the TEG is too low, the bidirectional Ćuk converter will be turned off and the battery will be switched and connected to the external load directly to avoid the system operating in the pure battery discharging mode.

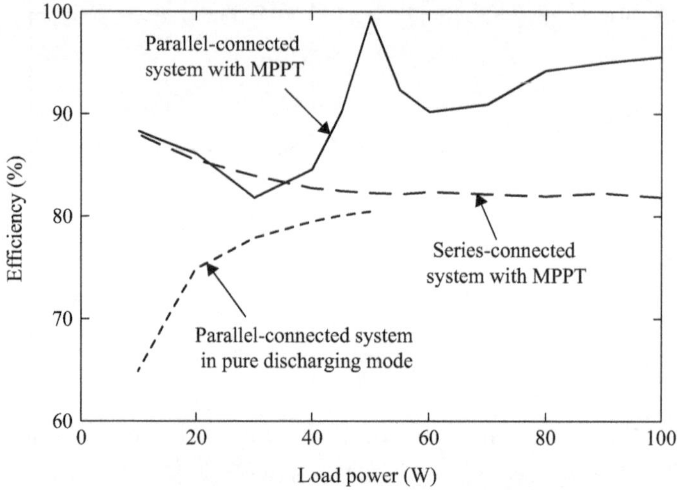


*Figure 7.25 Comparison of efficiencies under different conditions*

## References

Bari S., Hossain S.N., 'Waste heat recovery from a diesel engine using shell and tube heat exchanger'. *Applied Thermal Engineering.* 2013; 61: 355–363.

Bell L.E., 'High power density thermoelectric system', *Proceedings of International Conference on Thermoelectrics*, Australia, Jul 2004, pp. 1–5.

Chau K.T., Chan C.C. 'Emerging energy-efficient technologies for hybrid electric vehicles'. *Proceedings of the IEEE.* 2007; 95(4): 821–835.

Elsheikh M.H., Shnawah D.A., Sabri M.F.M., *et al.*, 'A review on thermoelectric renewable energy: Principle parameters that affect their performance'. *Renewable and Sustainable Energy Reviews*. 2014; 30: 337–355.

Gao J., Sun K., Ni L., *et al.*, 'A thermoelectric generation system and its power electronics stage'. *Journal of Electronic Materials.* 2012; 41(6): 1043–1060.

Hatami M., Ganji D.D., Gorji-Bandpy M., 'A review of different heat exchangers designs for increasing the diesel exhaust waste heat recovery'. *Renewable and Sustainable Energy Reviews.* 2014; 37: 168–181.

He W., Zhang G., Zhang X., Ji J., Li G., Zhao X., 'Recent development and application of thermoelectric generator and cooler'. *Applied Energy.* 2015; 143: 1–25.

LaGrandeur J., Crane D., Eder A., 'Vehicle fuel economy improvement through thermoelectric waste heat recovery', *Proceedings of Diesel Engine Emissions Reduction Conference*, Chicago, Illinois, USA, Aug 2005, pp. 1–7.

LaGrandeur J., Crane D., Hung S., Mazar B., Eder A. 'Automotive waste heat conversion to electric power using skutterudite, TAGS, PbTe and BiTe'. *Proceedings of International Conference on Thermoelectrics*, Vienna, Austria, Aug 2006, pp. 343–348.

Liu Y.H., Chen J.H., Huang J.W., 'A review of maximum power point tracking techniques for use in partially shaded conditions'. *Renewable and Sustainable Energy Reviews*. 2015; 41: 436–453.

Lyden S., Haque M.E., 'Maximum power point tracking techniques for photovoltaic systems: a comprehensive review and comparative analysis'. *Renewable and Sustainable Energy Reviews*. 2015; 52: 1504–1508.

Saidur R., Rezaei M., Muzammil W.K., Hassan M.H., Paria S., Hasanuzzaman M., 'Technologies to recover exhaust heat from internal combustion engines'. *Renewable and Sustainable Energy Reviews*. 2012; 16: 5649–5659.

Stabler F., 'Automotive applications of high efficiency thermoelectrics', *Proceedings of DARPA/ONR/DOE High Efficiency Thermoelectric Workshop*, San Diego, California, USA, Mar 2002, pp. 1–26.

Wang Y., Dai C., Wang S., 'Theoretical analysis of a thermoelectric generator using exhaust gas of vehicles as heat source'. *Applied Energy*. 2013; 112: 1171–1180.

Yang J., 'Potential applications of the thermoelectric waste heat recovery in the automotive industry', *Proceedings of International Conference on Thermoelectric*, Clemson, South Carolina, USA, Jun 2005, pp. 155–159.

Yu C., Chau, K.T., 'Thermoelectric automotive waste heat energy recovery using maximum power point tracking'. *Energy Conversion and Management*. 2009; 50: 1506–1512.

Zhang X., Chau K.T., 'An automotive thermoelectric–photovoltaic hybrid energy system using maximum power point tracking'. *Energy Conversion and Management*. 2011; 52: 641–647.

*Chapter 8*

# Review of battery charging strategies for electric vehicles

*Weixiang Shen*[1]

This chapter provides the comprehensive review of charging strategies for the major batteries currently used in electric vehicles (EVs) and plug-in hybrid EVs (PHEVs), including lead acid, nickel cadmium (NiCd), nickel-metal hydride (NiMH) and lithium-ion (Li-ion) batteries. It first reviews charging algorithms for a single battery and discusses future development of charging algorithms. Then, it presents battery balancing methods for pack charging. Finally, it introduces charging infrastructure.

## 8.1 Introduction

With more stringent regulations on emissions and constraints on energy resources, EVs and PHEVs have been considered as sustainable transportation to tackle climate change and energy crisis (Chan and Chau, 2001). The major energy sources for EVs and PHEVs have been identified to be batteries due to their technological maturity and reasonable cost. The batteries that are widely used in modern EVs and PHEVs include lead acid, NiCd, NiMH and Li-ion) (Chan, 2007; Shafiei and Williamson, 2010). Hence, the focus is given to the review on charging strategies for these four types of batteries in EVs, where PHEVs are lumped together with EVs.

Battery charging strategies mainly involve three areas: charging algorithms, battery balancing methods for pack charging and charging infrastructure. The charging algorithm is the core of a battery charger. It defines the charging profile for a battery charger in terms of current or voltage over time and a condition to terminate a charging process. The charging profile depends on battery charging characteristics. For example, the charging profile of constant current (CC) is normally used to charge NiCd/NiMH batteries, whereas the charging profile of constant voltage (CV) is normally used to charge lead acid and Li-ion batteries. In addition to these basic charging algorithms only with the CC or the CV, many advanced charging algorithms have been developed and adapted to the types of the batteries as well as the requirements of the applications, including constant current

[1]Swinburne University of Technology, Australia

constant voltage (CC/CV), multistep constant current (MSCC), two-step constant current constant voltage (TSCC/CV), constant voltage constant current constant voltage (CVCC/CV), pulse current (PC) and pulse voltage (PV) (Shen *et al.*, 2012). With regard to the termination of a charging process, battery state of charge (SOC) is an ideal indicator to stop charging a battery when the SOC reaches a preset value of 100%. However, the SOC is not a directly measured parameter. The accurate estimation of the SOC is a very challenging task. Even if the SOC can be estimated accurately, the backup methods are still required to stop a charging process using the directly measured parameters, such as voltage, current or temperature of a battery. Different techniques have been proposed to stop the charging process in terms of battery electrochemical characteristics (Young *et al.*, 2013). For example, the voltage drop ($\Delta V$) is normally incorporated with the CC to stop charging NiCd/NiMH batteries, whereas the cut-off current is normally incorporated with the CV to stop charging lead acid and Li-ion batteries. There are many other termination techniques, such as timer, temperature cut-off, temperature change rate ($dT/dt$), cut-off voltage for charge, cut-off voltage for discharge and voltage change rate ($dV/dt$). Section 8.2 discusses the charging algorithms for a single battery and their termination techniques in details.

The battery balancing techniques are required to charge a battery pack. In the battery pack, there is an imbalance between the voltage and capacity of the serially-connected cells. With the increasing number of charging and discharging cycles, such imbalance can deteriorate, causing the weaker cells and stronger cells in the string of the pack. The weaker cell fully discharges earlier or takes longer timer to be fully charged than the other cells while the stronger cell cannot be fully discharged or suffers from overcharging. Battery balancing methods have been developed to equalize the cells for pack charging. The battery balancing methods can be categorized into four groups: battery sorting, overcharge for balancing, passive balancing and active balancing. Battery sorting can mitigate cell imbalance in the serially connected cells during the initial stage of building a battery pack. During battery sorting, all cells are screened and identified into the group of the cells with the similar electrochemical characteristics and the cells in the same group are built into a battery pack. Overcharge for balancing can be used to equalize lead acid and NiCd/NiMH batteries. In this balancing method, the fully charged cells are overcharged while continuing to charge the cells which have not reached the fully charged states. The excessive energy in the fully charged cells is converted into heat accompanying with gas emission. Passive balancing has been developed to equalize lead acid, NiCd/NiMH and Li-ion batteries. It simply uses the shunt resistor to convert the unbalanced energy in the stronger cells into heat. Active balancing has also been developed to equalize lead acid, NiCd/NiMH and Li-ion batteries. It uses the electric circuits to transfer the unbalanced energy in the stronger cells into the weaker cells or the stronger cell to the pack or the pack to the weaker cell. With aid of the balancing techniques, the charging algorithms for a single cell can be extended to charge a battery pack. The details of the battery balancing methods for pack charging are discussed in Section 8.3.

The charging infrastructure deploys chargers which implement the charging algorithms to recharge a battery pack in EVs. It includes the nature of power transfer for chargers (conductive, inductive or wireless), the power type and level of chargers (AC or DC and Levels 1, 2 or 3) and the location of chargers (on-board or off-board) (Khaligh and Dusmez, 2012; Yilmaz and Krein, 2013). Based on the nature of power transfer, EV chargers can be realized by conductive or inductive or wireless. The conductive chargers transfer power with the hard-wired connection between the power supply and the chargers. They can be used to build stationary charging stations. The inductive chargers transfer power magnetically without the hard-wired connection. They can be used to build contactless charging stations. The wireless chargers transfer power based on magnetic resonant coupling. They can be used to build roadway charging stations. Based on the power demand, EV chargers can be classified into two categories: three AC power levels and three DC power levels. The three AC power levels are AC Level 1 (1.92 kW max), AC Level 2 (19.2 kW max) and AC Level 3 (>19.2 kW). The three DC power levels are DC Level 1 (35 kW max), DC Level 2 (90 kW) and DC Level 3 (240 kW). All these levels of chargers are proposed to meet different charging requirements of EVs, which can fully charge EV batteries from a few minutes to several hours. Depending on the location of the chargers, EV chargers can be classified as on-board or off-board chargers. The on-board chargers reside inside EVs, which are constrained by weight, space and cost. The off-board chargers reside outside EVs, which are less constrained by size and weight. Instead of charging EV battery packs, battery swapping technique, which mechanically swaps the discharged batteries with the fully charged batteries, can be also used to refuel EVs. The charging infrastructure is briefly introduced in Section 8.4. Conclusions are provided in Section 8.5.

## 8.2 Charging algorithms for a single battery

When a battery is run out of its energy source or its voltage reaches cut-off voltage or the battery SOC reaches 20% or lower, it must stop discharging and needs to be recharged. When the battery is charged, the charging algorithm must be adapted to the charging characteristics of different types of batteries. The typical charging characteristics of the major four EV batteries including lead acid, NiCd/NiMH and Li-ion are shown in Figure 8.1.

In addition to battery charging characteristics, battery charging process is affected by the amplitudes of charging current or voltage or combination of both as well as environmental temperature surrounding a battery. The key criterion to the charging algorithm is to recharge the batteries to its full capacity without causing extended overcharge or excessive temperature. In principle, every type of battery needs the specific charging algorithm adapting to its own charging characteristics. Practically, one charging algorithm may charge one or more battery types. Thus, the charging algorithms will be discussed together with the most suitable battery types which they conventionally charge. For the convenience, the basic terms for charging performance evaluation and characterization are first defined in the following section.

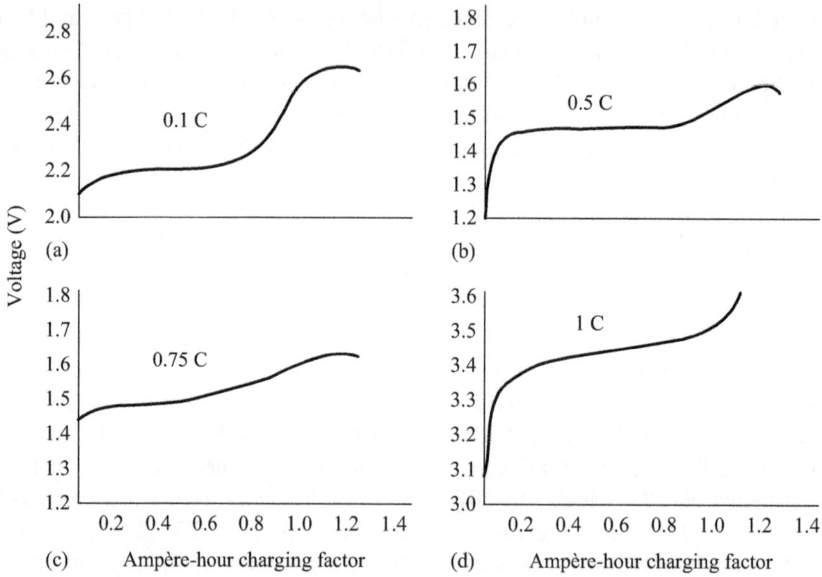


*Figure 8.1 Charging characteristics (cell voltage versus SOC) of different types of batteries (a) lead acid, (b) NiCd, (c) NiMH and (d) Li-ion*

### *8.2.1 Basic terms for charging performance evaluation and characterization*

Various terms are defined for battery charging performance evaluation and characterization (Berndt, 1997). These terms are summarized below.

#### 8.2.1.1 Cell and pack

A single cell is a complete battery with two current leads and separate compartment holding electrodes, separator and electrolyte. A pack of battery consists of many single cells connected in series and parallel with battery balancing and battery management system. More than one pack of battery may situate inside an EV.

#### 8.2.1.2 Nominal Ampère-hour capacity

Ampère-hour (Ah) is the total charge that can be discharged from a fully charged battery. The nominal capacity ($C_n$) is the rated Ah capacity of a fully charged battery under the conditions specified by a manufacturer.

#### 8.2.1.3 C-rate

C-rate is used to represent a charging or discharging rate. It equals the ratio of the nominal capacity to the duration of charge or discharge in hours. For a 2.3 Ah Li-ion battery, 1 C charging or discharging rate represents 2.3 A. Likely, 0.1 C and 10 C represent 0.23 A and 23 A, respectively. For a 95 Ah NiMH battery, 1 C charging or discharging rate represents 95 A. For a 10 Ah lead acid battery, 0.1 C charging or discharging rate represents 1 A.

#### 8.2.1.4 Cut-off voltage for discharge or charge

Cut-off voltage for discharge ($V^{dis}_{cut-off}$) is the minimum allowable voltage and can be interpreted as the 'empty' state of the battery at the discharging rate and temperature defined by a manufacturer. It is normally used to stop discharging process in the period of CC discharging. Cut-off voltage for charge ($V^{cha}_{cut-off}$) is the maximum allowable voltage defined by a manufacturer. It is a temperature-dependent parameter. It is normally used to stop charging process in the period of CC charging.

#### 8.2.1.5 Cut-off current

Cut-off current ($I^{cha}_{cut-off}$) is the predefined current which is normally very small (e.g. 0.01 C for Li-ion batteries) and used to stop the charging process during the period of CV charging.

#### 8.2.1.6 SOC

SOC is defined as the ratio of the remaining capacity ($C_r$) to the nominal capacity ($C_n$) of a battery,

$$SOC = \frac{C_r}{C_n} \tag{8.1}$$

or

$$SOC(t) = SOC(t_0) - \frac{1}{C_n}\int_{t_0}^{t} \eta i(\tau)\, d\tau \tag{8.2}$$

where $SOC(t_0)$ is the initial SOC. $\eta$ is charging or discharging efficiency. $i(t)$ is the current which takes negative if it is charging or positive if it is discharging. The SOC is a critical parameter for battery management in EVs and the key to the healthy and safe operation of batteries.

#### 8.2.1.7 State of health (SOH)

SOH is defined as the ratio of the capacity of the aged battery ($C_{aged}$) to the nominal capacity of the new battery. It is an important parameter to indicate the degree of performance degradation of a battery and to estimate battery remaining lifetime.

$$SOH = \frac{C_{aged}}{C_n} \tag{8.3}$$

#### 8.2.1.8 Cycle life

Cycle life is the number of charging and discharging cycles when the battery capacity fails to 80% of nominal capacity of the new battery. The actual cycle life is affected by the charging and discharging rates and other conditions such as temperature.

#### 8.2.1.9 Charge acceptance

Charge acceptance describes the share of the current that can actually be accepted by a battery and could be retrieved by subsequently discharging. It is determined by the balance between the current which converts electrical energy into chemical energy stored in the battery and the current which contributes to the internal losses of the battery. Charge acceptance is not a constant parameter, but depends largely on the SOC. It decreases with the increase of the SOC.

#### 8.2.1.10 Ampère-hour efficiency

Ampère-hour efficiency ($\eta_{Ah}$) is the ratio of the discharged capacity ($C_{dis}$) to the charged capacity of a battery ($C_{cha}$) under the specific charging and discharging conditions.

$$\eta_{Ah} = \frac{C_{dis}}{C_{cha}} \tag{8.4}$$

#### 8.2.1.11 Ampère-hour charging factor

Ampère-hour charging factor ($\alpha_{Ah}$) is the ratio of charged capacity ($C_{cha}$) to the discharged capacity ($C_{dis}$) under the specific charging and discharging conditions.

$$\alpha_{Ah} = \frac{C_{cha}}{C_{dis}} \tag{8.5}$$

If the battery is fully charged, the discharged capacity is the nominal capacity, namely $C_{dis} = C_n$, then

$$\alpha_{Ah} = \frac{C_{cha}}{C_n} \tag{8.6}$$

#### 8.2.1.12 Energy efficiency

Energy efficiency ($\eta_{en}$) is the ratio of the discharged energy ($E_{dis}$) to the charged energy of a battery ($E_{cha}$) under the specific charging and discharging conditions.

$$\eta_{en} = \frac{E_{dis}}{E_{cha}} \tag{8.7}$$

#### 8.2.1.13 Watt-hour charging factor

Watt-hour charging factor ($\alpha_{wh}$) is the ratio of the charged energy of a battery ($E_{cha}$) to the discharged energy ($E_{dis}$) under the specific charging and discharging conditions.

$$\alpha_{wh} = \frac{E_{cha}}{E_{dis}} \tag{8.8}$$

If the battery is fully charged, then

$$\alpha_{wh} = \frac{E_{cha}}{C_n \times V_n} \tag{8.9}$$

where $V_n$ is the normal voltage of the battery.

#### 8.2.1.14 Trickle charging (TC)

TC means continuous charge at low charging current ($I^{cha}_{trickle}$) approximately equivalent to the internal losses of a battery and suitable to maintain the battery in the fully charged state. The current required for TC amounts to 0.1 C or 0.01 C depending on battery types.

### *8.2.2 CC charging for NiCd/NiMH batteries*

The CC charging is a simple charging algorithm. It is commonly used to charge NiCd/NiMH batteries (Berndt, 1997). During the charging process, a regulated charging current (e.g. $I_{CC} = 0.5$ C for NiCd and 0.75 C for NiMH) is used to charge the batteries. The period of the CC charging is terminated at any time ($t_{end}$) if the battery voltage starts to drop. Then, the CC charging is switched to TC (e.g. $I^{cha}_{trickle} = 0.1$ C). Figure 8.2 shows the voltage profiles when the CC is applied to charge NiCd/NiMH batteries.

The charging time of the CC charging depends on the selection of the amplitude of the regulated charging current ($I_{CC}$). Generally, the higher current ($I_{CC}$) takes the shorter time to fully charge the battery with the lower efficiency, whereas the lower current ($I_{CC}$) takes the longer time to fully charge the battery with the higher efficiency. The regulated charging current ($I_{CC}$) is selected within the recommended limit to avoid any excessive rise of battery temperature. To implement a CC charger, a temperature sensor is required to monitor battery temperature for safety concerns and a voltage sensor is used to stop charging process. In addition, the temperature and timer are also used to stop charging process if the temperature exceeds the predetermined threshold or the timer with the programmed

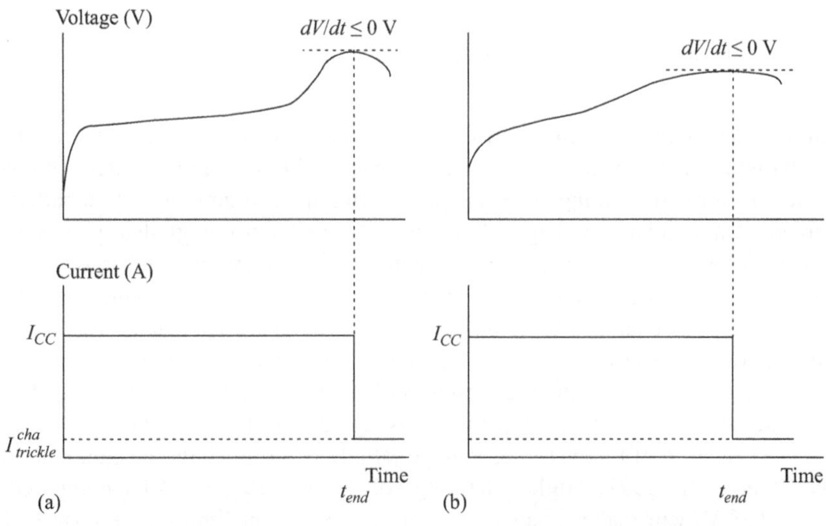


*Figure 8.2 Voltage and current profiles at CC charging for (a) NiCd and (b) NiMH batteries*

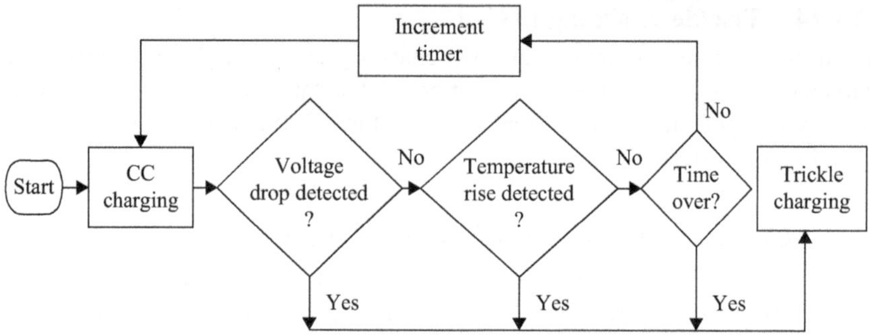


*Figure 8.3 Charging process of a CC charger for NiCd/NiMH batteries*

time duration expires. Figure 8.3 shows the charging process of a CC charger for NiCd/NiMH batteries (Nicolai and Wuidart, 1994).

There are a number of advantages and drawbacks for the CC charging for NiCd/NiMH batteries. The primary advantages are as follows:

- CC regulation is relatively inexpensive and easily implemented;
- Fast charging is possible at a high charging current with the ease of Ah computation (adding up currents over times);

The primary drawbacks are as follows:

- Single high CC charging rate with unregulated voltage may produce overcharge and result in short battery life;
- Charging current towards to the end of charge must be reduced adaptively to charge acceptance.

### 8.2.3 *CV charging for lead acid batteries*

The CV charging is another simple charging algorithm. It is the most common charging method for lead acid batteries (Berndt, 1997). A more descriptive term of the CV charging is actually current-limited CV charging because a real charger can only provide the limited charging current ($I_{\max}$). During the charging process, a battery is charged at a regulated voltage ($V_{CV}$), the charging current gradually decreases because of low charge acceptance of the battery with progressing recharge. The end of charge occurs at the time ($t_{end}$) when a CC is attained that remains unchanged (e.g. the current remains stable for about 3 h). The charging voltage requires the temperature compensation. The temperature coefficient is about −5 mV/°C. Figure 8.4 shows the voltage and current profiles when the CV is applied to charge lead acid batteries with different preset charging voltages (e.g. $V_{CV}^1$ and $V_{CV}^2$).

The selection of the charging voltage largely determines the desired charging time (Rand *et al.*, 2004). Higher charging voltage (e.g. $V_{CV}^2 = 2.40$ V compared to $V_{CV}^1 = 2.23$ V) can make a battery to stay in the current limit of the charger for longer time, leading to a significantly higher charging input during the early hours of charge and delivering a full charge in a shorter time. This can be realized using a

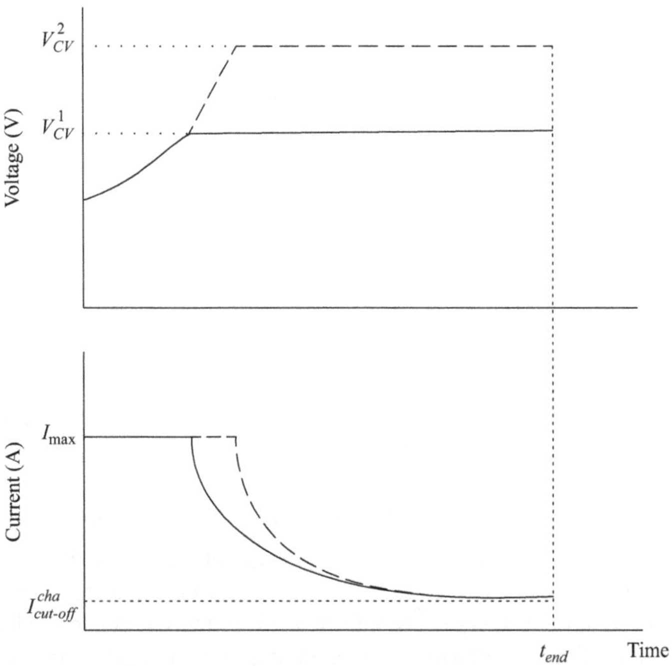


*Figure 8.4 Voltage and current profiles of CV charging at different voltage settings for lead acid batteries*

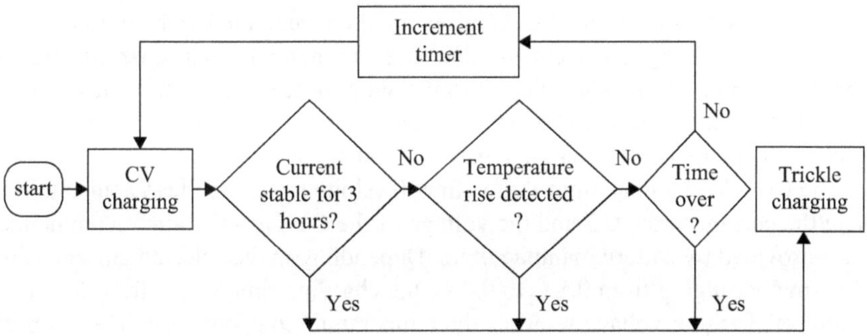


*Figure 8.5 Charging process of a CV charger for lead acid batteries*

higher current limit of the charger but care must be taken not to generate harmful amounts of heat due to ohmic ($I^2R$) effects. The CV charging algorithm has no charging current regulation. To implement a CV charger, the safety measures are taken to protect the battery. A temperature sensor is required to monitor temperature of the battery for safety concerns and a current sensor is required to terminate the charging process. Figure 8.5 shows the charging process of the CV charger for lead acid batteries.

The main advantages of the CV charging are as follows:

- CV regulation is relatively inexpensive and easy to be implemented;
- Overcharge is minimized with most of the current going into a battery converting into electrical energy;
- Charging voltage and current limit of a charger can be varied to support fast or slow charging.

On the other hand, there exist some disadvantages:

- There is a long duration charging 'tail', which creates long charging time in most cases;
- Ampère-hour counting must be measured by current integration due to changing current in charging process;
- Temperature compensation may be necessary with the preset high current limit of the charger if there is no environmental control.

### *8.2.4 CC/CV charging for lead acid and Li-ion batteries*

The CC/CV charging is a hybrid charging algorithm of the CC and the CV charging algorithms, with some good aspects of both. It starts with a high charging rate in the CC mode at the initial stage and ends at the appropriately regulated voltage in the CV mode which limits overcharge. It is most popular charging algorithm for lead acid and Li-ion batteries, and has become a benchmark with which the performance of the newly developed battery charging algorithms for lead acid and Li-ion batteries are compared. In the CC/CV charging algorithm, a regulated charging current ($I_{CCCV}$) is first applied to charge the batteries until the battery voltage rises to a preset voltage ($V_{CCCV}$). Then, the charging voltage is regulated to the preset voltage ($V_{CCCV}$) and the charging current will decrease exponentially. The charging process stops at the time ($t_{end}$) when the charging current reaches a predefined cut-off current ($I^{cha}_{cut-off}$). Figure 8.6 shows the voltage and current profiles when the CC/CV charging is applied to charge a lead acid battery.

The CC/CV charging algorithm is first developed to charge lead acid batteries, where the current of the CC and the voltage of the CV are set at the recommended values provided by battery manufacturers. Depending on the selected current of the CC normally ranging from 0.1 C to 0.4 C, the charging time varies from 6 to 12 h. The preset charging voltage requires the temperature compensation. The temperature coefficient is about −5 mV/°C (Berndt, 1997).

The CC/CV charging is then extended to charge Li-ion batteries with some modifications of voltage and current settings due to their different charging characteristics. Li-ion batteries generally have higher charge acceptance and higher nominal voltage than lead acid batteries. As a result, the charging current of the CC for Li-ion batteries is much higher than that for lead acid batteries, which can be chosen from 0.5 C to 3.2 C for the charging time in the range of 0.5–2.5 h (Kim *et al.*, 2008).

The selection of the voltage of the CV depends on the types of Li-ion batteries. For the four most commercially used Li-ion batteries, the typical voltage settings are 4.2 V for lithium cobalt oxide, lithium manganese oxide (LMO) and lithium

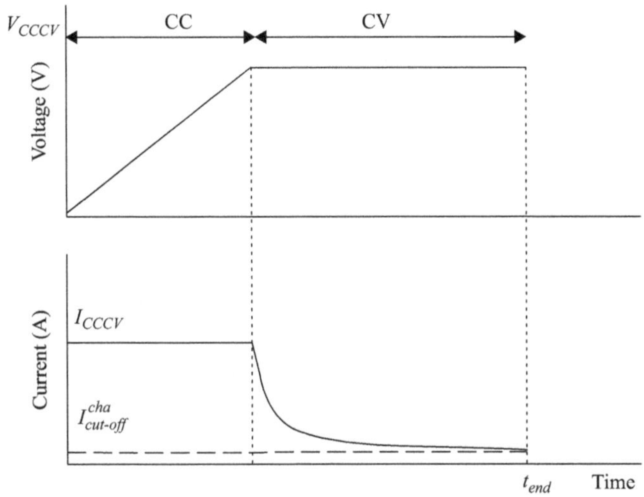


*Figure 8.6 Voltage and current profiles of CC/CV charging for lead acid batteries*

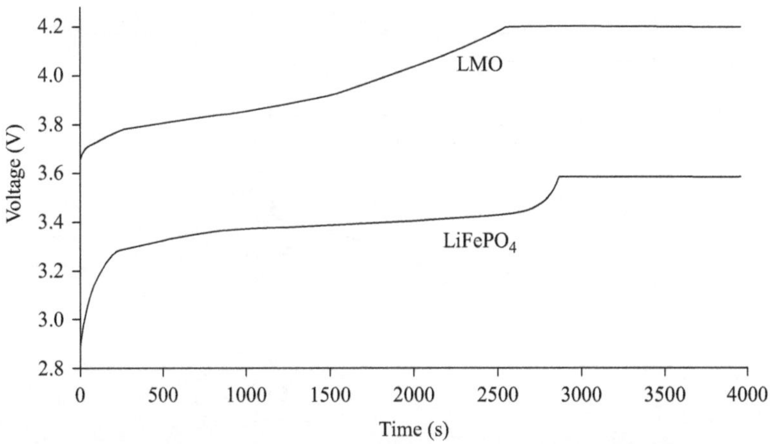


*Figure 8.7 Voltage profiles for CC/CV charging algorithm with the CC of 1 C at voltage of 3.6 V for a $LiFePO_4$ battery and at voltage of 4.2 V for a LMO battery*

nickel manganese cobalt oxide and 3.6 V for lithium iron phosphate ($LiFePO_4$) (Chen *et al.*, 2012). Figure 8.7 shows the CC/CV charging voltage profiles at the CC of 1 C and the CV of 4.2 V for a LMO battery and the CV of 3.6 V for a $LiFePO_4$ battery, respectively.

To implement a CC/CV charger, the precaution measures are taken to protect the batteries. Figure 8.8 shows the charging process when the CC/CV is used to charge lead acid and Li-ion batteries (Panasonic Inc., 2003, 2005). It shows that the

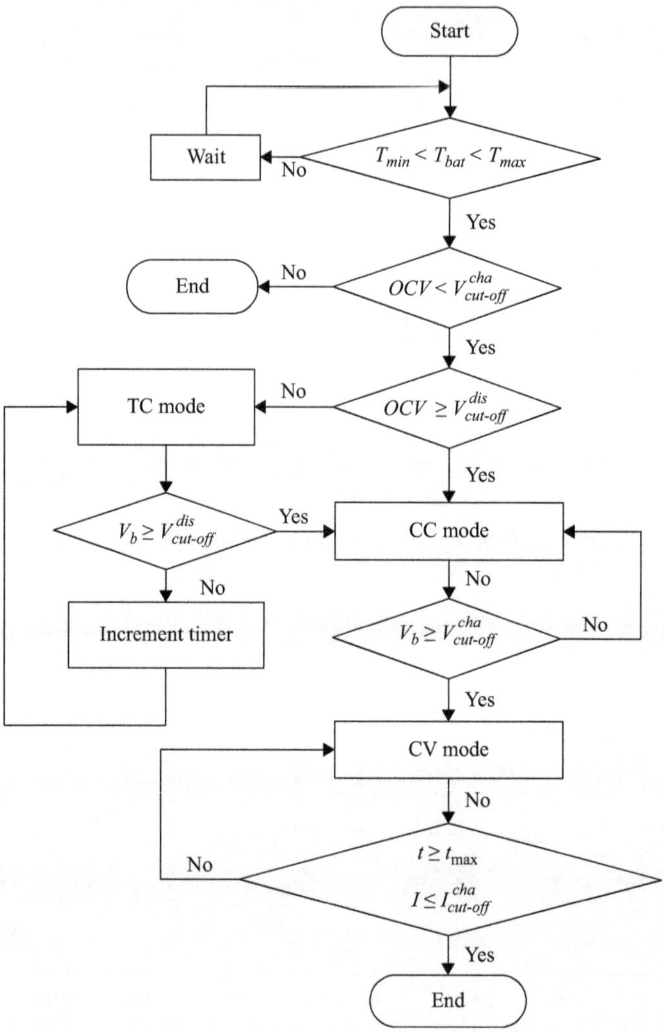


*Figure 8.8 Charging process of a CC/CV charger for lead acid and Li-ion batteries*

charging process of the CC/CV consists of three steps. First, the battery initial conditions, such as temperature and open circuit voltage (OCV), are checked if they are in the normal range. If the OCV is less than the cut-off voltage for discharge ($V_{cut-off}^{dis}$) (e.g. 1.8 V for lead acid and 2.5 V for LMO), the battery is charged by TC with small current (e.g. 0.1 C for Li-ion and 0.01 C for lead acid) until the battery voltage ($V_b$) rises to $V_{cut-off}^{dis}$.

Second, once the battery voltage exceeds $V_{cut-off}^{dis}$, the CC mode starts to charge the battery at the regulated charging current ($I_{CCCV}$). Third, when the battery voltage reaches to the cut-off voltage for charge (e.g. $V_{cut-off}^{cha} = 2.40$ V for lead acid and 4.2 V for LMO), the charging process switches to the CV mode, the battery is charged at the regulated voltage ($V_{CCCV}$) and the charging current is reduced exponentially. The charging period is terminated by either cut-off current ($I^{cha}_{cut-off}$) which is typically 0.01 C for Li-ion batteries and 0.001 C for lead acid batteries or the maximum charging time ($t_{\max}$). Depending on the charging current in the CC mode, the total charging time is varied. In general, the lower the charging current of the CC mode is, the higher the charging efficiency and longer the charging time and the battery life are. Three sensors are required to measure and monitor battery voltage, current and temperature, ensuring that they are all within the maximum allowable values. The main advantage of the CC/CV charging algorithm is that no model information is required to charge the battery. Furthermore, the CC control and the CV control can be easily realized with very simple integrated circuits (ICs) or a microcontroller at low cost. The main drawback is that there is a long 'tail' in the CV charging period which prolongs the charging time.

Recently, many variants of the CC/CV charging algorithm have been developed to charge Li-ion batteries. One of these variants uses a simple double-loop controller to achieve the charging profile similar to the conventional CC/CV charging profile (DL-CC/CV), as shown in Figure 8.9 (Tsang and Chan, 2011). The DL-CC/CV is implemented by using the negative and positive feedbacks of the battery voltage ($V_B(s)$) to the battery reference voltage ($V_{Br}(s)$) and the output of the PID controller. The large error ($e_B(s) = V_{Br}(s) - V_B(s)$) creates a similar charging process to the CC charging while the minimization of $e_B(s)$ creates a charging process equivalent to the CV charging. Since only the battery terminal voltage is required to measure, the need for a current sensor is eliminated. As a result, the performance of the CC/CV charging can be achieved with the requirement of only voltage and temperature sensors at low cost.

The other two variants use the advanced control schemes to realize the charging profile similar to that of the conventional CC/CV charging algorithm. The advanced control schemes are the fuzzy logic control (Hsieh *et al.*, 2001) and the grey-predicted control (Chen *et al.*, 2008). For the former, a fuzzy-controlled active SOC controller is used to realize CC/CV (FL-CC/CV). For the latter, a grey-predicted technique is used to realize CC/CV (GP-CC/CV). The essence of these two charging algorithms is to replace the regulated terminal voltage in the charging process with the OCV (e.g. $V_{OC} = 4.2$ V) so that the appropriate charging current in the CV mode is dynamically determined based on the SOC and the charging current in the CV charging period under the conventional CC/CV. As a result, the charging current of the CV mode is larger at higher current part and smaller at the

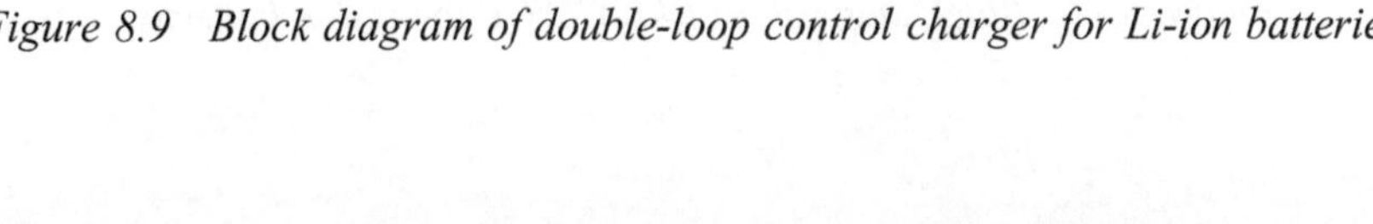


*Figure 8.9 Block diagram of double-loop control charger for Li-ion batteries*

lower current part than the charging current in the CV mode of the conventional CC/CV, eliminating the long 'tail' of the CV charging as shown in Figure 8.10. Thus, more capacity is able to be charged into the battery in the period of the CV mode and the shorter time is required to complete the CV mode. The FL-CC/CV and the GP-CC/CV have a shorter charging time and a higher charging efficiency. Due to their complexity and the requirement of high computation power, a microcontroller is more suitable for the realization of both charging algorithms.

Another variant of the CC/CV charging algorithm is implemented by applying the principle of the phase-locked loop (PLL) control to the charging process (Chen, 2004). The PLL process naturally coincides with the requirement of the charging process of the CC/CV. Figure 8.11 shows the block diagram of PLL-based CC/CV charging algorithm (PLL-CC/CV). The working principle of the PLL-CC/CV can be explained in the three steps. First, the battery voltage ($V_b$) is measured and sent to a difference amplifier. The output of the amplifier ($V_o$) is conditioned to adapt the input voltage scale of the voltage control oscillator (VCO). Second, the VCO generates a feedback phase ($P_o$) that represents the battery voltage and sends to a phase

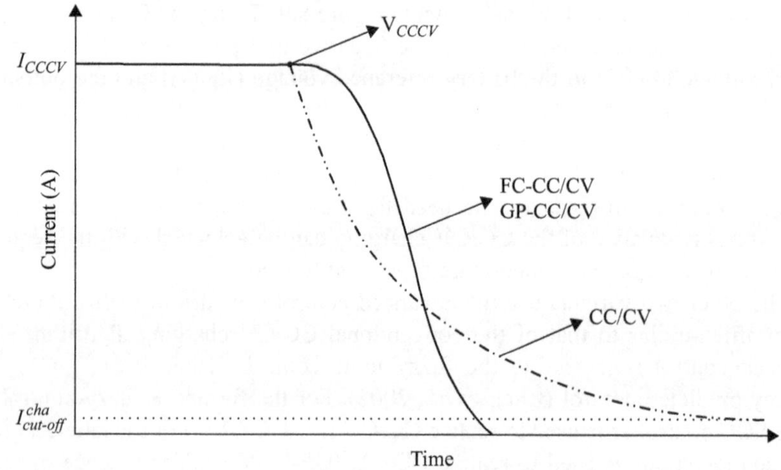


*Figure 8.10 Charging current profiles of FL-CC/CV, GP-CC/CV and CC/CV for Li-ion batteries*

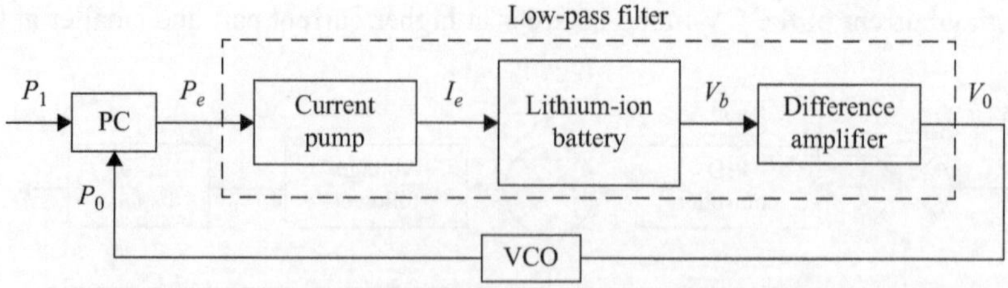


*Figure 8.11 Block diagram of PLL-CC/CV for Li-ion batteries*

comparator (PC). The PC then compares the input reference phase ($P_i$) to produce the phase error ($P_e$). Finally, this phase error is sent to the current pump to produce a suitable current to charge Li-ion batteries.

In this PLL-CC/CV arrangement, the auto-tracking process, i.e. the frequency-tracking process, is corresponding to the regulated current charging or bulk charging which is similar to the CC mode of the CC/CV. The auto-locking process from the phase-tracking to the phase-locked state is corresponding to the variable current charging and float charging which is similar to the regulated voltage charging in the CV mode of the CC/CV. Figure 8.12 shows the flowchart for the charging process of the PLL-CC/CV.

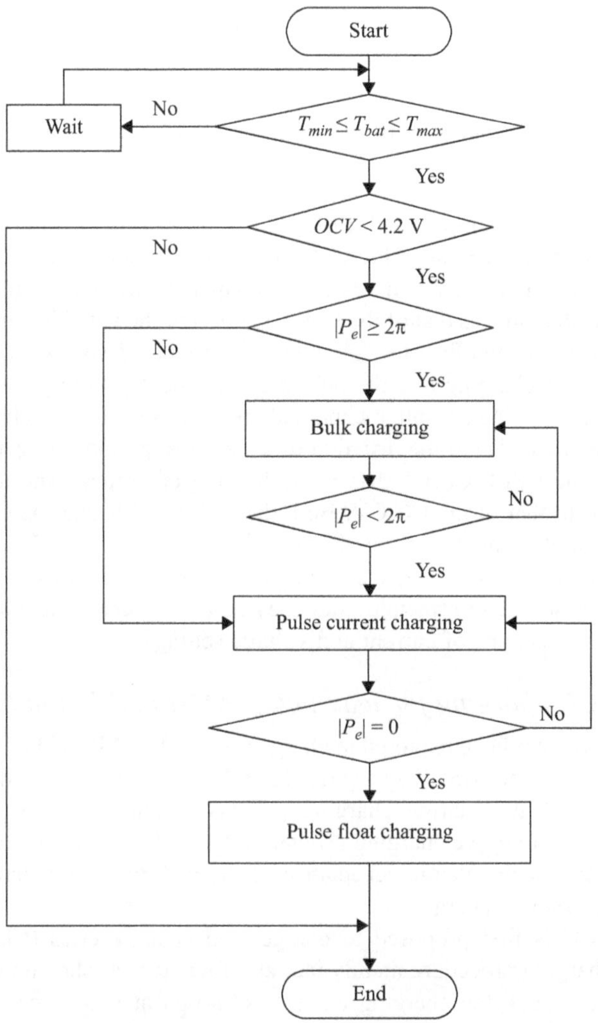


*Figure 8.12 Flowchart for charging process of PLL-CC/CV for Li-ion batteries*

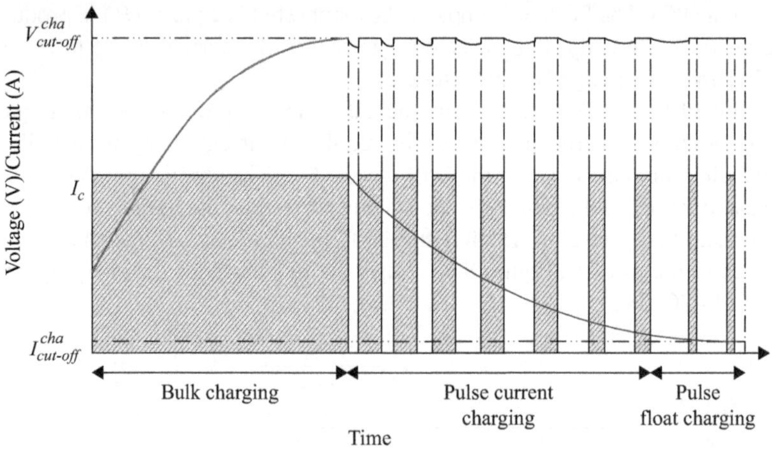


*Figure 8.13 Charging profile of CPPL-CC/CV for Li-ion batteries*

Later, a current-pump based PLL-CC/CV (CPLL-CC/CV) is proposed to improve the charging process of the PLL-CC/CV (Chen *et al.*, 2008). The charging process of the CPLL-CC/CV consists of the regulated current charging (CC mode) which remains the same as that of the PLL-CC/CV, but the variable current charging and float charging of the PLL-CC/CV under the regulated voltage are modified to the pulsed current charging and the pulsed current float charging, respectively, as shown in Figure 8.13. As a result, the internal pressure of the battery charged by a PC is smaller than a CC due to the insertion of a short rest period. Comparing with the PLL-CC/CV, the CPLL-CC/CV has high charging efficiency. The total charging time is similar to that of the CC/CV. Both the PLL-CC/CV and the CPLL-CC/CV can be easily implemented by using the specific IC with the PLL function.

All the variants of the CC/CV charging algorithms are only explored to charge Li-ion batteries so far. In principle, they may be able to charge lead acid batteries with some modifications of current and voltage settings.

### *8.2.5 MSCC charging for lead acid, NiMH and Li-ion batteries*

MSCC algorithm has been explored to charge lead acid, NiMH and Li-ion batteries. The main idea of the MSCC charging algorithm is to use the reduced current amplitude in multistep during charging process to minimize the possibility of overcharge and to improve charging efficiency. The adoption of the reduced CC is due to the fact that the charge acceptance of the batteries is gradually decreased with the progressing recharge.

The MSCC is first proposed to charge lead acid batteries (Chan and Chu, 1990). The charging procedure mainly includes four steps as shown in Figure 8.14. In the first step (S1), the charging current is chosen at C/6 before reaching the gassing voltage of 2.45 V. Then, the charging current is cut-off so that the cell voltage is allowed to decay until it reaches a predefined value of 2.2 V. In the

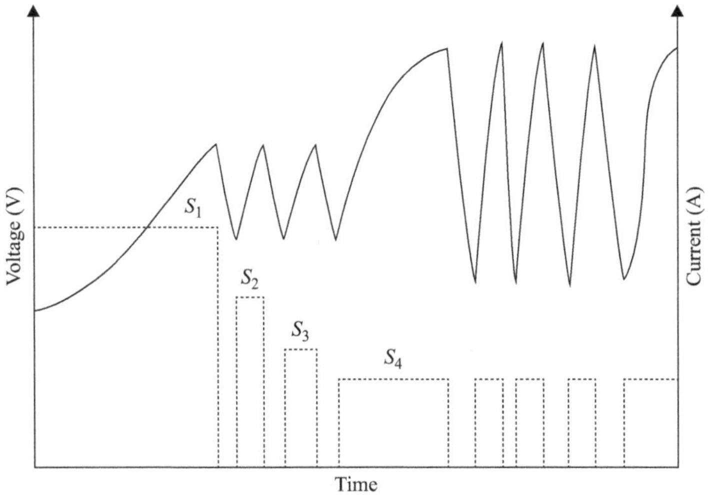


*Figure 8.14 Charging profile of MSCC for lead acid batteries*

second step (S2) and the third step (S3), the charging currents are taken as 75% and 50% of that in S1, respectively, to charge the battery up to the gassing voltage and then decay down to the predefined voltage. In the fourth step (S4), the charging current is chosen at 25% of that in S1 to continue charging the battery over the gassing voltage until no rise in the cell voltage is detected within 15 min which leads to the termination of S4.

Later, the experiments of the MSCC with two, three, four and six steps are investigated for lead acid batteries (Ikeya *et al.*, 1998). It is found that the MSCC with four and six steps can further shorten the charging time with higher energy efficiency and longer cycle life than the MSCC with two steps.

The MSCC is then explored to charge NiMH batteries (Ikeya *et al.*, 2002). The experiments of MSCC with one, two and three steps are investigated to seek the optimal charging algorithms for NiMH batteries with the improvement of charging efficiency, charging time and cycle life. It shows that the MSCC with three steps results in the shortest charging time less than 2 h with reasonable charging efficiency and cycle life. Figure 8.15 shows the charging process with the respective charging currents in three steps of 1.0 C, 0.5 C and 0.14 C.

The first two steps are controlled by the preset cut-off voltage for charge ($V^{cha}_{cut-off}$) and the last step is regulated by the changing rate of the battery temperature, if $dT/dt$ is equal to a predefined value ($\alpha_T$) (e.g. $\alpha_T = 0.13$ °C/min), the charging process will be terminated.

Recently, the MSCC has been explored to charge Li-ion batteries with five steps, as illustrated in Figure 8.16. The amplitude of charging current in each step must be appropriately determined to charge the battery. Four approaches are proposed to optimize charging currents in the MSCC charging algorithms. The first approach uses a fuzzy logic controller to determine a charging current

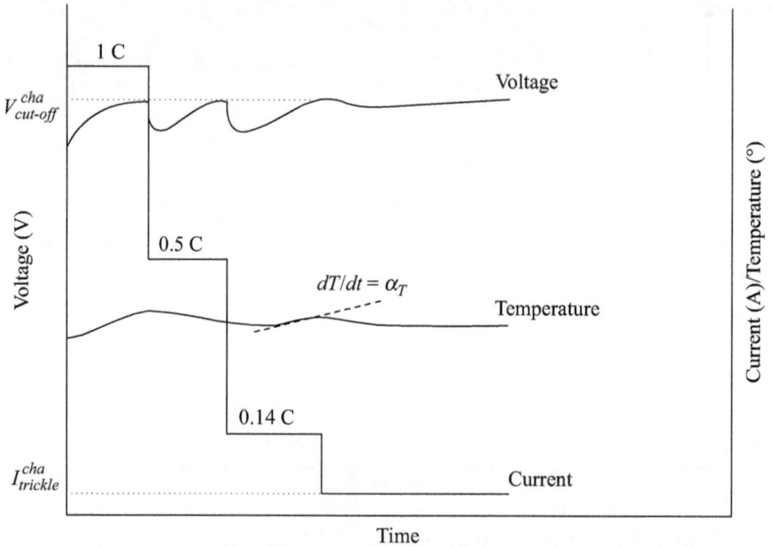


*Figure 8.15 Charging profile of MSCC with three steps for NiMH batteries*

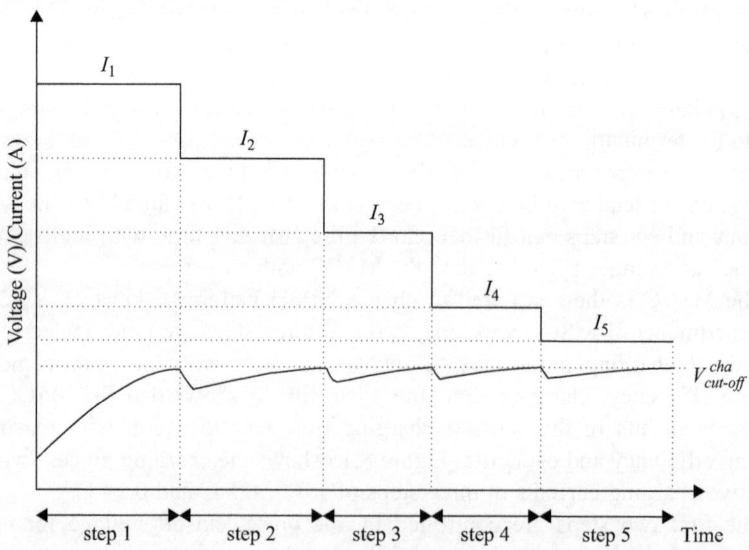


*Figure 8.16 Charging profile of MSCC with five steps for Li-ion batteries*

(Huang *et al.*, 2009), where the inputs of the fuzzy controller are the temperature and the change of the temperature, and the output of the controller is the charging current. The second approach uses the consecutive orthogonal array (or Taguchi method) to search an optimal charging current profile (Liu and Luo, 2010; Liu *et al.*, 2011). The third approach applies an ant colony system to optimize a charging

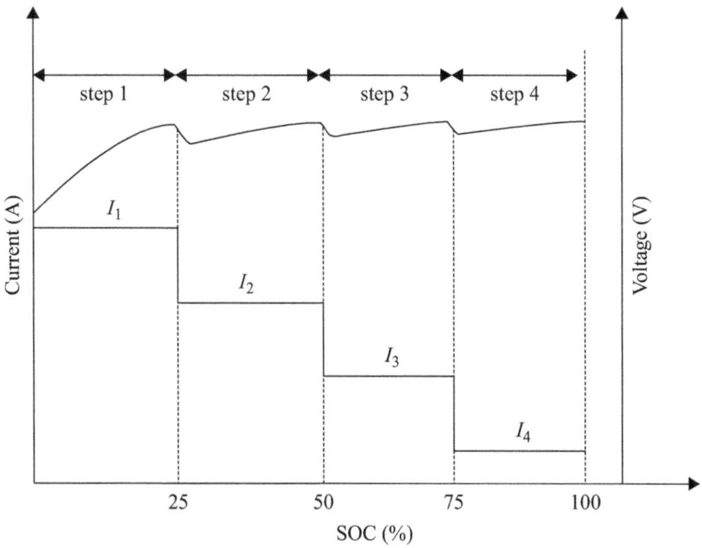


*Figure 8.17 Charging profile of MSCC with four steps for Li-ion batteries*

current profile (Liu *et al.*, 2005). The fourth approach applies an integer linear programming to search an optimal charging current profile (Dung and Yen, 2010).

The MSCC with four steps has also been explored to charge a Li-ion battery recently (Vo *et al.*, 2015). In this approach, Taguchi method is employed to search an optimal charging current pattern and an adaptive switching gain sliding mode observer is employed to estimate the SOC which controls and terminates the charging process. The total charging period is equally divided into four steps, where each step has the range of the SOC equal to 25%. In each step, the Li-ion battery is charged by a pre-set current. During charging process, the SOC is estimated in real time. When the estimated SOC reaches the predetermined SOCs, such as 25%, 50% and 75%, the charging process will be shifted to next step and a new pre-set charging current will be applied to charge the battery accordingly. The charging process will continue until the estimated SOC reaches 100%, namely, the battery at this state is considered as fully charged and the charging process terminates. Figure 8.17 shows the charging current profile for MSCC with four steps. This charging algorithm has been used to charge a Li-ion battery at different SOCs of 0, 17%, 37%, 67% and 84%, respectively, and their experimental results are shown in Figure 8.18.

It shows that this charging algorithm can charge Li-ion batteries with any SOCs, where the estimated SOC is used to switch the steps and stop charging process. To be safe, the temperature and cut-off voltage for charge can also be used as the backup methods to stop the charging process together with the estimated SOC.

As a summary, MSCC charging algorithms can be developed based on the flowchart as shown in Figure 8.19. In this flowchart, the blocks highlighted by the shaded area may vary from one approach to another as various methods can be

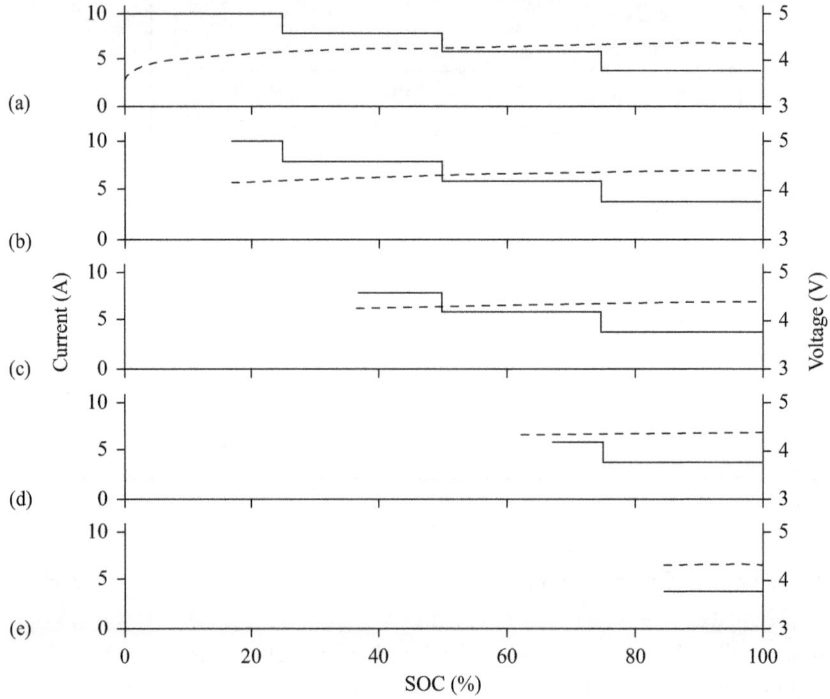


*Figure 8.18 Experimental results of charging profiles of MSCC with four steps at different SOCs of a Li-ion battery: (a) SOC = 0, (b) SOC = 17%, (c) SOC = 37%, (d) SOC = 67% and (e) SOC = 84%*

chosen to optimize charging profiles in terms of the 'best value' which can be the highest charging efficiency or the shortest charging time or compromise between both, the rest of the blocks remain the same. Generally, the MSCC algorithm has higher charging speed and higher charging efficiency than the CC/CV charging algorithm. A microcontroller or a computer is required to implement this charging algorithm due to its complexity.

### 8.2.6 TSCC/CV charging for Li-ion batteries

TSCC/CV charging has been developed to charge a high power $LiFePO_4$ battery within approximately 20 min (Anseán *et al.*, 2013). The TSCC/CV charging algorithm combines the high CC, low CC/CV into one charging profile. It is developed on the basis of the evolution of internal resistance during the charging process which is divided into three steps, referred to as CC-I, CC-II and CV-I. In the step of CC-I, the battery is charged at a regulated CC ($I^1_{CC-I}$) (e.g. $I^1_{CC-I} = 4C$) up to the cut-off voltage for charge (e.g. $V^{cha}_{cut-off} = 3.6$ V). The charging current of 4 C is selected because it corresponds to initial low internal resistance and high charge acceptance as well as the maximum charging current recommended by the manufacturer.

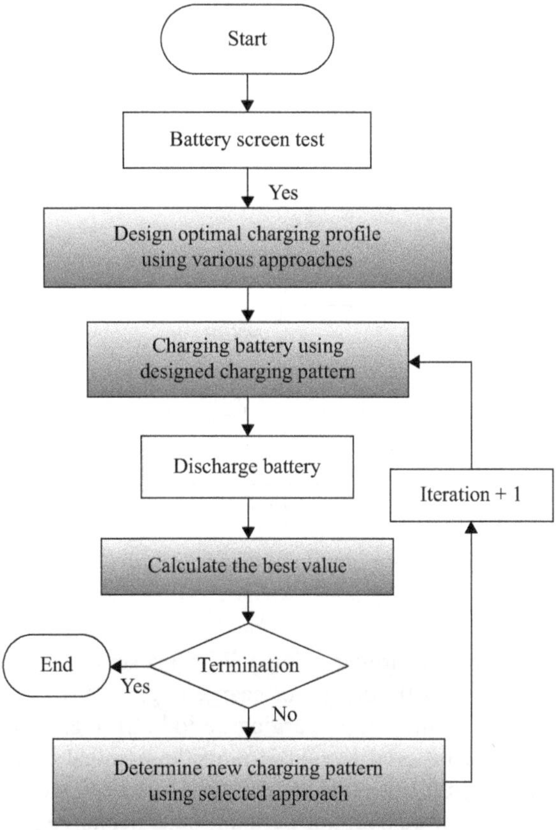


*Figure 8.19 Flowchart to implement MSCC with various approaches for charging profile optimization*

In the step of CC-II, the battery is charged at another regulated CC ($I^2_{CC-II}$) (e.g. $I^2_{CC-II} = 1$ C), since the current in CC-II is much lower than that in CC-I, the battery voltage drops below 3.6 V so that the charging process can continue until the battery voltage reaches $V^{cha}_{cut-off}$ again. In the last step of CV-I, the battery is charged at a regulated CV of 3.6 V only for a short duration of 5 min because it avoids an acceleration of the aging process that occurs at high SOC values over longer periods of time. The extreme high charging current is used to charge the battery within the safe voltage initially, significantly shortening charging time. Figure 8.20 shows the charging profiles of the TSCC/CV charging algorithm.

### *8.2.7 CVCC/CV charging for Li-ion batteries*

CVCC/CV has been designed to charge a Li-ion battery. It is also called the boost charging due to initial high charging voltage (Notten *et al.*, 2005). The CVCC/CV is required to fully discharge a battery before charging. It consists of three charging steps, including the regulated high CV, regulated CC and regulated low CV. For the

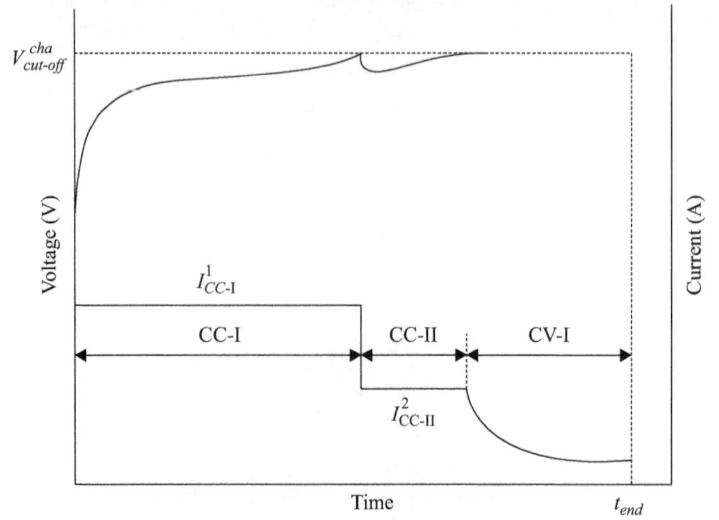


*Figure 8.20 Charging profile of TSCC/CV for Li-ion batteries*

CVCC/CV, the initial high charging voltage ($V_b^{max}$) is set to 4.3 V which is 0.1 V higher than the normal cut-off voltage for charge (e.g. $V^{cha}_{cut-off} = 4.2$ V). The battery is charged at $V_b^{max}$ in the boost charging period ($t_b$) (e.g. $t_b = 5$ min) and the charged capacity can be restored around 30% of the nominal capacity of the battery. It shows that a significant amount of charge has been stored in the battery within a relatively short period $t_b$. If this period is extended to 10 min, about 60% of the nominal capacity of the battery can be charged into the battery. After this period, the charging algorithm is then switched to the conventional CC/CV. Figure 8.21 shows the charging profile of the CVCC/CV. Due to the initial high charging voltage, the CVCC/CV can charge the battery faster than the standard CC/CV. However, the discharging circuits are required to fully discharge the battery before charging. This will increase the number of components and cost. Discharging a battery before charging also makes this charging algorithm inefficient.

### 8.2.8 Pulse charging for lead acid, NiCd/NiMH and Li-ion batteries

The pulse charging has been considered as the promising candidate for fast charging algorithm for lead acid, NiCd/NiMH and Li-ion batteries. The effects of pulse charge on Li-ion batteries are evaluated using an electronic network model (Jongh and Notten, 2002). Simulation results provide some insight into the effect of the pulse on internal process, such as diffusion, migration, electrochemical reactions and heat generation. Experimental approach is also used to investigate the effect of pulse charge on the cycle life of a Li-ion battery (Li *et al.*, 2001). Many pulse charging algorithms have been developed. They can be classified into two groups, namely, constant voltage pulse charging (CV-PC) and constant current pulse charging (CC-PC).

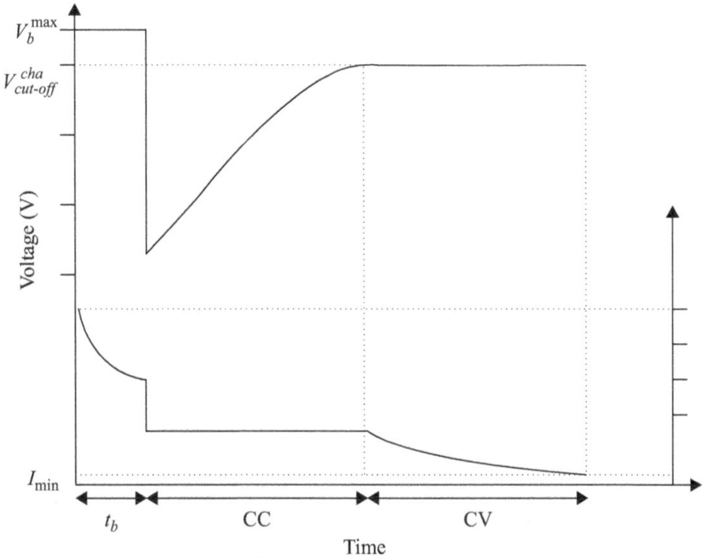


*Figure 8.21 Charging profile of CVCC/CV for Li-ion batteries*

The CV-PC is realized by keeping the constant amplitude of PV in the entire charging process while changing frequency of the pulse (FCV-PC) (Chen, 2007) or duty cycle of the pulse (DCV-PC) (Chen, 2009). The basic idea of the FCV-PC is to adjust the frequency of the pulse within a certain range and observe the response of the charging current. The optimal frequency ($f_{opt}$) of the pulse charging is obtained when the battery receives the highest charging current ($I_{opt}$) at the minimum impedance. Figure 8.22 shows the flowchart of the FCV-PC for a Li-ion battery. The DCV-PC is very similar to the FCV-PC. The only difference is that the DCV-PC changes the duty cycle of the pulse, instead of the frequency of the pulse, to achieve the highest charging current at the lowest impedance. The prototypes of implementing these two charging algorithms are made, demonstrating that the charging time is shorter than that of the conventional CC/CV (Chen, 2007, 2009) with higher charging efficiency and longer cycle life.

On the other hand, the CC-PC is realized by keeping the constant amplitude of PC in the entire charging process while monitoring the battery voltage to ensure that it is always lower than the cut-off voltage for charge. The charging profile can be varied by changing the amplitude and width of the PC and the relaxation period between the pulses (Purushothama *et al.*, 2005; Purushothama and Landau, 2006). Figure 8.23 shows the charging profile of the CC-PC. Simulation study has been conducted to optimize the selection of the charging time ($t_c$) and the rest time ($t_r$) for the given amplitude of the PC. The charging time is optimized when the maximum concentration is reached and the relaxation period is determined such that it provides sufficient time to a reset concentration. As a result, electrochemical reaction inside the battery neither produces heat nor causes the accumulation

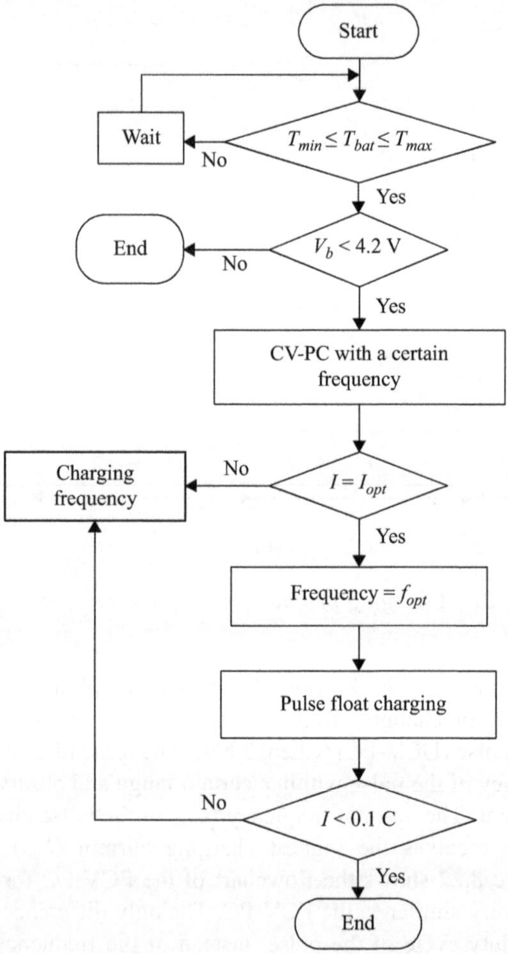


*Figure 8.22 Flowchart of FCV-PC for a Li-ion battery*

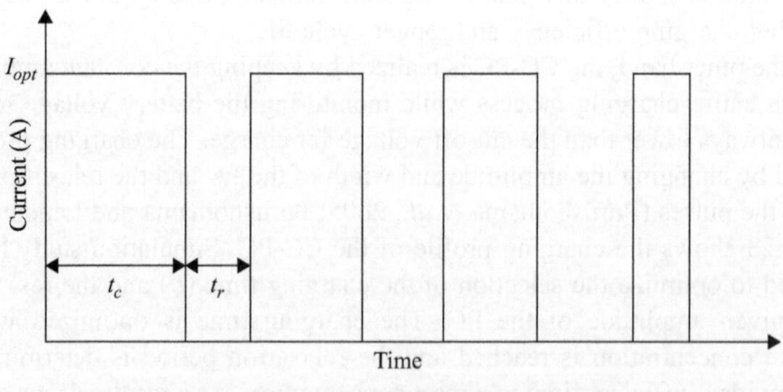


*Figure 8.23 Charging profile of CC-PC*

pressure. Since this charging algorithm is designed based on the link between the pulse charging current profile and the chemical reaction process inside the battery, in principle, it can charge the battery faster and more efficient. However, it is hard to be applied in a real charger since the concentration change inside a battery is not measureable.

### 8.2.9 Charging termination techniques

Charging termination is a critical part of a charging algorithm. Ideally, the battery SOC is the best indictor to safely stop charging process when the SOC reaches a preset value of 100%. However, the SOC cannot be measured directly but has to be estimated. Due to the complexity of the SOC estimation, the estimated SOC may not be accurate and the backup methods are always required to stop charging process by using other directly measured parameters such as voltage, current and temperature of a battery. The followings are some typical techniques to terminate charging process (Berndt, 1997; Hussein and Batarseh, 2011; Young *et al.*, 2013).

- Timer. It can be used for any types of batteries. When a preset timer expires, the charging process is stopped. This method is very simple but can be advantageous if the charger fails to detect a full charge state by other means. Normally, it is used as a backup together with other termination methods.
- Temperature cut-off. It can be used for any types of batteries. The charging process will be stopped if the absolute temperature of a battery rises to a certain threshold value. This method is very simple and normally is used as a backup or safety measures with the support of other termination methods.
- Temperature rise. When the temperature rise during the charging process exceeds the preset safe value, the charging process will be terminated. This termination method is suitable for NiCd/NiMH batteries, particularly for NiMH batteries, because when they are fully charged some gases are produced to result in sudden rise in temperature.
- Temperature change rate ($dT/dt$). If the temperature change rate is over the safety threshold value, the charging process will be terminated. This termination method is suitable for NiMH batteries.
- Cut-off current for charge. When the charging current reaches the preset small current, the charging process stops. This method is normally incorporated with a CV charging algorithm and normally used for lead acid and Li-ion batteries.
- Cut-off voltage for charge. When the battery voltage reaches a threshold value, the charging process will be terminated. This method is normally incorporated with a CC charging method to stop charging lead acid and Li-ion batteries.
- Voltage change rate ($dV/dt$). The charging process stops if the battery voltage does not change versus time, or even if it starts to drop (a negative value of $dV/dt$). This termination method is usually used to stop charging NiCd/NiMH batteries.
- Voltage drop ($-\Delta V$). The charging process will be terminated if a preset value of the voltage drop is reached. Upon completion of the charging process, the voltage drop occurs in NiCd/NiMH batteries because the internal resistance of the batteries drops when the batteries are fully charged, particularly for NiCd batteries.

### 8.2.10 Comparisons of charging algorithms and new development

The charging algorithms have been reviewed together with most widely used batteries in EVs, including lead acid, NiCd/NiMH and Li-ion batteries. The CC/CV is the conventional approach to charge lead acid and Li-ion batteries while the CC is the conventional approach to charge NiCd/NiMH batteries. The MSCC is the advanced approach to charge lead acid, NiCd/NiMH and Li-ion batteries. These charging algorithms vary in charging time, charging efficiency, implementation complexity and cost, impact on cycle life and sensors required. They range from the simplest charging algorithms, such as the CC, the CV and the CC/CV, to the most creatively complicated charging algorithms, such as MSCCs with ant colony or consecutive orthogonal array or Taguchi method. Table 8.1 summarizes the major aspects of these charging algorithms together with their most suitable battery types which these charging algorithms conventionally charge (Shen *et al.*, 2012). It should be noted that all the charging algorithms discussed in sections 8.2.2–8.2.8 have constant amplitude of voltage or current or combination of both during the charging process which simplifies their implementation in a real charger.

There is new devolvement of the charging algorithms on the concept of variable amplitude of voltage or current or combination of both during the charging process. Under this new concept, the charging voltage or current is not predefined. It will be dynamically determined adaptively to the battery states during the entire charging process. One of these charging algorithms is developed to minimize the charging time for a Li-ion battery, where different constraints on internal battery states are considered such as the over-potential in the negative electrode and the final SOC (Klein *et al.*, 2011). The simulation results based on electrochemical model suggest a decrease in charging time by 50% compared to the CC/CV. Another one of these charging algorithms is developed to compromise the charging time and the charging loss (Hu *et al.*, 2013). The battery equivalent circuit model is used to analyse the influences of the maximum charging voltage, battery temperature and battery aging on charging results for both lithium nickel–manganese–cobalt oxide and lithium iron phosphate batteries. The simulation results show that within the same charging time, the newly developed charging algorithm has the advantage of energy-saving in terms of the battery aging over the CC/CV. The third one is developed to charge a Li-ion battery with the goal to minimize the losses of a battery in the charging process (Chen *et al.*, 2015). In this algorithm, the battery is charged in the defined range from the initial state (e.g. SOC = 0) to the final state (e.g. SOC = 80%) for the given charging time. With the given status of the start and end of the charge, many possible paths in terms of charging profiles can be chosen to charge the battery. The objective is to find the optimal charging profile to minimize the energy loss in the charging process. The experimental results demonstrate that the charging algorithm can reduce the charging loss in the charging process compared to the CC charging algorithm. However, the former two approaches are still in the stage of theoretical investigation with the battery model while the later one is hardly to be implemented into a real charger due to their complexity of the formation of the charging profiles.

*Table 8.1 Major characteristics of charging algorithms and their most conventional batteries to charge*

| Charging algorithms | Analogue or digital | Charging time | Charging** efficiency | Implementation complexity, cost | Cycle life | Sensors required | Battery types |
|---|---|---|---|---|---|---|---|
| CC | Analogue | Short | Low | Low | Short | $V$, $T$ | NiCd/NiMH |
| CV | Analogue | Long | Medium | Low | Short | $I$, $T$ | Lead acid, Li-ion |
| CC/CV | Analogue or digital | Long | Medium | Medium | Short | $V$, $I$, $T$ | Lead acid, Li-ion |
| DL-CC/CV* | Analogue | Long | Medium | Low | Short | $V$, $T$ | Li-ion |
| FL-CC/CV* | Digital | Medium | Medium | High | Medium | $V$, $I$, $T$ | Li-ion |
| GP-CC/CV* | Digital | Medium | Medium | High | Medium | $V$, $I$, $T$ | Li-ion |
| PLL-CC/CV* | Analogue | Long | Medium | Medium | Medium | $V$, $I$, $T$ | Li-ion |
| CPLL-CC/CV* | Analogue | Long | Medium | Medium | Medium | $V$, $I$, $T$ | Li-ion |
| MSCC* | Digital | Medium | Medium | High | Long | $V$, $I$, $T$ | Lead acid, Li-ion, NiCd/NiMH |
| TSCC/CV | Digital | Short | Medium | Medium | Long | $V$, $I$, $T$ | High power Li iron phosphate |
| CVCC/CV | Digital | Short | Low | Medium | Short | $V$, $I$, $T$ | Li-ion |
| FCV-PC* | Digital | Short | High | High | Long | $V$, $I$, $T$ | Li-ion |
| DCV-PC* | Digital | Short | High | High | Long | $V$, $I$, $T$ | Li-ion |
| CC-PC | Analogues or digital | Short | High | High | Medium | $V$, $I$, $T$ | Lead acid, Li-ion, NiCd/NiMH |

*Notes*: $V$: voltage sensor, $I$: current sensor, $T$: temperature sensor.

*These charging algorithms are only explored to charge Li-ion batteries so far, they may be able to extend to charge lead acid batteries.

** Charging efficiency refers to overall charging efficiency including Ampère-hour charging efficiency and energy charging efficiency.

## 8.3 Balancing methods for battery pack charging

A battery pack supplies power to EVs. The EV battery pack generally consists of hundreds or thousands of cells connected in series and parallel to provide sufficient operating voltage, power and energy for EVs. Refuelling of EVs is not just to charge a single cell but a battery pack. Due to manufacturing tolerances, temperature differences and aging effects, the cells in a battery pack may differ in their capacities, internal resistances, self-discharge rates and degradation rates. Charging a battery pack consisting of the serially connected cells may experience an unbalance charging, where the stronger cells in the pack can be overcharged or the weaker cells in the pack can be over-discharged. After many cycles of charging and discharging the battery pack, it may lead to the capacity losses and the reduced cycle life of the cells in the pack and the pack itself. Battery balancing methods have been developed to solve this unbalance charging problem among the cells in the pack. Battery balancing methods can be divided into four main categories: battery sorting, overcharge for balancing, passive balancing and active balancing.

Battery sorting can alleviate cell imbalance in a battery pack during the initial stage of building the pack. The purpose of battery sorting is to carry out the screening tests for the cells to make up a battery pack. The experimental data of the cells are obtained to cluster cells into the group with the similar electrochemical characteristics in terms of the characteristic parameters of cell or the profiles of voltage or the profiles of temperature during charging and discharging processes, etc. Then, the cells in the same group will be used to build the battery pack. The sorted battery pack which has consistent characteristics of the cells can be charged as if it is a single battery. However, such consistence may not be able to maintain along their lifetime due to different operational environments, aging speeds and self-discharge rates of each single cell. Battery sorting can only be useful if it is complemented by one of the other three balancing methods, namely overcharge for balancing, passive balancing and active balancing (Park *et al.*, 2012).

Overcharge for balancing can be used to equalize a low voltage battery pack for lead acid and NiCd/NiMH batteries (Kutkut *et al.*, 1998) as Li-ion batteries cannot be overcharged. In this method, the fully charged cells are overcharged while continuing to charge the cells which have not reached the fully charged states. The excessive energy in the fully charged cells is converted into heat accompanying gas emissions. The passive balancing has been developed to equalize lead acid, NiCd/NiMH and Li-ion batteries. It uses shunt resistor across each cell to dissipate the excessive energy. Since the energy in the passive balancing will be converted into heat in the shunt resistor, it creates an inefficient charging process and adds the extra burden on battery thermal management. To minimize the energy losses and create an efficient charging process, the active balancing has been developed to equalize lead acid, NiCd/NiMH and Li-ion batteries. Instead of wasting the excessive energy in the stronger cells, the active balancing methods transfer the unbalanced energy in the stronger cells to the weaker cells or the stronger cells to the pack or the pack to the weaker cells (Rahimi-Eichi *et al.*, 2013).

### 8.3.1 Battery sorting

Many methods have been developed for battery sorting (Li *et al.*, 2014). Among them, two methods have been adopted to select the cells with similar electrochemical characteristics to build a battery pack for EVs.

One is to determine the similarity of the cells by observing the directly measured parameters of cell together with the identified parameters from a battery equivalent circuit model (Kim *et al.*, 2012). In this method, two screening processes in sequence have been developed, namely the capacity screening process and the resistance screening process. The capacity screening process is to match the capacities of the cells. The battery capacities of 20 Li-ion cells with their nominal capacities of $C_n$ (e.g. $C_n = 1.3$ Ah) have been tested. The CC/CV charging algorithm is recommended to charge the cells and the five CCs at the discharging rates of 3 C, 2 C, 1 C, 0.5 C and 0.2 C are used to discharge the cells. Then, the average values of capacities under five discharging rates are calculated. The cells with the average capacity fall into the predefined range from $C_a^u$ to $C_a^l$ are selected (e.g. $C_a^l = 1.2815$ Ah and $C_a^u = 1.2864$ Ah), namely the cells with no. 5, no. 6, no. 13, no. 16 and no. 19, as shown in Figure 8.24. For these five cells, the OCVs are obtained at the SOCs of 30%, 50% and 70% which are calculated by Ah counting. The results are listed in Table 8.2. It shows that the OCVs are almost the same at these specific SOCs of the five cells. The relationship between the OCVs and the SOCs is checked because such relationship is crucial to the accurate estimation of the SOC for the battery pack in EVs.

The resistance screening process is to match DC internal resistance in the battery equivalent circuit model as shown in Fig 8.25. It consists of an OCV, an

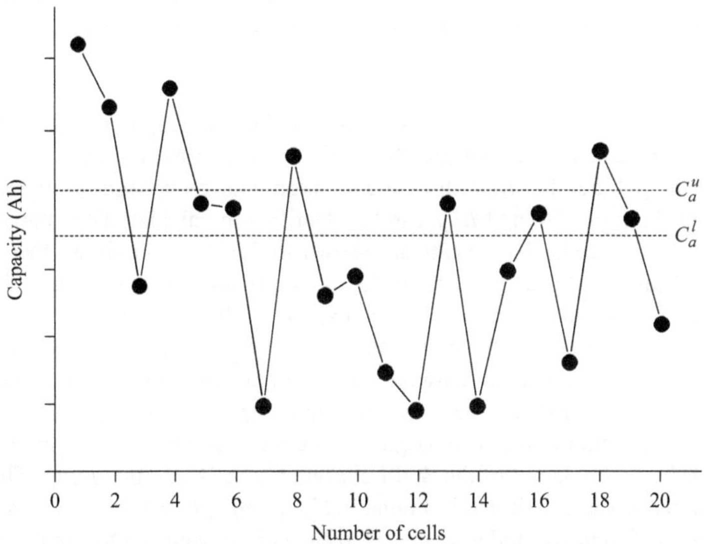


*Figure 8.24 Capacities of 20 Li-ion cells for capacity screening process*

Table 8.2 OCVs (V) of five cells at SOCs of 30%, 50% and 70%

| Battery SOCs | No. 5 | No. 6 | No. 13 | No. 16 | No. 19 |
|---|---|---|---|---|---|
| 30% | 3.6738 | 3.6745 | 3.6751 | 3.6752 | 3.6741 |
| 50% | 3.8353 | 3.8360 | 3.8358 | 3.8369 | 3.8348 |
| 70% | 3.9760 | 3.9756 | 3.9773 | 3.9771 | 3.9752 |

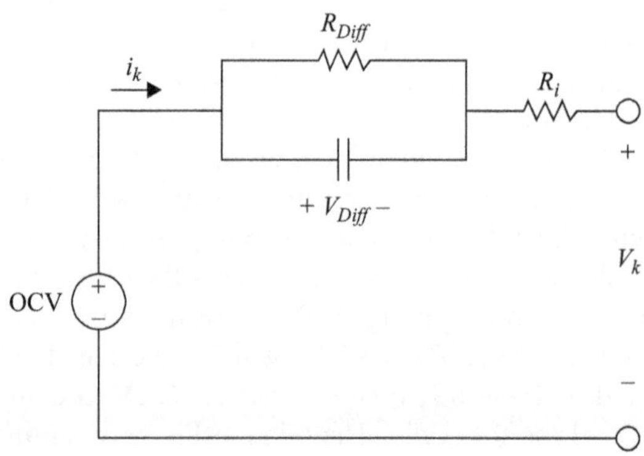


Figure 8.25 Battery equivalent circuit model

internal resistor $R_i$ and a $RC$ pair connected in series, which approximately reflects the dynamics of a battery.

Two resistances $R_i$ and $R_{Diff}$ are considered to be critical for the circuit, which are defined as DC internal resistance ($R_{DC}$) and can be obtained as

$$R_{DC} \approx R_i + R_{Diff} \tag{8.10}$$

The PC discharging (e.g. $I_{PC}^{dis} = 4$ A) and the PC charging (e.g. $I_{PC}^{cha} = 4$ A) are used to test the five cells at the specific SOCs of 30%, 50% and 70%. Figure 8.26 shows the amplitude of current, the time duration and the voltage responses of the tests. From the figure, $R_i$ and $R_{Diff}$ can be identified to calculate DC internal resistances of the five cells. The results are shown in Table 8.3. It shows that the DC internal resistance of the cells no. 5, no. 13 and no. 16 are almost the same. Thus, the cells no. 5, no. 13 and no. 16 are grouped together to build a battery pack.

The other is to cluster the cells based on machine learning approach. In this method, one type of neutral network: a self-organizing map (SOM) has been adopted (He *et al.*, 2014), where the available capacity, internal resistance and temperature variation during discharging process are selected as the input vectors of the SOM and the output of the SOM clusters the cells into the group. The SOM model for cell sorting is shown in Figure 8.27, where $x_i$ ($i = 1, 2, 3$) represents the input vectors of each cell and $w_i$ connects to each of the input vectors to the neurons which perform clustering. This SOM contains two layers: an input layer and a competition layer with a regular two-dimensional grid of mapping units.

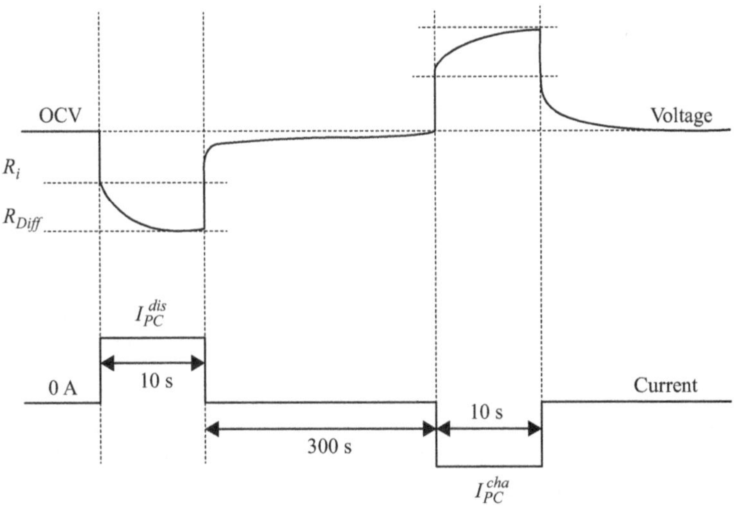


*Figure 8.26 Voltage responses at a specific SOC for PC discharging and PC charging*

*Table 8.3 DC internal resistances (Ω) of five cells at SOCs of 30%, 50% and 70%*

| Battery SOCs | No. 5 | No. 6 | No. 13 | No. 16 | No. 19 |
|---|---|---|---|---|---|
| 30% | **0.07775** | 0.06969 | **0.07796** | **0.07757** | 0.08321 |
| 50% | **0.07427** | 0.06641 | **0.07418** | **0.07462** | 0.08050 |
| 70% | **0.07789** | 0.06981 | **0.07802** | **0.07812** | 0.08334 |

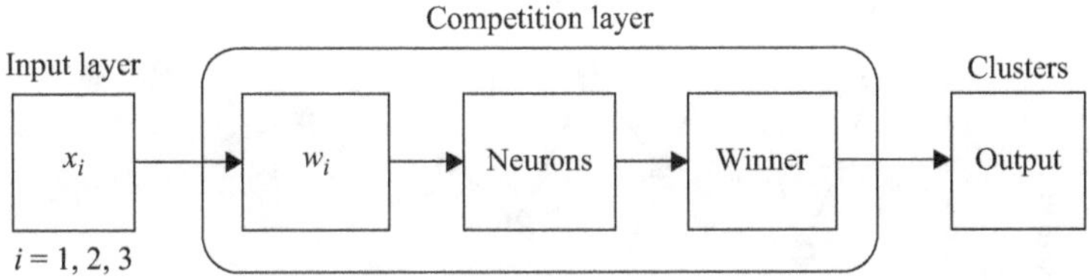


*Figure 8.27 SOM model for cell sorting*

Every unit (or neuron) $i$ is represented by a prototype vector $w_i$ of the same dimension as the input vectors $x_i$. The units are connected to the adjacent ones through a neighbourhood relation. The available capacity, internal resistance and temperature variation of 12 cells are chosen as the input vectors $x_i$ of the SOM while the outputs of the SOM are three clusters which classify 12 cells into three groups. The temperature variation has been introduced into the input vector of the SOM to achieve temperature consistency among the cells in the pack and further enhance the battery pack performance. Distances between $x_i$ and all the prototype vectors $w_i$ are calculated (Vesanto and Alhoniemi, 2000). The best matching unit, which is represented by $w_b$, is the mapping unit with prototype closest to $x_i$

$$||x_i - w_b|| = \min\{||x_i - w_i||\} \tag{8.11}$$

After the calculation, the prototype vectors are updated. The best matching unit and its topological neighbours are moved closer to the input vectors in the input space. The updating rule for the prototype vector of unit $i$ is

$$w_i(t+1) = w_i(t) + \alpha(t)[x_i - w_i(t)] \tag{8.12}$$

where $\alpha(t)$ is an adaptation coefficient. The updating procedure repeats until all the 12 cells are clustered successfully. The steps to implement this method are explained as follows.

- Conduct the experiments for 12 $LiFeO_4$ cells with each of them having the nominal capacity of 2.3 Ah under the constant discharging rate of 1 C for eight times. In each time, the cells are fully charged by the CC/CV charging algorithm with the CC of 1 C and the CV of 3.6 V.
- Collect the experimental data of battery available capacity, internal resistance at the fully charged state and temperature variations between the highest temperature and initial environmental temperature during discharging process.
- Take average values of battery capacities, internal resistances and temperature variations from eight tests for each cell, as shown in Figure 8.28.
- Regularize the averaged battery capacity, internal resistance and temperature variation into the range of [0, 1].
- Apply the SOM to cluster the 12 cells into three groups. The results are shown in Figure 8.29.

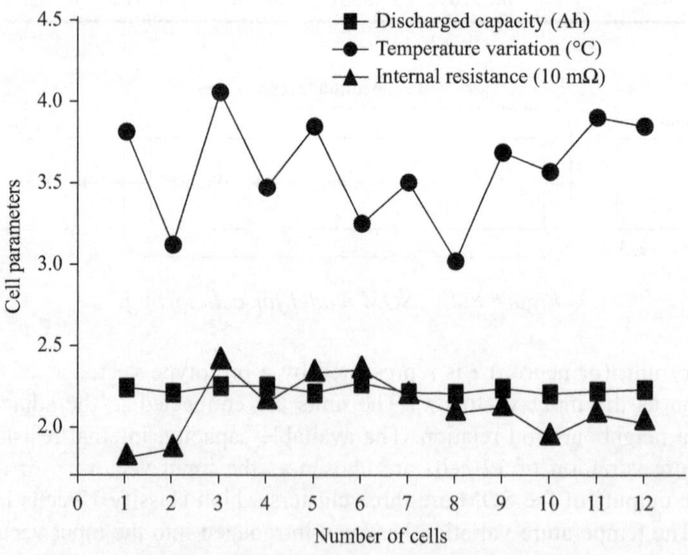


*Figure 8.28 Average values of available capacity, temperature variation and internal resistance for 12 $LiFeO_4$ cells*

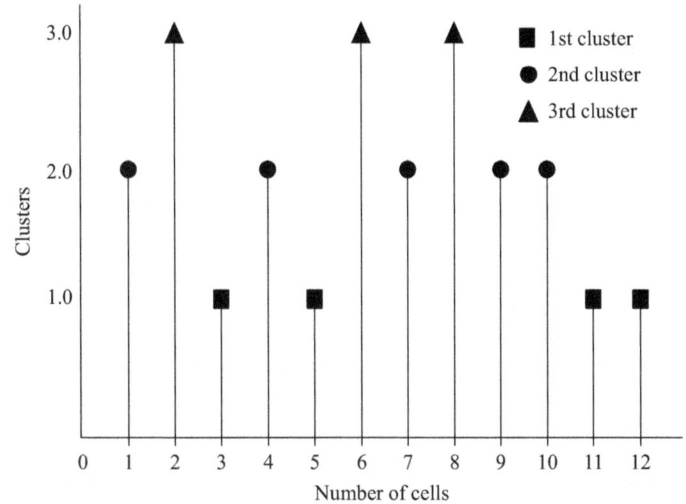


*Figure 8.29 Clustering results of 12 cells*

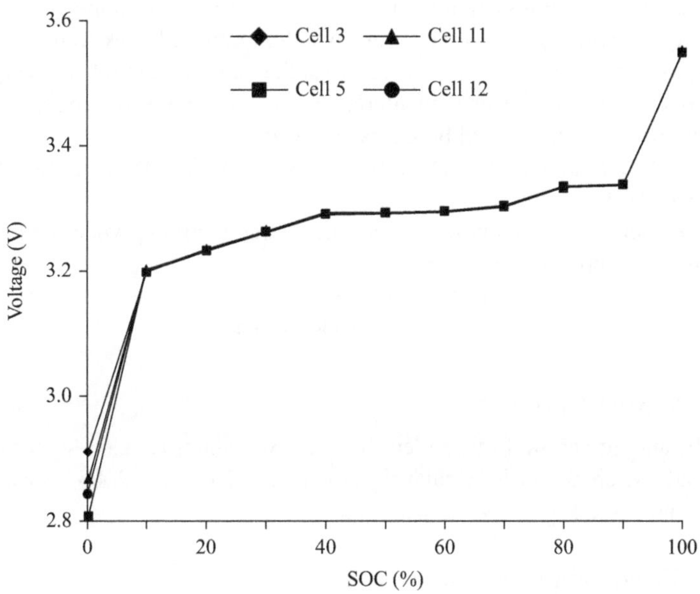


*Figure 8.30 OCVs versus SOCs of cells in sorted battery pack*

The cells in the first cluster are connected in series to build a battery pack. The experimental results of the relationship between the OCVs and the SOCs for each cell in the sorted battery pack are shown in Figure 8.30. It indicates that the cells in the same cluster have almost the same relationship of the OCVs versus the SOCs. This makes the accurate estimation of the SOC for the battery pack in EVs much easy.

### *8.3.2 Overcharge for balancing*

Four methods are used to overcharge lead acid and NiCd/NiMH battery packs for balancing, where the battery pack consists of the serially connected cells (Berndt, 1997):

- Charge lead acid and NiCd/NiMH battery packs with TC after the end of charge.
- Charge lead acid battery packs with Ah charging factor of 1.07–1.2 or watt-hour charging factor of 1.2–1.5, depending on their cycle life. The more the cycle life of the battery is, the higher the charging factor is.
- Charge NiCd/NiMH battery packs with Ah charging factor of 1.2–1.4 or watt-hour charging factor of 1.4–1.65, depending on their life cycle. The more the cycle life of the battery is, the higher the charging factor is.
- Charge lead acid and NiCd/NiMH battery packs with the small amplitude of PC after the end of charge.

During the overcharging process, five methods are taken to prevent lead acid and NiCd/NiMH batteries from excessive overcharge which may cause over-heating, explosion or permanent damaging to the batteries (Berndt, 1997; Quinn and Hartley, 2013):

- The gas combination system in the batteries is specially designed to recombine the oxygen with hydrogen into water which is returned back to the batteries.
- The gassing valve which is equipped with the batteries can release oxygen and hydrogen generated inside the batteries. It prevents explosion from the increase of internal pressure caused by excessive overcharge.
- The pack charging current is decreased as the pack voltage reaches the pre-defined level.
- The temperature is monitored to terminate pack charging when the maximum allowable temperature exceeds.
- The timer is set to terminate pack charging when the predefined capacity (or charge) has been charged back to the battery pack.

### *8.3.3 Passive balancing*

Passive balancing can be further identified as fixed shunt resistor, switched shunt resistor and switched transistor balancing methods (Cao *et al.*, 2008; Lozano *et al.*, 2014; Qi and Lu, 2014).

#### **8.3.3.1 Fixed shunt resistor**

The fixed shunt resistors are connected in parallel with each individual cell in a battery pack, as shown in Figure 8.31. During charging process, the current is partially bypassed from the cells through the resistor in order to limit cell voltage. This method is continuously bypassing the current and continuously dissipates energy into heat for the cells. It is the simplest balancing method, but it is very inefficient.

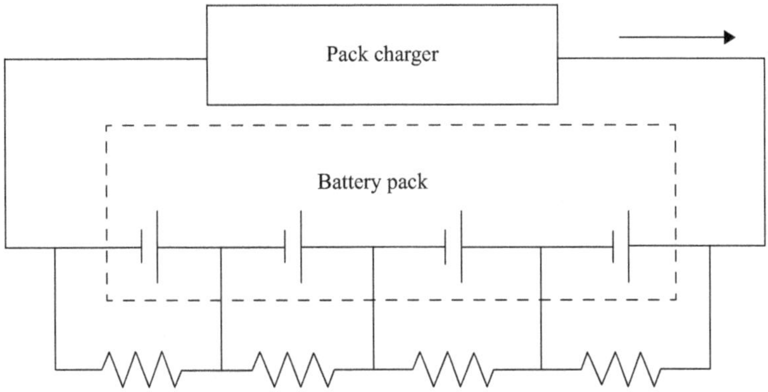


*Figure 8.31 Fixed shunt resistor balancing method*

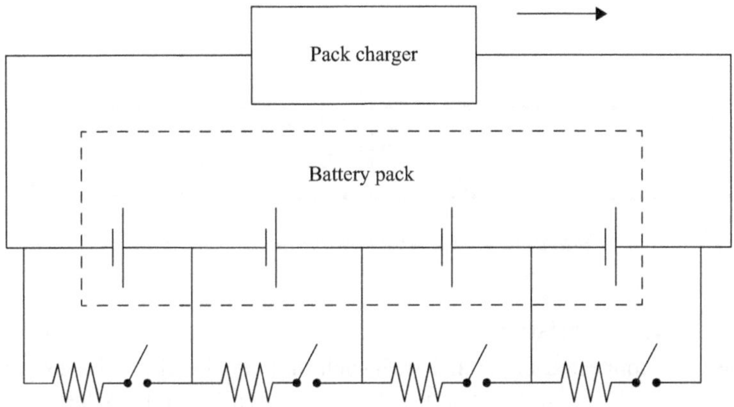


*Figure 8.32 Switched shunt resistor balancing method*

#### 8.3.3.2 Switched shunt resistor

A switch in series with a resistor is connected in parallel with each individual cell in a battery pack, as shown in Figure 8.32. There are two operation modes. The first mode turns on all switches when it is necessary. If one cell has higher voltage, more current will flow through the resistor instead of charging the cell and vice versa. With the properly selected resistance value, the weaker cells will catch up with the stronger cells effectively. The second mode involves voltage sensing and complex switch control. The voltages of each cell are sensed. The duty cycles of the switches can be regulated according to the sensed voltage levels of each cell. The higher the cell voltage is and the more the energy will be dissipated. It is a simple and reliable balancing method, but it is still inefficient.

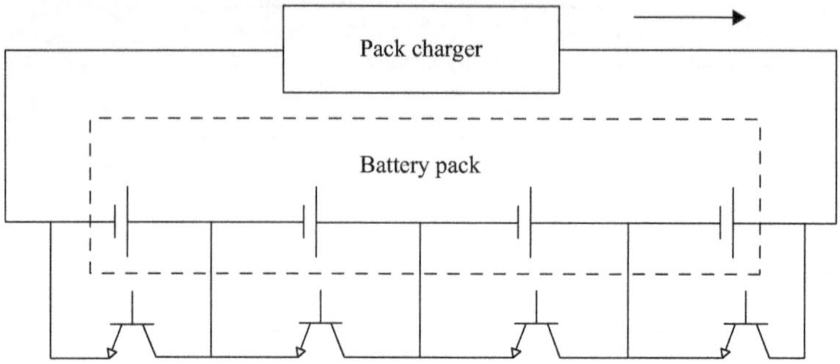


*Figure 8.33 Shunt transistor balancing method*

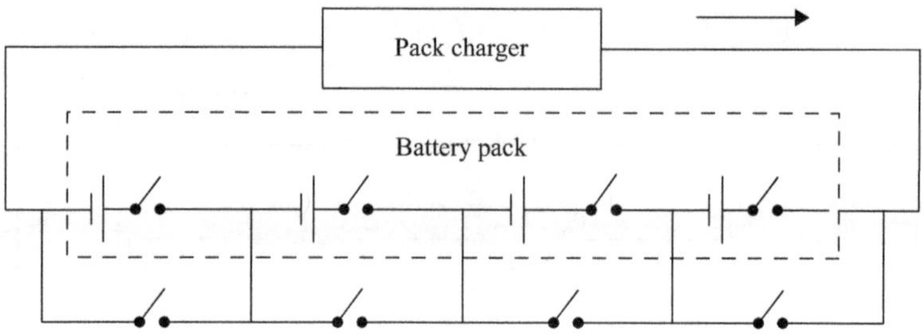


*Figure 8.34 Cell bypass balancing method*

#### 8.3.3.3 Shunt transistor

A transistor is connected in parallel with each individual cell in a battery pack, as shown in Figure 8.33. During charging process, when the cell reaches the cut-off voltage for charge, the current is proportionally bypassed around the cell to the corresponding transistor so that the cell will be charged at almost CV. In this method, the current is only shunted at the end of the charging process. Compared to the switched shunt resistor working in the first mode, it has less energy loss. Compared to the switched shunt resistor working in the second mode, it does not need intelligent control, and therefore the cost is lower.

### *8.3.4 Active balancing*

Active balancing can be further divided into the six groups based on energy flow, including cell bypass, cell to cell, cell to pack, pack to cell, cell to energy storage tank to cell, cell to pack or pack to cell.

#### 8.3.4.1 Cell bypass

In this balancing method, each cell is individually controlled by two switches, as shown in Figure 8.34. When one cell reaches its cut-off voltage for charge, the cell is completely shunted by using two switches associated with this cell. The charge finishes until the last cell in the pack is fully charged. The advantages of this method are relatively low cost and high efficiency. It is easy to be modularized. The main disadvantage is that it can only be used for low power battery pack since the higher the voltage of the battery pack is (i.e. the more the number of cells is), the higher the currents flowing through the switches are, the less the efficiency is due to the on-resistance of the switches.

#### 8.3.4.2 Cell to cell

Many methods have been developed to transfer energy from cell to cell. They can be implemented based on the switched capacitor, the double-tiered switching capacitors, the Cûk converter, the pulse width modulation controlled converter, the quasi-resonant converter and the resonant converter. Figure 8.35 shows the switched capacitor balancing method.

This method has two working modes. In the first mode, each capacitor is set in parallel with its corresponding upper cell, therefore the capacitor is set to the cell voltage, delivering or demanding energy from these cells. In the second mode, the capacitors are set in parallel with their corresponding lower cells, transferring or demanding energy from these cells in order to reach the new voltage. After many cycles of this process, the voltage of all cells will be equalized. The main advantages of this method are high efficiency, low complexity and the possibility of low and high power applications. No sensing or closed-loop control is needed. The main disadvantage is the speed, as the lower the voltage difference between adjacent cells is, the lower the balancing current is, and therefore the lower the balancing speed is.

#### 8.3.4.3 Cell to pack

Many methods have been developed to transfer energy from cell to pack. They can be implemented based on the shunt inductor, boost shunting, multiple transformers, multi-secondary windings transformer and switched transformer. Figure 8.36 shows the shunt inductor balancing method. In this method, the balancing circuit works in two modes. In the first mode, when a cell is detected to have a higher voltage than the other cells of the pack, this cell will be selected in parallel with the inductor by

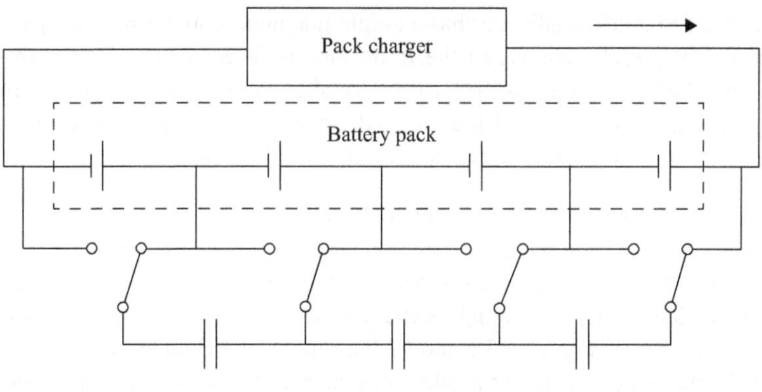


*Figure 8.35 Switched capacitor balancing method*

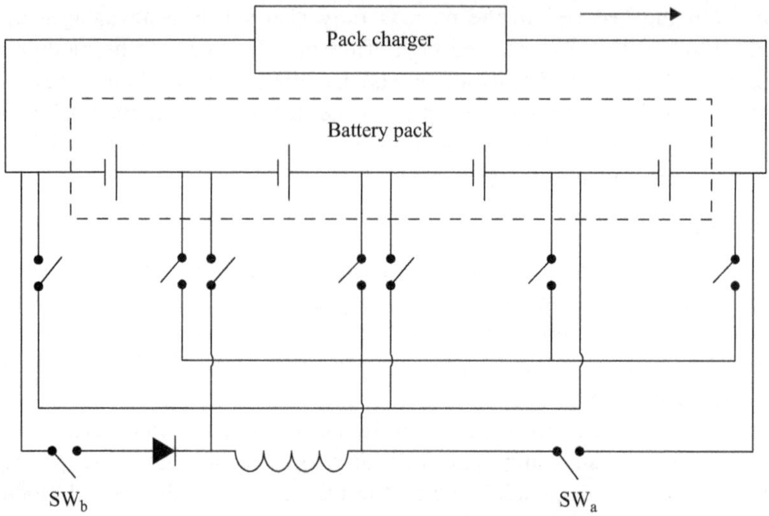


*Figure 8.36 Shunt inductor balancing method*

activating the corresponding switches of the cell, i.e. the cell is shunted by the inductor, and the extra energy of the cell will be transferred into the inductor. In the second mode, the switches corresponding to the higher voltage cell will be turned off while the switches $SW_a$ and $SW_b$ will be activated to connect the inductor to the whole battery pack, with the aim of transferring the extra energy from the higher voltage cell to the pack. This method is good for high power applications, but it is very slow since only one cell is being balanced at every instant.

#### 8.3.4.4 Pack to cell

Many methods have been developed to transfer energy from pack to cell. They can be implemented based on the multi-secondary windings transformer, voltage multiplier, full-bridge converter, multiple transformer and switched transformer. Figure 8.37 shows the multi-secondary windings transformer balancing method. In this method, a shared transformer has a single magnetic core with many secondary sides connecting each cell. When the imbalance is detected, the current from the pack is switched into the transformer primary side and induces the current in each of the secondary sides. Most of the induced current will be provided to the cell(s) with the lowest voltage (lowest reactance) via the diode(s).

#### 8.3.4.5 Cell to energy storage tank to cell

Three methods have been developed to transfer energy from the strongest cell to the energy storage tank and then to the weakest cell. They can be implemented based on the single switched capacitor, single switched inductor and pulse width modulation controlled converter. Figure 8.38 shows the single switched capacitor balancing method. In this method, the controller only selects the switches that connect the capacitor in parallel with the strongest cell (i.e. the highest voltage cell). Once the

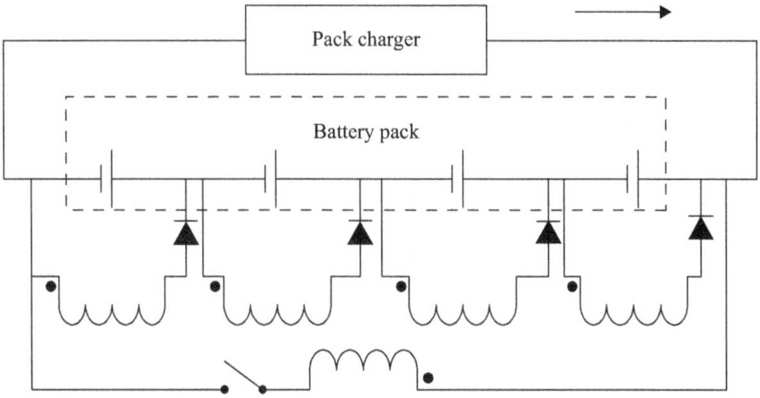


*Figure 8.37* *Multi-secondary windings transformer balancing method*

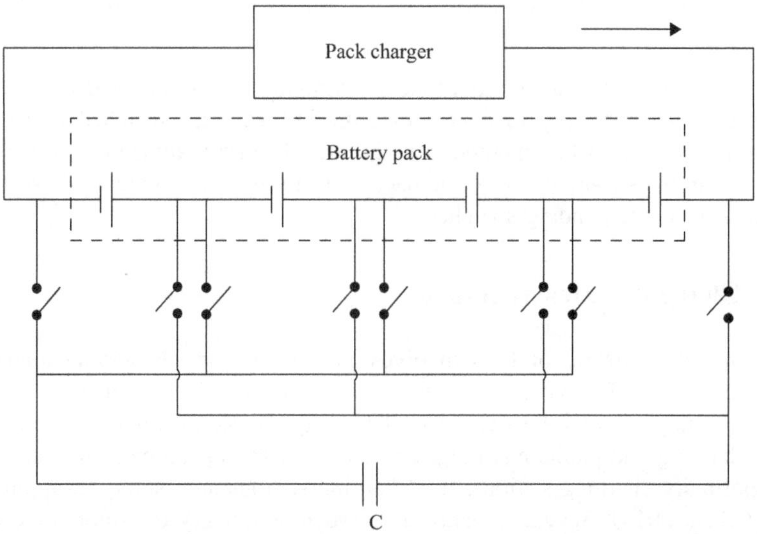


*Figure 8.38* *Single switched capacitor balancing method*

capacitor reaches the cell voltage, the controller selects the switches that connect the capacitor in parallel with the weakest cell (i.e. the lowest voltage cell), transferring the excessive energy in the strongest cells through the capacitor to the weakest cell. This method has relatively low cost and high efficiency for high power applications. But, it is slow balancing as only one cell is balanced at the same time.

#### 8.3.4.6 Cell to pack or pack to cell

Three methods have been developed to transfer energy from the cell to pack or pack to cell. They can be implemented based on the bidirectional switched

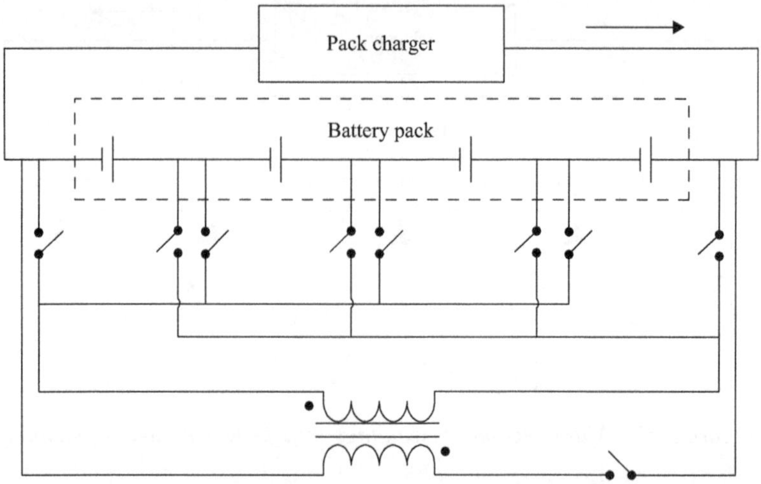


*Figure 8.39 Bidirectional switched transformer balancing method*

transformer, bidirectional multi-secondary winding transformer and bidirectional multiple transformer. Figure 8.39 shows the bidirectional switched transformer balancing method. In this method, a single transformer is utilized to transfer the energy from the strongest cell to the pack or from the pack to the weakest cell by selecting the corresponding switches.

## 8.4 Charging infrastructure

Charging infrastructure of EVs involves two main areas: battery chargers and charging stations. The battery chargers can be categorized as conductive chargers, inductive chargers or wireless chargers. The charging stations can be categorized as home charging and public charging. The public charging can be further classified as opportunity charging stations, fast charging stations and battery swapping stations (Chau, 2014). Several organizations, such as Society of Automotive Engineers (SAE), International Electromechanical Commission (IEC) and CHAdeMO EV standards, have established a few international standards for EV charging infrastructure.

### *8.4.1 Battery chargers*

Battery chargers can be the on-board charger which is installed inside EVs or the off-board charger which is installed outside EVs. Due to the constraint in weight, space and cost, the on-board charger needs to be light and compact. It is designed at low charging rate for a slow charger. On the other hand, the off-board charger has less constrained by size and weight. It is designed at high charging rate for a fast charger. In principle, both on-board and off-board chargers can be implemented as conductive, inductive or wireless chargers.

#### 8.4.1.1 Conductive chargers

The conductive charger for EVs has the advantages of maturity, simplicity and low cost. It simply uses plugs and sockets to conduct electrical energy with physical metallic contacts. According to SAE J1772 standard (SAE International, 2010), three AC levels (AC Levels 1, 2 and 3) and three DC levels (DC Levels 1, 2 and 3) are proposed for EV conductive chargers. Tables 8.4 and 8.5 show the AC and DC charging power levels and the approximate charging times for EV battery packs, respectively.

AC Levels 1 and 2 are designed for a single phase on-board chargers and AC Level 3 are mainly dedicated for three phase off-board chargers. All DC Levels 1, 2 and 3 are dedicated for off-board chargers (Williamson *et al.*, 2015).

The AC Level 1 charger is for the opportunity or slow charging. It uses a single phase grounded outlet, where a standard connector can be used to connect into the EV AC port at home or business sites. It allows EV owners to charge their vehicles wherever a suitable power source is available. The AC Level 2 charger is for the primary or normal charging. It requires the dedicated equipment and connector installed at home or public sites. The AC Level 3 charger and all DC Levels 1, 2 and 3 chargers are intended for fast charging in commercial and pubic applications. The fast charger can be installed in highway rest areas and city refuelling points, analogous to gas stations for conventional vehicles. It requires the dedicated DC connector. The fast charger is rarely feasible for residential areas due to its extremely high power demand.

*Table 8.4 AC charging power levels and charging times for EV battery pack*

| Power level types and nominal voltage | Expected output power (kW) | Charging time (h) | Vehicle technology |
|---|---|---|---|
| AC Level 1, 120 V, 1-phase | 1.44 | 4–11 | PHEVs (5–15 kWh) |
| | 1.92 | 11–36 | EVs (16–50 kWh) |
| AC Level 2, 240 V, 1-phase | 4 | 2–5 | PHEVs (5–15 kWh) |
| | 8 | 3–6 | EVs (16–30 kWh) |
| | 19.2 | 2–3 | EVs (30–50 kWh) |
| AC Level 3, 1-phase or 3-phase | 50 | 0.4–1 | EVs (20–50 kWh) |
| | 100 | 0.2–0.5 | |

*Table 8.5 DC charging power levels and charging times for EV battery pack*

| Power level types and supplied voltage range | Expected output power (kW) | Charging time (h) | Vehicle type |
|---|---|---|---|
| DC Level 1, 200–450 V | $\leq 36$ | 0.5–1.4 | EVs (20–50 kWh) |
| DC Level 2, 200–450 V | $\leq 90$ | 0.2–0.5 | |
| DC Level 3, 200–600 V | $\leq 240$ | 0.1–0.2 | |

#### 8.4.1.2 Inductive chargers

Inductive chargers use a varying magnetic field to transfer electrical energy across air gap to a battery pack in EVs. Different from conductive chargers where energy transfer occurs through metal-to-metal contact or cable, inductive chargers transfer energy magnetically without any physical contact. Their working principle is similar to that of a high-frequency transformer. The primary winding is mounted in the charger coupler while the secondary winding is embedded in the inlet of the EV. The main AC supply with the frequency of 50 or 60 Hz is rectified and converted to a high frequency AC power of 10–50 kHz within the charger (Musavi *et al.*, 2012). The EV inlet receives this high-frequency AC power from the charger coupler by magnetic induction. This high-frequency AC power is then converted to DC power to charge battery. The whole process is free from any metallic contacts between the charger and the EV, hence offering galvanic isolation, along with other advantages, such as durability, elimination of contactor wear caused by excessive use and safe under rainy, snowy and dirty conditions. The main drawback is that high leakage reluctance of the large air gap results in low magnetizing inductance and thus high magnetizing current, causing high winding losses and low charging efficiency. This type of the chargers has been explored in AC power Levels 1 and 2 (SAE International, 1995).

To promote the concept of park-and-charge, the inductive charger is further extended to charge the EV without necessity of plugging in a charger. Under this concept, the primary coil is installed on the floor of a parking lot while the secondary coil is installed on the EV. When the vehicle is properly parked over the primary coil, the secondary coil is coupled with primary coil and the power transfer is activated to charge the batteries. It offers the advantages of more convenience for users and larger tolerance for industry standardization.

#### 8.4.1.3 Wireless chargers

Wireless chargers use magnetic resonant coupling which is the near field wireless transmission of electrical energy between a transmitter's coil and a receiver's coil that are highly resonant at the same frequency. The transmitter's coil is spread underneath an area on the roadway and the power transfer occurs at multiple locations within the area, called the charging zone, and the receiver's coil residing inside EVs picks up the power wirelessly for battery charging.

The wireless charger is the most ideal way to charge EV batteries. It can charge batteries while the EV is moving on the road, the so-called move-and-charge. Such a technique improves easiness of charging and reduces the cost of EVs. The ability to charge the EV while it is moving enables the small size of the battery, achieving a long driving range without any waste time at the charging stations. Different from the magnetic induction which is adopted by inductive chargers, the coupled magnetic resonance which is adopted by wireless chargers enables high current operation in the primary circuit without suffering from high losses while the resonance at the secondary can boost the efficiency of the power transfer (Kurs *et al.*, 2007), namely, the power transmitters beneath the roadway and the receivers mounted on EVs at the same resonant frequency can wirelessly transform power efficiently with high power density.

The wireless move-and-charge needs much research in both technological and biological aspects, such as the varying displacement between the transmitter and the receiver, the misalignment between the transmitter and the receiver, and the biological response of humans to strong magnetic field. Nevertheless, this charging concept has a promising future as it can potentially solve the fundamental problem of EVs: short driving range per charge.

## *8.4.2 Home charging*

Charging at home is the most preferable way for EV owners to charge their EVs by simply plugging in the on-board charger to the outlet that is installed nearby the parked car. Most of EV owners will be able to meet their daily driving range requirements by charging the batteries in EVs overnight with an AC Level 1 charger. Since the AC Level 1 charger belongs to the slow charger, it may take about 6–8 h to fully charge the EV batteries with the charger power ratings of a few kilowatts. In general, home charging favours an effective utilization of electricity since EVs are usually charged at night or off-peak periods. Thus, power utilities are willing to impose an incentive tariff and rebate to attract EV owners to charge their EVs during off-peak periods.

The basic requirement of home charging is the availability of a garage or a parking lot that is fed with electricity. For those houses with private garage, an indoor socket outlet can be installed for recharging. For those apartments and multi-storey buildings with car parks attached, an outdoor socket outlet can be installed. The outdoor socket outlets should have individual protection circuits, and can be independently operated.

## *8.4.3 Public charging*

Public charging for EVs is similar to gasoline refuelling for conventional vehicles. It involves the deployment of public charging stations which can be easily accessed by EV drivers to recharge their EVs when necessary. Public charging can address range anxiety.

### 8.4.3.1 Opportunity charging stations

EVs may be used both day and night or desired to travel long distance per day. During the daily trip, there has always been an opportunity that EVs are likely to park for half an hour to several hours in parking lots, shopping centres, hotels, theatres, restaurants, airports, schools and supermarkets, etc. EV drivers can make use of this opportunity to charge the batteries in EVs, hence the name opportunity charging.

EV drivers should not always expect to fully charge up their EVs within the parking period. They should consider this is chance to increase the battery usable capacity for the upcoming travel or simply as an additional reserve. Typically, the driving range of EVs may be extended by about 40 km for an hour of opportunity charging.

### 8.4.3.2 Fast charging stations

Fast charging is known as rapid charging or quick charging, aiming to recharge EV batteries within a short period comparable to the time for gasoline refuelling of conventional vehicles. The time necessary for fast charging is about 20 min for charging 80% of battery nominal capacity. If there are sufficient fast charging stations on the way, the total travelling distance of EVs can be greatly extended. Thc key to fast charging stations is the off-board charger which can provide 35 kW or even higher. The corresponding voltage and power ratings are 200–600 V and 36–240 kW, respectively. As both voltage and power ratings are so high, such charging stations have to be installed in supervised stations or service centres.

Although fast charging enables EVs to have driving range similar to that of conventional vehicles, it creates adverse impacts on power systems, particularly when high power demand superimposes on the peak-hour consumption.

#### 8.4.3.3 Battery swapping stations

Instead of charging the batteries, another way to refuel EV batteries is to mechanically swap the discharged batteries with fully charged batteries. In such a case, all these batteries should be owned by the service station or battery companies while the EV driver is only a battery borrower. The discharged batteries will either be charged at the service station or centrally collected and charged. Since the battery swapping process involves mechanical replacement and battery recharging, it is also named as mechanical refuelling or mechanical recharging. These battery swapping stations combine the merits of both slow charging and fast charging, namely, slowly recharging the EV batteries at off-peak periods while quickly refuelling the EVs by swapping within a very short time. With the use of robotic machinery, the whole battery swapping process can be carried out within a few minutes, directly comparable to the existing refuelling mechanism for conventional vehicles.

There are many obstacles to practically implementing battery swapping. First, the initial cost to set up this swapping station is very high, involving expensive robotic machinery to swap the battery and a large number of costly batteries for necessary operation. Second, due to the need to store both discharged and fully charged batteries, the necessary space to build a battery swapping station is much larger than that for a charging station. Third, the EV batteries need to be standardized in physical dimensions and electrical parameters before the possible implementation of automatic battery swapping.

## 8.5 Conclusions

The chapter reviews the charging algorithms along with the termination methods to stop charging process and discusses future development of charging algorithms for lead acid, NiCd/NiMH and Li-ion batteries which are widely adopted in EVs and PHEVs. Due to the simplicity and low cost, the CC/CV charging algorithm is still the most favourable choice to charge lead acid and Li-ion batteries whereas the CC charging algorithm is the most favourable choice to charge NiCd/NiMH batteries. In order to charge EV battery packs safely and effectively, battery balancing techniques should be implemented in EV battery packs which should be built from the cells with the similar electrochemical characteristics selected through battery sorting process.

Charging infrastructure plays a critical role in the development and commercialization of EVs. The key to the charging infrastructure is a battery charger. There are different types of the battery chargers. The fast charger in public can charge batteries within a few minutes comparable to the time required for refuelling conventional vehicles but have an adverse effect on power systems. The slow charger at home may take a few hours to fully charge the battery. But, it can charge the battery efficiently, and charging EVs slowly overnight can benefit to the power systems for load levelling. The conductive charger is mature, simple and cheap. The inductive charger is a safe, durable and proven technology. The wireless charger is emerging and convenient. All these chargers have their own merits which can meet various charging needs in EVs and PHEVs.

## Acknowledgements

I would like to express sincere thanks to the team members of Electric Vehicle Research Group at Swinburne University of Technology, especially to Mrs Fengxian He for her contribution to the preparation of figures in this chapter. I must also express my indebtedness to my wife, Haihong Jiang for her patience and support along the way.

## References

Anseán, D., González M., Viera, J.C., García, V.M., Blanco, C., Valledor, M. (2013) Fast charging technique for high power lithium iron phosphate batteries: a cycle life analysis. *Journal of Power Sources*, vol. 239, pp. 9–15.

Berndt, D. (1997) *Maintenance-Free Batteries: Lead Acid, Nickel/Cadmium, Nickel/Metal Hydride. A Handbook of Battery Technology.*Taunton, Somerset, England, 2nd edition, Research Studies Press LTD.

Cao, J., Schofield, N. and Emadi, A. (2008) Battery balancing methods: a comprehensive review. *Proceedings of IEEE Conference on Vehicle Power and Propulsion,* Harbin, China, pp. 1–6.

Chan, C.C. (2007) The state of the art of electric, hybrid, and fuel cell vehicles. *Proceedings of the IEEE*, vol. 95, no. 4, pp. 704–718.

Chan, C.C. and Chau, K.T. (2001) *Modern Electric Vehicle Technology*. Oxford: Oxford University Press.

Chan, C.C. and Chu, K.C. (1990). A microprocessor-based intelligent battery charger for electric vehicle lead acid batteries. *Proceedings of the 10th International Electric Vehicle Symposium*, Hong Kong, China, pp. 456–66.

Chau, K.T. (2014). Pure electric vehicles. In *Alternative Fuels and Advanced Vehicle Technologies for Improved Environmental Performance – Towards Zero Carbon Transportation*, ed. R. Folkson. Woodhead Publishing, March 2014, pp. 655–684.

Chen, L.R. (2004) PLL-based battery charge circuit topology. *IEEE Transactions on Industrial Electronics*, vol. 51, no. 6, pp. 1344–1346.

Chen, L.R. (2007) A design of an optimal battery pulse charge system by frequency-varied technique. *IEEE Transaction on Industrial Electronics*, vol. 54, no. 1, pp. 398–405.

Chen, L.R. (2009) A design of duty-varied voltage pulse charger for improving lithium-ion battery-charging response. *IEEE Transactions on Industrial Electronics*, vol. 56, no. 2, pp. 480–487.

Chen, L.R., Chen, J.J., Chu, N.Y. and Han, G.Y. (2008) Current pumped battery charger. *IEEE Transaction on Industrial Electronics*, vol. 55, no. 6, pp. 2482–2488.

Chen, L.R, Hsu, C.M, and Liu, C.S. (2008) A design of a grey-predicted lithium-ion battery charge system. *IEEE Transactions on Industrial Electronics*, vol. 48, no. 3, pp. 3692–3701.

Chen, X.P., Shen, W.X., Vo, T.T., Cao, Z.W., and Kapoor A. (2012) An overview of lithium-ion batteries for electric vehicles. *Proceedings of IEEE Conference on Power and Energy,* Ho Chi Minh City, Vietnam, pp. 230–235.

Chen, Z. Xia, B., Chris Mi, C.T., and Xiong, R. (2015) Loss minimisation-based charging strategy for lithium-ion battery. *IEEE Transactions on Industry Applications*, DOI: 10.1109/TIA.2015.2417118.

Dung, L.R. and Yen, J.H. (2010) ILP-based algorithm for lithium-ion battery charging profile. *Proceedings of IEEE International Symposium on Industrial Electronics*, Bari, Italy, pp. 2286–2291.

He, F.X., Shen, W.X., Song, Q., Kapoor, A., Honnery, D. and Dayawansa, D. (2014) Clustering $LiFePO_4$ cells for battery pack based on neural network in EVs. *Proceedings of IEEE Conference and Expo on Transportation Electrification Asia-Pacific*, Beijing, China, pp. 1–5.

Hsieh, G.C., Chen, L.R., and Huang, K.S. (2001) Fuzzy-controlled lithium-ion battery charge system with active state of charge controller. *IEEE Transactions on Industrial Electronics*, vol. 48, no. 3, pp. 585–593.

Hu, X.S., Li, S.B., Peng, H., and Sun, F.C. (2013) Charging time and loss optimisation for LiNMC and $LiFePO_4$ batteries based on equivalent circuit models. *Journal of Power Sources*, vol. 239, pp. 449–457.

Huang, J.W., Liu, Y.H., Wang, S.C., and Yang, Z.Z. (2009) Fuzzy-control-based five-step lithium-ion battery charger. *Proceedings of IEEE Conference on Power Electronics and Drive Systems*, Taipei, Taiwan, pp. 1547–1551.

Hussein, A. and Batarseh, I. (2011) A review of charging algorithms for nickel and lithium battery chargers. *IEEE Transactions on Vehicular Technology*, vol. 60, pp. 830–838.

Ikeya, T., Sawada, N., Murakami, J., *et al.* (2002) Multi-step constant-current charging method for an electric vehicle nickel/metal hydride battery with high-energy efficiency and long cycle life. *Journal of Power Sources*, vol. 105, pp. 6–12.

Ikeya, T., Sawada, N., Takagi, S., *et al.* (1998) Multi-step constant-current charging method for electric vehicle, valve-regulated lead acid batteries during night time for load-levelling. *Journal of Power Sources*, vol. 75, pp. 101–107.

Jongh, P.E.De and Notten, P.H.L (2002) Effect of current pulses on lithium intercalation batteries. *Solid State Ionics*, vol. 148, pp. 259–268.

Khaligh, A and Dusmez, S. (2012) Comprehensive topological analysis of conductive and inductive charging solutions for plug-in electric vehicle. *IEEE Transactions on Vehicular Technology*, vol. 61, no. 8, pp. 3475–3489.

Kim, J., Shin, J., Chun, C., and Cho, B.H. (2012) Stable configuration of a li-Ion series battery pack based on a screening process for improved voltage/SOC balancing. *IEEE Transactions on Power Electronics,* vol. 27, no. 1, pp. 411–424.

Kim, B.G, Tredeau, F.P., and Salameh, Z.M. (2008) Fast chargeability lithium polymer batteries, *Proceedings of IEEE Conference on Power and Energy Society General Meeting: Conversion and Delivery of Electrical Energy in the 21st Century*, Pittsburgh, PA, USA, pp. 1–5.

Klein, R., Chaturvedi, N.A., Christensen, J., Ahmed, J., Findeisen, R, and Kojic, A. (2011) Optimal charging strategies in lithium-ion battery. *Proceedings of American Control Conference*, San Francisco, CA, USA, pp. 382–387.

Kurs, A., Karalis, A., Moffatt, R., Joannopoulos, J.D., Fisher, P. and Soljačić, M. (2007) Wireless power transfer via strongly coupled magnetic resonance, *Science*, vol. 317, pp. 83–86.

Kutkut, N.H., Wiegman, H.L.N., Divan, D.M. and Novotny, D.W. (1998) Charge equalisation for an electric vehicle battery system. *IEEE Transaction on Aerospace and Electronics Systems*, vol. 34, no. 1, pp. 235–246.

Li, J., Murphy, E., Winnick, J. and Kohl, P.A. (2001) The effects of pulse charging on cycling characteristics of commercial lithium-ion batteries. *Journal of Power Sources*, vol. 102, pp. 302–309.

Li, X., Wang, T., Pei, L., Zhu, C., and Xu, B.L. (2014) A comparative study of sorting methods for lithium-ion batteries. *Proceedings of IEEE Conference and Expo on Transportation Electrification Asia-Pacific*, Beijing, China, pp. 1–6.

Liu, Y.H., Hsieh, C.H. and Luo, Y.F. (2011) Search for an optimal rapid charging pattern for Li-ion batteries using consecutive orthogonal arrays. *IEEE Transactions on Industrial Electronics*, vol. 26, no. 2, pp. 654–661.

Liu Y.H. and Luo, Y.F. (2010) Search for an optimal rapid charging pattern for Li-ion batteries using Taguchi approach. *IEEE Transactions on Industrial Electronics*, vol. 57, no. 12, pp. 3963–3971.

Liu, Y.H., Teng, J.H. and Lin, Y.C. (2005) Search for an optimal rapid charging pattern for Li-ion batteries using ant colony system algorithm. *IEEE Transactions on Industrial Electronics*, vol. 52, no. 5, pp. 1328–1336.

Lozano, J.G., Cadaval, E.R., Montero, M.M., and Martinez, M.A.G. (2014) Battery equalisation active methods, *Journal of Power Sources*, vol. 246, pp. 934–949.

Musavi, F., Edington, M. and Eberle, W. (2012) Wireless power transfer: a survey of EV battery charging technologies. *Proceedings of IEEE Conference on Energy Conversion Congress and Exposition*, Raleigh, NC, USA, pp. 1804–1810.

Nicolai, J. and Wuidart, L. (1994) From nickel–cadmium to nickel–metal–hydride fast battery charger. STMicroelectronics Inc. Application Note. [Online]. Available: http://www.st.com/stonline/books/pdf/docs/2074.pdf

Notten, P.H.L., Op het Veld, J.H.G., Van Beek, J.R.G. (2005) Boostcharging Li-ion batteries: a challenging new charging concept. *Journal of Power Source*, vol. 145, no. 1, pp. 89–94.

Panasonic Inc. (2003) Methods of Charging, the Valve-Regulated Lead-Acid Battery, Application Note [online].

Panasonic Inc. (2005) Lithium-Ion Charging Datasheet. Available: http://www.panasonic.com/industrial/includes/Panasonic_LiIon_Charging.pdf

Park, S.H., Park, K., Kim, H.S., Moon, G.W., and Youn, M.J. (2012) Single magnetic cell-to-cell charge equalisation converter with reduced number of transformer windings. *IEEE Transactions on Power Electronics*, vol. 27, pp. 2900–2911.

Purushothama, B.K. and Landau, U. (2006) Rapid charging of lithium-ion batteries using pulsed current. *Journal of the Electrochemical Society*, vol. 153, no. 3, pp. A533–A542.

Purushothama, B.K., Morrison, P.W. and Landau, U. (2005) Reducing mass-transport limitations by application of special pulsed current modes. *Journal of the Electrochemical Society*, vol. 152, no. 4, pp. J33–J39.

Qi, J., and Lu, D.C. (2014) Review of battery cell balancing techniques. *Proceedings of Australasian Universities Power Engineering Conference*, Perth, Australia, pp. 1–6.

Quinn, D.D., and Hartley, T.T. (2013) Design of novel charge balancing networks in battery packs. *Journal of Power Sources*, vol. 240, pp. 26–32.

Rahimi-Eichi, H., Ojha, U., Baronti, F. and Chow, M.Y. (2013) Battery management system: an overview of its application in the smart grid and electric vehicles. *IEEE Industrial Electronics Magazine*, vol. 7, no. 2, pp. 4–16.

Rand, D.A.J., Moseley, P.T., Garche, J. and Parker, C.D. (2004) *Valve-Regulated Lead-Acid Batteries*, Amsterdam: Elsevier.

SAE International (1995) Electric Vehicle Inductive Coupling Recommended Practice, J1773.

SAE International (2010) Surface Vehicle Recommended Practice J1772, SAE Electric Vehicle and Plug in Hybrid Electric Vehicle Conductive Charge Coupler.

Shafiei, A. and Williamson, S. (2010) Plug-in hybrid electric vehicle charging: current issues and future challenges, *Proceedings of IEEE Conference on Vehicle Power and Propulsion*, Lille, France, pp. 1–8.

Shen, W.X., Vo, T.T., and Kapoor, A. (2012) Charging algorithms of lithium-ion batteries: an overview. *Proceedings of IEEE Conference on Industrial Electronics and Applications*, Singapore, pp. 1567–1572.

Tsang, K.M., Chan, W.L. (2011) Current sensorless quick charger for lithium-ion batteries. *Energy Conversion and Management*, vol. 52, pp. 1593–1595.

Vesanto, J. and Alhoniemi, E. (2000) Clustering of the self-organising map. *IEEE Transactions on Neural Networks*, vol. 11, pp. 586–600.

Vo, T.T., Shen, W.X., Kapoor, A. (2014) New charging strategy for lithium-ion batteries based on the integration of Taguchi method and state of charge estimation. *Journal of Power Source*, vol. 273, pp. 413–422.

Williamson, S.S., Rathore, A.K and Musavi F. (2015) Industrial electronics for electric transportation: current state-of-the-art and future challenges. *IEEE Transactions on Industrial Electronics*, vol. 62, no. 5, pp. 3021–3032.

Yilmaz, M. and Krein, Philip T. (2013) Review of battery charger topologies, charging power levels, and infrastructure for plug-in electric and hybrid vehicles. *IEEE Transactions on Power Electronics, vol.* 28, no. 5, pp. 2151–2169.

Young, K., Wang, C.S., Wang, L.Y. and Strunz, K. (2013) Electric vehicle battery technologies. In *Electric Vehicle Integration into Modern Power Networks*, ed. R. Garcia-Valle, New York: Springer-Verlag.

# Chapter 9
# Wireless power transfer systems for electric vehicles

*Chi-Kwan Lee*[1] *and Wen-Xing Zhong*[1]

Cutting the annoying power cables, wireless power transfer technology based on the magnetic resonance and near-field coupling of two-loop resonators has tremendously transformed the way that electrical and electronic devices receive their power. This emerging technology certainly has makes our lives more convenient, comfortable, and productive than any generation before us. By introducing wireless power transfer systems for electric vehicles, the obstacles of transferring energy to the vehicles can be resolved. The driver only needs to park the car and leave. Charging the vehicle's battery becomes an easy task. In dynamic wireless charging application like roadway-powered electric vehicle, the technology also enables battery charging while driving. The electric vehicle is possible to run continuous without stopping.

Starting from Tesla's principles of wireless power transfer, this chapter introduces the basic concepts, study method, design principles, and latest developments in wireless power transfer technology.

## 9.1 Introduction

As a matter of fact, wireless power transfer technology has already been developed by Tesla a century ago (Lomas, 1999; Sarkar *et al.*, 2006). As pioneered by Tesla, the technology can be radiative or nonradiative depending on the energy transfer mechanisms. Radiative power can be emitted from an antenna and propagates through a medium (such as vacuum or air) over long distance (i.e. many times larger than the dimension of the antenna) in the form of an electromagnetic wave. Due to the omni-directional nature of the radiative power emission, the energy efficiency of power transmission is very low. Nonradiative technique relies on the near-field magnetic coupling of conductive loops and resonance techniques. In the

[1]Department of Electrical and Electronic Engineering, The University of Hong Kong, Hong Kong, China

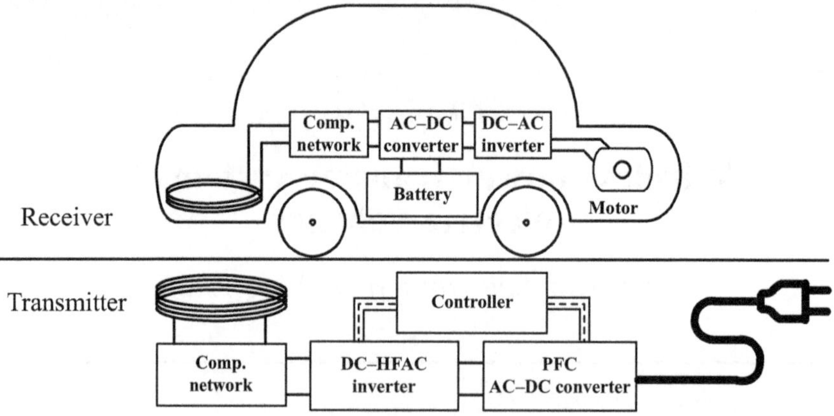


*Figure 9.1 Block diagram of a wireless power transfer system for electric vehicle*

modern applications such as charging an electric vehicle like trains or buses, energy efficiency higher than 90% is possible.

Figure 9.1 depicts a wireless power transfer system for an electric vehicle (Li and Mi, 2015; Madawala and Thrimawithana, 2011). It includes several stages to charge an electric vehicle wirelessly. The main difference between a wireless charger and a conventional conductive or wired charger is that a transformer is replaced by a set of couple coils. The energy is coupled from the transmitting coil to the receiving coils by using an alternating magnetic field. To avoid the magnetic flux goes inside the chassis causing high eddy current loss, the couple coils consist of ferrite structure as a magnetic flux guide and metal plates (such as aluminum) as magnetic shield (Nagatsuka *et al.*, 2010; Budhia *et al.*, 2010). The frequency of the alternating magnetic field typically from ten to hundreds of kHz in order to meet the size, power, and efficiency criteria with the use of latest state-of-the-art power electronics devices.

At first, the utility AC power is converted to a DC power source by an AC–DC converter with power factor correction (PFC). Then, the power inverter converts the DC power to a high-frequency AC (HFAC) voltage or current to drive the transmitting coil through a compensation network. Compensation networks are made of tuned LC circuits. The operating point (i.e. frequency, voltage, and/or current) of the power inverter is designed to match with the electrical characteristics of the networks to maximize the power transfer and efficiency. A high-frequency transformer could be inserted between the power inverter and the transmitting coil to provide extra isolation and protection. The vehicle equipped a receiving coil, an AC voltage, is induced. The energy will be rectified by using a high-frequency fed AC–DC converter. The second-stage power conversion consists of a DC–DC converter charging the battery. At last, DC–AC power inverter converts the DC power into desired voltage and current format for the motor (Lukic

*Figure 9.2 A commercial wireless electric vehicle charging system. The vehicle adapter is installed underneath the vehicle. A parking pad is installed on the floor and a control panel is mounted on the wall*

and Pantic, 2013; Wang *et al.*, 2005). Figure 9.2 is a photo of a commercial wireless electric vehicle charging system. The charging system consists of three components. The vehicle adapter is installed underneath the vehicle. A parking pad is installed on the garage floor or parking space. A control panel is mounting on the wall, which supplies power to the parking pad and also provides alignment guidance and charging status to the driver.

## 9.2 Tesla's early work of nonradiative wireless power transfer

As the inventor of a series of technologies that have affected human society since the 20th century, Tesla's work on tuned circuits, wireless power, and radio circuits has shared some common themes. In a study of Tesla's contributions (Marincic, 1982), quotations are cited from a 1943 technical article (Wheeler, 1943) that Tesla is entitled to either distinct priority or independent discovery of:

1. The idea of inductive coupling between the driving and the working circuits.
2. The importance of tuning both circuits, that is, the idea of an "oscillation transformer."
3. The idea of a capacitance loaded open secondary circuit.

Obviously, these three aspects of discovery have formed the founding principles for both nonradiative and radiative wireless transfer. In particular, his discovery of using both tuned circuits as an "oscillation transformer" indicates that both the transmitter and receiver circuits are tuned to operate in the resonance mode. The "oscillation transformer" concept goes beyond pure magnetic induction principle, and more precisely, refers to the use of magnetic resonance between two magnetically coupled coil-resonators. Tesla's idea of a capacitance loaded open secondary circuit is highlighted in Marincic (1982) and Wheeler (1943). These open resonant circuits were used at the transmitting and the receiving sides. Electrical energy can be transmitted through the globe just as through a good conductor.

Despite the lack of modern equipment such as radio-frequency power amplifier or other forms of high-power power supply with tens of Mega-Hertz frequency range a century ago, Tesla's early work still influences recent wireless applications. According to Tesla (1914), Tesla designed his own "high-frequency" AC generator and managed to test his apparatus at hundreds of kilohertz. A drawing of a magnifying transmitter (resonant transformer) patented by Tesla in 1914 for transmitting electrical energy to a distance through the natural media is shown in Figure 9.3 (https://commons.wikimedia.org/wiki/File:US_patent_1119732_Nikola_Tesla_1907_Apparatus_for_transmitting_electrical_energy.png). Magnifying transmitter is an air-core, multiple-resonant transformer that is used to generate a very high-frequency and high voltage signal. This system was the basis for Tesla's later Wardenclyffe Tower project for wireless telecommunications and electrical power transmission.

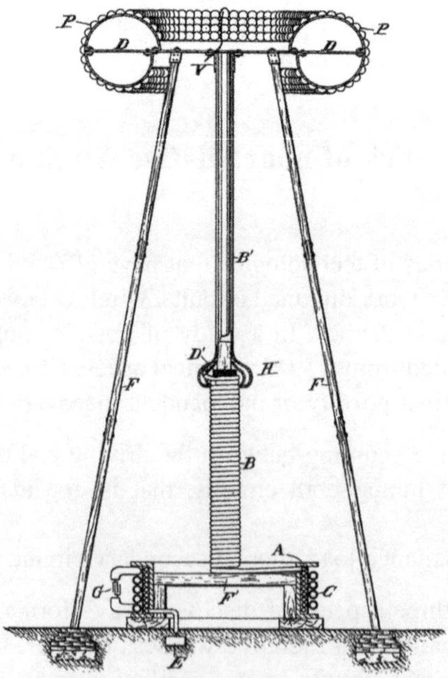


*Figure 9.3 A drawing of Tesla's magnifying transmitter (https://commons.wikimedia.org/wiki/File:US_patent_1119732_Nikola_Tesla_1907_Apparatus_for_transmitting_electrical_energy.png)*

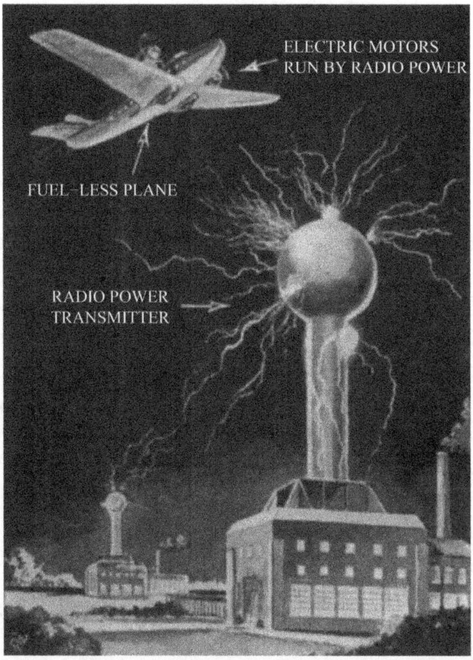


*Figure 9.4* *Tesla envisioned a world wireless system consisting of a worldwide network of dome-shaped antennas which would transmit both power and information over areas of hundreds of miles to power homes, factories, and vehicles (https://commons.wikimedia.org/wiki/File: Tesla_wireless_power_future_1934.png)*

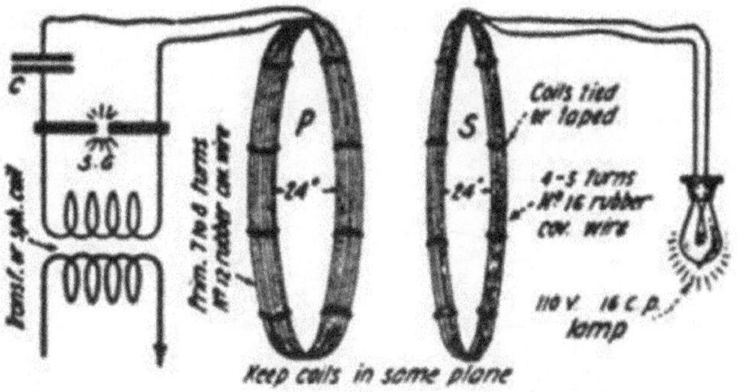


*Figure 9.5* *An illustration of wireless power transfer experiment by Nikola Tesla in 1921 (https://commons.wikimedia.org/wiki/File: Tesla_wireless_power_circuit.jpg)*

Figure 9.4 is a drawing of Tesla's vision of the future of wireless power (https://commons.wikimedia.org/wiki/File:Tesla_wireless_power_future_1934.png). He envisioned a world wireless system consisting of a worldwide network of dome-shaped antennas which would transmit both power and information over areas of hundreds of miles to power homes, factories, and vehicles. A diagram of Tesla's wireless power transfer experiment in 1921 is shown in Figure 9.5 (https://commons.wikimedia.org/wiki/File:Tesla_wireless_power_circuit.jpg). The setup consists of a primary (transmitter) coil, resonant circuit, a secondary (receiver) coil and a lamp. A resonant circuit is formed by the capacitor (*C*) and primary coil (*P*) at the transmitter. The capacitor (*C*) repeatedly charges up, then discharges through the spark gap (*S-G*), creating a radio-frequency current in the primary coil to generate an alternating magnetic field. An electric current is induced in the secondary coil powering the lamp. Tesla demonstrated optimal energy transfer could be achieved using a pair of magnetically coupled coils operating at the resonance frequency. With this setup, he is able to light lamps wirelessly at a distance of tens of feet.

## 9.3 Basic principles for wireless power transfer using near-field coupling technique

### 9.3.1 Basic circuit model

Near-field coupling technique has been applied extensively in AC machines, which were also pioneered by Tesla. Energy transfer via coupled windings is the basic principle used in electric machines. Using a cage induction machine as an example, energy is transferred from the excited stator winding across the air gap to the rotor cage. Therefore, wireless power system can be mathematically described by electric circuit theory for magnetically coupled circuits (Lee *et al.*, 2012; Zhong *et al.*, 2012, 2013). Assuming a general system with $n$ magnetically coupled coils as shown in Figure 9.6. The mathematical model using circuit equation can be

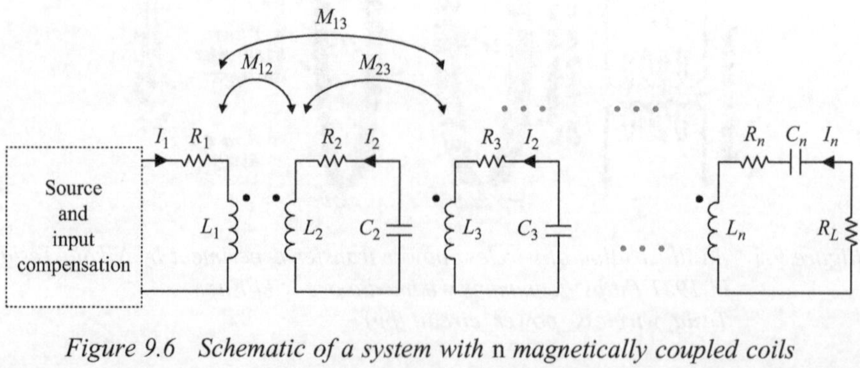


*Figure 9.6 Schematic of a system with* n *magnetically coupled coils*

obtained as:

$$
\begin{bmatrix}
j\omega M_{12} & R_2 + j\left(\omega L_2 - \frac{1}{\omega C_2}\right) & j\omega M_{23} & \cdots & \cdots & j\omega M_{2n} \\
j\omega M_{13} & j\omega M_{23} & R_3 + j\left(\omega L_3 - \frac{1}{\omega C_3}\right) & \cdots & \cdots & j\omega M_{3n} \\
\vdots & \vdots & \vdots & \ddots & \vdots & \vdots \\
j\omega M_{1(n-1)} & \cdots & \cdots & & R_{n-1} + j\left(\omega L_{n-1} - \frac{1}{\omega C_{n-1}}\right) & j\omega M_{(n-1)n} \\
j\omega M_{1n} & \cdots & & & j\omega M_{(n-1)n} & R_n + R_L + j\left(\omega L_n - \frac{1}{\omega C_n}\right)
\end{bmatrix}
\cdot
\begin{bmatrix}
\mathbf{I}_1 \\ \mathbf{I}_2 \\ \vdots \\ \mathbf{I}_{n-1} \\ \mathbf{I}_n
\end{bmatrix}
=
\begin{bmatrix}
V_1 \\ 0 \\ \vdots \\ 0 \\ 0
\end{bmatrix}
\tag{9.1}
$$

where $M_{ij} = k_{ij}\sqrt{L_iL_j}(i,j = 1,2,\ldots,n; i \neq j)$ is the mutual inductance between winding-$i$ and winding-$j$, $L_i$ is the self-inductance of winding-$i$, $k_{ij}$ is magnetic coupling coefficient between winding-$i$ and winding-$j$. $C_i$ is the compensating capacitance of winding-$i$, $R_i$ is the resistance in resonator-$i$ (including the resistance of winding-$i$ and the equivalent series resistance (ERS) of the capacitor $C_i$), and $R_L$ is the load resistance which is connected to winding-$n$. $I_1$ is the AC excitation current source and it can be expressed by:

$$I_1 = I_S \sin(\omega t) \tag{9.2}$$

where $\omega$ is angular frequency of the current source. $I_i$ is the current in winding-$i$. It should be noted that the model (9.1) represents a general system that is not restricted to resonance-frequency operation.

## 9.3.2 Power flow analysis

The power flow in each path can be worked out in order to understand how the power flows will change under frequency variation. Using a system with three magnetically coupled coils as an example, the relationships between the winding currents can be determined by solving (9.1) as:

$$\frac{\mathbf{I}_2}{\mathbf{I}_1} = \frac{-\omega M_{12}R_{out} + j(\omega^2 M_{13}M_{23} - \omega M_{12}X_3)}{R_2X_3 + R_{out}X_2 + j(X_2X_3 - R_2R_{out} - \omega^2 M_{23}^2)} \tag{9.3}$$

$$\frac{\mathbf{I}_3}{\mathbf{I}_1} = \frac{-\omega M_{13}R_2 + j(\omega^2 M_{12}M_{23} - \omega M_{13}X_2)}{R_2X_3 + R_{out}X_2 + j(X_2X_3 - R_2R_{out} - \omega^2 M_{23}^2)} \tag{9.4}$$

where $X_2 = \omega L_2 - 1/(\omega C_2)$, $X_3 = \omega L_3 - 1/(\omega C_3)$, and $R_{out} = R_3 + R_L$. Now assuming the input current is 1 A, with the help of (9.3) and (9.4), the complex power flows can be expressed as:

$$\overline{\mathbf{S}}_{12} = P_{12} + jQ_{12} = -j\omega M_{12}\mathbf{I}_1\mathbf{I}_2^*$$

$$= \frac{\begin{pmatrix} \omega^2 M_{12}^2 X_3 R_2 X_3 + \omega^2 M_{12}^2 R_{out}^2 R_2 \\ + \omega^4 M_{12}^2 M_{23}^2 R_{out} \\ - \omega^3 M_{12}M_{13}M_{23}R_2X_3 \\ - \omega^3 M_{12}M_{13}M_{23}R_{out}X_2 \end{pmatrix} + j\begin{pmatrix} \omega^3 M_{12}M_{13}M_{23}X_2X_3 \\ + \omega^4 M_{12}^2 M_{23}^2 X_3 - \omega^5 M_{12}M_{13}M_{23}^2 \\ - \omega^3 M_{12}M_{13}M_{23}R_2R_{out} \\ - \omega^2 M_{12}^2 M_{out}^2 X_2 - \omega^2 M_{12}^2 X_3X_2X_3 \end{pmatrix}}{(R_2X_3 + R_{out}X_2)^2 + (X_2X_3 - R_2R_{out} - \omega^2 M_{23})^2} \tag{9.5}$$

$$\overline{\mathbf{S}}_{23} = P_{23} + jQ_{23} = -j\omega M_{23}\mathbf{I}_2\mathbf{I}_3^*$$

$$= \frac{\begin{pmatrix} \omega^3 M_{12}M_{23}M_{13}R_2X_3 \\ + \omega^4 M_{12}^2 M_{23}^2 R_{out} \\ - \omega^3 M_{12}M_{23}M_{13}R_{out}X_2 \\ - \omega^4 M_{13}^2 M_{23}^2 R_2 \end{pmatrix} + j\begin{pmatrix} \omega^5 M_{12}M_{23}^3 M_{13} + \omega^3 M_{12}M_{23}M_{13}X_2X_3 \\ - \omega^3 M_{12}M_{23}M_{13}R_{out}R_2 \\ - \omega^4 M_{12}^2 M_{23}^2 X_3 \\ - \omega^4 M_{13}M_{23}^2 M_{13}X_2 \end{pmatrix}}{(R_2X_3 + R_{out}X_2)^2 + (X_2X_3 - R_2R_{out} - \omega^2 M_{23}^2)^2} \tag{9.6}$$

$$\overline{\mathbf{S}}_{13} = P_{13} + jQ_{13} = -j\omega M_{13}\mathbf{I}_1\mathbf{I}_3^*$$
$$= \frac{\begin{pmatrix} \omega^2 M_{13}^2 R_2^2 R_{out} \\ +\omega^4 M_{23}^2 M_{13}^2 R_2 \\ +\omega^2 M_{13}^2 R_{out} X_2^2 \\ -\omega^3 M_{12} M_{23} M_{13} R_2 X_3 \\ -\omega^3 M_{12} M_{23} M_{13} R_{out} X_2 \end{pmatrix} + j \begin{pmatrix} \omega^2 M_{13}^2 R_2^2 X_3 + \omega^2 M_{13}^2 X_2^2 X_3 \\ +\omega^3 M_{12} M_{23} M_{13} R_2 R_{out} \\ +\omega^5 M_{12} M_{23}^3 M_{13} \\ -\omega^3 M_{12} M_{23} M_{13} X_2 X_3 \\ -\omega^4 M_{23}^2 M_{13}^2 X_2 \end{pmatrix}}{(R_2 X_3 + R_{out} X_2)^2 + (X_2 X_3 - R_2 R_{out} - \omega^2 M_{23}^2)^2} \tag{9.7}$$

$$\overline{\mathbf{S}}_3 = P_3 + jQ_3 = |\mathbf{I}_3|^2 (R_{out} + jX_3)$$
$$= \frac{\omega^2 M_{13}^2 R_2^2 + (\omega^2 M_{12} M_{23} - \omega M_{13} X_2)^2}{(R_2 X_3 + R_{out} X_2)^2 + \left(X_2 X_3 - R_2 R_{out} - \omega^2 M_{23}^2\right)^2} (R_{out} + jX_3) = \overline{\mathbf{S}}_{23} + \overline{\mathbf{S}}_{13} \tag{9.8}$$

where $\overline{\mathbf{S}}_{ij}(i, j = 1, 2, 3)$ represents the complex power generated in coil-$j$ due to the current in coil-$i$ and $\overline{\mathbf{S}}_3$ represents the total complex power in coil-3.

## 9.4 Magnetic resonant

Due to the large leakage inductance and low coupling coefficient in the wireless power transfer coupling systems, it is required to use a compensation network to reduce the voltage-to-current rating in the coil and power converter. Commonly, external capacitors are connected in both primary and secondary sides to improve the system performance. Figure 9.7 shows four basic compensation topologies.

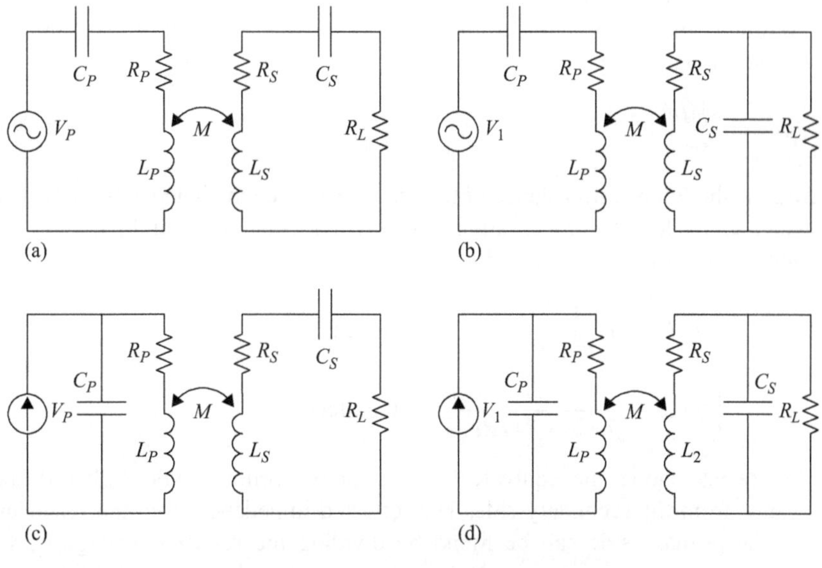


*Figure 9.7 Four basic compensation topologies: (a) SS, (b) SP, (c) PS, and (d) PP*

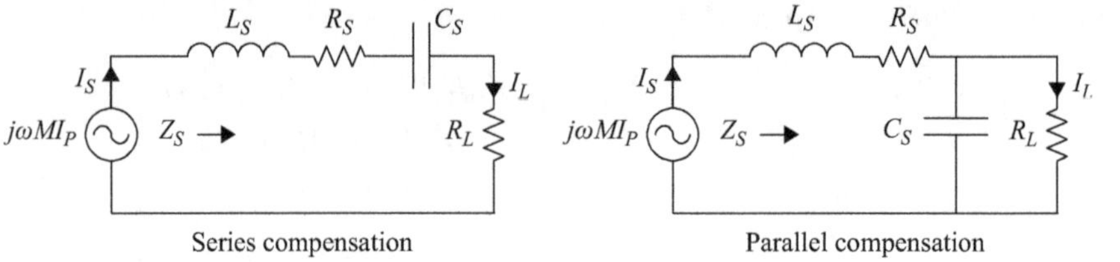


*Figure 9.8 The secondary side equivalent circuits of a wireless power transfer system*

Depending on how the capacitors are connected to the coils, there are four basic compensation topologies, which are called series-series (SS), series-parallel (SP), parallel-parallel (PP), and parallel-series (PS) (Wang *et al.*, 2005; Stielau and Covic; Zhang and Mi, 2015).

### *9.4.1 Compensation in secondary*

The secondary side equivalent circuits of a wireless power transfer system are shown in Figure 9.8. $L_P$ and $L_S$ are used to represent the inductance of primary coil and secondary coil, respectively. The capacitor added at the secondary side is $C_S$ which can be placed in-series or in-parallel with the secondary coil. $R_L$ is the equivalent resistor of the load. Resistors $R_P$ and $R_S$ represent the power losses of the primary and secondary coils, respectively. They are frequency-dependent because of the high-frequency skin and proximity effects in the windings and core losses in the magnetic material. In the below analysis, $R_P$ and $R_S$, are initially ignored because they are normally much smaller than the load resistor $R_L$, therefore, simplified expression can be obtained to describe the characteristic of different type of compensation topologies.

Assuming the current in the primary winding is $I_P$, the current flowing through the secondary winding is:

$$I_S = \frac{j\omega MI_P}{Z_S} \tag{9.9}$$

where $Z_S$ is the lumped impedance of the secondary side which depends on the load resistance and selected compensation topology. $Z_S$ using series and parallel compensation topologies are expressed as:

$$Z_S = \begin{cases} j\omega L_S + \left(\dfrac{1}{\omega C_S}\right) + R_L & \text{(Series)} \\ j\omega L_S + \dfrac{1}{j\omega C_S + (1/R_L)} & \text{(Parallel)} \end{cases} . \tag{9.10}$$

Figure 9.9 shows the equivalent circuit of the primary side with reflected impedance from the secondary side. The reflected impedance from the secondary side to the primary side can be found by dividing the reflected voltage by the

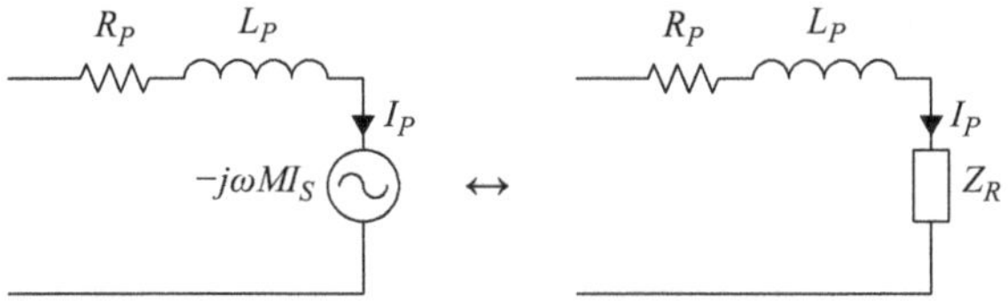


*Figure 9.9 The primary side equivalent circuit*

primary current. Thus, the reflected impedance can be expressed as:

$$Z_R = \frac{-j\omega MI_S}{I_P} = \frac{\omega^2 M^2}{Z_S}. \tag{9.11}$$

The reflected resistance and reactance can be derived as:

$$\operatorname{Re} Z_R = \begin{cases} \dfrac{\omega^4 C_S^2 M^2 R_L}{(\omega^2 C_S L_S - 1)^2 + \omega^2 C_S^2 R_L^2} & \text{(Series)} \\[2ex] \dfrac{\omega^2 M^2 R_L}{R_L^2(\omega^2 C_S L_S - 1)^2 + \omega^2 R_L^2} & \text{(Parallel)} \end{cases} \tag{9.12}$$

and

$$\operatorname{Im} Z_R = \begin{cases} \dfrac{-\omega^3 C_S M^2 (\omega^2 C_S L_S - 1)}{(\omega^2 C_S L_S - 1)^2 + \omega^2 C_S^2 R_L^2} & \text{(Series)} \\[2ex] \dfrac{-\omega^3 M^2 \left[C_S R_L^2(\omega^2 C_S L_S - 1) + L_S\right]}{R_L^2(\omega^2 C_S L_S - 1)^2 + \omega^2 L_S^2} & \text{(Parallel)} \end{cases}. \tag{9.13}$$

If the system works at the resonant frequency of secondary side, that is $\omega_O = 1/\sqrt{C_S L_S}$, (9.12) and (9.13) become the following simple equations:

$$\operatorname{Re} Z_R = \begin{cases} \dfrac{\omega_O^2 M^2}{R_L} & \text{(Series)} \\[2ex] \dfrac{M^2 R_L}{L_S^2} & \text{(Parallel)} \end{cases} \tag{9.14}$$

and

$$\operatorname{Im} Z_R = \begin{cases} 0 & \text{(Series)} \\[2ex] \dfrac{-\omega_O M^2}{L_L} & \text{(Parallel)} \end{cases}. \tag{9.15}$$

It can be observed that the reflected resistance is proportional to the square of the mutual inductance, implying that the output power drops rapidly with

*Table 9.1 Secondary impedance, load voltage, and current*

| Topology | Impedance ($Z_S$) | Load voltage ($V_L$) | Load current ($I_L$) |
|---|---|---|---|
| Series | $j\omega L_S + \frac{1}{j\omega C_S} + R_L$ | $I_S R_L$ | $I_S$ |
| Parallel | $j\omega L_S + \frac{1}{j\omega C_S + (1/R_L)}$ | $V_S$ | $\frac{V_S}{R_L}$ |

*Table 9.2 Reflected impedance at the secondary resonant frequency $\omega_O$*

| Topology | Reflected resistance | Reflected reactance | Output power ($I_P$ = constant) | Type of sources |
|---|---|---|---|---|
| Series | $\frac{\omega_O^2 M^2}{R_L}$ | 0 | $P \propto \frac{1}{R_L}$ | Voltage |
| Parallel | $\frac{M^2 R_L}{L_S^2}$ | $-\frac{M^2 \omega_O}{L_S}$ | $P \propto R_L$ | Current |

transmission distance. The influence of the load resistance to the reflected resistance of the series- and parallel-compensated secondary is different. In a series-compensated secondary, the reflected resistance is inversely proportional to the load resistance. When the primary side has a constant current, the load resistance decreases, then the power increases. It makes the secondary output like a voltage source. In contrast, the reflected resistance is proportional to the load resistance in a parallel-compensated secondary. It makes the secondary side output like a current source because the load resistance decreases, then the power decreases.

The advantage of a series-compensated secondary is that there is no reflected reactance at the secondary resonant frequency. On the other hand, the parallel-compensated secondary reflects a capacitive reactance at the secondary resonant frequency. But it can be turned out because it is independent of the load.

The secondary impedance, load voltage, and current for series- and parallel-compensated networks are summarized in Table 9.1. The reflected resistance and reactance at the resonant frequency of secondary side are also given in Table 9.2 and depend on the compensations used.

### 9.4.2 Compensation in primary

The primary winding can either be parallel- or series-compensated. Although the primary is parallel-compensated, an inductor is inserted to change the voltage source converter to a current source converter as shown in Figure 9.10. On the other hand, a voltage source converter can be connected directly to the couple coil when the primary coil is series-compensated as shown in Figure 9.11.

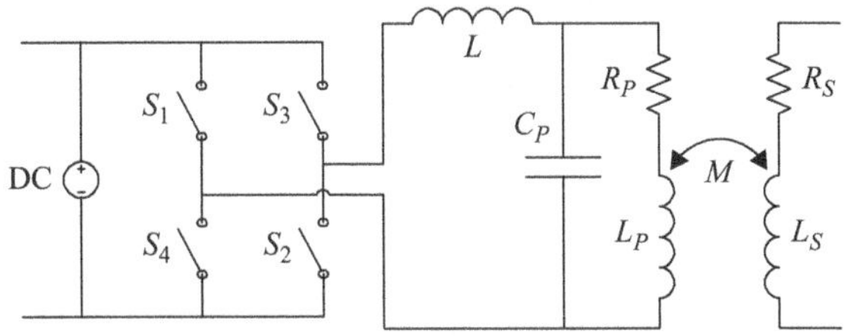


*Figure 9.10 Current source converter for parallel-compensated primary coil*

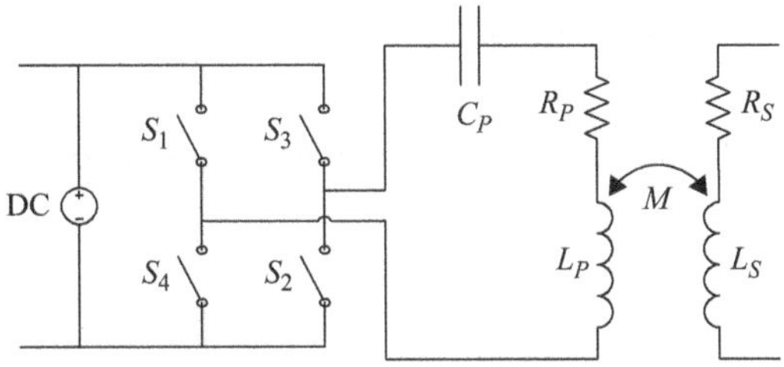


*Figure 9.11 Voltage source converter for series-compensated primary coil*

In order to reduce the power rating of the converter, the primary side capacitor is usually tuned to make the input voltage and current in phase at certain coupling and load conditions (Hu *et al.*, 2000; Wang *et al.*, 2004). Practically, the primary side compensation network is often tuned to make the primary side has a small portion of reactive power which enables the converter to operate in zero voltage switching (ZVS) and/or zero current switching (ZCS) conditions (Kutkut and Klontz, 1997; Zhang *et al.*, 2014). The power converter employs soft switching, the switching loss in the power devices are substantially reduced, thus offering higher efficiency than normal hard switching converters.

#### 9.4.2.1 SS compensation topology

The compensation capacitance for the primary coil, in order to achieve a zero phase angle, depends on the compensation topologies being used. In SS compensation, the reflected impedance on the primary side only consists of resistive component. When primary coil is operated at or near the resonant frequency of the secondary side where $1/\sqrt{L_P C_P} = 1/\sqrt{L_S C_S} = \omega_0$, the compensation capacitance for the primary coil can be obtained as:

$$C_P = \frac{C_S L_S}{L_P}. \tag{9.16}$$

It is important to note that the primary compensation capacitance is a constant value regardless of the coupling and load conditions. This is certainly an advantage of using SS compensation methods.

#### 9.4.2.2 SP compensation topology

The lumped impedance $Z_P$ of the primary side using SP compensation is:

$$Z_P = \frac{M^2 R_L}{L_S^2} - j\frac{\omega_0 M^2}{L_S} + j\left(\omega_0 L_P - \frac{1}{\omega_0 C_P}\right). \tag{9.17}$$

To achieve a zero phase angle at the primary side, $-(\omega_0 M^2/L_S) + \omega_0 L_P - (1/\omega_0 C_P) = 0$. Therefore, the compensation capacitance for the primary can be obtained as:

$$C_P = \frac{L_S}{\omega_0^2 L_P L_S - \omega_0^2 M^2} = \frac{C_S L_S^2}{L_P L_S - M^2}. \tag{9.18}$$

Since the mutual inductance is related to the magnetic coupling coefficient by:

$$k = \frac{M}{\sqrt{L_P L_S}}, \tag{9.19}$$

(9.18) can be rewritten as:

$$C_P = \frac{C_S L_S}{L_P} \times \left(\frac{1}{1-k^2}\right). \tag{9.20}$$

For SP compensation topology, the compensation capacitance for the primary coil varies with the change of magnetic coupling coefficient (i.e. The distance between the primary and secondary coils).

#### 9.4.2.3 PS compensation topology

The calculation of the primary side capacitance for parallel-compensated primary topologies is more complicated. A secondary side load quality factor $Q_S$ is defined, which is the ratio between reactive and real power. For series-compensated secondary, $Q_S$ is:

$$Q_S = \frac{\omega_0 L_S}{R_L}. \tag{9.21}$$

The lumped impedance $Z_P$ of the primary side using PS compensation topology can be obtained as:

$$Z_P = \frac{((\omega_0^2 M^2/R_L) + j\omega_0 L_P)(1/j\omega_0 C_P)}{(\omega_0^2 M^2/R_L) + j\omega_0 L_P + (1/j\omega_0 C_P)} = \frac{1}{(\omega_0^2 M^2/R_L)^2 + (\omega_0 L_P - (1/\omega_0 C_P))^2} \times \left[\frac{\omega_0^2 M^2 L_P}{C_P R_L} - \frac{\omega_0 M^2}{C_P R_L}\left(\omega_0 L_P - \frac{1}{\omega_0 C_P}\right) - j\frac{\omega_0^3 M^4}{C_P R_L^2} - j\frac{L_P}{C_P}\left(\omega_0 L_P - \frac{1}{\omega_0 C_P}\right)\right]. \tag{9.22}$$

To achieve a zero phase angle at the primary side, the imaginary part of $Z_P$ is kept zero, $(\omega_0^3 M^4/C_P R_L^2) + (L_P/C_P)(\omega_0 L_P - (1/\omega_0 C_P)) = 0$. Therefore,

$$\frac{1}{C_P} = \frac{\omega_0^4 M^4}{L_P R_L^2} + \omega_0^2 L_P$$
$$C_P = \frac{C_S L_S}{(M^4/L_P C_S L_S R^2) + L_P}. \tag{9.23}$$

Substituting (9.19) and (9.21) into (9.23), the compensation capacitance for the primary can be rewritten as:

$$C_P = \frac{C_S L_S}{L_P} \times \frac{1}{((L_S/C_S R_L^2) \times (M^4/L_P^2 L_S^2) + 1)} = \frac{C_S L_S}{L_P} \times \frac{1}{(Q_S^2 k^4 + 1)}. \tag{9.24}$$

#### 9.4.2.4 PP compensation topology

In the case of PP compensation topology, the secondary load quality factor $Q_S$ is defined as:

$$Q_S = \frac{R_L}{\omega_0 L_S}. \tag{9.25}$$

The lumped impedance $Z_P$ of the primary side using PP compensation topology can be obtained as:

$$Z_P = \frac{((M^2 R_L/L_S^2) - j(\omega_0 M^2/L_S) + j\omega_0 L_P)(1/j\omega_0 C_P)}{(M^2 R_L/L_S^2) - j(\omega_0 M^2/L_S) + j\omega_0 L_P + (1/j\omega_0 C_P)}. \tag{9.26}$$

The Im $Z_P$ is kept zero,

$$\text{Im } Z_{IN} = \frac{1}{C_P}\left[\left(\frac{M^2}{L_S} - L_P\right)\left(\omega_0 L_P - \frac{1}{\omega_0 C_P} - \frac{\omega_0 M^2}{L_S}\right) + \left(\frac{-M^4 R_L^2}{\omega_0 L_S^4}\right)\right].$$

Therefore,

$$\frac{1}{C_P} = \frac{M^4 L_S R_L^2}{L_S^4 (L_S L_P - M^2)} + \frac{\omega_0^2 (L_S L_P - M^2)}{L_S}$$
$$C_P = \frac{(L_P L_S - M^2) L_S^4}{M^4 L_S R_L^2 + \omega_0^2 L_S^3 (L_P L_S - M^2)^2} = \frac{(L_P L_S - M^2) C_S L_S^2}{(M^4 C_S R_L^2/L_S) + (L_P L_S - M^2)^2}. \tag{9.27}$$

Substituting (9.19) and (9.25) into (9.27), the compensation capacitance for the primary can be rewritten as:

$$C_P = \frac{C_S L_S}{L_P} \times \left[\frac{(1 - (M^2/L_P L_S))}{(C_S R_L^2/L_S) \times (M^4/L_P^2 L_S^2) + (1 - (M^2/L_P L_S))^2}\right]$$
$$= \frac{C_S L_S}{L_P} \times \left[\frac{(1 - k^2)}{Q_S^2 k^4 + (1 - k^2)^2}\right]. \tag{9.28}$$

*Table 9.3 Compensation capacitances for primary to achieve zero phase angle*

| Topology | Primary capacitance ($C_P$) | Load quality factor ($Q_S$) |
|---|---|---|
| SS | $\frac{C_S L_S}{L_P}$ | |
| SP | $\frac{C_S L_S}{L_P} \times \frac{1}{1-k^2}$ | |
| PS | $\frac{C_S L_S}{L_P} \times \frac{1}{Q_S^2 k^4 + 1}$ | $\frac{\omega_0 L_S}{R_L}$ |
| PP | $\frac{C_S L_S}{L_P} \times \frac{1-k^2}{Q_S^2 k^4 + (1-k^2)^2}$ | $\frac{R_L}{\omega_0 L_S}$ |

For PS and PP compensation topologies, the compensation capacitance for the primary is affected by both the magnetic coupling coefficient and load conditions.

The compensation capacitances for different types of compensation topologies are summarized in Table 9.3.

## 9.5 Influence of the load resistance

The reflected resistance exhibited at the resonant frequency is very important to determine the performance of the wireless power transfer system. The power transferred from the primary to the secondary is the reflected resistance multiplied by the square of primary current as given by:

$$P = (\mathrm{Re}Z_R)I_P^2. \tag{9.29}$$

Selecting the appropriated compensation topology and identifying the optimal operating frequency in the secondary side, a boundary frequency $\omega_B$ is defined and it can be calculated as:

$$\omega_B = \frac{R_L}{L_S}. \tag{9.30}$$

When $\omega > \omega_B$ (i.e. $(\omega^2 M^2/R_L) > (M^2 R_L/L_S^2)$), the reflected resistance using a series-compensated secondary is higher than parallel compensation. This implies that the reflected resistance can be the maximum value of $\mathrm{Re}Z_R$ in (9.12) under this condition. A system with series-compensated secondary could have a higher output power and efficiency than parallel-compensated system with the same primary current $I_P$. Conversely, parallel-compensated secondary could achieve a better result than series-compensated system while $\omega < \omega_B$ (i.e. $(\omega^2 M^2/R_L) < (M^2 R_L/L_S^2)$). It is because the maximum reflected resistance can be located at a relatively lower frequency resulting a lower transmission loss (Liu *et al.*, 2008).

As far as the efficiency is concerned, the power losses of the couple coils are included in the following study. In the primary side, the active power is dissipated in the coil $R_P$ and reflected resistance $ReZ_R$, no matter what type of the compensation topology is chosen. Then the power absorbed by $ReZ_R$ can be further divided into two components, namely, the power loss in the secondary couple coil $R_S$, and the power absorbed by the load $R_L$. The general expression of the efficiency of couple coils $\eta$ is given by:

$$\eta = \eta_P \eta_S \tag{9.31}$$

where $\eta_P$ is the efficiency of the primary side and it is equal to

$$\frac{\operatorname{Re} Z_R}{\operatorname{Re} Z_R + R_P} \tag{9.32}$$

and $\eta_S$ represents the efficiency of the secondary side which has different forms depending on the secondary compensation topologies.

### 9.5.1 Series-compensated secondary

In the series compensation, $\eta_S$ is a simple expression because the load current flowing through the load is as same as the current in the secondary coil, that is:

$$\eta_S = \frac{R_L}{R_L + R_S}. \tag{9.33}$$

The reflected resistance $ReZ_R$ including the power loss in the secondary coil is:

$$\operatorname{Re} Z_R = \frac{\omega^2 M^2}{(R_S + R_L)}. \tag{9.34}$$

Therefore, the efficiency of the couple coils with series-compensated secondary can be expressed in (9.35) by substituting (9.32)–(9.34) into (9.31). It can be derived as:

$$\eta_{SC} = \frac{1}{1 + ((R_S + R_L)R_P/\omega^2 M^2)} \frac{R_L}{R_L + R_S}. \tag{9.35}$$

### 9.5.2 Parallel-compensated secondary

The efficiency analysis of parallel-compensated secondary is more complicated than the series-compensated secondary. The efficiency of the secondary side, $\eta_S$, can be derived as shown in (9.36).

$$\eta_S = \frac{R_L}{R_L + R_S + R_S R_L^2 \omega^2 C_S^2} \tag{9.36}$$

The reflected resistance $\operatorname{Re} Z_R$ including the power loss in the secondary coil is:

$$\operatorname{Re} Z_R = \frac{\omega^2 M^2 (R_L + R_S + R_S R_L^2 \omega^2 C_S^2)}{R_S^2 + \omega^2 (L_S + C_S R_S R_L)^2} \tag{9.37}$$

By combining (9.32), (9.36), and (9.37) into (9.31), the efficiency of the couple coils with parallel-compensated secondary can be derived as:

$$\eta_{SP} = \frac{\omega^2 M^2 R_L}{\omega^2 M^2 \left(R_S + R_L + R_S R_L^2 \omega^2 C_S^2\right) + \left[R_S^2 + \omega^2 (L_S + C_S R_S R_L)^2\right] R_P}. \tag{9.38}$$

Two couple coils are used to demonstrate this theory. The inductive parameters of the couple coils are given in Table 9.4.

Figure 9.12 shows the efficiency at different operating frequencies when the load resistance is 25 Ω. Under this load condition, the boundary frequency is calculated as $f_B = R_L/2\pi L_S = 50$ kHz. The dashed line is the result of series-compensated secondary while tracking the resonant condition of secondary side at different frequencies. It can be observed that series compensation (dashed line) is always better than parallel compensation (solid lines) when the system is operated above the boundary frequency. Conversely, the maximum efficiency of parallel compensation is better than series compensation when the system operated below the boundary frequency, as shown by the dots in Figure 9.12. When couple coils are operated at the boundary frequency, the efficiency of using series compensation or parallel compensation is similar.

*Table 9.4 Inductive parameters of the couple coils*

| | $L_P$, $L_S$ | $M$ | $K$ | $R_P$, $R_S$ | $R_L$ |
|---|---|---|---|---|---|
| Value | 79.58 μH | 15.96 μH | 0.2 | 0.125 Ω | 25 Ω |

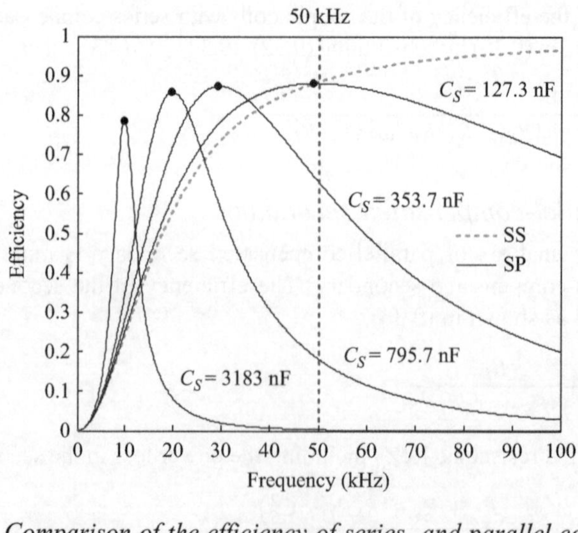


*Figure 9.12 Comparison of the efficiency of series- and parallel-compensated secondary*

## 9.6 Transmission distance

The analysis of mutual coupling between two couple coils has been well established. It has been shown in Waffenschmidt and Staring (2009) and Imura and Hori (2011) that the magnetic coupling coefficient $k_{12}$ between two resonator coils can be expressed as:

$$k_{12} = \frac{1}{\left[1 + 2^{2/3}\left(d/\sqrt{r_1 r_2}\right)^2\right]^{3/2}}. \tag{9.39}$$

where $d$ is the transmission distance. $r_1$ and $r_2$ are the radii of the transmitter and receiver coils, respectively. If the transmission distance $d$ is comparable with the radii of the transmitter and receiver coils $r_1$ and $r_2$ (i.e. $d \geq r_1$ and $d \geq r_2$), a simplified equation can be obtained as shown in the following equation:

$$k_{12} \approx \frac{1}{2\left(d/\sqrt{r_1 r_2}\right)^3}. \tag{9.40}$$

It is important to note that the real-power energy efficiency is proportional to the square of the magnetic coupling coefficient which is:

$$\eta \propto \frac{k_{12}^2}{2}, \tag{9.41}$$

implying that the efficiency drops rapidly with transmission distance. These limitations of wireless power transfer have been addressed in Imura *et al.* (2011) and Pinuela *et al.* (2013). If identical coils are used and $d \gg r$, it has been pointed out in Imura and Hori (2011) and Pinuela *et al.* (2013) that the magnetic coupling coefficient and the efficiency will decrease exponentially with transmission distance $d$. A typical graphical relationship of the power transmission efficiency versus transmission distance is shown in Figure 9.13.

So far, most of these modern wireless power applications (i.e. mobile phones and electric vehicles) are of the "short-range" type in which the diameters of the

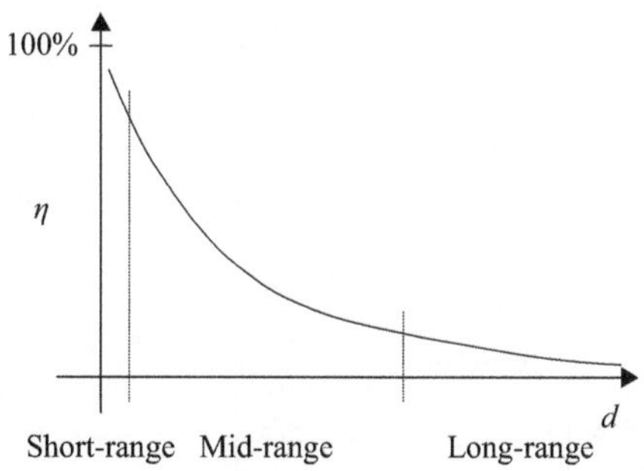


*Figure 9.13 Typical exponential decay curve of the efficiency η as a function of transmission distance* d *for wireless power transfer*

transmitter and receiver coils are designed much larger than the transmission distance to achieve high power transmission efficiency. For wireless electric vehicle charging, the typical working distance between the transmitter and receiver coils is 100–150 mm. The diameters of the transmitter and receiver are in between 400 mm and 500 mm. When the transmission distance is larger than the diameters of the transmitter and receiver, which is termed "mid-range" wireless power transfer.

At the boundary of the short-range and mid-range transmission at which the transmitter coil dimension and the transmission distance is the same, it has been shown that high system efficiency can be achieved by designing the transmitter and receiver coils with high $Q$ factor (Mur-Miranda *et al.*, 2010). In Pinuela *et al.* (2013), a system efficiency of 77% has been reported in transferring power of 100 W over a distance of 300 mm when the dimension of the transmitter coil and the transmission distance are the same. In this experiment, a transmitter coil (with a diameter of 300 mm and a $Q$ factor of 1,270) and a receiver coil (with a diameter of 200 mm and a $Q$ factor of 1,100) are used. However, the system efficiency drops significantly when the transmission distance exceeds the transmitter coil dimension. The quality factor is $Q = \omega L/R$ where $\omega = 2\pi f$ is the angular frequency, $L$ is the inductance, and $R$ is the resistance of the loop resonator at operating frequency. An operating frequency of 10 MHz was adopted in Kurs *et al.* (2007) in order to increase the transfer efficiency over "mid-range" distance. In this system, a coil radius $r$ of 300 mm, a transmission distance $d$ of 2,400 mm (i.e. a ratio of $d/r$ of 8) and a $Q$ of about 1,000 is used. An efficiency of 40% was reported.

## 9.7 Transmission efficiency and energy efficiency of the system

A general power transfer network is shown in Figure 9.14 helping to differentiate the transmission efficiency $\eta_T$ and energy efficiency $\eta_E$ of a system (Hui *et al.*, 2014). The transmission efficiency $\eta_T$ is the ratio of the output power $P_3$ and available power from the output of the power source for Port-1 $P_2$, and does not include the power loss in the power source.

The transmission efficiency $\eta_T$ is defined as:

$$\eta_T = \frac{P_3}{P_2}. \tag{9.42}$$

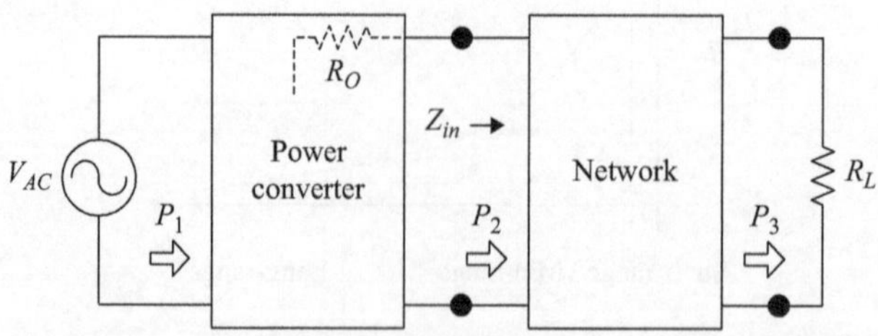


*Figure 9.14 Schematic of a power transfer network*

The maximization of transmission efficiency $\eta_T$ aims at minimizing the power losses in the network. Using an SS-compensated wireless power transfer system as an example, the design objective is to maximize the transmission efficiency function as shown in the following equation:

$$\eta_T = \frac{I_S^2 R_L}{I_P^2 R_P + I_S^2 (R_S + R_L)} \tag{9.43}$$

where $R_P$ and $R_S$ represent the power losses in the primary and secondary couple coils, respectively. The power loss in the couple coils includes the energy loss in the ESR of the capacitors, winding resistance, and magnetic core. The $I_P$ and $I_S$ are the current in the primary and secondary couple coils, respectively. The $R_L$ is the load resistance. Figure 9.15 shows typical variations of primary, secondary, and transmission efficiency of an SS-compensated wireless power transfer system. Increase in the load power by reducing the load resistance, the power losses in the secondary coil are increased accordingly. However, the reflected resistance in the primary coil given by (9.14) is inversely proportional to the load resistance. As a result, the efficiency of the primary coil increases. There is an optimum load resistance for maximum transmission efficiency $\eta_T$.

The system energy efficiency $\eta_E$ refers to the ratio of the output power $P_3$ and total input power $P_1$ from the power source. Its calculation including the power loss in the source $\eta_E$ is defined as:

$$\eta_E = \frac{P_3}{P_1} = \frac{P_2}{P_1} \times \frac{P_3}{P_2} = \eta_C \times \eta_T \tag{9.44}$$

where $\eta_C$ is the efficiency of the power source (i.e. power converter). A system obtained a high transmission efficiency does not necessarily imply high system

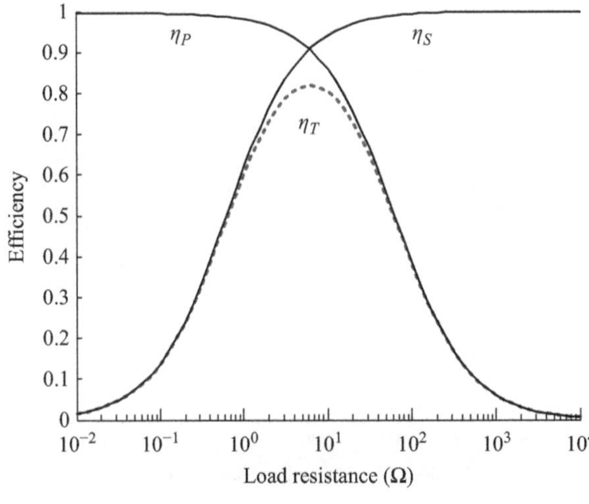


*Figure 9.15 Variations of primary, secondary, and transmission efficiency of an SS-compensated wireless power transfer system*

energy efficiency because a significant amount of power could be lost in the power converter. To achieve an ultimate performance of the system from the power source to the load, we need to not only maximize the transmission efficiency of the network but also need to minimize the power loss in the power converter. The system design optimization requires engineering trade-offs to balance the system's cost, size, weight, reliability, and performance.

## 9.8 Transducer power gain and maximum power transfer of the system

The transducer power gain $G_T$ of the network is defined as:

$$G_T = \frac{P_3}{P_{2,\max}} \tag{9.45}$$

where $P_3$ is the power delivered to the load and $P_{2,\max}$ is the maximum available power of the power source. $P_{2,\max}$ is obtained from the source when the impedance $Z_{in}$ connected it is the complex conjugate of the source impedance, a consequence of the maximum power theorem. In Figure 9.14, maximum output power $P_2$ can be achieved if $Z_{in} = R_O$. Under this condition, it is important to note that same amount of power will be dissipated in source resistance $R_O$ (i.e. $P_1 \geq 2 \times P_2$). As a result, the system energy efficiency $\eta_E$ that includes the power loss in the power source cannot excess 50%. As explained in (9.42), the maximum energy efficiency $\eta_E$ is governed by $\eta_E \leq (1/2) \times \eta_T$.

Typical variations of transducer power gain and energy efficiency of an SS-compensated wireless power transfer system are plotted in Figure 9.16. When the load resistance is about 100 Ω, the system achieves a high transducer power gain

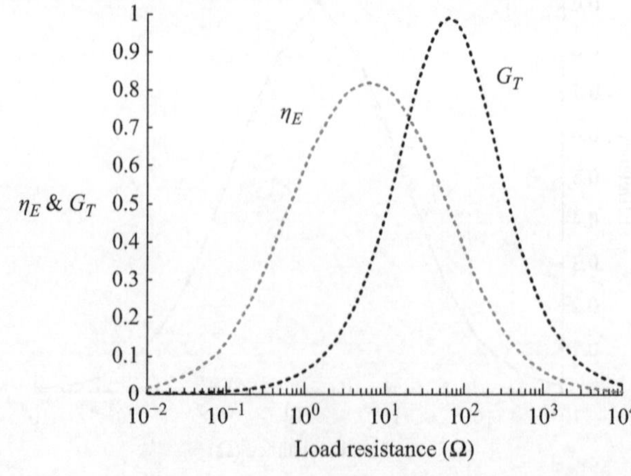


*Figure 9.16 Variations of transducer power gain $G_T$ and system energy efficiency $\eta_E$ of an SS-compensated wireless power transfer system*

between the transmitter and receiver coils (i.e. $G_T = 0.95$). But, the overall system energy efficiency is only 45%. The objective of the maximization of transducer power gain is to maximize the power delivery from the transmitter to the load based on the maximum power transfer principle. This implies that the transmission distance is of the primary concern, not the system energy efficiency. It is suitable for relatively low power application such as wireless charging of wireless sensor nodes (Xie *et al.*, 2013; Porto *et al.*, 2015).

## 9.9 Frequency-splitting phenomenon

A phenomenon recently observed in the wireless power transfer research is called "frequency splitting" (Sample *et al.*, 2011; Huang *et al.*, 2014). Frequency splitting occurs when the conditions for the maximum power theorem cannot be met at the resonance frequency of the resonators within the over-coupled region. The transmission power gain $G_T$ as a function of the mutual coupling coefficient and operating frequency is plotted in Figure 9.17. The parameters of the wireless power transfer system are tabulated in Table 9.5. The mutual coupling coefficient $k_{12}$ is inversely proportional to the transmission distance $d$. So a decreasing $k_{12}$ means an

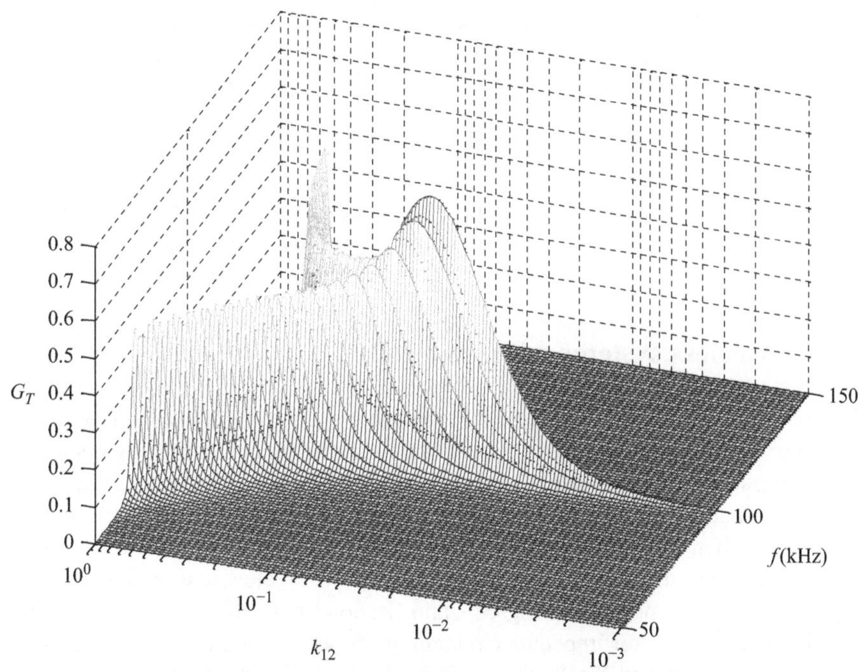


*Figure 9.17 A plot of $G_T$ as a function of mutual coupling coefficient $k_{12}$ and operating frequency $f$ for the example with parameters shown in Table 9.5*

Table 9.5 Parameters for an SS-compensated wireless power transfer system

| Frequency | $L_1$ | $L_2$ | $R_S$ | $R_L$ |
|---|---|---|---|---|
| 100 kHz | 100 μH | 100 μH | 1 Ω | 5 Ω |

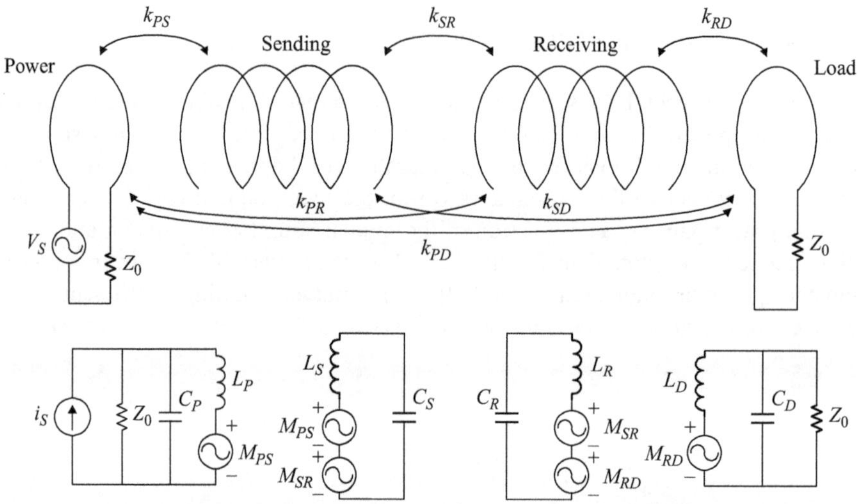


Figure 9.18 Wireless power system with four coils and its equivalent circuit

increasing $d$. It can be seen from Figure 9.17 that, within the over-coupled region, maximum $G_T$ occurs at two frequencies in this example. Beyond the critical coupling point, $G_T$ reduces exponentially with an increasing $d$ (i.e. a decreasing $k_{12}$).

## 9.10 Wireless systems with four coils

The recent interests in the four-coil systems were sparked by the work reported in Kurs *et al.* (2007). This structure involves a power driving coil, a sending resonator, a receiving resonator, and a load coil as shown in Figure 9.18. Although the coupled mode theory was used in Kurs *et al.* (2007) to explain its operation, several references (Chen *et al.*, 2010; Cheon *et al.*, 2011) have transformed the analysis using electric circuit theory that is easily understood by electrical engineers.

The use of the power driving coil and the load coil offers two extra mutual coupling coefficients for impedance matching (assuming that the mutual coupling of the driving loop and the load loop is negligible). Beside the mutual coupling coefficient between the sending coil and the receiving coil ($k_{SR}$), the two extra coefficients are the mutual coupling coefficient between the power driving coil and the sending coil ($k_{PS}$), and that between the receiving coil and the load coil ($k_{RD}$).

Cheon *et al.* (2011) contain a detailed circuit analysis of this four-coil system and step-by-step explanations on how to match impedances in various stages in order to maximize power transfer. The four-coil system provides three mutual coupling coefficients, $k_{PS}$, $k_{SR}$, and $k_{RD}$, which can be utilized to maximize the power transfer if the following condition can be met as shown in the following equation:

$$\frac{k_{PS}k_{RD}}{k_{SR}} = 1 \tag{9.46}$$

In the mid-range wireless power transfer applications (i.e. the transmission distance is larger than the coil radii, $d \gg r$), the mutual coupling coefficient between the sending coil and the receiving coil ($k_{SR}$) should be small. Compared with basic two-coil systems, the two extra mutual coupling coefficients ($k_{PS}$ and $k_{RD}$) in the four-coil systems provide extra freedom for extending the transmission distance by minimizing $k_{SR}$ through the use of (9.46). For example, if $k_{SR} = 0.01$, by keeping $k_{PS} = 0.1$ and $k_{RD} = 0.1$, the conditions of (9.46) can be met. Therefore, the two extra coefficients allow a small $k_{SR}$ to be selected. However, the impedance matching requirement in order to satisfy (9.46) also implies that such system has its overall energy efficiency not higher than 50%. So far, this design approach has been proved to be an effective means in extending the transmission distance at the expense of system energy efficiency.

Frequency splitting phenomenon has been observed in the four-coil systems (Sample *et al.*, 2011). The reason for such phenomenon is similar to that of the two-coil systems when the concept of impedance matching is involved. Within the over-coupled range, near-constant power transfer can be achieved. Therefore, the power receiving load can be placed within a certain range, which becomes an important feature for medical implants (Waters *et al.*, 2012).

Most of the recent mid-range wireless power research based on the four-coil systems adopts the maximum power transfer approach by matching the load impedance with the source impedance. Although this approach is not preferably from a system energy efficiency standpoint, it could achieve a longer transmission distance.

## 9.11 Conclusion

Starting from Tesla's principles of wireless power transfer a century ago, this chapter outlines basic concepts, study method, design principle, and latest development of the wireless power transfer technology. The most important fundamental concepts and electrical characteristics of a range of wireless power transfer design are summarized. The differences between the system energy efficiency and the transmission efficiency are also highlighted. The chapter also covers the four-coil wireless power transfer system which uses impedance matching technique for maximizing the transmission distance.

The challenges in wireless power transfer for electric vehicles include low-loss couple coils design, high-efficient switched-mode power supplies design, new dynamic power, and frequency control strategies for making the wireless power systems adaptive to the presence and movements of objects. To cope with a number of challenges in the future, research requires the combined efforts of the professionals and researchers in the areas of power electronics, control, radio frequency, high-frequency magnetics, material science, and electromagnetic compatibility.

## References

Budhia, M., Covic, G., and Boys, J. (2010) A new IPT magnetic coupler for electric vehicle charging systems, *Proc. IECON 2010 – 36th Annual Conference on IEEE Industrial Electronics Society*, Glendale, AZ, USA, 2487–2492.

Chen, C., Chu, T.Lin, C., and Jou, Z. (2010) A study of loosely coupled coils for wireless power transfer, *IEEE Transactions on Circuits and System – Part II: Express Briefs*, 57(7), 536–540.

Cheon, S., Kim, Y., Kang, S., Lee, M., Lee, J., and Zyung, T. (2011) Circuit-model-based analysis of a wireless energy-transfer system via coupled magnetic resonances, *IEEE Transaction on Industrial Electronics*, 58(7), 2906–2914.

https://commons.wikimedia.org/wiki/File:US_patent_1119732_Nikola_Tesla_1907_Apparatus_for_transmitting_electrical_energy.png

https://commons.wikimedia.org/wiki/File:Tesla_wireless_power_future_1934.png

https://commons.wikimedia.org/wiki/File:Tesla_wireless_power_circuit.jpg

Hu, A., Boys, J., and Covic, G. (2000) Frequency analysis and computation of a current-fed resonant converter for ICPT power supplies, *Proc. International Conference on Power System Technology*, Perth, Australia, Vol. 1, 327–332.

Huang, R., Zhang, B., Qiu, D., and Zhang, Y. (2014) Frequency splitting phenomena of magnetic resonant coupling wireless power transfer, *IEEE Transactions on Magnetics*, 50(11).

Hui, S., Zhong,W., and Lee, C. (2014) A critical review of recent progress in mid-range wireless power transfer, *IEEE Transactions on Power Electronics*, 29(9), 4500–4511.

Kurs, A., Karalis, A., Moffatt, R., Joannopoulos, J., Fisher, P., and Soljacic, M. (2007), Wireless power transfer via strongly coupled magnetic resonances, *Science*, 317(5834), 83–86.

Imura, T. and Hori, Y. (2011) Maximizing air gap and efficiency of magenticresonant coupling for wireless power transfer using equivalent circuit and Neumann formula, *IEEE Transactions on Industrial Electronics*, 58(10), 4746–4752.

Kutkut, N., and Klontz, K. (1997) Design considerations for power converters supplying the SAE J-1773 electric vehicle inductive coupler, *Proc. Applied Power Electronics Conference and Exposition*, Atlanta, GA, USA, Vol. 2, 841–847.

Lee, C., Zhong, W., and Hui, S. (2012) Effects of magnetic coupling of nonadjacent resonators on wireless power domino-resonator systems, *IEEE Transactions on Power Electronics*, 27(4), 1905–1916.

Li, S., and Mi, C. (2015) Wireless power for electric vehicle applications, *IEEE Journal of Emerging and Selected Topics in Power Electronics*, Austin, TX, USA, 3(1),4–17.

Liu, X., Ng, W., Lee, C., and Hui, S. (2008) Optimal operation of contactless transformers with resonance in secondary circuits, *Proc. 2008 Twenty-Third Annual IEEE Applied Power Electronics Conference and Exposition*, 645–650.

Lomas, R. (1999) The man who invented the twentieth century – Nikola Tesla – forgotten genius of electricity, *Headline*, 146.

Lukic, S., and Pantic, Z. (2013) Cutting the cord: static and dynamic inductive wireless charging of electric vehicles, *IEEE Electrification Magazine*, 1(1), 57–64.

Madawala, U., and Thrimawithana, D. (2011) A bidirectional inductive power interface for electric vehicles in V2G systems, *IEEE Transactions on Industrial Electronics*, 58(10), 4789–4796.

Marincic, A. (1982) Nikola Tesla and the wireless transmission of energy, *IEEE Transactions on Power Apparatus and Systems*, PAS-101(10), 4064–4068.

Mur-Miranda, J., Fanti, G., Feng, Y., *et al.* (2010) Wireless power transfer using weakly coupled magnetostatic resonators, *Proc. IEEE Energy Conversion Congress and Exposition*, Atlanta, GA, USA, 4179–4186.

Nagatsuka, Y., Ehara, N., Kaneko, Y., Abe, S., and Yasuda, T. (2010) Compact contactless power transfer system for electric vehicles, *Proc. 2010 International Power Electronics Conference (IPEC)*, Sapporo, Japan, 807–813.

Pinuela, M., Yates, D., Lucyszyn, S., and Mitcheson, P. (2013) Maximising DC to load efficiency for inductive power transfer, *IEEE Transactions on Power Electronics*, 28(5), 2437–2447.

Porto, R., Brusamarello, V., Muller, I., and Sousa, F. (2015) Design and characterization of a power transfer inductive link for wireless sensor network nodes, *Proc. IEEE International Instrumentation and Measurement Technology Conference (I2MTC)*, Pisa, Italy, 1261–1266.

Sample, A., Meyer, D., and Smith, J. (2011) Analysis, experimental results and range adaptation of magnetically coupled resonators for wireless power transfer, *IEEE Transactions on Industrial Electronics*, 58(2), 544–554.

Sarkar, T.K., Mailloux, R., Oliner, A.A., Salazar-Palma, M., and Sengupta, D.L. (2006) History of wireless: Chapter 8. *Nikola Tesla and His Contributions to Radio Development*, Hoboken, NJ: Wiley-IEEE Press.

Stielau, O., and Covic, G. (2000) Design of loosely coupled inductive power transfer systems, *Proc. 2000 International Conference on Power System Technology*, Perth, Australia, 1, 85–90.

Tesla, N. (1914) Apparatus for Transmitting Electrical Energy, U.S. Patent No. 1,119,732. Washington, DC: U.S. Patent and Trademark Office.

Waffenschmidt, E., and Staring, T. (2009) Limitation of inductive power transfer for consumer applications, *European Conference on Power Electronics and Applications, 2009. EPE'09*, Barcelona, Spain, 1–10.

Wang, C., Covic, G., and Stielau, O. (2004) Power transfer capability and bifurcation phenomena of loosely coupled inductive power transfer systems, *IEEE Transaction on Industrial Electronics*, 51(1), 148–157.

Wang, C., Stielau, O., and Covic, G. (2005) Design considerations for a contactless electric vehicle battery charger, *IEEE Transactions on Industrial Electronics*, 52(5), 1308–1314.

Waters, B., Sample, A., Bonde, P., and Smith, J. (2012) Powering a ventricular assist device (VAD) with the free-range resonant electrical energy delivery (FREE-D) system, *Proceedings of IEEE,* 100(1), 138–149.

Wheeler, L. (1943) *Tesla's Contribution to High Frequency*, Electrical Engineering, NY, 355.

Xie, L., Yi, S., Hou, Y., and Lou, A. (2013) Wireless power transfer and applications to sensor networks, *IEEE Wireless Communication*, 20(4), 140–145.

Zhang, W., and Mi, C. (2015) Compensation topologies for high power wireless power transfer systems, *IEEE Transactions on Vehicular Technology*. DOI: 10.1109/TVT.2015.2454292.

Zhang, W., Wong, S., Tse, C., and Chen, Q. (2014) Analysis and comparison of secondary series- and parallel-compensated inductive power transfer systems operating for optimal efficiency and load-independent voltage-transfer ratio, *IEEE Transactions on Power Electronics*, 29(6), 2979–2990.

Zhong, W., Lee, C., and Hui, S. (2012) Wireless power domino-resonator systems with noncoaxial axes and circular structures, *IEEE Transactions on Power Electronics*, 27(11), 4750–4762.

Zhong, W., Lee, C., and Hui, S. (2013) General analysis on the use of Tesla's resonators in domino forms for wireless power transfer, *IEEE Transactions on Industrial Electronics*, 60(1), 261–268.

# Chapter 10

# Move-and-charge technology for electric vehicles

*Chun T. Rim*[1]

Recently, roadway-powered electric vehicles (RPEVs), also called move-and-charging EVs, are widely studied, being attractive candidates for future transportations because they do not rely on large and heavy batteries but directly and efficiently get power while moving along a road. The inductive power transfer systems (IPTSs) used for the wireless powering of RPEVs are widely reviewed in this chapter. The development history of the IPTS is tracked from the origin to recent RPEVs. Throughout the history of the RPEV, the size, weight, efficiency, air-gap, lateral tolerance, and cost of the IPTS have been substantially improved, and now RPEVs are becoming commercialized. Important milestones of the developments of the IPTS and RPEV are introduced in this chapter, focusing on on-line electric vehicles (OLEV) that have been commercialized firstly in the world in 2013.

## 10.1 Introduction to the wireless power transfer technologies for EVs

EVs are becoming more attractive than ever due their clean, calm, and low operation cost. Various EVs, including pure battery EV (PEV), hybrid EV (HEV), plug-in hybrid EV (PHEV), battery replace EV (BREV), and roadway powered EV (RPEV), which is also called move-and-charge EV, are being developed. One of the challenges to the commercialization of the EV is the battery, which is still heavy, bulky, and expensive. The charging of battery is an obstacle in the commercialization of EVs because they should be frequently recharged after operating for a short range due to the low energy density of batteries. The currently available quick charging time of a few 10 minutes is still too long for drivers accustomed to rapid fueling and deteriorates battery lifetime severely, requiring quite expensive and large-size charging facilities. Extremely quick charging less than 5 min make the problems worse because the lifetime of a battery tends to be rapidly degraded and a large and expensive charging infra is inevitable.

[1]Department of Nuclear & Quantum Engineering, Korea Advanced Institute of Science and Technology, Korea

Once the power supply rails for transmitting power to RPEV are fully deployed under the road, RPEVs do not require battery for their traction because they directly get required power from a road while they are moving on it. Hence, RPEVs are most free from the battery-related problems among EVs and quite promising candidates for future transportation of small cars, passenger cars, buses, trucks, and trains, even in competition with internal combustion engine vehicles (ICEVs). The biggest challenge of the RPEV in commercialization is to transfer high power from the road in an efficient, economic, and safe way. The power transfer can either be wired or wireless. Even though wired electric buses are no longer widely used in urban areas now, it is not amazing that still the highest speed train is powered through pantographs. Wired power transfer is gradually being replaced with wireless one as hundreds of kilowatts of power become available. Various wireless power transfer systems (WPTSs) have been widely developed for RPEVs.

In this chapter, a history of the WPTSs for RPEVs is described from its advent developments to recent technologies. Important technical issues in the developments of IPTSs, the majority of WPTSs, are explained, and major milestones of the developments of RPEVs are summarized, focusing on the developments of OLEVs that have been recently commercialized. A recent trend of interoperability between the stationary and in-motion charging is added at the end.

## 10.2 Basic principles of WPTSs for RPEV

### *10.2.1 Configuration of the WPTS*

The WPTS for RPEV should be able to deliver high power efficiently through a moderate air-gap for avoiding collisions between the RPEVs and the road. This is quite crucial for the RPEVs in practice because the air-gap is ever changing due to the weight of passengers and dynamic vibration of the vehicles. The WPTS are composed of two subsystems: one is the roadway subsystem for providing power, which includes a rectifier, a high-frequency inverter, a primary capacitor bank, and a power supply rail, and another is the on-board subsystem for receiving power, which includes a pick-up coil, a secondary capacitor bank, a rectifier, and a regulator for battery, as shown in Figure 10.1(a). The roadway subsystem should be so robust and cheap that it may withstand severe road environments for a long time and should be economic to install over a long distance, whereas the on-board subsystem should be compact in size and light in weight so that it may be adopted into the RPEV. These different requirements make the development of RPEVs more difficult than stationary charging EVs.

WPTSs, in general, can be classified into IPTSs, coupled magnetic resonance systems (CMRSs) (Huh *et al.*, 2011c, 2013; Lee *et al.*, 2013), and capacitive power transfer systems (CPTSs) (Hanazawa *et al.*, 2012). These three systems were regarded as quite different from one another; however, CMRS are found to be just a special form of IPTS whose quality factor $Q$ is extremely high, and resonating repeaters extend power delivering distance (Lee *et al.*, 2013). Moreover, it is no longer true that CMRS are appropriate for a long-distance power delivery whereas

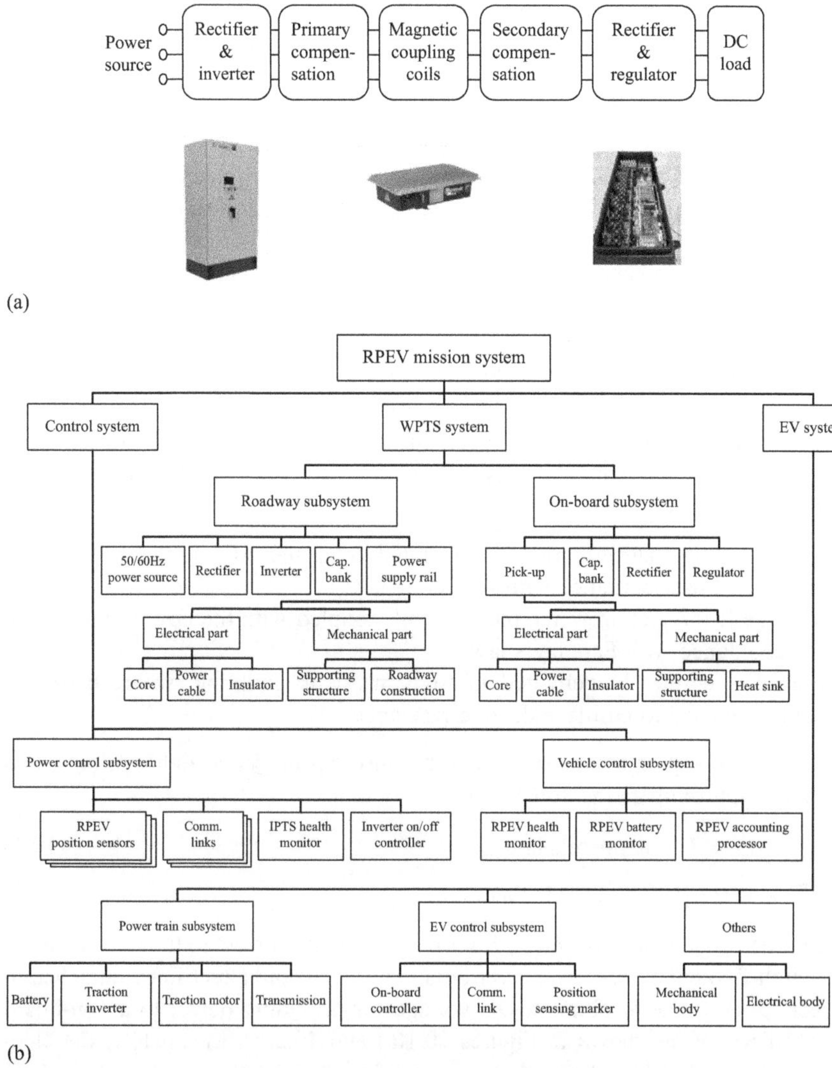


*Figure 10.1 Overall configurations of RPEV and IPTS: (a) The configuration of a WPTS for RPEV and (b) The breakdown structure of RPEV. © 2015 IEEE. Reprinted with permission from Choi et al. (2015c)*

IPTS are adequate for a short-distance high-power delivery because, recently, a new world record of 5-m-off long-distance power delivery has been demonstrated by using IPTS without a very high $Q$ factor (Park *et al.*, 2012, 2015). Considering the difficulties in maintaining resonance for multiple resonant repeaters with very high $Q$ factor and the bulky configuration of pick-ups, conventional CMRSs are not quite appropriate candidates for RPEV, in general. Therefore, IPTSs will be dealt with in detail in the subsequent sections, though CMRSs and CPTSs are not completely excluded from review of papers. More important viewpoint is probably core-less or with-core structure of the power supply rail because this changes the cost and voltage stress of the power supply rail significantly.

The RPEV mission system, as shown in Figure 10.1(b), includes not only the WPTS system but also the control system and EV system. The control system is unique and crucial for RPEVs because it senses and identifies the EV and then appropriately turns on and off the inverters. Moreover, it monitors the health of the IPTSs and RPEVs and provides accounting service and communication links.

### 10.2.2 Fundamental principles of the IPTS

Now the fundamental principles of IPTSs, instead of all WPTSs, for RPEVs will be briefly explained. The IPTSs are governed by Ampere's law and Faraday's law among four Maxwell equations, as shown in Figure 10.2(a). It can be briefly explained as follows:

- Time-varying magnetic flux is generated from the AC current of a power supply rail in accordance with Ampere's law.
- Voltage is induced from the pick-up coil, coupled with the power supply rail, in accordance with Faraday's law.
- Power is wirelessly delivered through magnetic coupling, where capacitor banks are used to nullify inductive reactance.

The governing equations of IPTS for sinusoidal magnetic field, voltage, and current are approximated as follows:

$$\nabla \times \mathbf{H} = \mathbf{J} \quad \text{(Ampere's law)}. \tag{10.1}$$

$$\nabla \times \mathbf{E} = -j\omega\mathbf{B} \quad \text{(Faraday's law)}. \tag{10.2}$$

In order to provide moderately high-frequency AC current to the power supply rail, a high-frequency switching inverter is introduced, and a rectifier is attached to the pick-up coil side to obtain a DC voltage for providing power to an on-board battery of RPEV, as shown in Figures 10.1(a) and 10.2(b). Meanwhile, the electromagnetic field (EMF) for pedestrians should be under the constraints such as ICNIRP guidelines (ICNIRP Guidelines, 1998, 2010). This is why passive and active EMF cancellation techniques have been so developed.

It is quite often misunderstood that the purpose of resonating LC circuits in the IPTS, as shown in Figure 10.2(b), is to amplify power or energy by $Q$ factor. Actually, resonance is not mandatory if the required power delivery is small so that the voltage drop of leakage inductances of coils may not be large due to small current. Moreover, the switching frequency of the inverter $f_s$ for IPTS is not necessarily exactly tuned to the resonant frequency of LC circuits $f_r$. Very often, $f_s$ is intentionally increased a little bit higher than $f_r$ to guarantee the zero voltage switching (ZVS) of inverters (Huh *et al.*, 2011a, 2011b). The frequency deviation is typically 5–10% of $f_s$. Note that the inductance of a power supply rail tends to

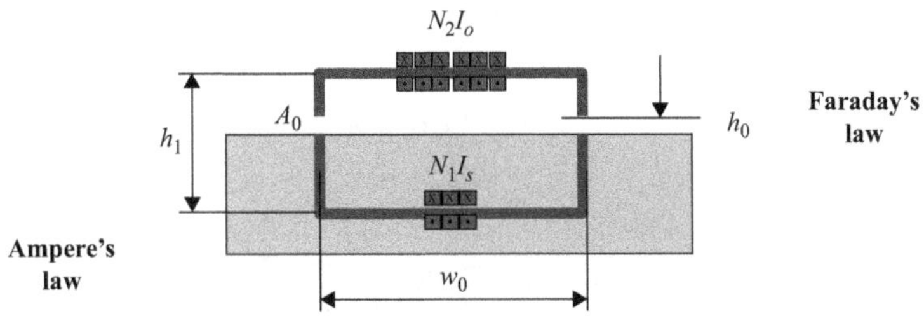


| | |
|---|---|
| $\nabla \times \mathrm{E} = -\dfrac{\partial \mathrm{B}}{\partial t}$ | $\oint_C \mathrm{E} \cdot dl = -\dfrac{d}{dt} \int_S \mathrm{B} \cdot dA$ |
| $\nabla \times \mathrm{H} = \mathrm{J} + \dfrac{\partial \mathrm{D}}{\partial t}$ | $\oint_C \mathrm{H} \cdot dl = \int_S \mathrm{J} \cdot dA + \dfrac{d}{dt} \int_S \mathrm{D} \cdot dA$ |

(a)

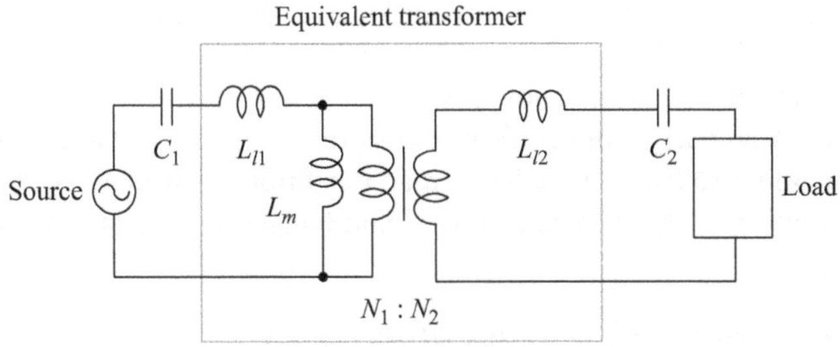


(b)

*Figure 10.2 The principle and an equivalent circuit of IPTS for RPEV: (a) The governing equations of the IPTS and (b) An example of equivalent circuit of IPTS*

decrease as vehicles are displaced on the power supply rail, and this change should be considered in the design of the inverter and primary-side compensation circuit. Depending on the source and load types, the compensating capacitors can be connected in series or parallel with the source or loads. Figure 10.2(b) shows a series–series compensation example, which is adequate for high-power application of IPTS whose output characteristic is an ideal voltage source.

### *10.2.3 Discussions on the requirements of IPTS*

As shown in Figure 10.3, the IPTSs for RPEVs are differentiated from conventional IPTSs for stationary charging because additional requirements such as larger lateral tolerance, higher air-gap, and lower construction cost should be considered for dynamic charging. Moreover, the IPTSs for RPEVs should survive from harsh road conditions such as extremely high and low temperatures, high humidity, and repetitive mechanical shocks for at least 10 years. Basically, the high voltage power rail under a road should be electrically and mechanically protected well, but

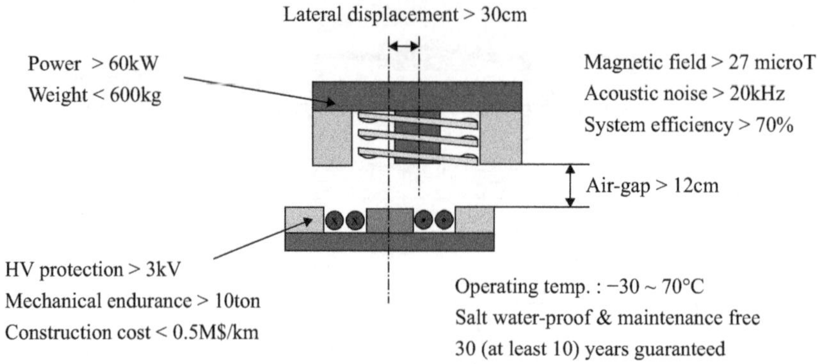


*Figure 10.3 An example of desired requirements of the IPTSs for RPEVs*

electricity under the conditions of a wet and fragile structure is not compatible with a road, in general. Therefore, it is quite challenging to build the roadway subsystem robustly. Fortunately, rain and snow are not so serious factors to be considered if the power rail is properly isolated from moisture. On the other hand, the on-board subsystem should also survive the harsh road conditions as well as vibrating RPEV operating conditions, where strict technical and legislative regulations for vehicles should be met.

### *10.2.4 Important design issues of the IPTS*

It is not a straightforward process to determine the design goal in general. Moreover, the design of the IPTS for RPEVs is not yet fully established. So the design procedure introduced here should be regarded as an example. The design goals of the IPTS are then summarized as follows:

- to increase magnetic coupling as much as possible so that higher induced voltage can be obtained
- to increase power efficiency for given power capacity, device ratings, and cost
- to make modules as compact as possible to accommodate for given space and weight
- not to increase or cancel out EMF
- to manage resonance frequency variation and coupling factor change due to misalignment of pick-up position, air-gap change, and even temperature change.

In order to meet the design goals, several important design issues of the IPTSs, which are unique for RPEVs, should be resolved as follows:

1. *Switching frequency*: The switching frequency of the inverter and rectifier of an RPEV governs overall performances. As the switching frequency $f_s$ increases, the coil and capacitor size decreases for the same required power, but the switching loss and core loss as well as the conduction loss of wires due

to the skin effect increase. In the case of the switching frequency lower than 20 kHz, acoustic noise may be a problem. As shown in Figure 10.2(b), another unique feature of RPEVs concerning frequency is the increased voltage stress in a distributed power supply rail $V_{l1}$ for higher switching frequency, which is proportional to the frequency and line current $I_s$ as follows:

$$V_{l1} = j\omega_s L_{l1} I_s. \tag{10.3}$$

$$\frac{\partial V_{l1}}{\partial x} = j\omega_s I_s \frac{\partial L_{l1}}{\partial x}. \tag{10.4}$$

Because of these restrictions, the switching frequency tends to be no less than 20 kHz but not far beyond 50 kHz. For the OLEVs, the frequency was finally selected as 20 kHz after examining for 20 kHz, 25 kHz, and 30 kHz. The higher the frequency, the less the efficiency.

The interoperability of an IPTS between RPEVs and stationary wireless EVs is an important issue that needs further study because the design of IPTSs for RPEVs is not a direct extension of that of stationary chargers. The design goals of the IPTSs of RPEVs include low infra cost, high power, continuity of power delivery, and low fluctuation during in-motion, which are not so important for stationary chargers.

2. *High power and large current*: In addition to the high operating frequency of RPEVs, high power of hundreds of kilowatts with large current of hundreds of ampere (Huh *et al.*, 2011a, 2011b) makes it difficult to handle cables, converters, and devices. For example, a Litz wire of 300 A rating with 20 kV insulation capability is not commercially mass produced. Capacitors and IGBTs, which have high voltage and large current ratings with an operating frequency higher than 20 kHz, become scarce.
3. *Power efficiency*: In order to make RPEVs competitive against ICEVs, the power efficiency or energy efficiency should not be too low. The bottom line of the overall power efficiency, which is defined from the AC power source to the DC output power for battery, would be about 50% (Huh *et al.*, 2011a, 2011b), considering grid loss and fuel cost. Fortunately, modern IPTSs have fairly good power efficiency larger than 80% (Huh *et al.*, 2011a, 2011b).
4. *Coil design*: In order to focus magnetic field from the power supply rail to the pick-up and mitigate leakage flux so that large induced voltage as well as low EMF level can be achieved, a novel coil design is crucial for the IPTS. This design is unique for the roadway subsystem in that the cost per km is critical; therefore, the roadway power supply rail should be low enough not to deteriorate the economic feasibility of overall RPEV solution, whereas the pick-up should be neither heavy nor thick so that it can be successfully implemented on the bottom of the RPEV.
5. *Insulation*: In order to guarantee stable operation of IPTS, a few kilovolt level insulations of the roadway subsystem and on-board subsystem should be provided. The insulation of a roadway power supply is of great importance because high voltage is induced along through the distributed power rail over a

road rather than over a point and is increased by the back EMF of the pick-up coil abruptly displaced on the power supply rail.

6. *Segmentation of power supply rail*: RPEVs typically require many IPTS along a road because a power supply rail cannot be infinitely deployed. It should be segmented so that each segment can be independently turned on and off (Choi *et al.*, 2013). The length of a power supply rail is an important design issue because it would be too expensive if the length is very short due to an increased number of inverters and switch boxes, whereas the power loss would be too large if the length is very long due to increased resistance.
7. *Roadway construction*: In order to minimize traffic obstruction, the roadway construction time of the IPTS should be as short as possible. Moreover, there are numerous difficulties in roadway work because of debris and dirt. Keeping all of the electric components of the IPTS clean during roadway construction for a few days or weeks is a serious problem in practice.

## 10.3 Advent of RPEV

### *10.3.1 Origin of the RPEV: "Transformers for electric railways"*

The concept of the RPEV stems from the first patent, "Transformer system for electric railways" by M. Hutin and M. Leblanc in France in 1894 (Hutin and Leblanc, 1894), where a large air-gap transformer was displaced under a train for electric power transfer, as shown in Figure 10.4.

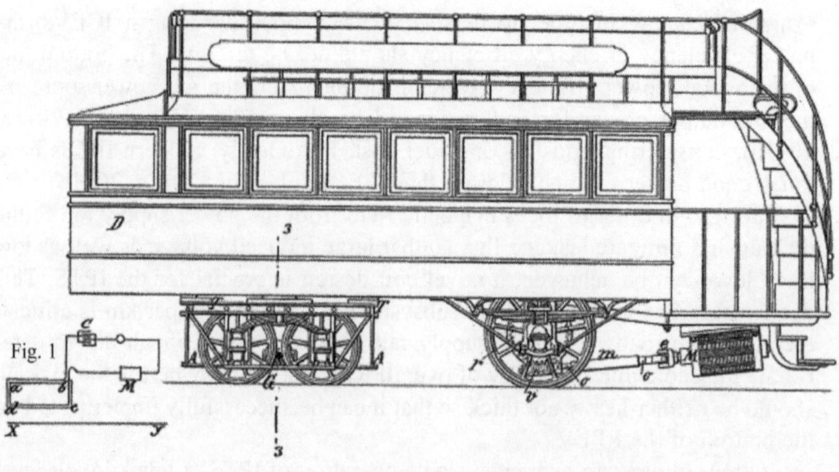


*Figure 10.4 The first patent of an RPEV in 1894. © 2015 IEEE. Reprinted with permission from Choi et al. (2015c)*

### *10.3.2 The first development of RPEVs*

During the oil crisis of the 1970s, interest in RPEVs increased in the United States, where several research teams started to investigate RPEVs for reducing the use of petroleum in highway vehicles. The first development of RPEVs was begun in 1976 by the Lawrence Berkeley National Laboratory in order to confirm the technical feasibility of RPEVs. A prototype IPTS, as shown in Figure 10.5, was built and tested for 8 kW wireless power transfer; however, it was not a fully operational system. In 1979, the Santa Barbara Electric Bus Project was started and another prototype RPEV was also developed.

After the two frontier projects of RPEV, the Partners for Advanced Transit and Highways (PATH) program began in 1992 to determine the technical viability of RPEV in the University of California, Berkeley (Bolger, 1994; California PATH Program, 1994). Throughout the PATH program, broad research and field tests on RPEV, as shown in Figure 10.6, including designs of IPTS, installations of an IPTS to a bus, road constructions of power supply rails, and environmental impact

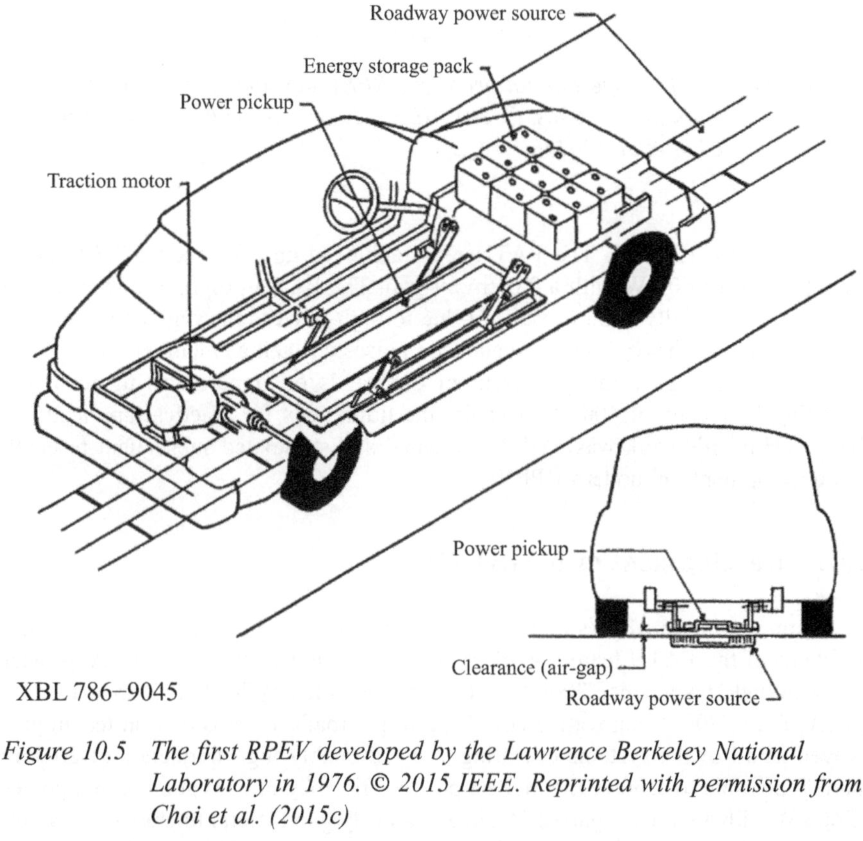


*Figure 10.5 The first RPEV developed by the Lawrence Berkeley National Laboratory in 1976. © 2015 IEEE. Reprinted with permission from Choi et al. (2015c)*

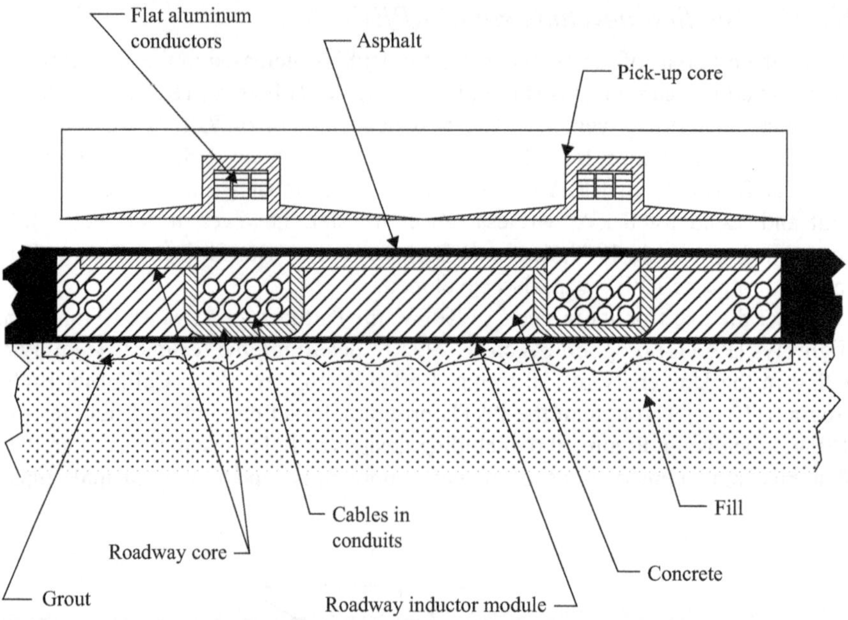


*Figure 10.6 The first operational prototype RPEV developed by the PATH team in the 1990s. © 2015 IEEE. Reprinted with permission from Choi et al. (2015c)*

studies, were performed. The PATH team achieved an efficiency of 60% at an output power of 60 kW with a 7.6 cm air-gap. The PATH program, however, had not been successfully commercialized due to high-power rail construction cost of around 1 M$/km, heavy coils, and acoustic noises, as well as relatively low power efficiency and large primary current of thousands of amperes owing to a low operating frequency of 400 Hz. Despite the limitations for practical applications, the PATH team's work was well documented and stimulated subsequent research and development on modern RPEVs.

## 10.4 Development of KAIST OLEVs

A stream of modern RPEVs is the OLEV, which has solved most of the remaining problems of the PATH team's work, as shown in Figure 10.7. The OLEV project has been undertaken since 2009 by a research team led by KAIST, Korea (KAIST OLEV Team, 2009). Innovative coil designs and roadway construction techniques as well as all of the systems operating at a reasonably high frequency of 20 kHz made it possible to achieve the highest power efficiency of 83% at an output power of 60 kW with a large air-gap of 20 cm and a fairly good lateral tolerance of 24 cm. Moreover, the power rail construction cost of the OLEV, which is responsible for

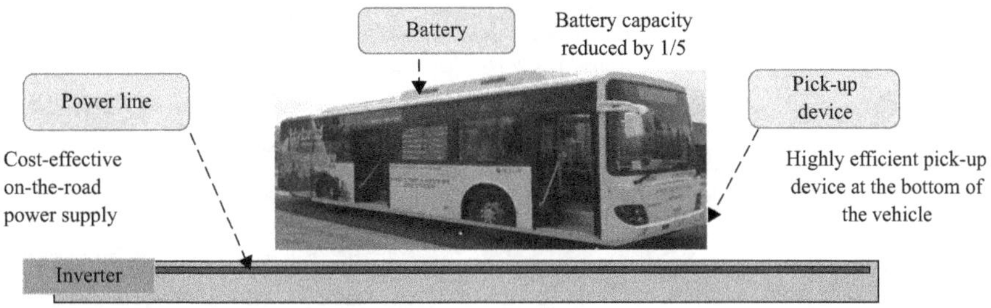


*Figure 10.7 The OLEV bus developed by KAIST*

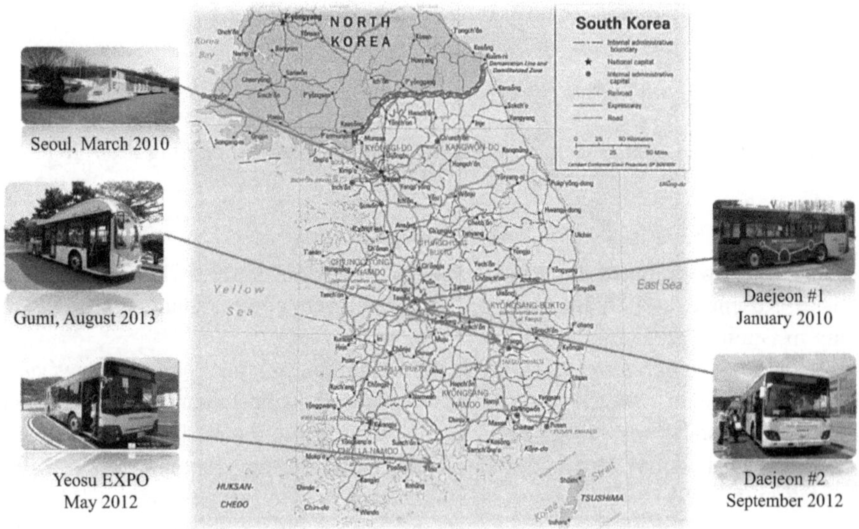


*Figure 10.8 Deployment status of OLEVs in Korea*

more than 80% of the total deployment cost for RPEVs, has been dramatically reduced to at least one-third of that of the PATH project. The primary current has been also reasonably mitigated as low as 200 A, and the battery size has been significantly reduced to 20 kWh, which can be further reduced by increasing the length of the power supply rails. The OLEV system has been successfully deployed at five locations now, as shown in Figure 10.8.

As shown in Figure 10.9, the first-generation (1G) concept-demonstration-car of OLEV, the second-generation (2G) OLEV buses, and the third-generation (3G) OLEV passenger car were developed and extensively tested at the test sites of KAIST since 2009, and three OLEV trains ($3^{+}$G) had been successfully deployed at the Seoul Grand Park, Korea since 2010. Two upgraded OLEV buses ($3^{+}$G) were deployed at the 2012 Yeosu EXPO, Korea, and other two OLEV buses ($3^{+}$G) have

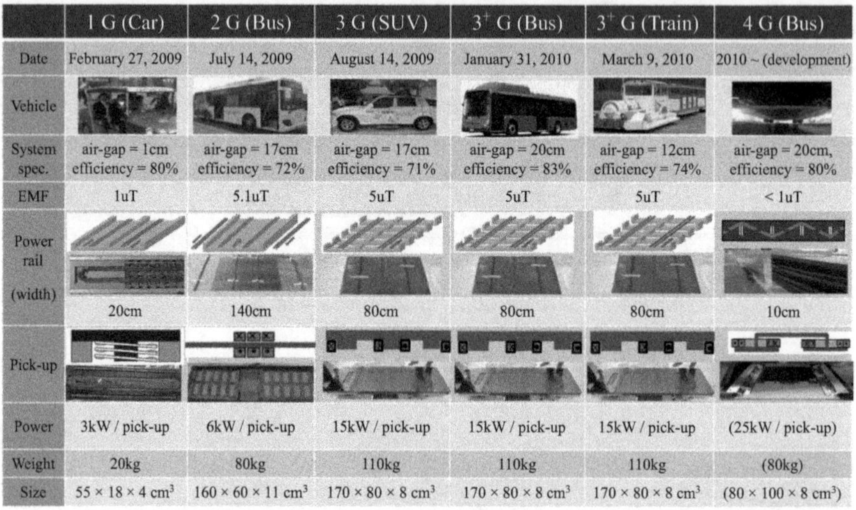


*Figure 10.9 A summary of the developments of OLEV including their IPTS*

been in full operation at the main campus of KAIST since 2012 (Choi *et al.*, 2015c). Recently, two OLEV buses ($3^{+}$G) were firstly commercialized at a 48 km route in Gumi, Korea.

The fourth-generation (4G) OLEV, showing more practical performances such as a very narrow rail width of 10 cm, a larger lateral displacement of 40 cm, lower EMF level, and lower power rail construction cost compared to the previous generation OLEV, has also been developed. Now, the development of the fifth-generation (5G) OLEV is in progress, where an ultra slim S-type power supply coil of 4 cm rail width is proposed for power rail construction with much less cost and time (Choi *et al.*, 2015a).

### *10.4.1 1G OLEV*

The 1G OLEV, which was announced on February 27, 2009, is a golf cart equipped with a mechanically controlled pick-up to automatically align to the power supply rail of 45 m within a 3 mm lateral displacement, as shown in Figure 10.10. The 1G OLEV adopted E-type cores for both the power supply rail and pick-up coil, as shown in Figure 10.9, and achieved an overall system efficiency of 80% at 20 kHz switching frequency with an output power of 3 kW at 1 cm air-gap (Choi *et al.*, 2015c).

### *10.4.2 2G OLEV*

The 2G OLEV, announced on July 14, 2009, focused on drastically improving the air-gap of the 1G OLEV without mechanical moving parts, and finally achieved a

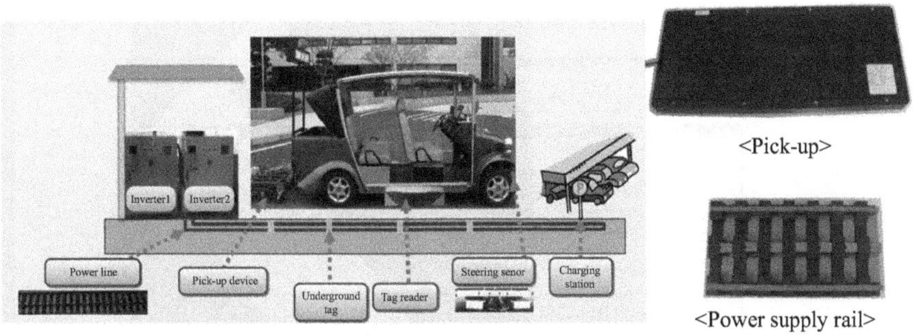


*Figure 10.10 The 1G golf cart platform of OLEV*

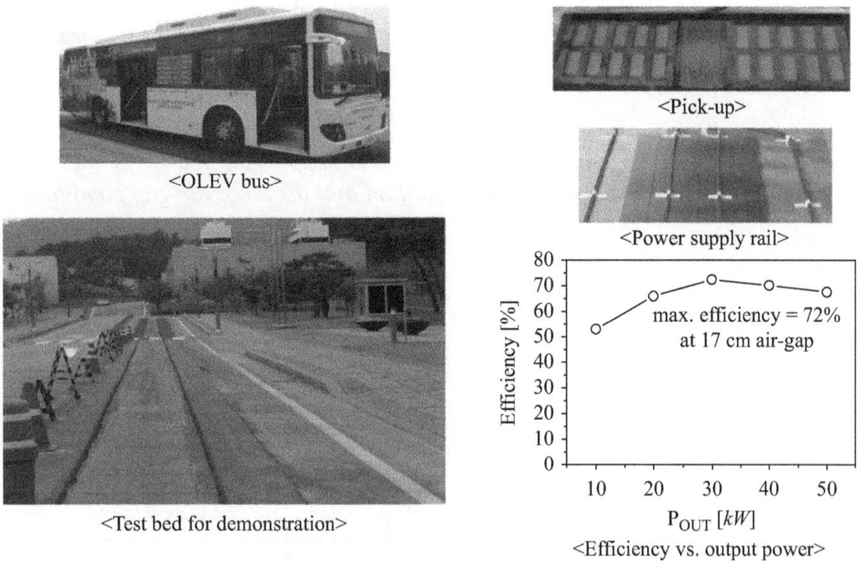


*Figure 10.11 The 2G OLEV bus at the KAIST Munji Campus with four test tracks of power supply rails (60 m each)*

17 cm air-gap, which meets road regulations, 12 cm in Korea and 16 cm in Japan. At the same time, it achieved a maximum efficiency of 72% and a maximum output power of 60 kW using 10 pick-ups. The efficiency may increase if a shorter power supply rail is adopted. Note that the efficiency may drop for a laterally displaced OLEV. The power supply rail, where the width is 1.4 m and the total length is 240 m, is paved with asphalt to provide the same friction force as normal roads, as shown in Figure 10.11 (Lee *et al.*, 2010). In order to realize the 17 cm air-gap, the U-type power supply rail and the flat pick-up coil of IPTS have been newly developed, where the name "U-type" stems from the cross-section of the power supply rail, as shown in Figure 10.12. A pair of return power cables is used to

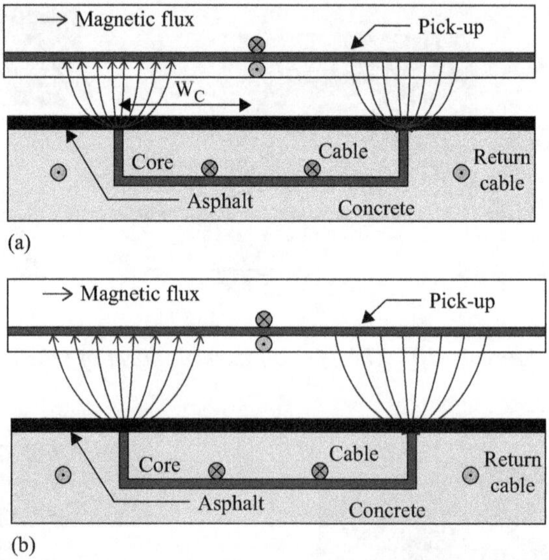


*Figure 10.12 Cross-section of the U-type power supply rail and I-type pick-up coil: (a) small air-gap condition and (b) large air-gap condition*

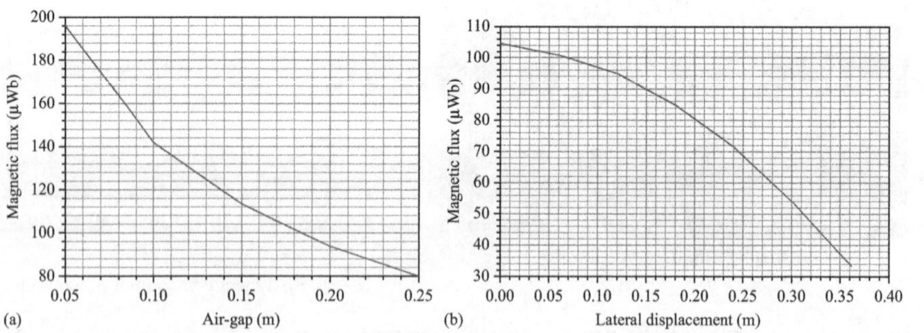


*Figure 10.13 Magnetic flux characteristics of the 2G IPTS: (a) along the air-gap and (b) along the lateral displacement. © 2015 IEEE. Reprinted with permission from Choi et al. (2015c)*

mitigate the EMF from the power supply rail for the ICNIRP guidelines (ICNIRP Guidelines, 1998, 2010). With the U-type power supply rail and I-type pick-up coil, it is found that an effective area of the magnetic path between the power supply rail and the pick-up coil increases as the air-gap increases due to fringe effect, as shown in Figures 10.12 and 10.13.

The lateral displacement, at which the output power drops to a half of the maximum output power and the induced voltage or magnetic flux density becomes

70.7% of their maximum, was achieved to be 23 cm, as shown in Figure 10.13(b). The cross-section of the core plates of the power rail, as shown in Figure 10.12, is significantly reduced by using the high operating frequency of 20 kHz, which is 50 times of the PATH team. The reason for selecting 20 kHz is that it is the lowest inaudible frequency, which mitigates the line voltage stress, as in (10.3) and (10.4). The upward magnetic flux generated from the pick-up coil was appropriately shielded by an aluminum plate with a 5 cm distance.

### *10.4.3 3G OLEV*

The 3G OLEV SUV (Sports Utility Vehicle), announced on August 14, 2009, as shown in Figure 10.9, newly adopted the W-type power supply rail and the flat pick-up with overlapped double coils for higher power and increased lateral tolerance. Thus, the upward magnetic leakage flux from the pick-up of 2G OLEV had been drastically mitigated by using the flat pick-up core, which prohibits the magnetic flux between the power supply rail and pick-up from leaking. A magnetic field shield, such as an aluminum plate, or additional space is no longer required. The overall efficiency and air-gap of the 3G OLEV were 71% and 17 cm, respectively, which were respectable, but slightly disappointing numbers; therefore, overall systems including a roadway rectifier & inverter, power supply rail, pick-up, and on-board regulator were redesigned to achieve a maximum efficiency of 83% with a 20 cm air-gap, and this new design was called the 3G$^{+}$ OLEV. Four 3G$^{+}$ OLEV buses, as shown in Figure 10.9, were built for test purposes, and six more 3G$^{+}$ OLEV buses were built for full operation purposes, whereas three 3G$^{+}$ OLEV train were deployed at the Seoul Grand Park, as shown in Figure 10.14.

The W-type power supply rail, as shown in Figure 10.15, has many cores of "W" shape, where the total magnetic resistance becomes three-quarters smaller

*Figure 10.14 The 3G$^{+}$ OLEV train on the 2.2 km road at the Seoul Grand Park, where 375 m was paved with 24 m power supply rails*

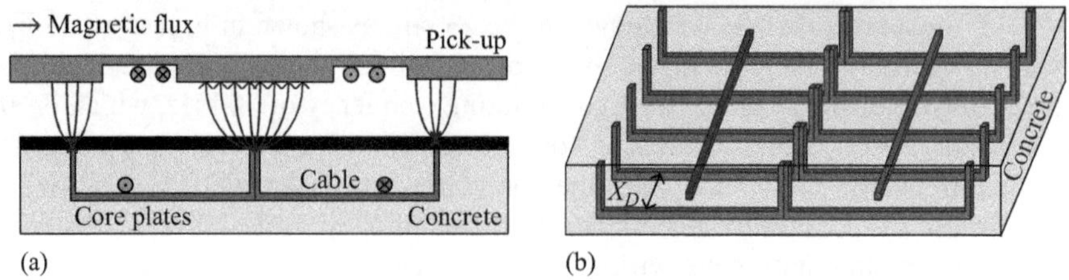


*Figure 10.15 Views of the W-type power supply rail and flat pick-up coil: (a) cross-section view and (b) bird's eye view of a bone structure. © 2015 IEEE. Reprinted with permission from Choi et al. (2015c)*

than that of the 2G OLEV for the same air-gap, and this eventually leads to the increase of the output power from 6 kW to 15 kW for each pick-up. Moreover, the width of the power supply rail is decreased to 70 cm, which is just a half of that of the 2G OLEV, and it can reduce the deployment cost of OLEV for commercialization. The EMF at 1.75 m distance meets the ICNIRP guideline because the two power cables of opposite polarity of magnetic flux cancel each other out.

One of the problems found during the development of 2G OLEV was the inherently weak structure of the power supply rail, where the core separates the concrete into two and undergoes severe mechanical stress from heavy vehicles. As a remedy for this mechanical weakness, bone structure cores of the power supply rail, as shown in Figure 10.15(b), have been proposed and registered as a patent (Suh *et al.*, 2011), where the bone structure cores are installed with the bone core-gap $X_D$. Despite the large gap between cores, the magnetic flux does not decrease considerably, as shown in Figure 10.16. This bone structure is now widespread used in other IPTS such as stationary charging systems.

In addition, during power rail road construction, concretes can percolate down through the bone structure cores. Therefore, the power supply rail, which is reinforced with two iron bars and in which power cables are protected by fiber reinforced plastic (FRP) pipes, has almost the same endurance of concrete, as shown in Figure 10.17. In practice, roadway construction of the power supply rail takes a few weeks, being affected by weather and debris, and this delay obstructs the commercialization of OLEV.

### *10.4.4 4G OLEV*

As shown in Figure 10.18, the 4G OLEV bus, announced in 2010, has an innovative I-type structure of a power supply rail with a maximum output power of 27 kW for a double pick-up coil having a 20 cm air-gap and 24 cm lateral displacement (Huh *et al.*, 2011a, 2011b). As shown in Figure 10.19(a), the I-type power supply rail, where the name "I-type" stems from its front shape, has only 10 cm width, which

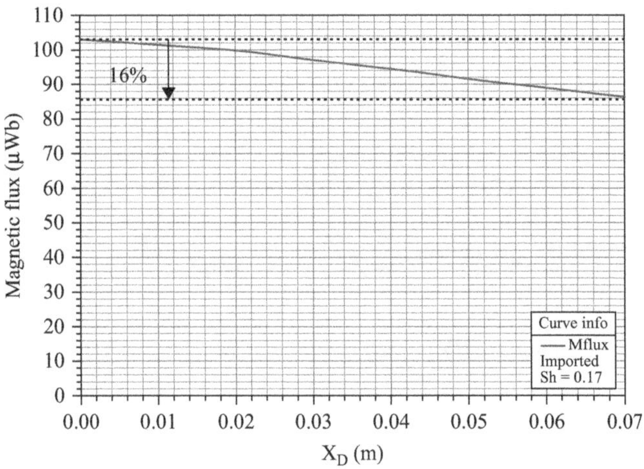


*Figure 10.16 Simulated magnetic flux characteristics for different bone core-gap $X_D$. © 2015 IEEE. Reprinted with permission from Choi et al. (2015c)*

*Figure 10.17 The W-type power supply rail of the 3G OLEV being tested. © 2015 IEEE. Reprinted with permission from Choi et al. (2015c)*

leads to a deployment cost reduction of 20% compared to the 3G OLEV. Unlike previous generations of OLEV, a very low EMF for pedestrians, as low as 1.5 μT at a distance of 1 m from the center of the power supply rail, can be obtained because the power supply rail has alternating magnetic poles along the road, as shown in Figure 10.19(b). As shown in Figure 10.20, the 4G OLEV adopted a module

*Figure 10.18 The 4G OLEV bus at KAIST Munji Campus. © 2015 IEEE. Reprinted with permission from Choi et al. (2015c)*

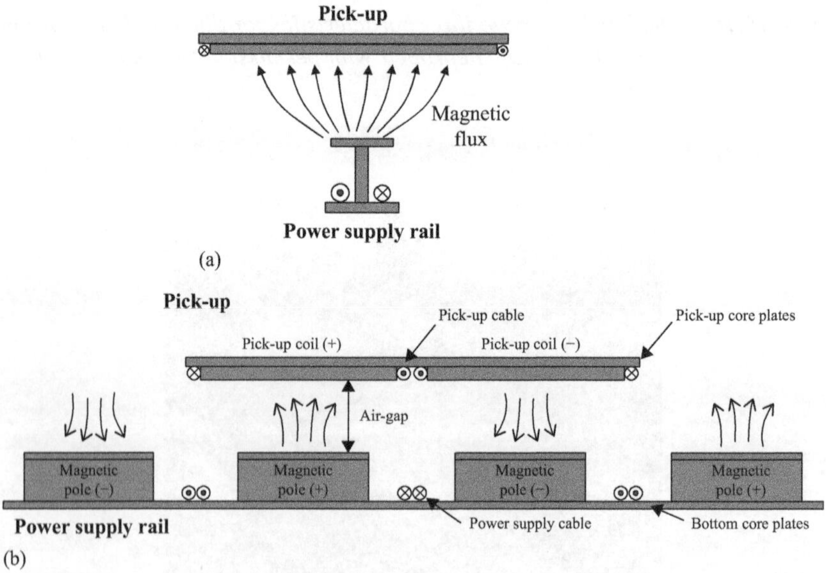


*Figure 10.19 The I-type power supply rail and double flat pick-up coil: (a) front view and (b) side view*

concept for the I-type power supply rail to reduce the deployment time within a few hours, which had been one of the drawbacks of the 3G OLEV. The I-type power supply module, which includes the power supply rail and capacitor banks inside, as shown in Figure 10.20, should be robust to high humidity and external mechanical impacts in accordance with the system requirements of Figure 10.3.

(a) (b)

*Figure 10.20 The I-type power supply module of the 4G OLEV: (a) a prototype module and (b) deployed modules at the 24 m test site. © 2015 IEEE. Reprinted with permission from Choi et al. (2015c)*

### *10.4.5 5G OLEV*

To further reduce the construction cost and increase the robustness of power supply rails, the 5G OLEV of ultra slim S-type core was recently proposed, which has an S-shape when viewed from the front, as shown in Figure 10.21 (Choi *et al.*, 2015a). The S-type power supply module has a width of only 4 cm, which has been decreased from the I-type width of 10 cm; hence, the S-type model leads to less construction cost and deployment time. Moreover, the S-type model makes it easier to fold itself, which means connecting power cables is no longer necessary after being deployed.

## 10.5 Generalized active EMF cancellation methods

In the IPTS of OLEV, the total EMF, which is the summation of EMF generated by the power supply rail and the pick-up coil, should be lower than the ICNIRP guideline for the safety of pedestrians. Among EMF cancellation methods, passive methods rely on the use of ferromagnetic materials, conductive materials, and selective surfaces for protecting against radio frequency interference, which is found to be adequate for OLEV. An active EMF cancellation design was applied to the RPEV of PATH team (California PATH Program, 1994), where a cancelling coil mitigates the EMF without RPEV. Another active EMF cancellation design was made for the pick-up coil of the OLEV (Kim *et al.*, 2013), where counter current is appropriately controlled. However, the EMF is generated from both the power supply rail and pick-up coil of the IPTS, as shown in Figure 10.22; therefore, they must be canceled together without accurate sensing of EMF and complicated control circuits, which should be avoided for practical use.

A generalized active EMF design principle, which can be extended to any IPTS, has been developed (Choi *et al.*, 2015b), where there are three design methods: independent self EMF cancellation (ISEC), 3 dB dominant EMF

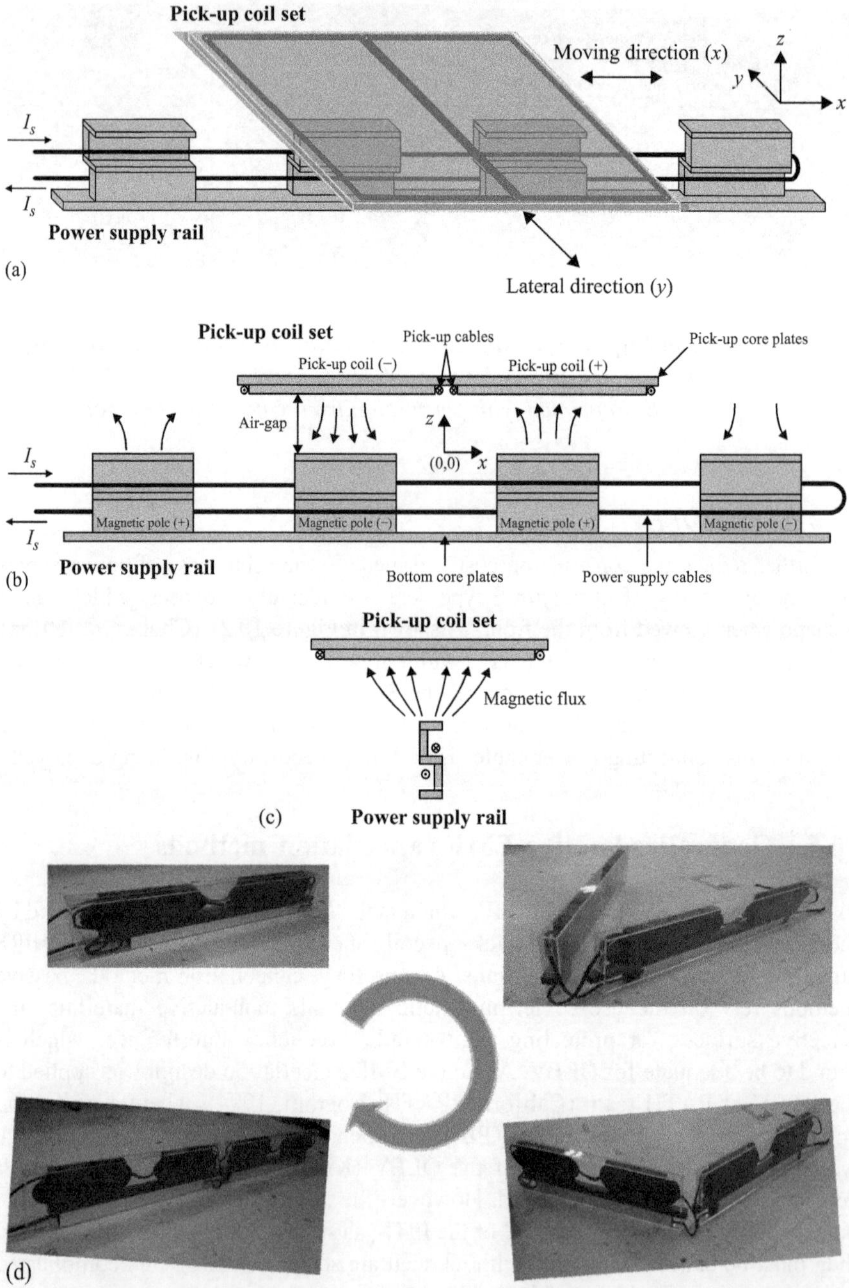


*Figure 10.21 Proposed ultra-slim S-type power rail and the flat pick-up coil of the IPTS: (a) bird's eye view, (b) side view, (c) front view and (d) deployment process.* 

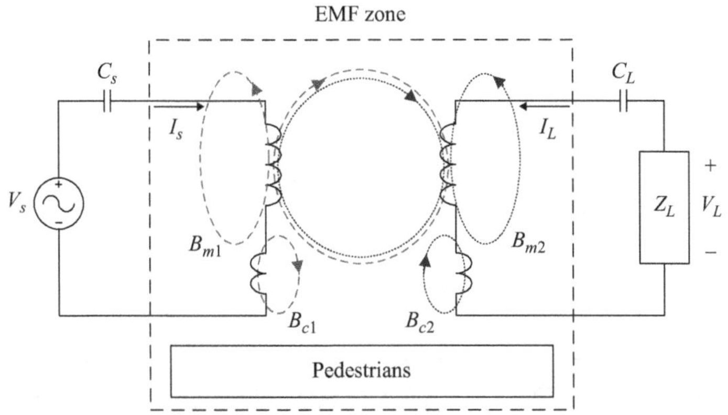


*Figure 10.22 ISEC method for primary and secondary sides, fetching its cancellation current from each main coil. © 2015 IEEE. Reprinted with permission from Choi et al. (2015c)*

cancellation (3DEC), and leakage-free EMF cancellation (LFEC). By adding an active EMF cancellation coil to each power supply rail and pick-up coil without EMF sensing and control circuits, the EMF generated from each main coil can be independently canceled by their corresponding cancellation coils, as shown in Figure 10.22.

Because the current of each cancellation coil is fetched from its corresponding main coil, the magnetic flux of a cancellation coil is in the phase of that of a main coil; hence, the EMF $\boldsymbol{B}_t$ becomes independent of both phase and load conditions as follows:

$$\mathbf{B}_t = (\mathbf{B}_{m1} + \mathbf{B}_{c1}) + (\mathbf{B}_{m2} + \mathbf{B}_{c2}) \cong 0,$$

$$\because \mathbf{B}_{c1} \cong -\mathbf{B}_{m1}, \quad \mathbf{B}_{c2} \cong -\mathbf{B}_{m2} \quad \text{and} \quad \mathbf{B}_k \equiv B_{kx}\mathbf{x}_0 + B_{ky}\mathbf{y}_0 + B_{kz}\mathbf{z}_0. \tag{10.5}$$

Together with ISEC, the 3DEC method is quite useful in practice for designing the IPTS because it makes it possible to completely isolate the primary and secondary cancellation designs from each other. For the full resonant IPTS that is very common in practice (Huh *et al.*, 2011a, 2011b), the primary and secondary currents are in quadrature, i.e., $B_{1x} \perp B_{2x}, B_{1y} \perp B_{2y}, B_{1z} \perp B_{2z}$. Then the total magnetic field $B_t$ can be constrained to the ICNIRP guidelines so far as a dominating magnetic field $B_1$ is less than 3 dB below the guideline as follows (Choi *et al.*, 2015b):

$$B_t \equiv |\mathbf{B}_t| = \sqrt{|\mathbf{B}_1|^2 + |\mathbf{B}_2|^2} = \sqrt{B_1^2 + B_2^2} \leq \sqrt{B_1^2 + B_1^2} = \sqrt{2}B_1 \leq B_{ref}.$$
$$\because B_1 \equiv |\mathbf{B}_1|, \quad B_2 \equiv |\mathbf{B}_2|, \quad B_2 \leq B_1 \tag{10.6}$$

A design example for the 3G W-type IPTS based on the ISEC and 3DEC is shown in Figure 10.23, where no cancellation coil was used for the power supply

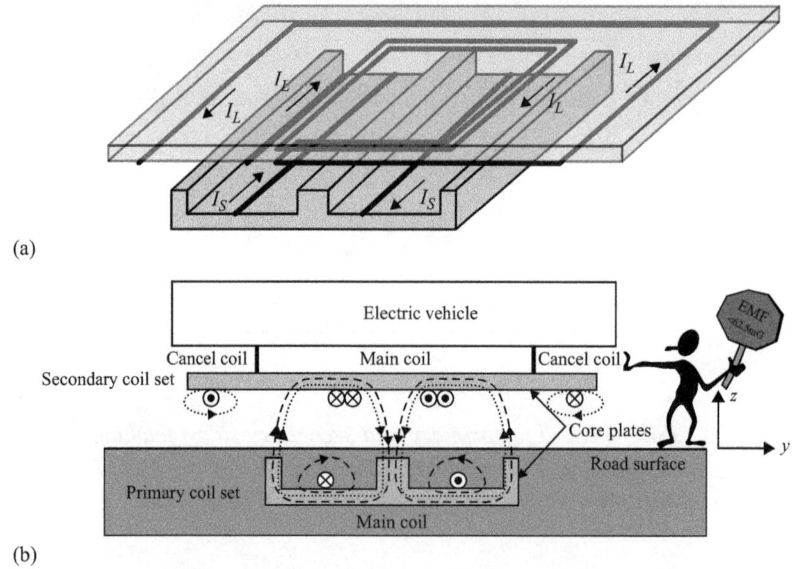


*Figure 10.23 An active EMF cancellation design example for the W-type IPTS: (a) bird's eye view and (b) front view*

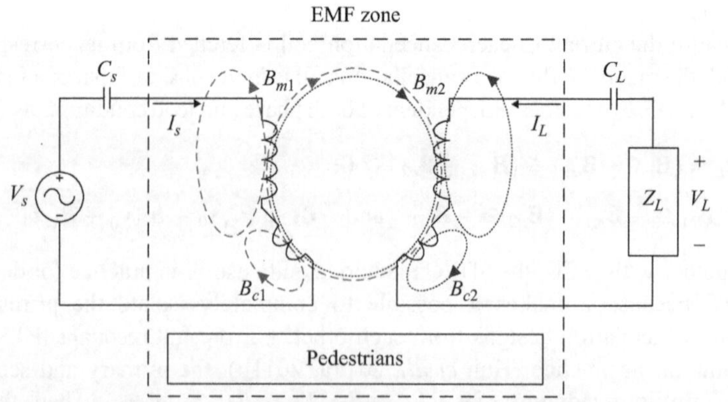


*Figure 10.24 A bad design case for cancellation coils, which are strongly coupled with main magnetic linkage. © 2015 IEEE. Reprinted with permission from Choi et al. (2015c)*

rail due to the low EMF generated from itself. Note that a large EMF is generated from the pick-up for high-power applications.

In case the EMF cancellation coils are involved in an unwanted magnetic linkage, as shown in Figure 10.24, the induced load voltage drops; therefore, it is

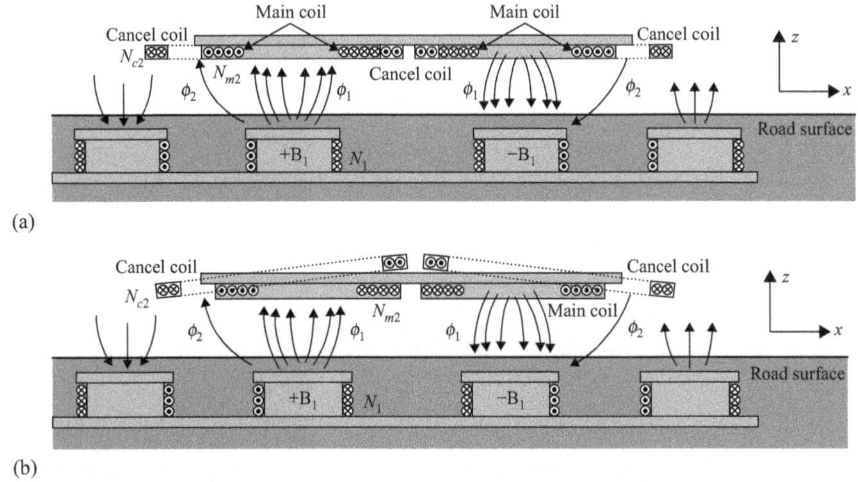


*Figure 10.25 The application of the LFEC design to a pick-up set: (a) not applied and (b) applied*

highly recommended to design the cancellation coils based on the LFEC method so that the magnetic linkage does not intersect them. As shown in Figure 10.25, it was verified that the LFEC method can improve the load voltage by 21%.

## 10.6 Research trends of other RPEVs

### *10.6.1 The Auckland University Research Team*

Since the 1990s, a research team in the Auckland University, New Zealand, known as the Auckland team, has been proposing various IPTSs for wireless charging. Among them, the circular coils that adopted ferrite bars instead of ferrite plates need to be addressed because of their compact structure, low weight, and low EMF, as shown in Figure 10.26. The circular type coils, however, are inadequate for high-power RPEVs because of their low-power transfer capability and small lateral tolerance. In principle, the lateral tolerance can be increased with a larger diameter of the circular coils, but it is not applicable in practice due to the limited space of the bottom of a car and increased EMF for pedestrians.

In 2010, the double-sided coils that consist of a rectangular core plate and a vertically wounded cable were proposed, as shown in Figure 10.27, which have high lateral tolerance and a high coupling factor. As mentioned in section 10.4.2, the double-sided coil has similar problems such as upward magnetic flux leakage because its structure is similar to the I-type pick-up, except the ends.

After that, single-sided polarized coils comprising horizontally wound cables on core plates, as shown in Figure 10.28, were introduced (Budhia *et al.*, 2011, 2013; Covic *et al.*, 2011; Zaheer *et al.*, 2012). A larger coupling coefficient as well as better

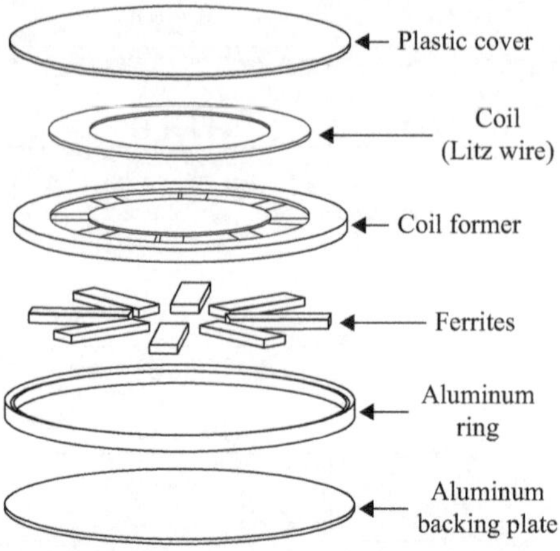


*Figure 10.26 The expanded view of a circular-type coil using ferrite bars by the Auckland team. © 2015 IEEE. Reprinted with permission from Choi et al. (2015c)*

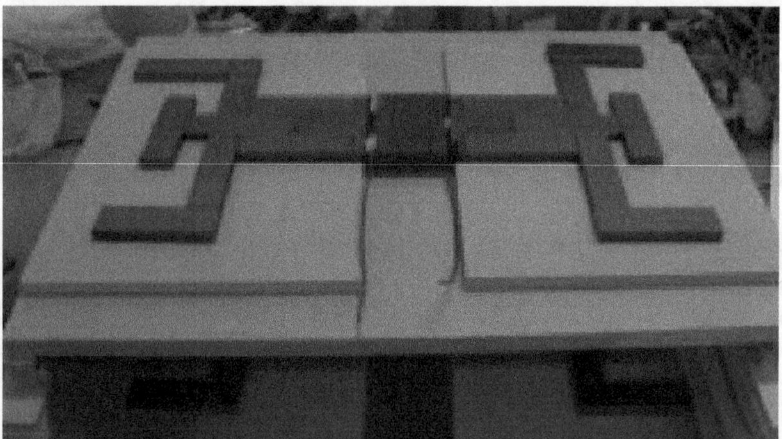

*Figure 10.27 The double-sided coils used by the Auckland team. © 2015 IEEE. Reprinted with permission from Choi et al. (2015c)*

lateral and longitudinal tolerances compared to the circular coil was obtained. In addition, the modified single-sided polarized coils, where another coil is added between the single-sided coils, as shown in Figure 10.28(c), or coils overlap each other, as shown in Figure 10.28(d), had improved performances, as shown in Figure 10.29.

In detail, Figure 10.29 shows the uncompensated power transfer when a circular type coil is used as the primary coil with the various secondary coils such as a

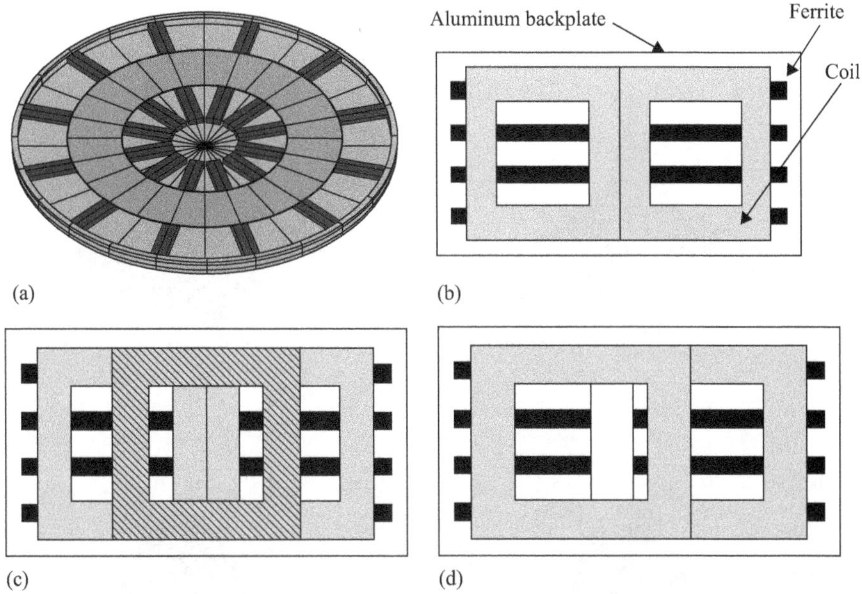


*Figure 10.28* *Coil structures created by the Auckland team: (a) circular coil, (b) a single-sided polarized coil, (c) a single-sided polarized coil adding a coil and (d) a single-sided polarized coil overlapping each other. © 2015 IEEE. Reprinted with permission from Choi et al. (2015c)*

circular type coil, double-sided coil, and single-sided polarized coil. The Auckland team proposed an IPTS including many small power pads for RPEV, where the length of a power pad is much shorter than that of a vehicle to avoid unwanted energizing and loading, as shown in Figure 10.30. This scheme, however, requires numerous considerations such as increased control complexity and deployment in addition to the maintenance costs of the power pads in ground, which should survive from the harsh road environments, as addressed in Figure 10.3.

### *10.6.2 The Bombardier Research Team*

Bombardier has developed several IPTS systems for the stationary and dynamic charging of trams and buses since 2010. With the PRIMOVE project applying IPTS technologies to transportation sectors, Bombardier used an operating frequency of 20 kHz and a three-phase power system for their IPTS to obtain a higher power density without exceeding the EMF guidelines. For stationary and dynamic charging, a power transfer of 250 kW for PRIMOVE trams has been achieved in Augsburg, Germany. The air-gap between primary and secondary coils is about 6 cm, and the lateral displacement of trams is as low as a few

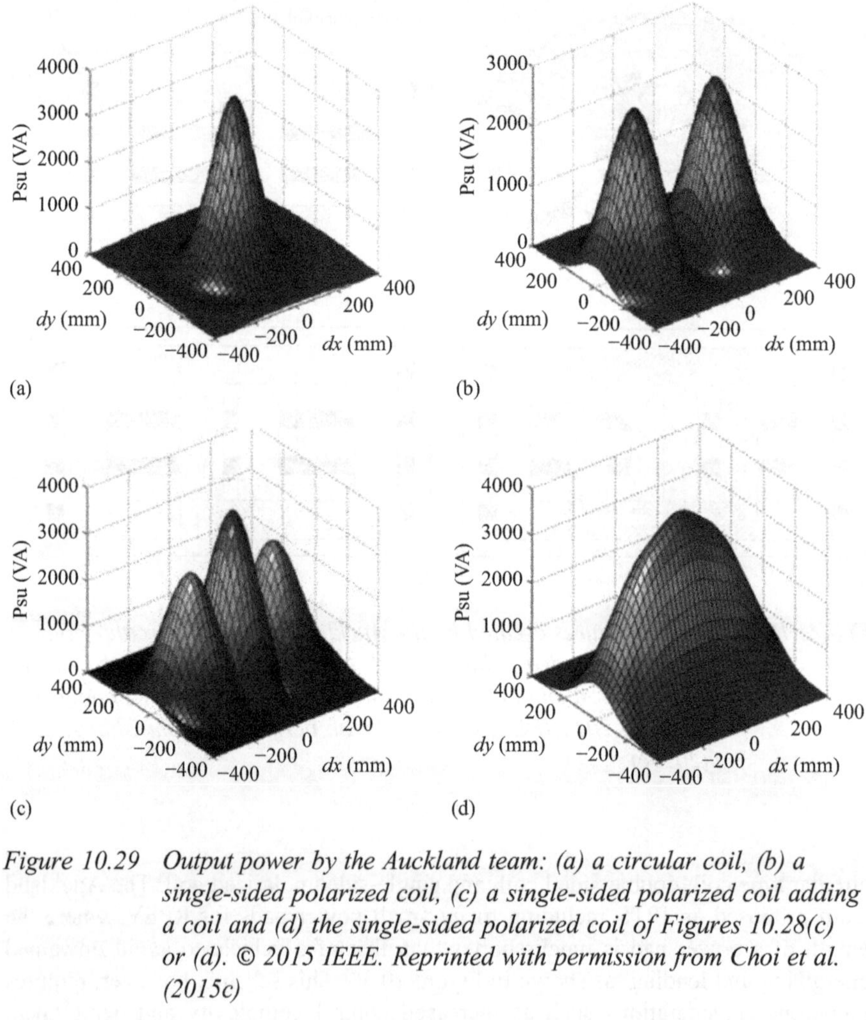


*Figure 10.29 Output power by the Auckland team: (a) a circular coil, (b) a single-sided polarized coil, (c) a single-sided polarized coil adding a coil and (d) the single-sided polarized coil of Figures 10.28(c) or (d). © 2015 IEEE. Reprinted with permission from Choi et al. (2015c)*

centimeters; hence, the design of an IPTS is much easier than that used on ordinary roadways.

Considering large-sized vehicles such as trams, it would be wise to use the three-phase power system for an IPTS because higher power can be transferred to pick-up coils than with a single-phase power system. These merits of high-power delivery capability and increased power efficiency may offset the initial investment costs for procuring more complex and expensive HF inverters and power supply coils. It is not clear, in general, that the multi-phase power systems for RPEVs (Covic *et al.*, 2007) are superior to a single-phase power system due to its high cost and complexity. There is not much open access information about the Bombardier team's works compared to that of the previous research teams. PRIMOVE buses of

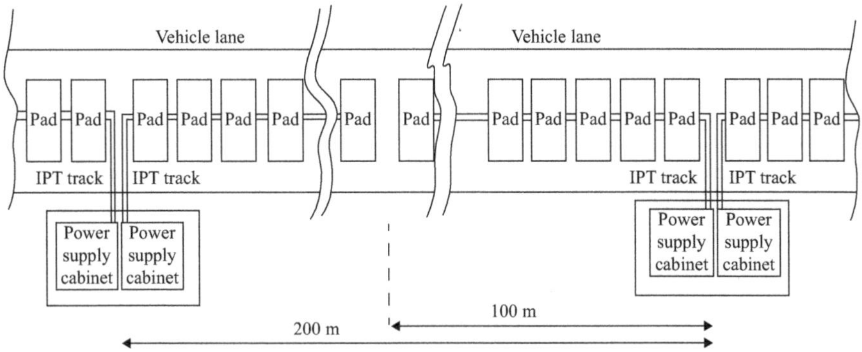


*Figure 10.30 The configuration of an IPTS using many small power pads by the Auckland team. © 2015 IEEE. Reprinted with permission from Choi et al. (2015c)*

a maximum power of 200 kW have been deployed and operated in Brunswick and Mannheim, Germany and in Bruges, Belgium since 2013.

### 10.6.3 The Endesa Research Team

Since April 2013, Endesa has been involved in the vehicle initiative consortium for transport operation & road inductive application (VICTORIA) project and has adopted the triple charging technologies such as conventional plug-in, stationary and dynamic charging using IPTS for RPEV (Mascarell, 2015). Hence, RPEV can be charged by a plug-in charger when parked at a bus terminal during night. At day time, the RPEV can be partially charged at wireless charging bus stops and bus lane equipped with power supply rails while moving on it to extend its driving range.

For the stationary and dynamic charging, Endesa did not adopt multiple small pads but a power supply rail to gain several advantages such as continuous power transfer, low construction cost and time, and simple control characteristics. With a rectangular pick-up, the U-type power supply rail, which had been used for the 2G OLEVs in 2011 developed by KAIST, was used for the interoperability between stationary and dynamic charging.

For the demonstration purpose, the RPEV of a maximum power of 50 kW was deployed and operated in a bus route of 10 km in Malaga Spain since December 2014. Moreover, a self-guided control system, which automatically controls a steering wheel to follow the center of a road, was adopted to minimize lateral displacements for its efficient power transfer.

### 10.6.4 The INTIS Research Team

The Integrated Infrastructure Solutions (INTIS) has been investigating the IPTS for stationary wireless electric vehicles (SWEVs) and RPEVs to provide engineering services and consulting from developments to field tests since 2011 (INTIS

Website, 2014). Now, INTIS has its own 25 m power supply rail in the test center in Lathen, Germany, and the test center can be used to evaluate the IPTS for SWEVs and RPEVs from a component level to a completed system level.

Based on a single-phase power system, the test power supply rail of the IPTS for SWEVs and RPEVs is available for tests up to a maximum power of 200 kW at an operating frequency up to 35 kHz while the double U-type power supply rail, which has two U-type power rails in parallel, is adopted in the test center. So far, INTIS has developed two IPTSs for SWEVs and RPEVs, which used the operating frequency of 30 kHz for the maximum power of 30 kW at an air-gap of 10 cm

## 10.7 Conclusion

The development history of IPTS for RPEV from its advent to its current status as a state-of-the-art technology has been introduced throughout this chapter. The size, weight, efficiency, air-gap, lateral tolerance, EMF, and cost of the IPTS have been substantially improved during a century, so RPEVs are becoming viable solutions for future transportation. The firstly commercialized OLEV is an especially strong candidate for the near-future widespread use of RPEVs in public transportation. IPTSs that are more economic, compact, efficient, robust, and easy to deploy and maintain will be welcome for future commercialization.

## References

J. G. Bolger (1994) Urban electric transportation systems: the role of magnetic power transfer. *IEEE WESCON94 Conference*, Anaheim, CA, USA, pp. 41–45.

M. Budhia, J. T. Boys, G. A. Covic, and C.-Y. Huang (2013) Development of a single-sided flux magnetic coupler for electric vehicle IPT charging systems. *IEEE Trans. Ind. Electron.*, 60, 318–328.

M. Budhia, G. A. Covic, J. T. Boys, and C.-Y. Huang (2011) Development and evaluation of single sided flux couplers for contactless electric vehicle charging. *IEEE Energy Conversion Congress and Exposition (ECCE)*, Phoenix, AZ, USA, pp. 614–621.

California PATH Program (1994) *Roadway Powered Electric Vehicle Project Track Construction and Testing Program Phase 3D*, California PATH Research Paper.

S. Y. Choi, B. W. Gu, S. Y. Jeong, and C. T. Rim (2015a) Ultra-slim S-type inductive power transfer system for roadway powered electric vehicles. *International Electric Vehicle Technology Conference & Automotive Power Electronics in Japan (EVTeC & APE Japan), 20144027.*

S. Y. Choi, B. W. Gu, S. W. Lee, W. Y. Lee, J. Huh, and C. T. Rim (2015b) Generalized active EMF cancel methods for wireless electric vehicles. *IEEE Trans. Power Electron.*, 29, 5770–5783.

S. Y. Choi, B. W. Gu, S. Y. Jeong, C. T. Rim (2015c) Advances in wireless power transfer systems for roadway-powered electric vehicles. *IEEE Journal of Emerging and Selected Topics in Power Electronics*, 3(1), 18–36.

S. Y. Choi, J. Huh, W. Y. Lee, S. W. Lee, and C. T. Rim (2013) New cross-segmented power supply rails for roadway powered electric vehicles. *IEEE Trans. Power Electron.*, 28, 5832–5841.

G. A. Covic, J. T. Boys, M. Kissin, and H. Lu (2007) A three-phase inductive power transfer system for roadway power vehicles. *IEEE Trans. Ind. Electron.*, 54, 3370–3378.

G. A. Covic, L. G. Kissin, D. Kacprzak, N. Clausen, and H. Hao (2011) A bipolar primary pad topology for EV stationary charging and highway power by inductive coupling. *IEEE Energy Conversion Congress and Exposition (ECCE)*, Phoenix, AZ, USA, pp. 1832–1838.

M. Hanazawa, N. Sakai, and T. Ohira (2012) SUPRA: supply underground power to running automobiles. *IEEE International Electric Vehicle Conference, IEVC'2012*, Greenville.

J. Huh, W. Y. Lee, G. H. Cho, B. H. Lee, and Chun T. Rim (2011a) Characterization of novel inductive power transfer systems for on-line electric vehicles (OLEV). *IEEE Applied Power Electronics Conference & Exposition (APEC)*, Fort Worth, TX, USA, pp. 1975–1979.

J. Huh, S. W. Lee, W. Y. Lee, G. H. Cho, and Chun T. Rim (2011b) Narrow-width inductive power transfer system for on-line electrical vehicles (OLEV). *IEEE Trans. Power Electron.*, 26, 3666–3679.

J. Huh, W. Y. Lee, S. Y. Choi, G. H. Cho, and C. T. Rim (2011c) Explicit static circuit model of coupled magnetic resonance system. *IEEE Energy Conversion Congress and Exposition (ECCE)-Asia*, Jeju, Korea, pp. 2233–2240.

J. Huh, W. Y. Lee, S. Y. Choi, G. H. Cho, and C. T. Rim (2013) Frequency-domain circuit model and analysis of coupled magnetic resonance systems. *J. Power Electron.*, 13, 275–286.

ICNIRP Guidelines (1998) Guidelines for limiting exposure to time-varying electric and magnetic fields (up to 300 GHz).

ICNIRP Guidelines (2010) Guidelines for limiting exposure to time-varying electric and magnetic fields (up to 100 kHz).

M. Hutin and M. Leblanc (1894) Transformer system for electric railways. Patent US 527857.

KAIST OLEV team (2009) *Feasibility Studies of On-Line Electric Vehicle (OLEV) Project*, KAIST, Daejeon, Korea, Internal Report.

J. Kim, J. H. Kim, and S. Y. Ahn (2013) Coil design and shielding methods for a magnetic resonant wireless power transfer system. *Proc. IEEE*, 101, 1332–1342.

E. S. Lee, J. Huh, X. V. Thai, S. Y. Choi, and C. T. Rim (2013) Impedance transformers for compact and robust coupled magnetic resonance systems. *IEEE Energy Conversion Congress and Exposition (ECCE)*, Denver, CO, USA, pp. 2239–2244.

S. W. Lee, C. B. Park, J. G. Cho, G. H. Cho, and Chun T. Rim (2010) Ultra slim U & W power supply and pick-up coil design for OLEV. *Korean Institute of Power Electronics (KIPE) Annual Summer Conference*, Pyeong-chang, Korea, pp. 353–354.

E. Mascarell (2015) *VICTORIA: Towards and Intelligent E-Mobility*. UNPLUGGED Final Event.

C. B. Park, S. W. Lee, and C. T. Rim (2012) 5 m-off-long-distance inductive power transfer system using optimum shaped dipole coils. *International Power Electronics and Motion Control Conference (IPEMC)*, Harbin, China, pp. 1137–1142.

C. B. Park, S. W. Lee, and C. T. Rim (2015) Innovative 5 m-off-long-distance inductive power transfer system with optimum shaped dipole coils. *IEEE Trans. Power Electron.*, 30, 817–827.

N. P. Suh, D. H. Cho, G. H. Cho, J. G. Cho, C. T. Rim, and S. H. Jang (2011) Ultra slim power supply and collector device for electric vehicle. Patent KR 1010406620000.

Technical Article eCarTec 2014 in INTIS Website, http://www.intis.de/intis/downloads_e.html.

A. Zaheer, D. Kacprzak, and G. A. Covic, (2012) A bipolar receiver pad in a lumped IPT system for electric vehicle charging applications. *IEEE Energy Conversion Congress and Exposition (ECCE)*, Raleigh, NC, USA, pp. 283–290.

Chapter 11

# Energy cryptography for wireless charging of electric vehicles

*Zhen Zhang*[1]

As one of the most epoch-making technologies, the wireless power transfer system shows increasing significance for various electrified devices, especially for electric vehicles (EVs), such as the battery charging for normal vehicular operations, and energy exchange or energy arbitrage for advanced vehicle-to-grid operations. Obviously, wireless charging technologies are changing our traditional usage pattern of the energy, thus promoting the pervasive application of sustainable energies into our daily life.

In this chapter, various wireless power transfer technologies, including acoustic, optical, microwave, capacitive, and inductive means, are first introduced. Then, the inductive resonant charging and dynamic charging are described, which are essential for EV wireless charging. After revealing the principle of energy cryptography, the realization, control, and experimentation of energy cryptography for EV wireless charging systems are discussed in detail.

## 11.1 Wireless power transfer

Literally, the wireless power transfer can fulfil delivering the energy from the power supply to the receptor in a cordless way. According to the transmission distance, the wireless power transfer can be broadly classified into the near-field and far-field transmissions. The near-field power transfer utilizes electromagnetic field couplings, such as the capacitive and inductive mechanisms, for relatively short-range transmission. In contrast, the far-field power transfer typically utilizes various schemes such as the acoustic, optical, and microwave means to deal with long-range transmission.

### *11.1.1 Acoustic*

The acoustic power transfer technique utilizes acoustic waves or mechanical vibrations to wirelessly transmit the energy. As depicted in Figure 11.1, the

[1]School of Electrical Engineering and Automation, Tianjin University, China

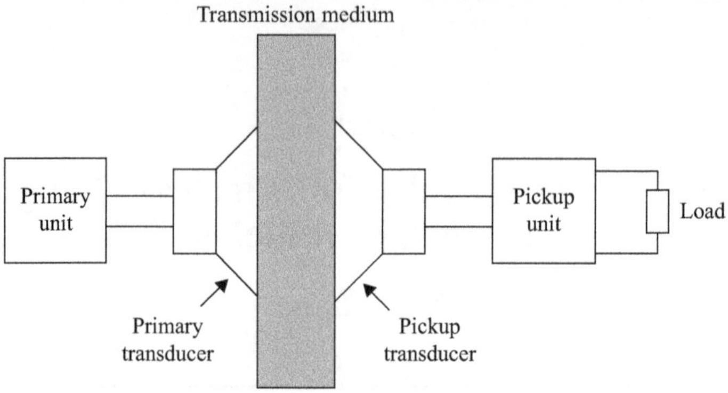


*Figure 11.1 Acoustic wireless power transfer*

acoustic wireless power transfer system mainly consists of the primary unit for the AC power supply, the primary and pickup transducers for the conversion between electricity and mechanical acoustic energy, and the pickup unit for the energization of the load.

Due to the ultrasonic frequency of the resulted waves or vibrations, the acoustic mechanism can offer long-range power transfer. In addition, the acoustic mechanism is applicable for various transmitting medium, such as the living tissue, metal material, and air. However, this infancy technique still faces with some challenges. In particular, the resulting spatial resonance restricts the placement of the receiver and thus limits its applications. Additionally, the corresponding theoretical analysis is still incomplete, which significantly impedes the further development of the acoustic power transfer technique. Besides, the transducer design deserves particular attention, in terms of the power level, efficiency, and influence on reflections (Roes *et al.*, 2013).

### *11.1.2 Optical*

As depicted in Figure 11.2, the laser can be utilized as the media to wirelessly transmit the energy to a photovoltaic cell. Due to its long transmission range, the optical wireless power transfer technique is mainly used for military weapons and aerospace applications. By comparing with other mechanisms, the laser shows salient advantages, such as the ultralong-distance transmission, zero-interference to radio-frequency applications, concentrated transmitting energy, and targeted transmission. Besides, the laser wireless power transfer has some inevitable drawbacks, including the low efficiency for the conversion between light and electricity, straight transmission, and hazardous laser radiation. In recent years, the optical wireless power transfer technique has also been utilized for the charging of

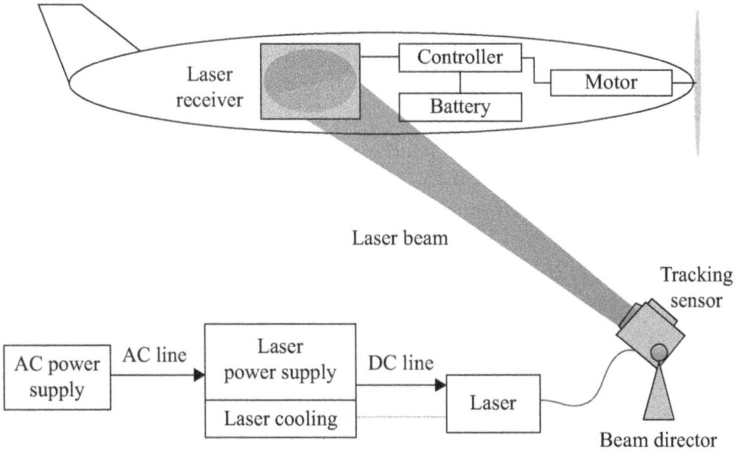


*Figure 11.2 Optical wireless power transfer*

consumer electronics and low-power sensors in industrial applications (Sahai and Graham, 2012).

### *11.1.3 Microwave*

As one of the far-field transmission mechanisms, the microwave is utilized for various applications, such as the low-power sensor network, space, and military applications. As shown in Figure 11.3, it depicts a typical configuration of the microwave power transfer system. Firstly, the microwave is generated by the microwave generator. Then, the radiation successively passes through the coax-waveguide adaptor and the waveguide circulator, aiming to reduce the outside exposure of the radiation. As the last step of the transmitting part, the radiation passes through the tuner and directional coupler device for signal separation based on the propagation direction. In such ways, the radiation can be transmitted over the air through antennae. In the receiving part, the microwave radiation acquired by the receiving antenna passes through a low-pass filter and a matching network. Lastly, the radiation is converted to DC power by a rectifier for further uses (Reddy *et al.*, 2013). In recent years, the microwave power transfer has attracted increasing attentions. For example, a rectenna architecture for the microwave wireless power transfer technique has been proposed, which can ensure the best possible energy conversion efficiency over a very wide range of input power levels (Marian *et al.*, 2012).

### *11.1.4 Capacitive*

In the capacitive wireless power transfer technique, the transmitter and receiver electrodes can form a capacitor, where an induced alternating potential of the

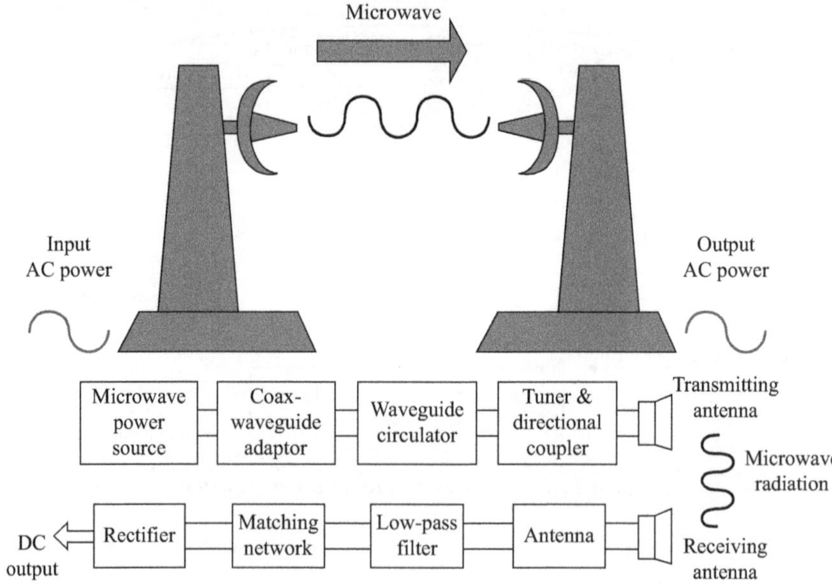


*Figure 11.3 Microwave wireless power transfer*

receiving plate can be generated by the alternating voltage of the transmitting plate via the electrostatic induction. The transmitting power is determined by the switching frequency and the capacitance between the plates. The capacitive wireless power transfer technique can be classified into the bipolar and unipolar systems. As depicted in Figure 11.4(a), the bipolar system comprises of two pairs of coupled transmitting and receiving plates, where the voltage has the phase difference of 180º for each transmitting plate by using the oscillator. Then, the receiving plates can induce the alternating potentials with the opposite phase, respectively. In such ways, the energy can be wirelessly transmitted between the plates in the bipolar way. Figure 11.4(b) shows the unipolar system, which has only one pair of coupled plates to transmit the energy, and the passive plates function to form the return path (Liu *et al.*, 2011).

The capacitive mechanism can offer the reduced requirement of alignment for plates and the confined energy field to decrease the unexpected interferences. However, since the high voltage on the plate may inevitably produce noxious gas and deteriorate the safety performance, the capacitive wireless power transfer technique cannot be used for high power applications.

### *11.1.5 Inductive*

As shown in Figure 11.5, the most basic form of inductive power transfer systems mainly consists of the primary and secondary units. First, the primary coil is energized by a high-frequency AC power source. In such ways, the energy can be wirelessly transferred from the primary coil to secondary coil via the induced

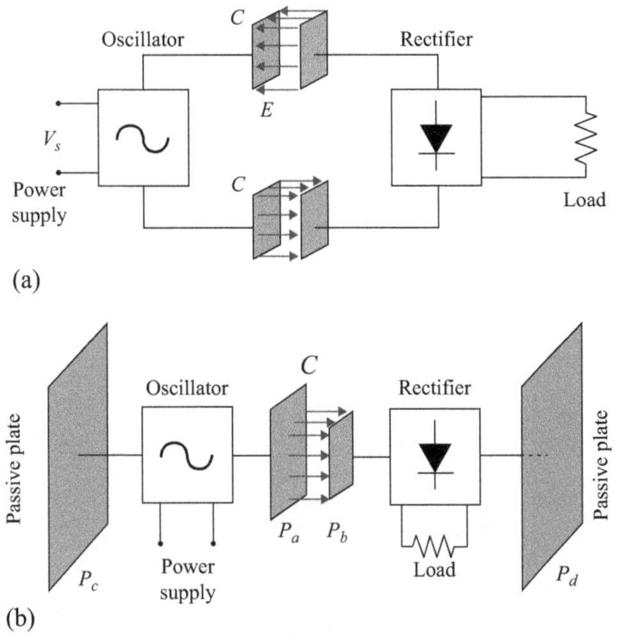


*Figure 11.4 Capacitive wireless power transfer: (a) bipolar and (b) unipolar*

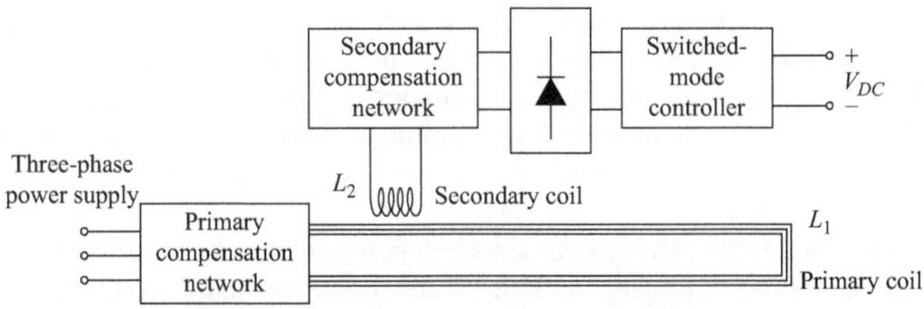


*Figure 11.5 Inductive power transfer*

electromagnetic field. Conceptually, the inductive power transfer is very similar to a transformer with a weak coupling effect. Meanwhile, the magnetic material such as the ferrite is also used to direct the magnetic flux and improve the coupling between the primary and secondary coils (Covic and Boys, 2013).

In addition, the magnetic resonant coupling (MRC) mechanism can be also utilized to further improve the transmission performance of inductive power transfer. The MRC-based wireless power transfer has been experimentally demonstrated by delivering the energy with the power of 60 W over the distance of 2 m (Kurs *et al.*, 2007), which is identified as the most promising solution for

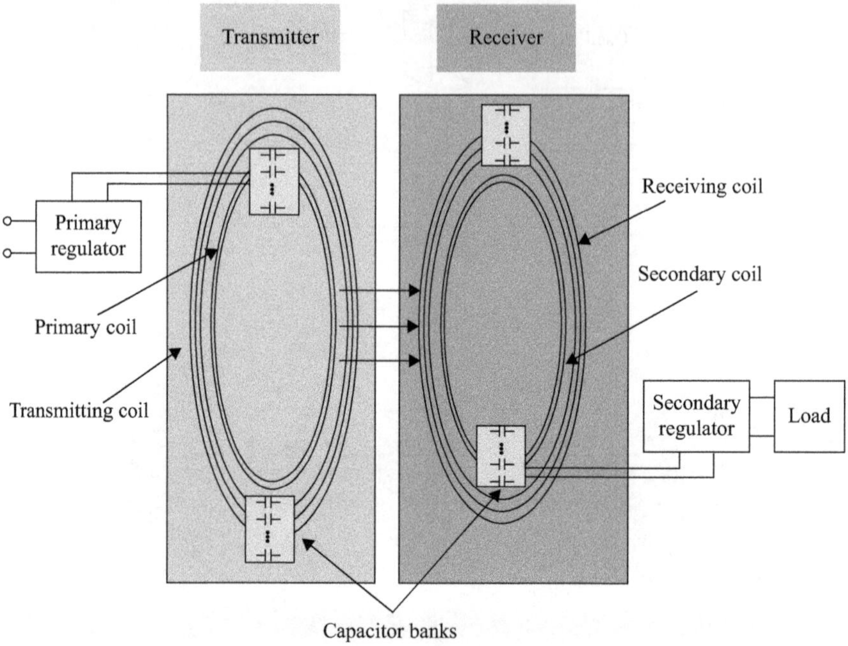


*Figure 11.6 MRC-based wireless power transfer*

mid-range and low-power applications. As depicted in Figure 11.6, the configuration of a basic four-coil MRC-based wireless charging system mainly consists of a high-frequency AC power supply, a transmitter, a receiver, and a regulator connected to the battery. The most difference from the inductive power transfer system is to utilize the parasitic capacitance of the transmitting and receiving coils to drive the system to work at the resonant state.

In previous studies, an MRC-based wireless power transfer system was utilized to wirelessly transfer the energy to multiple small receivers and EVs (Cannon *et al.*, 2009). Meanwhile, the equivalent circuit and Neumann formula was adopted to maximize the power transfer efficiency and air-gap length (Imura and Hori, 2011). In addition, the wireless domino-resonator system was proposed by adopting non-coaxial axes and circular structures (Zhong *et al.*, 2012). Also, an automated impedance matching system was proposed to improve the power transfer efficiency by matching the resonant frequency of the resonator pair with that of the power source (Beh *et al.*, 2013).

From the perspective of inductive power transfer technologies for EVs, previous studies have obtained fruitful achievements. For example, an inductive power transfer system was proposed with a narrow rail width for charging roadway-powered online EVs, which can provide energy across the entire width of the roadway surface (Huh *et al.*, 2011). For plug-in EVs and vehicle-to-grid services, a bidirectional inductive power interface was developed, which can facilitate simultaneous and controlled charging or discharging of multiple EVs (Madawala and Thrimawithana, 2011). In addition, a transient load detection model was proposed to detect load conditions for inductive power transfer systems (Wang *et al.*, 2013). Also, the mutual coupling effect between planar inductors was modeled for inductive power transfer systems (Raju *et al.*, 2014).

## 11.2 Wireless charging for EVs

As the greenest road transportation, EVs have to face with one fundamental problem impeding their further popularization, namely the short driving range. By significantly increasing the number of batteries for EVs, the driving range can be extended to over 400 km, but it will bring about other tough issues, such as the vehicle weight, volume, maintenance, and unaffordable cost. Rather than waiting for the breakthrough of battery technology, the wireless power transfer can make EVs harness the energy via the induced electromagnetic field and then charge their batteries over the air when running on the road. Accordingly, this roadway-powered EV technology can fundamentally get through problems of short driving range due to limited battery storage or high initial cost due to installation of a large number of batteries.

### *11.2.1 Inductive resonant charging*

Technically speaking, the inductive power transfer has been recognized as the preferred solution for EV wireless charging systems. In particular, the inductive resonant charging system can offer enhanced static and dynamic charging performance. Typically, it has four basic compensation networks as depicted in Figure 11.7, including the series–series (SS), series–parallel (SP), parallel–series (PS), and parallel–parallel (PP) topologies. By denoting the magnetic coupling coefficient as $k$, the secondary quality factor as $Q_s$, the switching frequency as $\omega$, and resonant frequency as $\omega_0$, the circuit parameters of these topologies can be calculated by using a mutual inductance coupling model as given in Table 11.1. It can be seen that there are three important features:

- In the SS topology, the series-compensated secondary unit has no reflected reactance when working at the resonant frequency. Thus, the series-compensated primary unit is independent of the coupling effect and the load.
- In the SP topology, the parallel-compensated secondary unit inevitably produces the reflected load-independent reactance when working at the resonant frequency. Thus, the series-compensated primary unit is dependent of the coupling effect.
- In the PS and PP topologies, the reflected impedance is affected by the load impedance. Thus, the parallel-compensated primary unit is dependent on both the coupling effect and the load.

Accordingly, the SS topology is widely adopted in inductive resonant charging for EVs, due to its independence on the mutual inductance and the load.

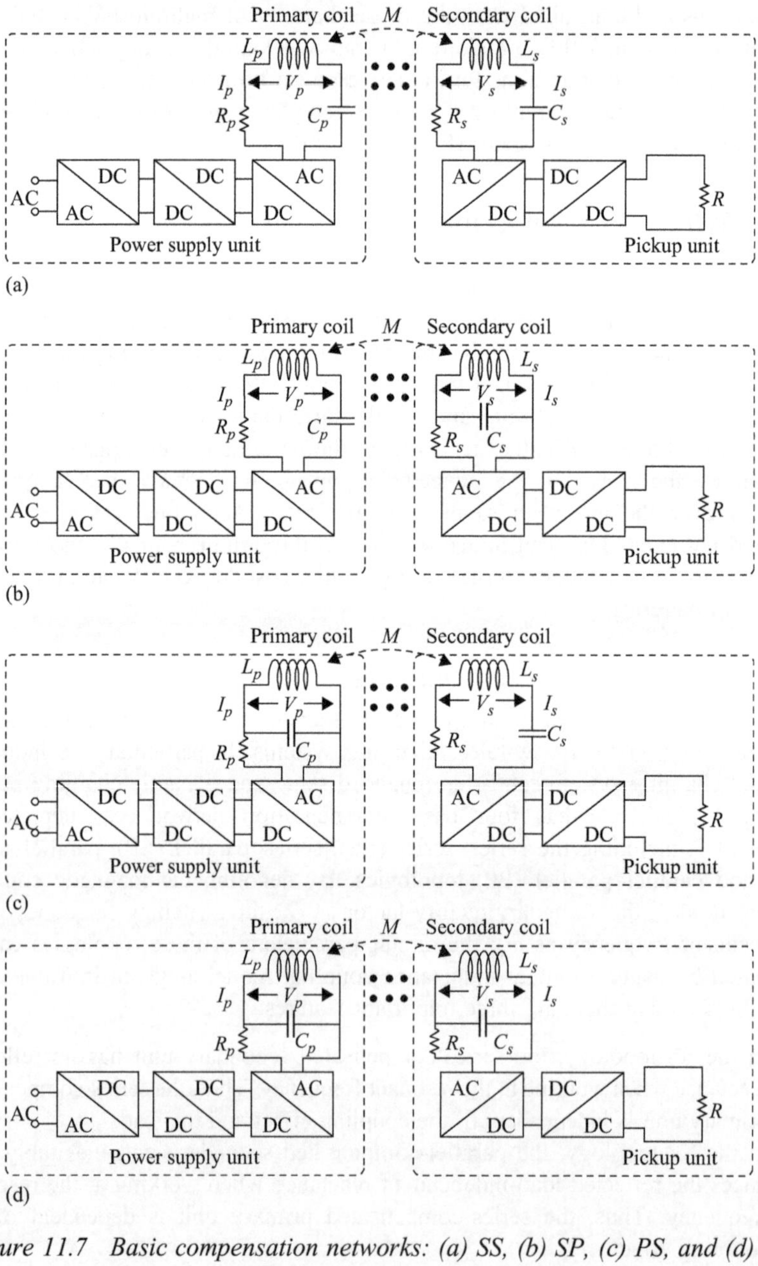


*Figure 11.7 Basic compensation networks: (a) SS, (b) SP, (c) PS, and (d) PP*

## 11.2.2 Dynamic charging

As shown in Figure 11.8, a typical wireless EV charging system mainly consists of several functional parts to fulfil the wireless energy delivery. By using the AC/DC converter with the function of power factor correction, the utility is

*Table 11.1 Comparative analysis of basic compensation networks*

| Topology | Primary capacitance | Normalized primary capacitance | Reflected resistance | Reflected reactance | Secondary impedance | Secondary quality factor | Load voltage | Load current |
|---|---|---|---|---|---|---|---|---|
| SS | $\frac{C_sL_s}{L_p}$ | 1 | $\frac{\omega_0^2M^2}{R}$ | 0 | $j\omega L_s+\frac{1}{j\omega C_s}+R$ | $\frac{\omega_0L_s}{R}$ | $I_sR$ | $I_s$ |
| SP | $\frac{C_sL_s^2}{L_pL_s-M^2}$ | $\frac{1}{1-k^2}$ | $\frac{M^2R}{L_s^2}$ | $-\frac{\omega_0M^2}{L_s}$ | $j\omega L_s+\frac{1}{j\omega C_s+\frac{1}{R}}$ | $\frac{R}{\omega_0L_s}$ | $V_s$ | $\frac{V_s}{R}$ |
| PP | $\frac{\left(L_pL_s-M^2\right)C_sL_s^2}{\frac{M^4C_sR}{L_s}+\left(L_pL_s-M^2\right)^2}$ | $\frac{1-k^2}{Q_s^2k^4+(1-k^2)^2}$ | $\frac{M^2R}{L_s^2}$ | $-\frac{\omega_0M^2}{L_s}$ | $j\omega L_s+\frac{1}{j\omega C_s+\frac{1}{R}}$ | $\frac{R}{\omega_0L_s}$ | $V_s$ | $\frac{V_s}{R}$ |
| PS | $\frac{C_sL_s}{\frac{M^4}{L_pC_sL_sR}+L_p}$ | $\frac{1}{Q_s^2k^4+1}$ | $\frac{\omega_0^2M^2}{R}$ | 0 | $j\omega L_s+\frac{1}{j\omega C_s}+R$ | $\frac{\omega_0L_s}{R}$ | $I_sR$ | $I_s$ |

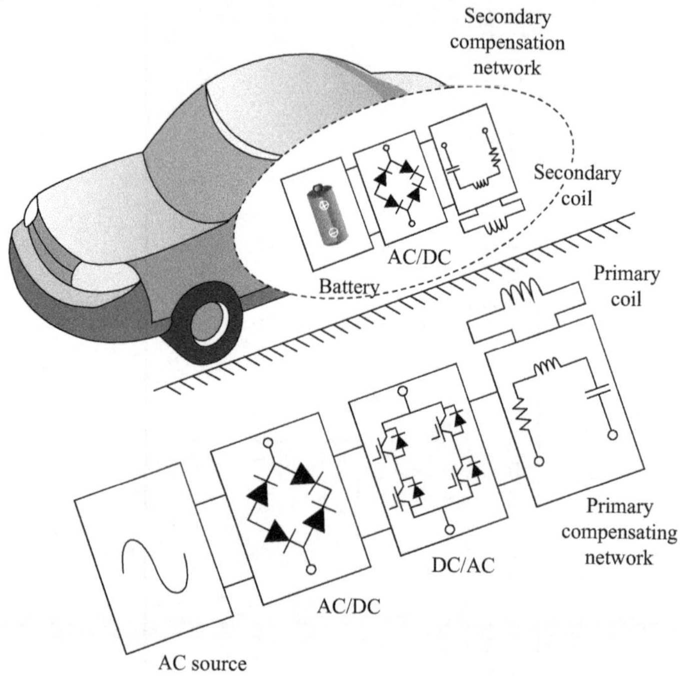


*Figure 11.8 Typical wireless EV charging system*

converted to a DC power source. Through a DC/DC converter, the expected voltage of DC bus can be obtained and then converted to a specified high-frequency AC power for the primary coil. So, an alternating electromagnetic field can be generated by the high-frequency current. Accordingly, the secondary coil can induce an AC voltage. By utilizing the compensation network, the transmission performance such as the power and efficiency can be improved significantly. Lastly, the acquired AC power is rectified and regulated to charge the EV battery. To sum up, the EV wireless charging consists of the three main parts as follows:

1. Primary and secondary coils;
2. Power converters for the conversion between AC and DC sources;
3. Compensation networks for primary and secondary circuits.

Apart from these three essential parts, the EV wireless charging system also needs assistive functions, such as the monitoring equipment, human machine interface, and sensor network.

Figure 11.9 shows the basic arrangement of EV dynamic charging system in which the power supply unit is embedded beneath the wireless power transfer lanes. By comparing with the static charging technique, the EV dynamic charging system

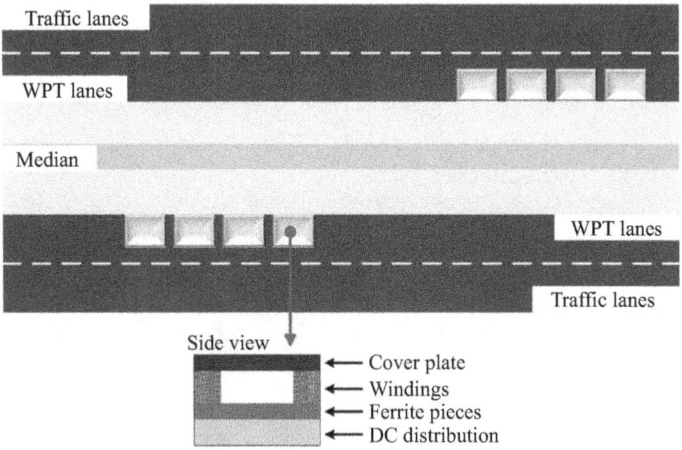


*Figure 11.9 Basic arrangement of EV dynamic charging system*

has to face with not only common issues as the static charging but also emerging unique challenges as follows:

- Design of AC or DC microgrid for the roadway-powered system;
- Impact of the vehicle speed on the roadway-powered system;
- Misalignment tolerance for wireless charging;
- Optimization and coordination of power tracks;
- Energy management for multiple vehicles with charging requests;
- Encrypted energy transmission channel;
- Bidirectional, secure, stable communications.

The EV dynamic charging technique has attracted increasing attentions in recent years. Undoubtedly, the dynamic charging system will be the most important development direction for EVs in the foreseeable future.

The wireless power transfer technique should effectively transmit the energy to moving targets, which is the key for EV dynamic charging systems. Figure 11.10 shows the basic structure of the EV dynamic charging system, where the power supply unit adopts multiple series-connected primary coils assembled along the moving trace of the charging target. The pickup unit can obtain the energy via the coupled magnetic fields between the secondary and primary coils. In addition, the cross-coupling effect between coils significantly depends on the corresponding mutual inductance, namely it can dramatically reduce to an extremely low level with the increasing distance. In order to deeply understand the EV dynamic charging system, the circuit-based analysis is employed which focuses on the mutual effect between the secondary coil and adjacent two primary coils.

Figure 11.11 shows the equivalent circuit model of the dynamic charging system when the charging target moves between the first and second primary coils, where $R_{p1}$,

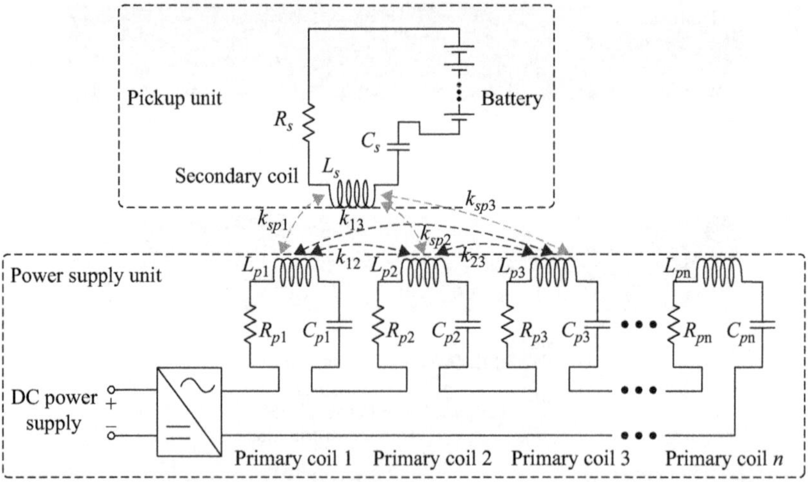


*Figure 11.10 Basic circuit diagram of EV dynamic charging system*

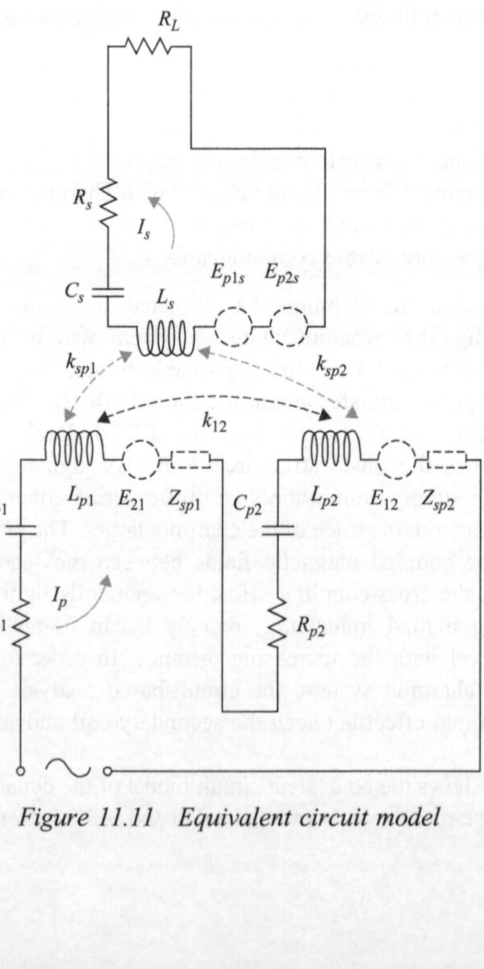


*Figure 11.11 Equivalent circuit model*

$R_{p2}$, and $R_s$ are the internal resistances of the first primary coil, second primary coil, and secondary coil, respectively; $R_L$ is the load resistance; $C_{p1}$, $C_{p2}$, and $C_s$ are the capacitances of the first primary coil, second primary coil, and secondary coil, respectively; and $L_{p1}$, $L_{p2}$, and $L_s$ are the inductances of the first primary coil, second primary coil, and secondary coil, respectively.

Denoting $\kappa_{12}$, $\kappa_{sp1}$, and $\kappa_{sp2}$ as the coupling coefficients between relevant primary and secondary coils, the corresponding mutual inductances can be expressed as:

$$L_{12} = \kappa_{12}\sqrt{L_{p1}L_{p2}} \tag{11.1}$$

$$L_{sp1} = \kappa_{sp1}\sqrt{L_{p1}L_s} \tag{11.2}$$

$$L_{sp2} = \kappa_{sp2}\sqrt{L_{p2}L_s} \tag{11.3}$$

where $L_{12}$ denotes the mutual inductance between the first and second primary coils, and $L_{sp1}$ and $L_{sp2}$ are the mutual inductances between the secondary coil and the first and second primary coils, respectively.

In the power supply unit, the current of the primary coil $I_p$ can be calculated as:

$$I_p = \frac{V_{in} - E_{12} - E_{21}}{Z_{p1} + Z_{p2} + Z_{sp1} + Z_{sp2}} \tag{11.4}$$

where $V_{in}$ is the input voltage, $E_{21}$ is the induced potential of the second primary coil on the first primary coil, and $E_{12}$ is the induced potential of the first primary coil on the second primary coil. In addition, $Z_{p1}$ is the impedance of the first primary coil, which is given by:

$$Z_{p1} = R_{p1} + \frac{1}{j\omega C_{p1}} + j\omega L_{p1} \tag{11.5}$$

where $\omega$ is the operating frequency, and $Z_{p2}$ is the impedance of the second primary coil, which is given by:

$$Z_{p2} = R_{p2} + \frac{1}{j\omega C_{p2}} + j\omega L_{p2} \tag{11.6}$$

Denoting $Z_{sp1}$ and $Z_{sp2}$ as the reflected impedances of the secondary coil to the first and second primary coils, they can be expressed as:

$$Z_{sp1} = \frac{j\omega L_{sp1} I_s}{I_p} \tag{11.7}$$

$$Z_{sp2} = \frac{j\omega L_{sp2} I_s}{I_p} \tag{11.8}$$

where $I_s$ is the current of the secondary coil. It can be calculated as:

$$I_s = \frac{j\omega (L_{sp1} + L_{sp2}) I_p}{Z_s + R_L} \tag{11.9}$$

where $Z_s$ is the impedance of the pickup unit as given by:

$$Z_s = R_{ps} + \frac{1}{j\omega C_s} + j\omega L_s \tag{11.10}$$

Then $I_p$ can be expressed as:

$$I_p = \frac{(Z_{p1} + Z_{p2})}{(Z_{p1} + Z_{p2})^2 - \omega^2 (L_{sp1} + L_{sp2})^2} (V_{in} - E_{12} - E_{21}) \tag{11.11}$$

Accordingly, the transferred power $P$ can be obtained as:

$$P = I_p{}^2 \mathrm{Re}(Z_{sp1} + Z_{sp2}) = \left[ \frac{(Z_{p1} + Z_{p2})}{(Z_{p1} + Z_{p2})^2 - \omega^2 (L_{sp1} + L_{sp2})^2} \right]^2$$

$$= \left[ \frac{\omega^2 (L_{sp1} + L_{sp2})^2}{(Z_s + R_L)^2} (V_{in} - E_{12} - E_{21}) \right]^2. \tag{11.12}$$

Thus, it can be observed that the energy transfer performance of EV wireless charging systems is synthetically affected by various factors, such as the operating frequency, mutual inductance, input voltage, and induced potential in primary coils, which are the key for the optimal design of EV wireless charging system, especially for dynamic charging application.

Besides, the power track is buried beneath the ground to induce the electromagnetic field for the dynamic wireless charging of EVs. Obviously, the design of power tracks is the key to the EV dynamic charging technique, which has been paid increasing attentions by many researchers. Previous studies have presented the circular- or rectangular-shaped power tracks instead of the conventional long loop tracks. Although these coil structures can offer a definite superiority of misalignment tolerance, the transmission performance may be significantly deteriorated when the vehicle runs over the gap of adjacent two tracks. On the other hand, the conventional loop track can reduce the number of the track gap, whereas it inevitably affects the initial installation, daily maintenance and charging flexibility of roadway-powered systems. Accordingly, the optimal design of the power track becomes particularly important for EV dynamic charging systems.

A comprehensive analysis was conducted for the optimal design of sectional power tracks (Zhang *et al.*, 2014), which reveals the relations among the track length, vehicle velocity, and the power transmission performance. As shown in Figure 11.12, specifically, the coupling coefficient dramatically decreases with the increasing vehicle velocity if the track length is fixed. Thus, the optimal design of power track should depend on the road situation, namely, the EV wireless charging system should adopt long scale of the track length for busy roads and short scale of the track length for relatively high-speed roads.

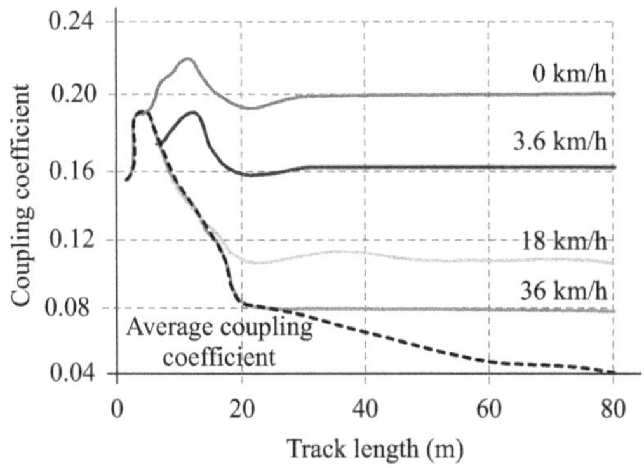


*Figure 11.12 Impact of vehicle speed and track length on power transmission performance*

## 11.3 Principle of energy cryptography

As aforementioned, a single power track adopting long scale for busy roads may have multiple vehicles with the charging request. It means that all involved vehicles may access the wirelessly transferred energy via the electromagnetic field induced by a single power track. In such ways, the energy is easily obtained by unauthorized vehicles, such as unpaid, unstable, and illegal vehicles. Then, the energy security issue has been increasingly concerned for EV wireless charging systems (Zhang *et al.*, 2015). Thus, the concept of energy cryptography, which is used to pack up the energy with security key, is becoming important. The presented energy cryptography aims to transfer the energy to authorized vehicles while prevents unauthorized vehicles from illegal access.

In wireless power transfer systems, the transmission performances greatly depend on the working frequency, which includes the transferred power, efficiency, and distance (Zhang *et al.*, 2014). As shown in Figure 11.13, the power can be efficiently transferred to the energy receptor with an optimal switching frequency, while the transmission power can be suppressed at an extremely low level as long as the frequency deviates from the optimal value even with only a slight deviation. Accordingly, this frequency sensitivity causes technical difficulties in tracing the maximum power for wireless charging systems.

The idea of energy cryptography is analogous to the anti-control method, aiming to convert the negative effect of frequency sensitivity to a positive usage. Specifically, the key is to purposely adjust the optimal frequency based on the predefined regulation by the power supply unit. The rule of frequency variation is confidential and unpredictable for all potential energy receptors. In such ways, the wirelessly transferred energy is split by the specified time slot and meanwhile packed with various frequencies. Accordingly, the rule of the frequency variation is the key to open the energy package. In other words, when the frequency is adjusted to vary with a predefined secure sequence, the unauthorized receptors cannot retrieve the energy

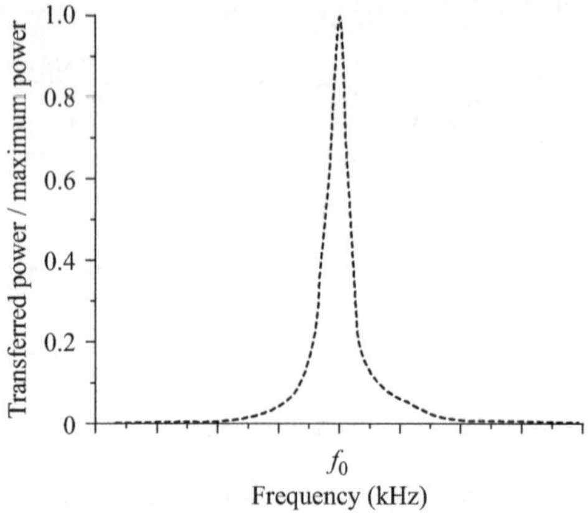


*Figure 11.13 Frequency sensitivity*

without knowledge of the sequence of frequency regulation, while the authorized receptors can receive the energy by adjusting the receptor circuit based on the acquired security key. Thus, the proposed energy encryption scheme essentially utilizes the anti-control of frequency regulation for wireless power transfer systems.

As shown in Figures 11.14 and 11.15, they depict the three-dimensional (3-D) electromagnetic field distributions on the horizontal plane of the pickup unit when the wireless power transfer system works in the resonant and nonresonant states, respectively. Figure 11.14(a) shows that the maximum flux density can reach around $3.5 \times 10^{-4}$ T in the resonant state, while Figure 11.15(a) shows that it dramatically reduces to around $5 \times 10^{-5}$ T in the nonresonant state. Meanwhile, both Figures 11.14(b) and 11.15(b) plot the two-dimensional (2-D) contours of their electromagnetic flux density distributions. Thus, the induced electromagnetic field can be purposely changed by regulating the frequency and capacitance to switch between the resonant and nonresonant states. Based on this unique characteristic, the corresponding transferred energy can be controlled for EV charging with security, which confirms the feasibility of energy encryption for EV wireless charging systems.

## 11.4 Realization of energy cryptography

Based on the above principle, the energy encryption for EV wireless charging system needs to focus on two key techniques, namely, the generation of the security key and the adjustment of the impedance.

### *11.4.1 Generation of security key*

Due to the unique characteristic of random-like behavior within an adjustable bounded domain, the chaotic sequence is utilized as the security key to encrypt the transferred energy. In fact, there are a variety of dynamic systems which can perform chaotic behaviors. In this chapter, two one-dimensional maps are introduced

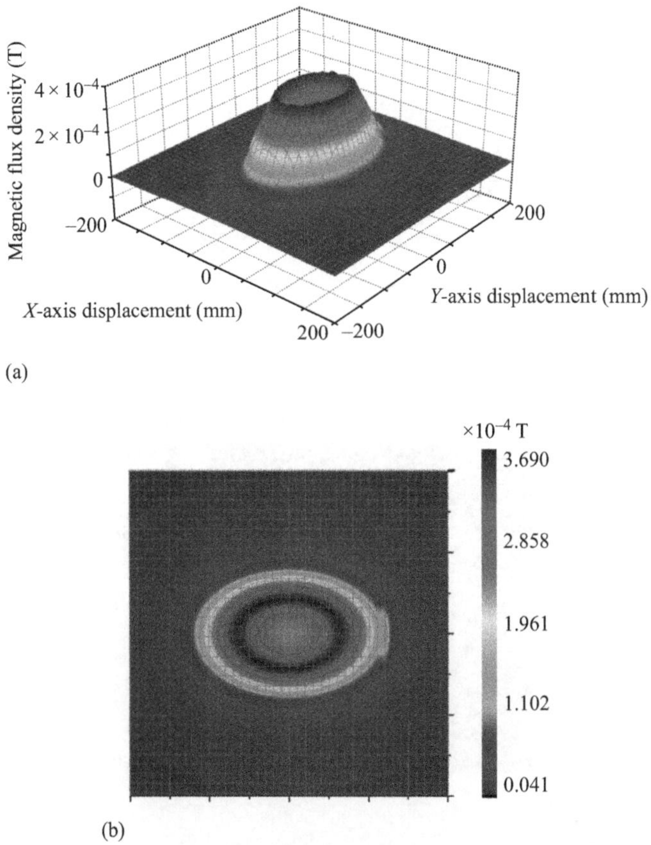


*Figure 11.14 Electromagnetic flux density distributions at resonant state: (a) 3-D and (b) 2-D contours*

for the generation of security key. First, the Logistic map can be utilized to generate the discrete-time chaotic series as given by (Chau and Wang, 2011):

$$\xi_{i+1} = A\xi_i(1 - \xi_i), \qquad A \in [0,\ 4] \tag{11.13}$$

where $\xi_i$ denotes the sequence and $A$ denotes the bifurcation parameter. Figure 11.16(a) depicts the bifurcation diagram where the phase portrait of $\xi_i$ and $\xi_{i+1}$ exhibits various topological structures along with the increase of $A$. Specifically speaking, the $\xi_i$ behaves as a constant value for $A \in [0, 1)$, a period-1 oscillation for $A \in [1, 3)$, a period-$n$ oscillation for $A \in [3, 3.57)$, and a chaotic oscillation for $A \in [3.57, 4]$. Consequently, $A = 3.9$ is selected to generate the random-like-but-bounded security key $\xi_i \in (0, 1)$ for the energy encryption. Accordingly, the encrypted switching frequency can be obtained as:

$$\omega = \delta_i \omega_0 \tag{11.14}$$

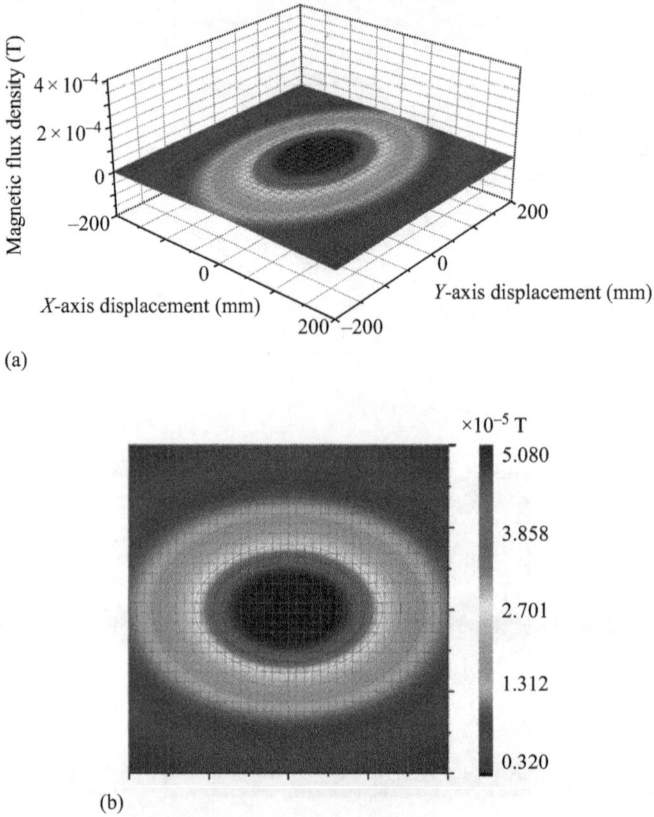


*Figure 11.15 Electromagnetic flux density distribution at nonresonant state: (a) 3-D and (b) 2-D contours*

where $\omega_0$ is the optimal frequency, and the chaotic security key $\delta_i$ can be expressed as:

$$\delta_i = a + (b - a)\xi_i, \qquad 0 < a < b \tag{11.15}$$

Then, the switching frequency can be regulated to chaotically vary within a regulating range of $(b - a)\omega_0$.

In addition, the Gaussian map exhibits random-like characteristics, which can be also utilized to generate the security key to regulate the switching frequency in an unpredictable way. The Gaussian map is a one-dimensional nonlinear iterated map, which is given by (Hilborn, 2004):

$$x_{n+1} = \exp\left(-\alpha x_n^2\right) + \beta \tag{11.16}$$

where $\alpha$ is the parameter to regulate the magnitude of chaotic oscillation and $\beta$ is the bifurcation parameter. Figure 11.16(b) shows the bifurcation diagram when $\alpha$ equals 6.2, where the sequence $x_n$ performs the chaotic oscillation within $(-0.3, 0.7)$ when $\beta$ varies from $-0.7$ to $-0.3$. Accordingly, the switching frequency

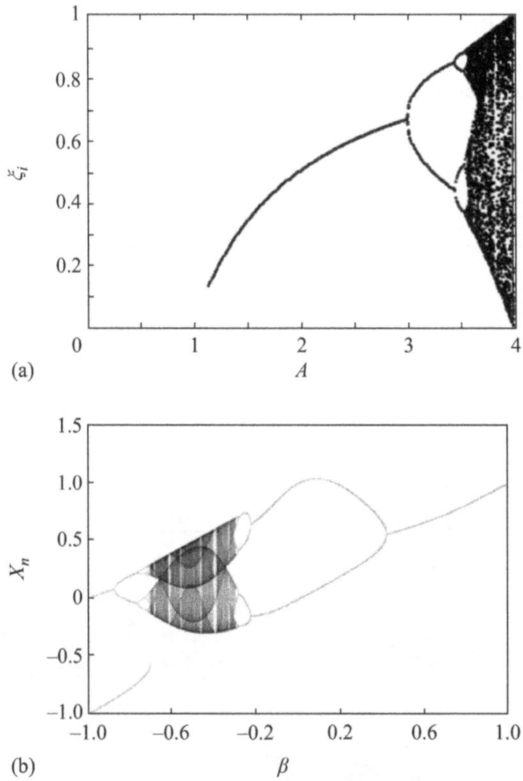


*Figure 11.16 Bifurcation diagrams: (a) Logistic map and (b) Gaussian map*

$\omega$ can be chaotically regulated within $((c-0.3)\omega_0,\ (c+0.7)\omega_0)$ by utilizing the security key $\mu$ which is given by:

$$\omega = \mu\omega_0 \tag{11.17}$$

Thus, the security key $\mu$ can be obtained according to $(c+x_n)$.

By properly choosing the constant coefficient $a$, $b$, and $c$, the regulating frequency range can be flexibly selected according to the power level and transmission distance.

## 11.4.2 Adjustment of impedance

With the chaotic security key, the primary and secondary capacitances can be accordingly regulated to achieve the optimal transferred power, which are respectively expressed as:

$$C_p = \frac{1}{\delta_i^2} \cdot \frac{1}{\omega_0^2 L_p} \tag{11.18}$$

$$C_s = \frac{1}{\delta_i^2} \cdot \frac{1}{\omega_0^2 L_s} \tag{11.19}$$

*Table 11.2 Key parameters of prototype*

| Circuit | Inductance (mH) | Capacitance (nF) | | | | | |
|---|---|---|---|---|---|---|---|
| Primary | $L_p$ | $C_{p1}$ | $C_{p2}$ | $C_{p3}$ | | | |
| | 0.09589 | 4.7 | 10 | 22 | | | |
| Secondary | $L_s$ | $C_{s1}$ | $C_{s2}$ | $C_{s3}$ | $C_{s4}$ | $C_{s5}$ | $C_{s6}$ |
| | 0.009372 | 1 | 2.2 | 4.7 | 47 | 100 | 220 |

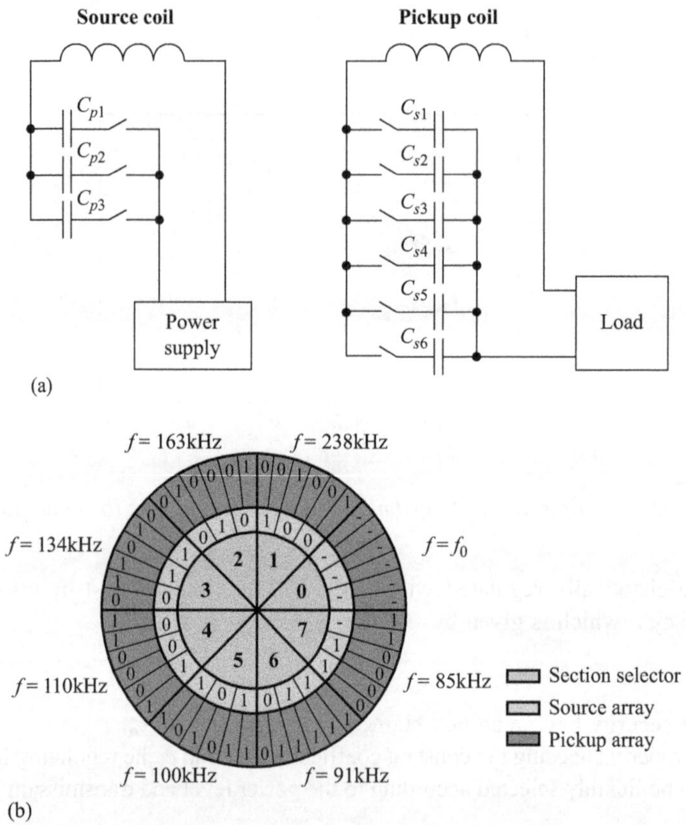


*Figure 11.17 Switched-capacitor arrays: (a) circuitry and (b) switching scheme*

In the inductive power transfer system, the primary and secondary capacitors need to be varied to meet the requirements of energy encryption and decryption. Then, the capacitor array is utilized to realize the variation of capacitance.

In order to elaborate the proposed capacitor array scheme, an exemplified wireless power transfer system is utilized to adjust the impedance. The corresponding key parameters are listed in Table 11.2. As shown in Figure 11.17(a), the capacitance of the primary and secondary circuits can be regulated by switching on and off the capacitors. For exemplification, the wireless system is designed to work in eight resonant coupling states, which can be selected by using the Logistic map as:

$$\varepsilon_i = floor(8\xi_i), \tag{11.20}$$

where $floor(\cdot)$ is the rounding function, $\xi_i$ is the chaotic Logistic sequence, and $\varepsilon_i$ is the section selector. Consequently, the capacitor arrays can be controlled according to the predefined switching table as depicted in Figure 11.17(b). For example, when $\varepsilon_i = 6$, the switch can be controlled based on the value of the corresponding section 6 in the counterclockwise direction. Then, the on–off states of the primary and resonant capacitor arrays are identical, namely, 0(OFF), 1(ON), and 1(ON) in sequence, while the state of the secondary capacitor array is 0(OFF), 1(ON), 1(ON), 1(ON), 1(ON), and 1(ON). Meanwhile, the switching frequency of the power supply should be regulated to 91 kHz. In addition, when $\varepsilon_i = 0$, it means that no regulation is required, namely, the switching state and frequency remain unchanged.

## 11.5 System control of energy cryptography

By encrypting the wirelessly transferred energy, the security performance of wireless power transfer systems can be improved significantly. After verifying the identification of EVs with the charging request, the packed energy can be effectively transferred from the power track to the authorized vehicle. Meanwhile, the unauthorized EV is rejected to access the energy field. As shown in Figure 11.18, the procedure of this energy encryption scheme for EV wireless charging systems is summarized in following steps:

**Step 1:** By arbitrarily setting the initial value of the chaotic system, like Logistic map or Gaussian map, a chaotic sequence can be generated based on a specified rule, which is utilized to generate the security key for the energy encryption of EV wireless charging systems.

**Step 2:** The working frequency of the primary side, namely the power supplier, is continually regulated to vary around the optimal frequency in a chaotic way. At this stage, the frequency of the transferred energy is confidential and unpredictable to the secondary sides, namely all EVs involved.

**Step 3:** Meanwhile, the capacitance in the primary side is regulated along with the varying frequency, aiming to make the primary circuit work in the resonant state and thus reducing the unnecessary reactive power. In such ways, the wirelessly transferred energy is successfully encrypted. Without the knowledge of the security key, the energy is hardly accessed by any EVs involved.

**Step 4:** The power supplier, namely the power track, works in the standby state awaiting the charging request from EVs. The supplier has predefined rules to verify the corresponding identification of EVs with the charging request. Then, the request from the unauthorized EV is rejected and the supplier

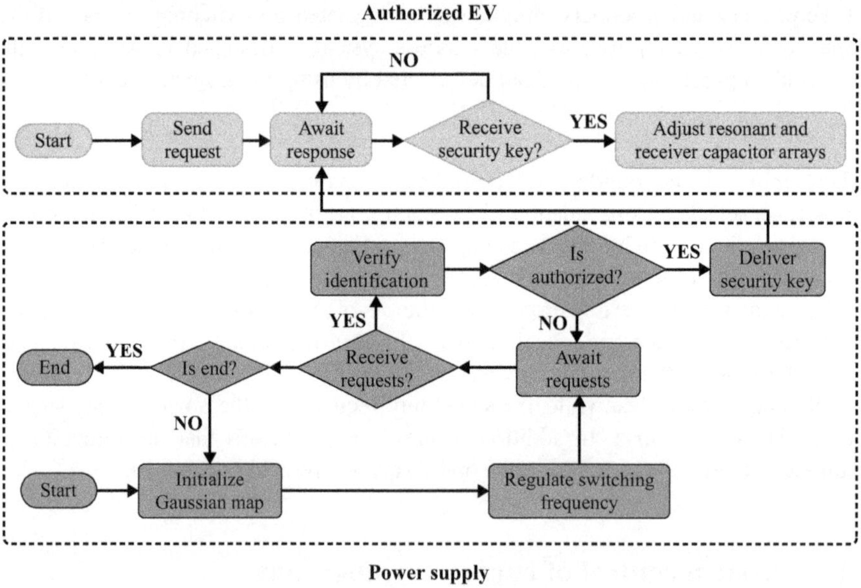


*Figure 11.18 Flowchart of energy-encrypted EV wireless charging system*

comes back to the standby state. For the authorized EV, the request is accepted and the supplier delivers the security key. Besides, the rule of authorization can be revised by the system administrator according to various situations.

**Step 5:** With the acquired security key, the secondary capacitance is simultaneously regulated via the capacitor array according to predefined rules, aiming to achieve the optimal transmission performance. Consequently, the encrypted energy can be decrypted by the synchronized regulation, which can effectively prevent unauthorized EVs from stealing the energy, thus significantly improving the security performance of EV wireless charging systems.

## 11.6 Experimentation of energy cryptography

In order to validate the energy encryption scheme for EV wireless charging systems, 3-D electromagnetic field analysis is performed by using JMAG and the system simulation is carried out by using MATLAB®/SIMULINK®. In the exemplified prototype, the power track adopts the length of 5 m and width of 1 m. The transmission clearance is set as 200 mm. Figure 11.19 illustrates that the presented power track can effectively offer an evenly distributed electromagnetic field on the horizontal plane of the EV pickup coil. In this study, the transmitted energy is encrypted by using a simple chaotic sequence, namely the Logistic map. The

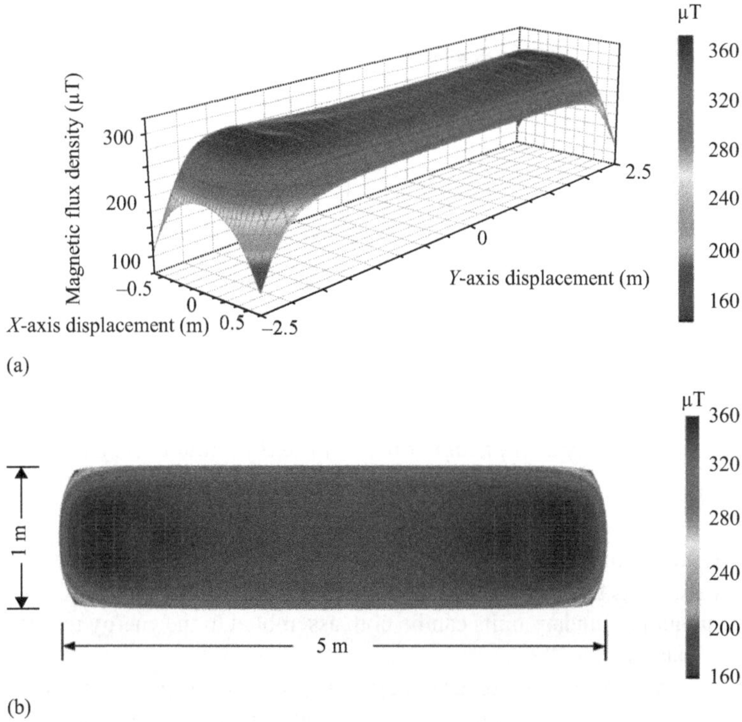


*Figure 11.19 Electromagnetic field analysis for energy-encrypted EV wireless charging system*

magnetic field is coupled by the receiver with and without tuning the resonant frequency for energy decryption. Accordingly, the corresponding power transfer to the load with and without energy decryption (equivalent to with and without authorization for EV charging) is depicted in Figure 11.20, which verifies that the proposed energy encryption for EV wireless charging system is functional. Besides, the efficiency can achieve around 92% for the power transfer of 8 kW.

Figure 11.21 shows a scale-down prototype, which aims to experimentally demonstrate the energy cryptography mechanism for EV wireless charging systems. The setup mainly consists of the primary, resonant, and secondary coils. Figure 11.22 depicts the dimensions of the primary and resonant coils, where both have a hollow circular shape with the inside diameter of 125 mm and the outside diameter of 215 mm, while the secondary coil adopts small-size design with the outside diameter of 90 mm and the inside diameter of 54 mm, aiming to facilitate testing the encryption performance at different positions with different magnetic flux densities. The primary and resonant coils directly face each other, and the secondary coil has an offset of 40 mm from the resonant coil. In this exemplified wireless power transfer system, the total transmission distance is 170 mm.

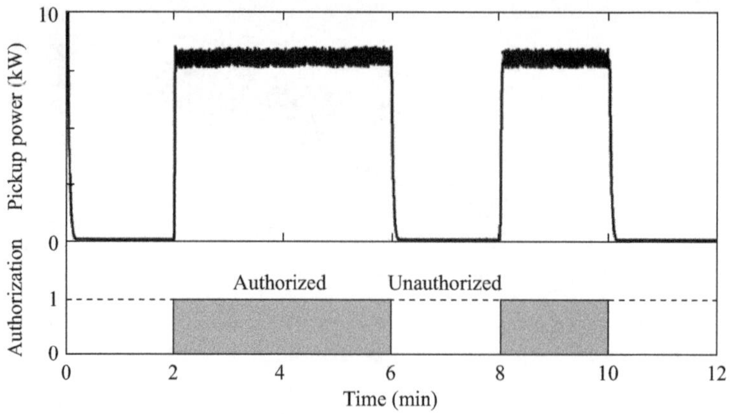


*Figure 11.20 Simulated pickup power with and without energy decryption*

Practically, the primary unit can be considered as the power supply mounted in a fixed place such as the EV charging panel in the parking lot or the roadway, while the resonant and secondary units can be both assembled in the energy receptor such as the EV chassis.

By utilizing the acquired security key from the power supply, the resonant and secondary circuits can be simultaneously regulated to work in the resonant state. Figure 11.23 shows the experimental measurement of the pickup voltage at the synchronized and unsynchronized states (namely, with and without appropriate decryption, respectively), where the pickup voltage can reach around 20 V at the synchronized state while the pickup voltage is suppressed at an extremely low level at the unsynchronized state. Figure 11.24 shows the experimental measurement of the pickup current at the synchronized and unsynchronized states, respectively. Similarly, the current can reach a high level, namely around 2 A, at the synchronized state and reduce to around 0 A at the unsynchronized state. From the experimentation, it shows that the received power of the load can effectively reach about 40 W, namely, the load voltage of 20 V and load current of 2 A. For the unauthorized receptor, the capacitor arrays of the resonant and secondary units cannot be appropriately regulated due to the absence of security key. Thus, the corresponding circuits work in the nonresonant state. Then, the pickup voltage and load current are suppressed to an insignificant level.

Technically speaking, the waveforms at the synchronized state successfully verify that the energy transmission channel is effectively established for the authorized receptor in the wireless power transfer system. On the other hand, the unsynchronized state verifies that the energy transmission channel is effectively switched off for the unauthorized receptor. Besides, by changing the position of the secondary coil, the feasibility of the proposed energy encryption strategy at different magnetic flux densities can be also verified by both simulation and experimental results. Additionally, the corresponding transmission efficiency is

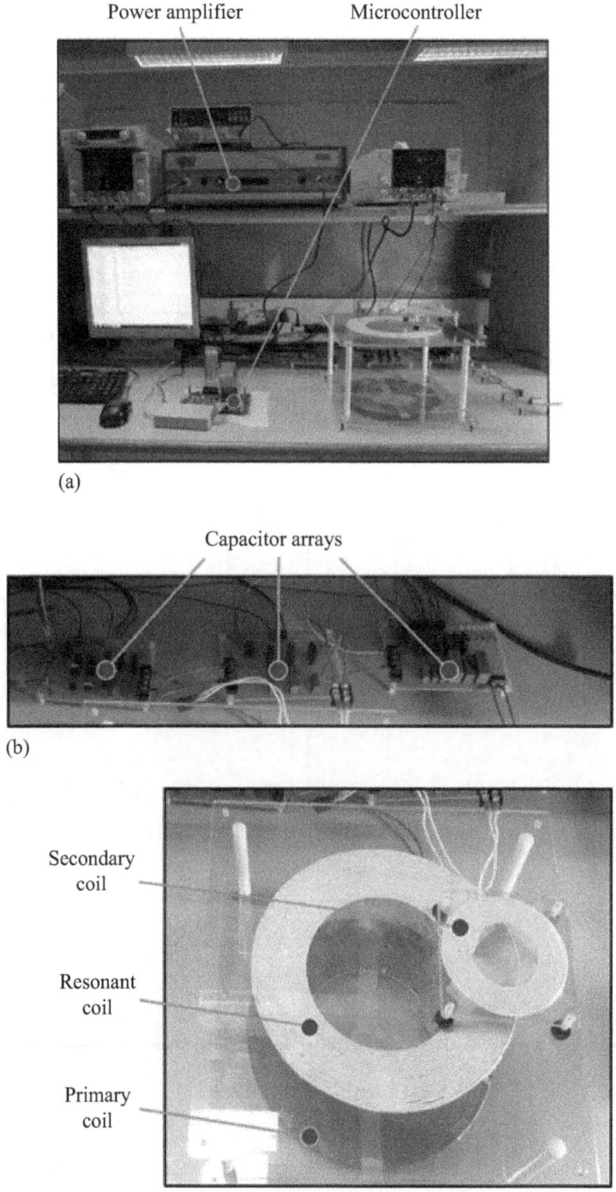


*Figure 11.21 Scale-down experimental prototype*

about 85%, where the main power loss is due to the conduction and switching losses of power switches and the internal resistance of coils. Finally, it can be observed that the experimental results well agree with the theoretical analysis and the simulation results. It verifies that the proposed energy strategy can effectively

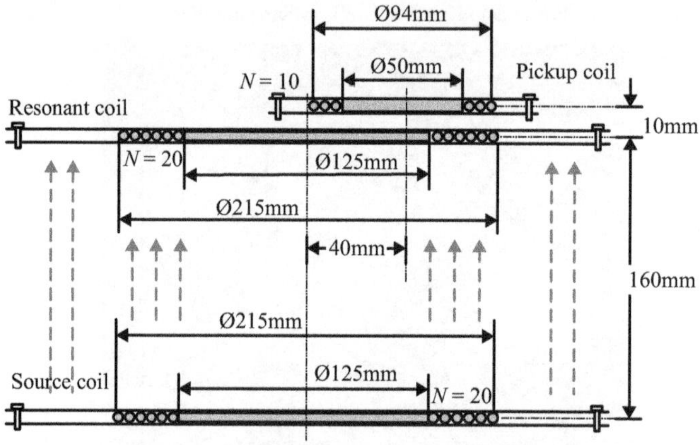


Figure 11.22 Geometry of experimental prototype

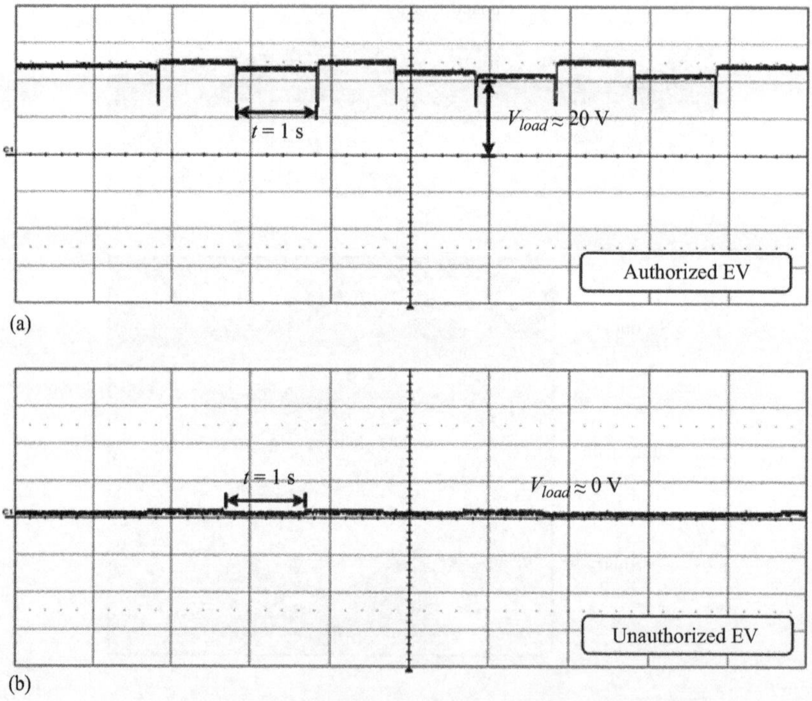


Figure 11.23 Experimental measurement of pickup voltage (X: 1 s/div, Y: 10 V/div): (a) synchronized state and (b) unsynchronized state

transfer the energy to the authorized receptor as well as to prevent the unauthorized receptor from stealing the energy. Thus, the energy cryptography shows great potentials to improve the energy security performance for EV wireless charging systems.

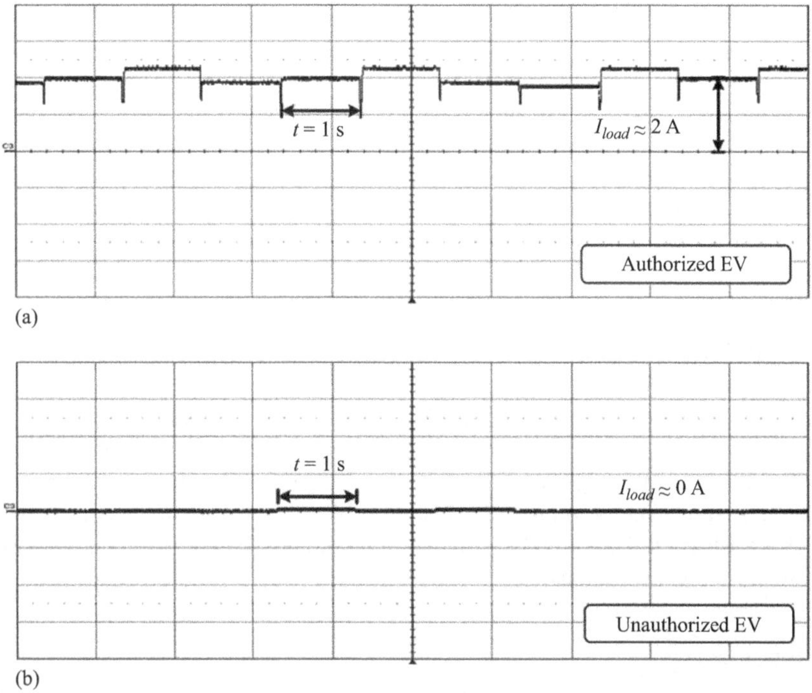


*Figure 11.24 Experimental measurement of pickup current (X: 1 s/div, Y: 1 A/div): (a) synchronized state and (b) unsynchronized state*

## 11.7 Conclusion

This chapter arouses attentions on the energy security issue for wireless power transfer systems, where the capacitor array method is utilized to improve the security performance of the wirelessly transferred energy. Nevertheless, the energy cryptography still needs further improvement to optimize the transmission performance, enhance the encryption security, and enrich the control strategies in the system level:

- The capacitor array should be further improved to enhance the encryption performance by properly increasing the section number, updating frequency, and algorithm complexity of the security key. Meanwhile, the associated negative impact on the transmission performance should be also taken into account in future studies, such as the additional power loss, increased circuit complexity, and dependence on the communication protocol. By systematically considering practical applications, it can achieve an optimal balance between the transmission performance and energy encryption.
- The capacitance can be adjusted in a continuous way instead of a discrete way, which can greatly increase the number of potential impedance values for the wireless power transfer circuit. For instance, the variable virtual capacitor

technique can be utilized to realize the continuously adjustable capacitance to supersede the capacitor array. Thus, there will be no need to divide the capacitance value into sections so that the security key can be directly used to represent the desired capacitance. The possibility of cracking the encryption strategy can be significantly reduced, therefore effectively improving the energy security performance.

- The system-level control and management policy of energy encryption should be formulated in future studies, which are essential to standardize the energy cryptography for EV wireless charging systems.

## Acknowledgements

This work was supported by Grants (Project No. SPF 201409176172) from The University of Hong Kong, Hong Kong S.A.R., China. The author would like to thank the editor-in-charge, Prof K. T. Chau, for his invitation and valuable suggestions to significantly improve the quality of this chapter, as well as my students, Mr B. Li, Mr R. X. Han, Mr B. N. Jia, Mr Y. Liu, and Mr S. Li for their hard work and important contributions.

## References

Beh, T.C., Kata, M., Imura, T., Oh, S., and Hori, Y. (2013) Automated impedance matching system for robust wireless power transfer via magnetic resonance coupling. *IEEE Transactions on Industrial Electronics*, 60, 3689–3698.

Cannon, B.L., Hoburg, J.F., Stancil, D.D., and Goldstein S.C. (2009) Magnetic resonant coupling as a potential means for wireless power transfer to multiple small receivers. *IEEE Transactions on Power Electronics*, 24, 1819–1825.

Chau, K.T. and Wang, Z. (2011) *Chaos in Electric Drive Systems – Analysis, Control and Application*, Singapore: Wiley-IEEE Press.

Covic, G.A. and Boys, J.T. (2013) Inductive power transfer. *Proceedings of the IEEE*, 101, 1276–1289.

Hilborn, R.C. (2004) *Chaos and Nonlinear Dynamics: An Introduction for Scientists and Engineers*, London, UK: Oxford University Press.

Huh, J., Lee, S.W., Lee, W.Y., Cho, G.H., and Rim, C.T. (2011) Narrow-width inductive power transfer system for online electric vehicles. *IEEE Transactions on Power Electronics*, 26, 3666–3679.

Imura, T. and Hori, Y. (2011) Maximizing air gap and efficiency of magnetic resonant coupling for wireless power transfer using equivalent circuit and Neumann formula. *IEEE Transactions on Industrial Electronics*, 58, 4746–4752.

Kurs, A., Karalis, A., Moffatt, R., Joannopoulos, J.D., Fisher, P., and Soljačić, M. (2007) Wireless power transfer via strongly coupled magnetic resonances. *Science*, 317, 84–86.

Liu, C., Hu, A.P., and Nair, N.K.C. (2011) Modeling and analysis of a capacitively coupled contactless power transfer system. *IET Power Electronics*, 4, 808–815.

Madawala, U.K. and Thrimawithana, D.J. (2011) A bidirectional inductive power interface for electric vehicle in V2G systems. *IEEE Transactions on Industrial Electronics*, 58, 4789–4796.

Marian, V., Allard, B., Vollaire, C., and Verdier, J. (2012) Strategy for microwave energy harvesting from ambient field or a feeding source. *IEEE Transactions on Power Electronics*, 27, 4481–4491.

Raju, S., Wu, R., Chan, M., and Yue, C.P. (2014) Modeling of mutual coupling between planar inductors in wireless power applications. *IEEE Transactions on Power Electronics*, 29, 481–490.

Reddy, M.V., Hemanth, K.S., and Venkat Mohan, CH. (2013) Microwave power transmission – a next generation power transmission system. *IOSR Journal of Electrical and Electronics Engineering*, 4, 24–28.

Roes, M.G.L., Duarte, J.L., Hendrix, M.A.M., and Lomonova, E.A. (2013) Acoustic energy transfer: a review. *IEEE Transactions on Industrial Electronics*, 60, 242–248.

Sahai, A. and Graham, D. (2012) Optical wireless power transmission at long wavelengths. *Proceedings of 2011 International Conference on Space Optical Systems and Applications*, Santa Monica, CA, USA, pp. 164–170.

Wang, Z., Li, Y., Sun, Y., Tang, C., and Lv, X. (2013) Load detection model of voltage-fed inductive power transfer system. *IEEE Transactions on Power Electronics*, 28, 5233–5243.

Zhang, Z., Chau, K.T., Qiu, C., and Liu, C.H. (2015) Energy encryption for wireless power transfer. *IEEE Transactions on Power Electronics*, 30, 5237–5246.

Zhang, W., Wong, S.C., Tse, C.K., and Chen, Q.H. (2014) An optimized track length in roadway inductive power transfer systems. *IEEE Journal of Emerging and Selected Topics in Power Electronics*, 2, 598–608.

Zhang, Y., Zhao, Z., and Chen, K. (2014) Frequency decrease analysis of resonant wireless power transfer. *IEEE Transactions on Power Electronics*, 29, 1058–1063.

Zhong, W.X., Lee, C.K., and Hui, S.Y.R. (2012) Wireless power domino-resonator systems with noncoaxial axes and circular structures. *IEEE Transactions on Power Electronics*, 27, 4750–4762.

# *Chapter 12*

# Review of battery management systems for electric vehicles

*Eric Ka-Wai Cheng*[1]

Battery management system (BMS) is a vital device for electric vehicles (EVs). As suggested by its name, it is to manage the battery operation and ensure the battery is safe to work and optimise its performance to the battery and the associated vehicles. It provides the monitoring and communicates with the other control and display units of a vehicle. The key function of BMS is to protect the battery cells, provide thermal management and determine the state of change (SoC) and state of health (SoH), so that it can be equipped to avoid the operation from hazardous and inefficient modes. The battery cells balancing is required in all BMS to ensure a healthy condition of operation for the battery cells. The chapter describes the structure of BMS and its major functions. Various control methods and circuits used for cell balancing and SoC calculation are shown.

## 12.1 What is BMS?

Energy storage is one of the most important devices in EVs or hybrid EVs (HEVs). Battery and supercapacitor are the two most popular energy storage devices. Their energy storage is large and it also governs the key performance of an EV. Supercapacitor is recently being used to support the battery system for electric and HEVs. Because it is an electric static energy storage device, its speed of operation is fast as compared to battery because energy storage in battery is through chemical reaction. Supercapacitor improves the transient power handling but its energy content is less as compared with battery. Therefore, the battery still provides the major energy storage of the vehicle. For battery-based vehicles, the supercapacitor is usually allocated by less than 5% of the total energy content because of the cost and the size of the supercapacitor. All these high-energy content devices require management system to ensure good operation.

[1]Department of Electrical Engineering, The Hong Kong Polytechnic University, Hong Kong, China

Besides the lead acid battery, the lithium-ion (Li-ion) based is today an acceptable battery for electric and HEVs. The common types of the Li-ion batteries are:

- $LiCoO_2$
- $LiMn_2O_4$
- $LiNi_{1/3}Mn_{1/3}Co_{1/3}O_2$
- $LiFePO_4$
- $Li_4Ti_5O_{12}$
- $LiNi_{0.8}Co_{0.15}Al_{0.05}O_2$

They all have high-energy storage content. Each of the Li-ion cell voltages is around 3.2–3.8 V depending on the types. Therefore, there are one to two hundreds of cells connected in series to make up the required voltage level for the vehicle. Each battery cell cannot be identical, and therefore, each cell may be present with different voltages that will imply problems of protection to the cells. The chemicals depreciation process is needed to be reduced by suitable battery conditioning. This can be done by proper charging and discharging profiles, and a management system. The battery lifetime will be reduced under wide operational temperature, too frequent charge and deep discharge cycle. Battery safety is the paramount concern and a good management system for battery will eliminate such fault and prolong the battery lifetime.

Therefore, the key function of BMS is to

- protect the battery and prolong its lifetime
- maintain the function of the battery for the EV application
- perform battery cell balancing
- determine the SoC
- determine the SoH
- communicate to other units and vehicle control unit (VCU)

The number of the data handling is massive in BMS. The measurements of the voltage, current and temperature of each of the cells are needed for BMS as the operation of BMS is highly data orientated. The communication of the data is usually using controller area network (CAN) (Lim and Anbuky, 2004) to deliver the data to other units such as microprocessor control unit (MCU). The whole BMS communicates with other system in a vehicle such as VCU through CAN as well.

## 12.2 BMS representation

The BMS consists of a number of functional blocks. The signals transferred from one block to others are continuously. They usually communicate using CAN-bus. Figure 12.1 shows the representation.

### *12.2.1 Battery module*

For vehicle design, the battery cells are connected in a module. Each module consists of 5–20 cells. A number of modules are then connected in series to make

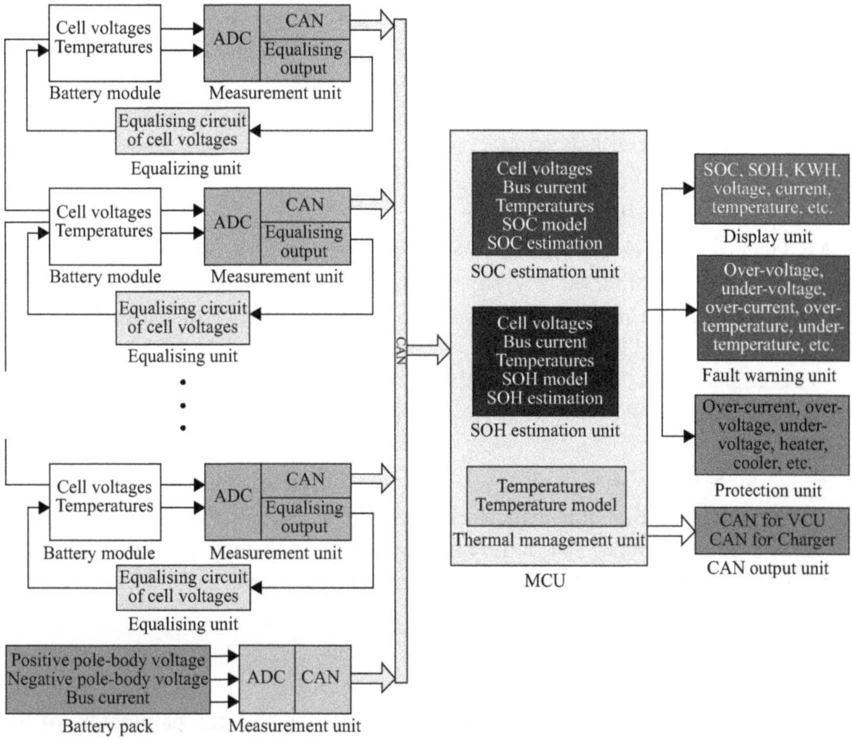


*Figure 12.1 BMS block representation*

up the total voltage requirement for the vehicle to form the battery pack. There are sensors installed into each of the battery module. Connection from the battery module includes the communication and power lines. Modular approach for battery cells can facilitate the mechanical protection, data communication, and thermal management, simplify the production and also make the power conditioning simple.

### *12.2.2 Measurement unit block*

Battery module consists of a measurement block which captures the basic parameters of individual cells including the voltage, current and temperature. The ambient temperature and the connector temperature may also be measured and converts all into digital format.

It is basically to report the battery status for other process. The measurement needs isolation because the cells are connected in series. A photo relay matrix may be used.

### *12.2.3 Battery equalisation: balancing unit*

Each of the battery cells are needed to be balanced in voltage in order to eliminate the risk of each cell under or over the rated voltage even the total voltage is correct.

Balancing circuit is basically a switched-mode power converter to transfer energy between the cells. A simply resistive dummy load is also being used to discharge the overvoltage cell rather than to transfer excess charge to other cells. The voltage of each cell, each module and the battery pack is measured, monitored and controlled to ensure the cell balancing.

### *12.2.4 MCU: estimation unit*

An MCU consists of the estimation unit for battery conditions. The voltage, current, temperature and the other battery parameters are used to calculate the SoC of the batteries. Another estimation is the SoH. Both of them are important parameters for the BMS.

The temperatures of the cells are also measured. The information is provided to the VCU to regulate the thermal control such as the cooling and heating system.

### *12.2.5 Display unit*

The display unit is like the vehicle dashpot which shows the current status of the vehicle battery units. The common parameters to display are the SoC, SoH, temperature, maximum temperature, maximum voltage of cells and minimum voltage of cells.

### *12.2.6 Fault warning block*

It is similar to the display block. Instead of only to display the parameters of the status, it will also compare the data with the setting, if any of the parameters is over the range of the setting, it will trigger the warning decision to give a warning procedure. This will be sent to the display unit and VCU for further action.

## 12.3 Data management and network

BMS has the abilities to record and manage many different types of data, but a large amount of the data needs BMS to interface or manage networks. The main job of a network is to move data from one area to another, which is the same as the common networks that are seen in households and companies. Many new BMS are smart systems that can decide what to do with data it receives instantly; however, it is still necessary for users to check up on the system every so often. A regular check up will allow users to see whether the batteries have been running consistently over time and ensure that there are no irregular spikes and transients during operation and charging. If there are irregularities, they can quickly locate the bad cell before harm is inflicted upon the system.

### *12.3.1 CAN-bus*

This transfer of data to users will not be possible without the installation of a network. For vehicles, when working with BMS, there are a few different types of network systems that are available to choose from. A few types of networks are

CAN-bus, local interconnect network (LIN), transmission control protocol/internet protocol (TCP/IP) and power line communication (PLC). A CAN-bus or CAN-bus system is commonly used in BMS because it is a system that allows all the parts within the BMS to communicate with each other. This communication is able to continuously occur with or without the appearance of a host computer. The concept of CAN-bus was originally invented to supplement the needs of vehicles. Most of the time a vehicle can have close to a hundred parts working simultaneously and to compute the feedback of all the parts is done best by a CAN-bus network. The network will work as intended during normal operation, and when an authorised, computer is attached so that it can still communicate with the user by providing current and past data.

### *12.3.2 LIN*

One drawback of the CAN-bus is that it is an expensive system to purchase and use. Therefore, a cheaper system was invented to complete many of the similar task. That system is called the LIN. A LIN can only work with up to 16 parts at once along with a master controller. The difference between the two networks is that a LIN system will only receive feedback when it is asked by the master controller and the communication is only between the two parts, while the CAN-bus allows for continuous communication between all parts. Although it may sound inferior, LIN system actually works extremely smoothly because it eliminates the possibility of data collision or jamming, which occasionally happens for CAN-bus systems. When a LIN system is used for vehicles, the vehicle has to be simplified by grouping similar parts together since only 16 slots are available. For projects involving a tighter budget and smaller design, a LIN system can complete the task.

### *12.3.3 TCP/IP*

Another method for BMS to communicate within itself and users is the TCP/IP system. This is the system that is currently used to support Internet use around the world, but it can also apply to BMS and it works in a similar fashion. When people use the Internet daily, it involves linking to some server with a wire or wirelessly and that server is linked together with many other servers allowing them to freely communicate. TCP/IP is the system that allows this large network to move data freely between each part. TCP/IP systems have become very advanced, allowing it to handle large amounts of data, moving the data to its designated area quickly, being able to work wirelessly and upgrade software components automatically. TCP/IP is a good selection for a system if the BMS is large and spread out over because the wireless uses can be an advantage.

### *12.3.4 Wireless and PLC*

Having a wireless system is certainly an attractive option because it can greatly reduce wire cluster within the system. However, being wireless will always include the risk of signal dropping, electromagnetic interference or server down time, so another option may be the PLC network. A PLC system has the benefit of not using any wires for communication because data can be transferred via the power cords. This technology is commonly used in many solar grid systems and intelligent home but its applications can be stretched out to many different fields. Since most electronic systems currently still need power supplied to them via a power cord, it would be ideal to add one more function to the existing piece. A PLC system would benefit a BMS that would like connection stability and minimum wiring.

Overall, the CAN is still the most common communication method in BMS.

## 12.4 SoC and SoH

The SoH and SoH are two states of batteries estimated in a BMS. The estimation is through the battery voltage, current, temperature measurement and other battery parameters and specification.

### *12.4.1 SoC*

The SoC is to measure the energy capacity as a percentage of the rated capacity. Therefore, it is equivalent to the fuel gauge of a petrol vehicle. The definition of the SoC, which can be found in many publications (Cheng *et al.*, 2011; Pattipati *et al.*, 2011), is to measure the charge accumulated in the battery:

$$SoC_{new} = SoC_{old} + \frac{ki\Delta t}{Q} \tag{12.1}$$

where $\Delta t$ is the sampling time, $k$ is Coulombic coefficient efficiency of the cell, $i$ is the cell current and is +ve for charging and −ve for discharging, and $Q$ is the capacity at the time of sampling.

SoC is usually presented in either fraction or percentage. It is an important parameter during charging and discharge of a vehicle. During discharge, it gives the remaining energy stored in the battery and therefore it can be used to estimate the remaining travelling distance and duration. During charging, it gives the charger necessary information to determine the charging current and termination of the charging process.

The SoC is an indicator to the BMS to execute the protection procedure. During an EV regeneration such as braking and deceleration, excess energy overcharge batteries when the SoC of the battery is high, BMS may channel excess energy to other units such as dumping resistors in order to protect the cells.

However, the SoC is not a simple product of the current and time, and cannot be measured directly like voltage and current, it highly depends on the electrochemical reaction of the battery cell that is highly non-linear and in terms of a number of parameters such as temperature, operating current and voltage, hence it varies with the vehicle driving condition such as vehicle speed. Sensors of current, voltage and temperature are usually installed in necessary positions for monitoring and its measured values are digitised using analog-to-digital (A/D) converter to deliver to an MCU for estimation. The primitive method to determine the SoC is to measure the current and time, and it is also known as Coulomb counting. However, the method may not work well with Li-ion battery because the middle range of the discharge curve is very flat and measurement error may result in large error in SoC estimation. Sample of SoC curve can be seen later in Figure 12.3. Also direct power measurement has not taken into account of the temperature and aging effect of the cells. Measurement error may be accumulated and therefore a rest process may be needed.

The key information of SoC is to provide the remaining energy content of a battery, rather than energy charged and delivered to the battery. A good estimation technique is therefore needed.

## *12.4.2 SoH*

SoH is defined as the state of health and is expressed as a percentage of the nominal capacity as a base value:

$$\text{SoH} = \frac{\text{Nominal capacity} - \text{Loss of capacity}}{\text{Nominal capacity}} \tag{12.2}$$

An ideal or new battery would have SoH of 100%. As a battery is aged, the capacity loses, and the remaining capacity that the battery can store is reduced. Usually when the SoH is reduced to less than 70%, a battery is considered to be not suitable for EVs. Most of the batteries may exhibit a decrease in SoH of 80% after thousands of cycles of operation.

The measurement of SoH imposes difficulty for HEV as the battery status of HEV does not involve reset condition in zero and fully charged, and also it generally operates between 30% and 80% of SoC. SoH also highly depends on the electrochemical status of the cell.

## *12.4.3 Estimation of SoC*

### 12.4.3.1 Coulomb-counting method

Coulomb-counting method is one of the most simplest and direct ways to approximate the SoC of battery using integration of electrical current passing through the battery system. With referring to the battery manufacturer's data or result of standard charging test, a fraction of remaining charge is directly obtained. This method could be implemented easily for most battery types, acquiring a pretty good result with stable loads and accurate current instrument, yet small offset error of current measurement could propagate with time; self-discharge and Peukert's law of battery (Leksono *et al.*, 2013) would lead to incorrect estimation of remaining energy stored in the battery system. This method usually cooperates with non-integral methods such as open circuit voltage (OCV) methods to calibrate the approximated charge level. Similarly, SoH could also be obtained by Coulomb-counting method that comparing the data of ordinary charge–discharge cycle with the battery's specifications.

### 12.4.3.2 OCV

OCV is the terminal voltage of the battery after it has settled down for a long duration. There is a strong relationship between the OCV and the SoC. Figure 12.2 shows an equivalent circuit model of a Li-ion cell (Petzl and Danzer, 2013; Michel and Heiries, 2015).

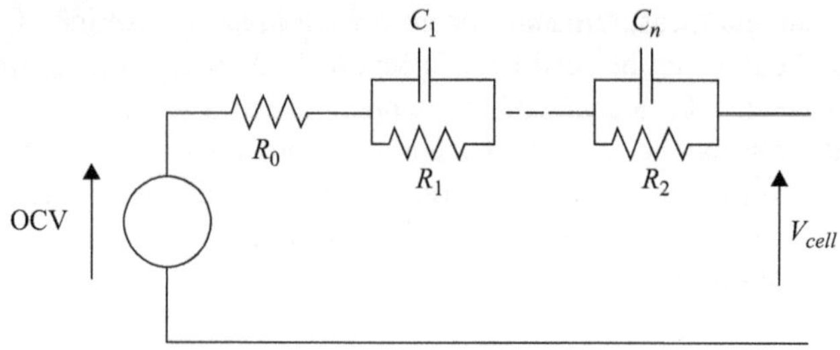


*Figure 12.2 Equivalent circuit model of a battery cell*

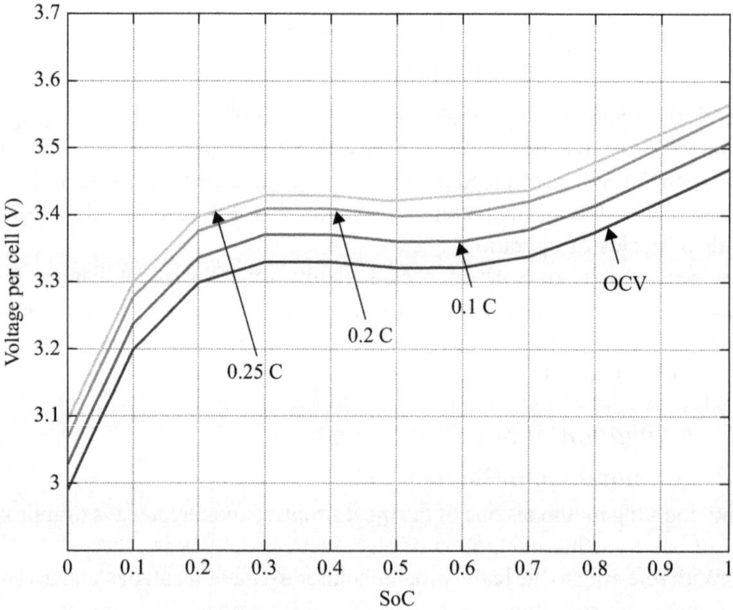


*Figure 12.3 A typical cell voltage versus SoC at different charging conditions*

It can be seen from the model, the capacitive component $C$ provides a delay in the measurement of the OCV. Hence, the measurement of OCV is difficult because the battery relaxation time is long. The relaxation time may take more than 6 h. The measurement is not practical to wait each measurement point. Estimation of OCV has been reported widely. The prediction method based on GM(1, 1) grey model has been proposed (Li *et al.*, 2015) which is robust and simple. The observability-Gramian-based (OGB) method has a wider load current operation range (Unterrieder *et al.*, 2012).

The OCV and the SoC has a non-linear correlation (Michel and Heiries, 2015). A typical voltage versus SoC is shown in Figure 12.3 with different $C$-rates. The model is derived from a 40 Ah Li-ion phosphate battery cell (Cheng *et al.*, 2011).

The sigma point Kalman filtering has been proposed to give improved estimation as compared with classical Kalman filter (Plett, 2006).

#### 12.4.3.3 Impedance-based methods

Impedance-based methods use prediction and experimental method to develop an impedance for the battery description. The models may require large amount of experimental data of voltage–current, voltage–charge, as well as time and environmental-varying characteristics of battery. Dynamic response of the battery could be represented by the impedance model of the battery, that is, an electrical circuit in which it is combined with the OCV. Some impedance-based methods such as electrochemical impedance spectroscopy (EIS) (Wang *et al.*, 2012) investigates the property of electrochemical system under various conditions to study the SoC. Other modelling methods (Chen and Rincon-Mora, 2006) using Thevenin and runtime electrical model are also popular circuit-based model for the estimation.

#### 12.4.3.4 Recursive and machine-learning approaches

Recursive and machine-learning approaches (Kim, 2008; Barbarisi *et al.*, 2006) further improve the robustness of physical modelling of battery in BMS by integration of recursive or intelligence functions that make use of system's historical data and system state. In BMS adopting artificial neural networks, actual batteries characteristics are learnt from time to time, modelling process is performed during monitoring operation and keep improving with time. Changes of battery characteristics are also measured as SoH. These are state-of-the-art SoC and SoH estimation techniques; along with the rapid development of processing units, the process of these complicated techniques could be implemented with a single chip. The adaptive battery model with predictor/corrector observer, and online subspace parameter estimation are shown good agreement with the measurement (Gould *et al.*, 2009). Related approaches have drawn much attention from battery and control engineers and scientists in recent years and are expected to be developing continuously in the near future.

A comparison of different modelling methods for SoC and SoH estimation of different battery types is given in Table 12.1.

For users' point of view, the remaining useful life (RUL) (Wei *et al.*, 2009) is another parameter that should be estimated and it is similar battery information to SoH. Estimation method using initial resistance for the estimation has been reported (Harris *et al.*, 2014).

## 12.5 Battery balancing

Battery balancing is an important part of the BMS as it provides the equal and enhanced safety operation for all battery cells. There are a number of topologies proposed in the literature. They can also be classified as passive and active balancing methods. The passive is simply resistive circuit to dump the excess energy. The active version is a switched-mode power converter–based circuits.

*Table 12.1 Comparison of different modelling methods for SoC and SoH estimation*

| Types of SoC estimation | Techniques | Battery types | Pros | Cons |
|---|---|---|---|---|
| Coulomb-counting | Current integration | Nearly all | Easy to implement, accurate result with stable loads | Calibration with other methods is required |
| OCV | Idle voltage measurement, electrical modelling | Lead-acid, Li-ion, and so on | Easy to implement | Limited to battery types with stable OCV–SoC relationship, limited accuracy without precise dynamic model |
| Impedance-based | Electrochemical modelling | Nearly all | Various SoH indicators are monitored | High instrumentation cost |
| Recursive/machine-learning approaches | Fuzzy logic, sliding mode observers, Kalman filters, artificial neural networks | All | Robust, accurate, self-tuning | Very difficult to implement, demanding system memory and processing power |

### *12.5.1 Resistive balancing*

Figure 12.4 shows a resistive balancing circuit. It is simply a switch to connect the battery cells to a resistor when unbalancing is detected. There are two schemes in general:

(a) Calculate the average voltage of the cells. If any cells have a higher than average values, the switch is triggered to discharge the cells.
(b) Set the maximum cell voltage, such as 4.3 V. If any cells reach the maximum voltage, the switch will be triggered to discharge the cells.

Resistive balancing is being used extensively in the commercial product. It provides a low cost, simple and robust solution. For application needs high efficiency and higher speed balancing, the active circuit should be used.

### *12.5.2 Classical switched-mode active balancing*

Active balancing is based on power electronic circuit to provide the power conversion between the battery cells. The classical circuit is a buck-boost converter that provides the power conversion between cells. Figure 12.5 shows the circuit. It is a combination of two buck-boost converters for both bi-directional power balancing. The circuit is easily extended to many cells as shown in Figure 12.6. The beauty of

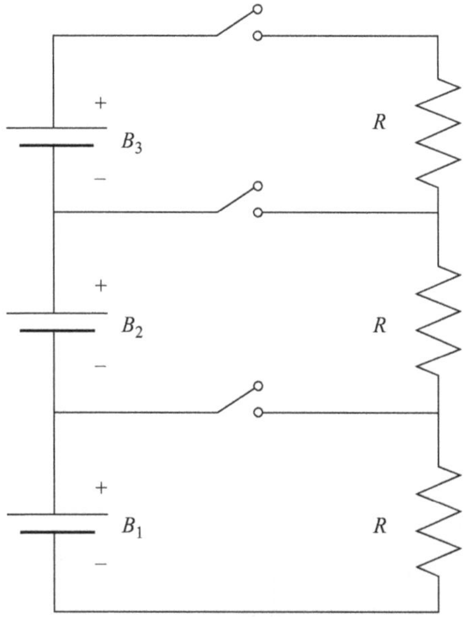


*Figure 12.4 Resistive battery balancing*

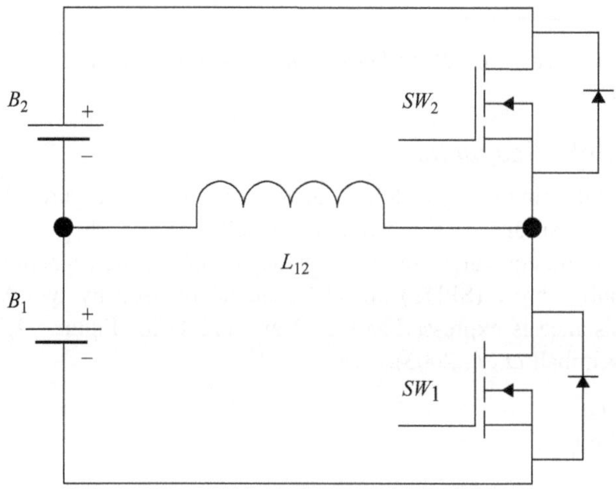


*Figure 12.5 Basic buck-boost converter balancing between two cells*

the design is that the switching signal for the complementary transistors, such as $SW_1$ and $SW_2$, can be 50% duty ratio without voltage detection and control.

The concept can be extended to many battery cells connected in series using a multi-phase converter (Mestrallet *et al.*, 2014) as shown in Figure 12.7. The number of transistors is higher than the basic topology.

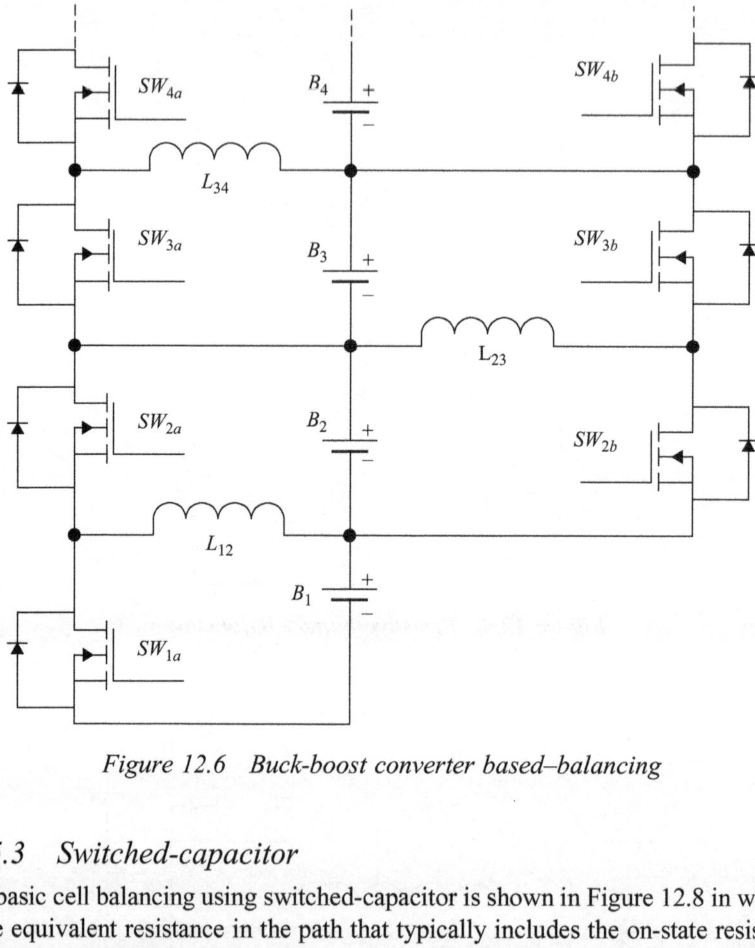


*Figure 12.6 Buck-boost converter based–balancing*

### *12.5.3 Switched-capacitor*

The basic cell balancing using switched-capacitor is shown in Figure 12.8 in which $r$ is the equivalent resistance in the path that typically includes the on-state resistance of the transistor and equivalent series resistance (ESR) of the capacitor $C$. The two single-port double-throw (SPDT) switches can be realised by two Mosfets. The equivalent resistance is expressed as (12.3) and (12.4) for Figure 12.8(a) and (b), respectively (Kimball *et al.*, 2005):

$$R_{SC} = \frac{1}{Cf_S} \tag{12.3}$$

$$R_{SC} = \frac{1 + e^{-1/(2rCf_S)}}{Cf_S[1 - e^{-1/(2rCf_S)}]} \tag{12.4}$$

### *12.5.4 SP balancing*

Figure 12.9 is an SP balancing circuit. Each battery cell is connected to the switched-capacitor $C$ for a portion of a switching cycle and then all the switched-capacitors are paralleled to equalise the voltages. The main disadvantage is that the number of switches required is four for each cell.

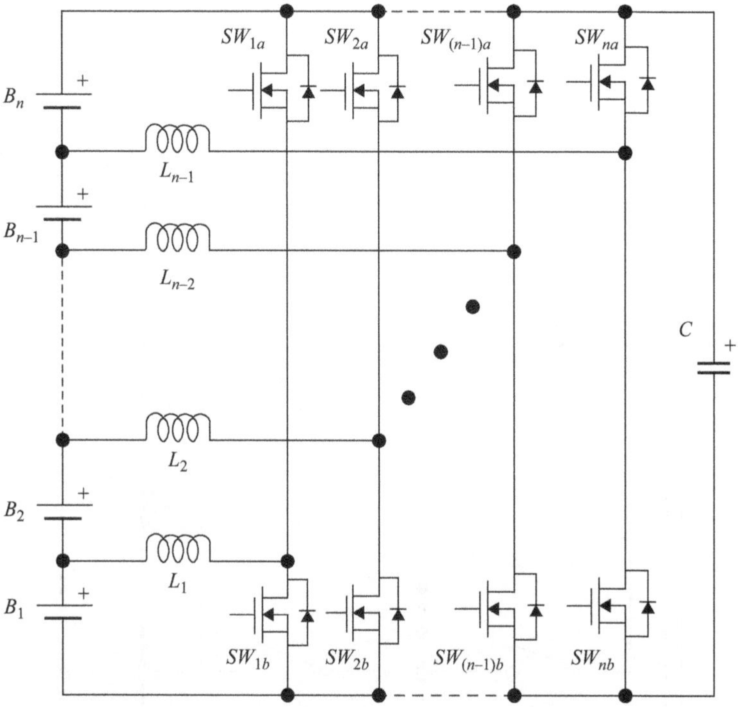


*Figure 12.7 Multi-phase buck-boost converter–based balancing circuit*

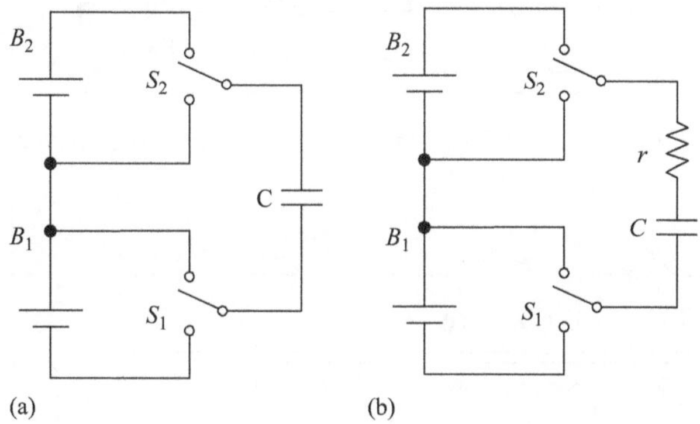


*Figure 12.8 Simple switched-capacitor balancing circuit for two cells: (a) ideal version and (b) realistic version with parasitic resistance*

## 12.5.5 Multi-winding balancing

The single core multi-winding battery balancing is shown in Figure 12.10 (Kutkut *et al.*, 1999). The transformer is needed to be designed according to the number of cells and therefore the flexibility is limited. The balancing is also dependent on the

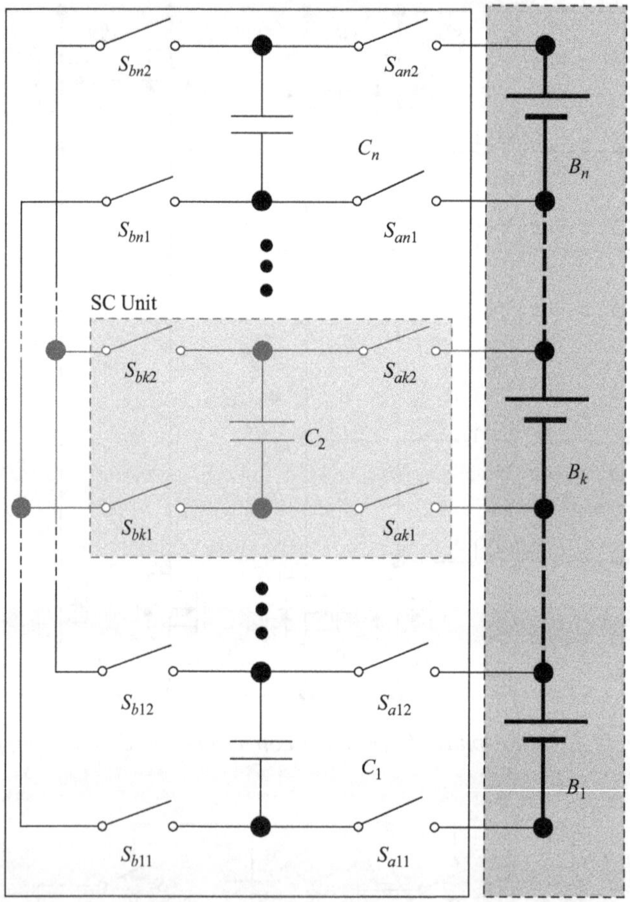


*Figure 12.9 SP switched-capacitor balancing circuit*

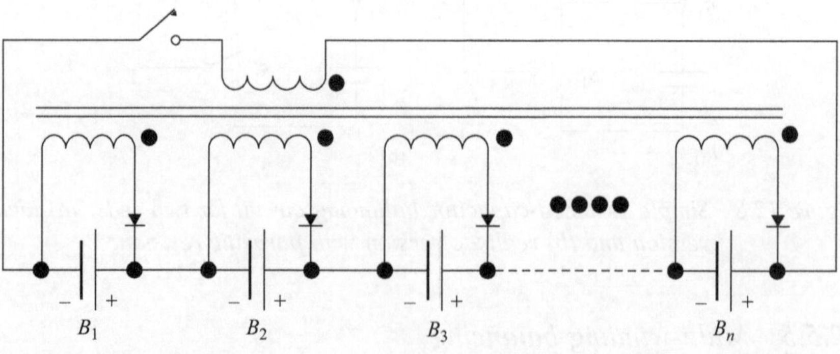


*Figure 12.10 Multi-winding transformer-based switched-capacitor balancing circuit*

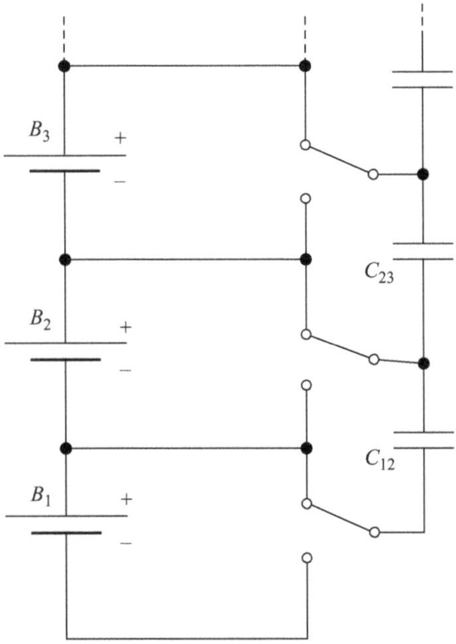


*Figure 12.11 Generalised Tier 1 switched-capacitor balancing circuit*

coupling between the windings, and the voltage gap between the cells is therefore reflected by the coupling coefficient among the windings.

### 12.5.6 Tier 1 balancing switched-capacitor

The basic circuit in Figure 12.8 can be extended to a generalised form as shown in Figure 12.11. Each battery cell needs one capacitor for balancing and the total number of the switched-capacitors is $n - 1$ for $n$ cells.

The circuit can be realised by Mosfet as shown in Figure 12.12(a). It shows a generalised version for any number of battery cells connected in series. The equivalent resistive model is shown in Figure 12.12(b) that is based on the principle in section 12.5.3.

### 12.5.7 Tier 2 balancing switched-capacitor

The Tier 1 circuit is slow because charge is needed to propagate from one to other in series. A second capacitor can be put between two capacitors to provide a faster path for balancing (Baughman and Ferdowsi, 2008). For a generalised Tier 2 circuit, the capacitor can be added to complete the generalised version as shown in Figure 12.13(a). Each Tier 1 capacitor has a Tier 2 capacitor to assist the balancing. The equivalent behavioural model using resistor is shown in Figure 12.13(b) which is derived from section 12.5.3.

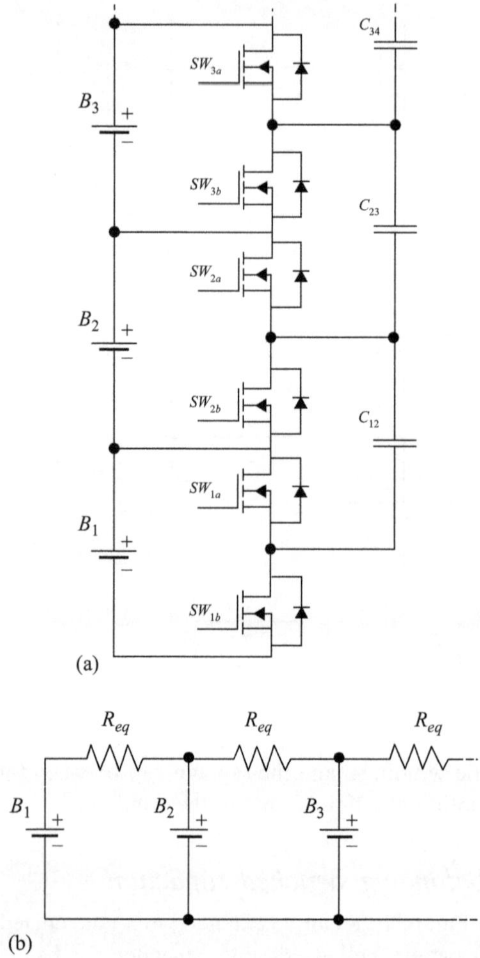


*Figure 12.12 Realisation of generalised Tier 1 switched-capacitor battery balancing circuit: (a) Mosfet realisation and (b) equivalent circuit*

### 12.5.8 Other switched-capacitor balancing

There are a number of variations of the switched-capacitor topology for battery balancing (Ye and Cheng, 2015). The common path for the switched-capacitor can be connected through the terminal connection of the battery bank that is shown in Figure 12.14(a) or connected to a common path in the mid-point of the battery bank. Figure 12.14(b) is a bilateral circuit and Figure 12.14(c) is a star circuit. For $n$ batteries balancing, the number of the balancing capacitor is $n - 1$ for circuits in (a) and (b) whereas it is $n - 1$ for circuit (c). The key advantage of the new topology is that the balancing speed is independent of the number of the battery cells.

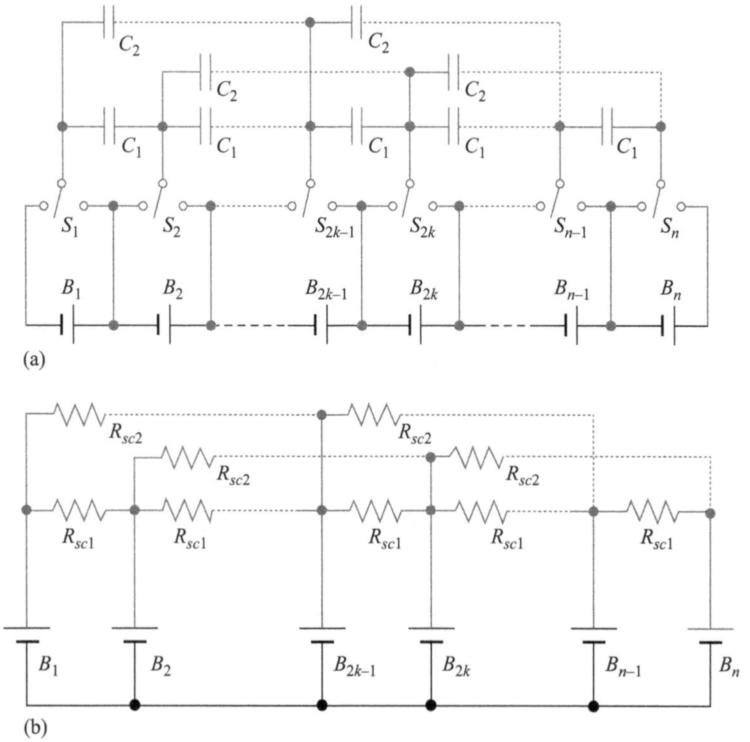


*Figure 12.13 Generalised Tier 2 switched-capacitor balancing circuit: (a) topology and (b) behavioural model*

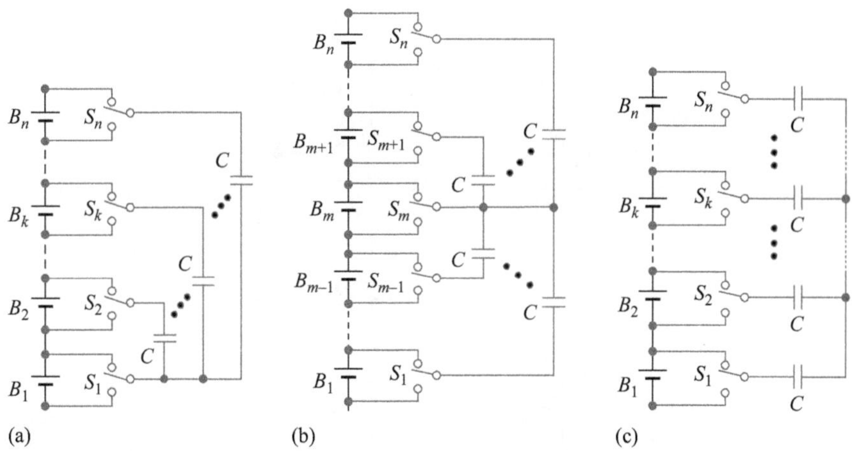


*Figure 12.14 Other variations for switched-capacitor balancing circuits: (a) lateral circuit, (b) bilateral circuit and (c) star circuit*

Supercapacitor is now also a common energy storage devices for EVs and HEVs. Their energy capacity is large. The balancing is also required for the capacitor. The above balancing circuits are also suitable for the applications.

### 12.5.9 Resonant version

Using switched-capacitor of the buck-boost switched-mode version has a disadvantage of conduction loss. Even the switched-capacitor version which uses a capacitor for energy transfer, but loss exists in the conducting. By inserting a small inductor as a resonant component with the capacitor, the $L$ and $C$ will provide a resonance and the energy can be transferred from one battery cell to other cell with a resonant energy transfer. The switching loss in the transistor can be reduced due to the zero-current switching (Cheng *et al.*, 2011), and the energy can be transferred more effectively to other set. Figure 12.15 shows an example (Ye and Cheng, 2012) using resonant tank for the balancing circuit.

For instance, if initially the voltage of $B_1$ is higher than the voltage of $B_2$. When the resonant tank formed by $L_{12}$ and $C_{12}$ is connected in parallel with battery

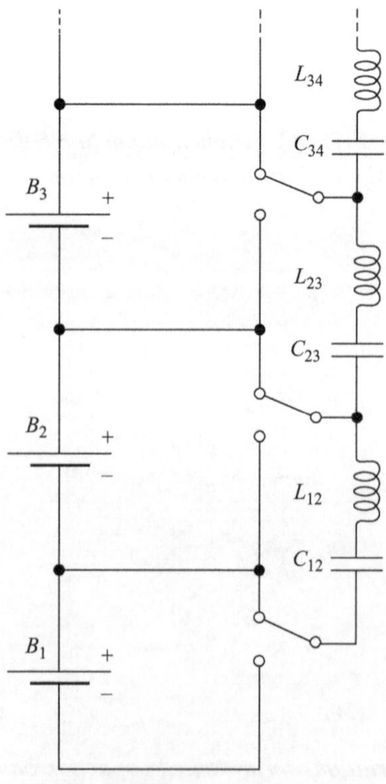


*Figure 12.15 Quasi-resonant switched-capacitor cell-balancing circuit*

cell $B_1$, the resonant equation is:

$$V_{C_{12}} = V_{B_1} + \Delta V_{12}\cos(\omega t + \phi). \tag{12.5}$$

$$i_r = I_m \sin(\omega t + \varphi) \tag{12.6}$$

where $\Delta V_{12}$ is the amplitude of $V_{C_{12}}$, $I_m$ is the amplitude of resonant current $i_r$, $\omega$ is resonant angular frequency and is equal to $1/\sqrt{L_{12}C_{12}}$, and $\varphi$ is the initial angle of the resonant state. After half a switching cycle, the resonant tank is connected to cell $B_2$. The resonant equation is:

$$V_{C_{12}} = V_{B_2} - \Delta V_{12}\cos(\omega t + \phi) \tag{12.7}$$

$$i_r = -I_m \sin(\omega t + \varphi) \tag{12.8}$$

The current in both cases are sinusoidal and the transistors are turned on and off under zero-current and the associated switching loss is small. The frequency of resonant current is usually slightly higher than the switching frequency in order to ensure zero-current switching during the turn-off condition of the transistor and this is easy to be achieved without using active control.

Figure 12.16 shows the battery balancing measurement of two battery cells of 0.6 Ah with initial voltage at 3.58 V and 3.38 V. The resonant tank is formed by a 10 μF capacitor and a 4.5 μH inductor. The switching frequency is 17.7 kHz. It takes 30 min for the balancing. The balancing time can be reduced by using larger capacitor and smaller inductor to increase the energy transfer per cycle.

### *12.5.10 Summary of balancing technology*

Although massive researches have been conducted in the areas, the commercial products are lacking behind the technology. Over 90% of the BMS in the market

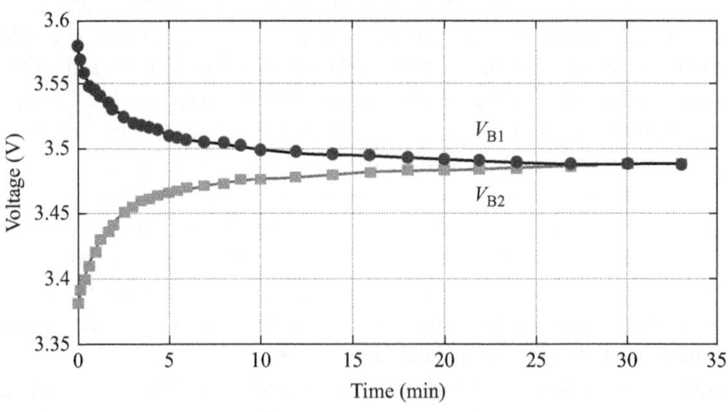


*Figure 12.16 Battery balancing process of two different cell voltages*

are passive resistive based because it is robust and low cost. Efficiency is important but it is unable to overwhelm the safety and cost.

For the various active balancing methods, the buck-boost converter based needs inductor; therefore, for battery module packaging point of view, the size will be a concern. However, the number of transistors required is less. For the switched-capacitor based balancing, it may be suitable for lower power and slow balancing speed and it is usually acceptable for EVs. The improved version, which is the resonant mode of switched-capacitor, its associated resonant inductor is small and the inductor is in the order of μH and therefore the size, weight and cost are not a concern.

One feature can be seen that the active balancing method usually does not require voltage control for the cells and only simple 50% duty ratio to the transistors is needed. The actual duty ratio is slightly less than 50% due to dead time needed between the complementary pair of the transistors. When the cells are near balancing, the balancing current between the cells will be reduced. The balancing speed is not needed to be high because most of the good Li-ion batteries should not deviate seriously and rapidly from others.

It is expected that higher power and higher energy content of batteries are increasingly important for EVs and HEVs with higher demand for bus and mini-bus and longer travelling per charge. The demand for battery balancing or higher efficiency balancing circuit is also increasing rapidly.

## 12.6 BMS standard

Another important aspect of BMS is the safety factor. BMS is no different, even the most advanced technology is unusable if it causes hazardous results. Currently, there are many international safety standards in place to ensure consumers and users that the product is safe for use. Since BMS is a system that is usually run by software, the standards in place are generally to ensure a minimum amount of software failures. BMS is also an electronics-based part, the major international standard they must be able to pass is IEC 61508 which ensures 'Functional Safety of Electrical/Electronic/Programmable Electronic Safety Related Systems' (Veritas, 2014). If a BMS can compile with this standard, it means that all the sensors function correctly, the data is accurate and the system is able to act according to the data received. Most safety organisations do not rule out the possibility of failure, but the product has to provide an acceptable failure rate and the system can shut down correctly in case of failures. The BMS also must be able to reduce the risk of electric shock, fire, energy hazards, heat hazards, mechanical hazards, chemical hazards and radiation in case of failure. IEC 61508 standard products can guarantee these tasks (Gall, 2008). IEC 61511 covers the frameworks and the initial design requirement which leads to the implementation, operation and maintenance in both hardware and software. Usually, the BMS should meet both IEC 61511 and IEC 61508.

Obviously, BMS is only one part of a larger system and the other parts usually have their own standards that they must compile. The IEC 62660 and ISO 12405 for Li-ion performance test are also useful for BMS design. For older systems with lead acid batteries, they have IEC/TR 61431/62060 or BS 6133. Although BMS has been applied for Li-ion batteries as the common implementation, but lead acid batteries also have the needs for balancing and parameter monitoring in order to maintaining their healthy operation. As BMS is used for EVs, then the system is subject to SAE J1939/J3009 or IEC 61982 and many more others. Vehicles usually have all their parts tested because failure could result in deaths, even the interfaces they use are tested. A CAN-bus interface will undergo ISO 11898 and its part 1, 2 and 3, while LIN interface will need ISO 9141. Certain BMS will have connection to charging system, and therefore, the charging system standard such as IEC 61851 is also needed to be considered when BMS is designed.

BMS is applied to many different fields of electrical system currently, so it is important to understand if the system that is paired with the BMS is safe along with the BMS. A safe BMS with all the correct standards may still cause a hazardous outcome if the equipment it used is unsafe. BMS development is still in the initial stage of R&D, new standards will emerge together with new technology of BMS. When the BMS is connected to vehicles and its interfacing parts, many standards for BMS are now needed for safe and successful operation of the BMS and the vehicles.

## 12.7 Conclusion

The BMS is now a major part in a vehicle. The development of BMS needs the technology in data processing, battery balancing, communication and control. Among all the technologies, the battery balancing and the prediction of the SOC and SoH are the key development recently. Various works have been done to develop methods in new circuits to make the balancing more efficient. It can be seen that the basic resistive balancing has been used extensively in the market. The newer balancing using switched-mode power conversion and the switched-capacitor based are also promising. The prediction of SoC and SoH is always difficult. The Coulomb-counting, OCV, impedance-based and recursive/machine-learning approaches are common and popular methods for the estimation.

According to market information, the demand for BMS is high. There is insufficient supply of BMS, especially for power battery application. The BMS is not only limited to EV and HEV, but also required for all energy storage applications such as solar power, hydro power and wind power generations. BMS governs the safety and the accurate operation of the battery and the vehicles. With the current rapid development of EVs, BMS is a major research and development of the EV and HEV technology.

## Acknowledgements

The support of the research team including Dr Xue Xiangdang, Mr Fong Yat Chi, Dr Xu Cuidong, Mr William Chen and Dr Ye Yuanmao is appreciated.

## References

Barbarisi, O., Vasca, F. and Glielmo, L. (2006) State of charge Kalman filter estimator for automotive batteries, *Control Engineering Practice*, 14(3) 267–275.

Baughman, A.C. and Ferdowsi, M. (2008) Double-tiered switched-capacitor battery charge equalization technique, *IEEE Transactions on Industrial Electronics*, 55(6), 2277–2285.

Bureau Veritas, (2014) *Functional and Safety Guide for Battery Management System (BMS): Assessment and Certification*, Bureau Veritas Certification, France.

Chen, M. and Rincon-Mora, G.A. (2006) Accurate electrical battery model capable of predicting runtime and *I–V* performance, *IEEE Transactions on Energy Conversion*, 21(2), 504–511.

Cheng, K.W.E. (2001) Zero-current-switching switched-capacitor converters, *IEE Proceedings – Electric Power Application*, 148(5), 403–409.

Cheng, K.W.E., Divakar, B.P., Wu, H.J., Ding, K. and Ho, H.F. (2011), Battery-management system (BMS) and SOC development for electrical vehicles, *IEEE Transactions on Vehicular Technology*, 60(1), 76–88.

Gall, H. (2008) Functional safety IEC 61508/IEC 61511 the impact to certification and the user. *IEEE/ACS International Conference on Computer Systems and Applications*, Doha, Qatar, pp. 1027–1031.

Gould, C.R., Bingham, C.M., Stone, D.A. and Bentley, P. (2009) New battery model and state-of-health determination through subspace parameter estimation and state-observer techniques, *IEEE Transactions on Vehicular Technology*, 58(8), 3905–3916.

Harris, A., O'Connor, P. and Cox, R.W. (2014) A time-domain approach for monitoring battery state of health (SOH) and remaining useful life (RUL). *IECON 2014 – 40th Annual Conference of the IEEE Industrial Electronics Society*, Dallas, TX, USA, pp. 3083–3087.

Huang, S.-J., Huang, B.-G. and Pai, F.-S. (2012) An approach to measurements of electrical characteristics of lithium-ion battery with open-circuit voltage function, *IET Power Electronics*, 5(9), 1968–1975.

Kim, I. S. (2008) Nonlinear state of charge estimator for hybrid electric vehicle battery, *IEEE Transactions on Power Electronics*, 23(4), 2027–2034.

Kimball, J.W., Krien, P.T. and Cahill, K.R. (2005) Modelling of capacitor impedance in switching converters, *IEEE Transactions Power Electronics*, 3(4), 136–140.

Kutkut, N.H., Wiegman, H.L.N., Divan, D.M. and Novotny, D.W. (1999) Design considerations for charge equalization of an electric vehicle battery system, *IEEE Transactions on Industry Applications*, 35(1), 28–35.

Leksono, E., Haq, I.N., Iqbal, M., Soelami, F.X.N and Merthayasa, I.G.N. (2013) State of charge (SoC) estimation on $LiFePO_4$ battery module using Coulomb counting methods with modified Peukert. *Joint International Conference on Rural Information & Communication Technology and Electric-Vehicle Technology (rICT & ICeV-T)*, Bandung, Indonesia, pp. 1–4.

Li, R., Wu, J.-F., Wang, H.-Y. and Li, G.-C. (2010) Prediction of state of charge of lithium-ion rechargeable battery with electrochemical impedance spectroscopy theory. *The 5th IEEE Conference on Industrial Electronics and Applications (ICIEA)*, Taichung, Taiwan, pp. 684–688.

Li, Z.-Y., Shang, Y.-L., Cui, N.-X. and Zhang, C.-H. (2015) The prediction method for battery open circuit voltage based on GM (1, 1) grey model. *Proceedings of 34th Chinese Control Conference*, Hangzhou, pp. 1902–1906.

Lim, D. and Anbuky, A. (2004), A distributed industrial battery management network, *IEEE Transactions on Industrial Electronics*, 51(6), 1181–1193.

Mestrallet, F., Kerachev, L., Crebier, J.C. and Collet, A. (2014) Multiphase interleaved converter for lithium battery active balancing, *IEEE Transactions on Power Electronics*, 29(6), pp. 2874–2881.

Michel, P.H. and Heiries, V. (2015) An adaptive sigma point Kalman filter hybridized by support vector machine algorithm for battery SoC and SoH estimation. *IEEE 81st Vehicular Technology Conference (VTC Spring)*, Glasgow, UK, pp. 1–7.

Pattipati, B., Sankavaram, C. and Pattipati, K.R. (2011) System identification and estimation framework for pivotal automotive battery management system characteristics, *IEEE Transactions Systems, Man and Cybernetics – Part C*, 41(6), 869–884.

Petzl, M. and Danzer, M.A. (2013) Advancements in OCV measurement and analysis for lithium-ion batteries, *IEEE Transactions on Energy Conversion*, 28(3), 675–681.

Plett, G.L. (2006) Sigma-point Kalman filtering for battery management systems of LIPB-based HEV battery packs. Part I: introduction and state estimation 35, *Journal of Power Sources*, 161, 1356–1368.

Unterrieder, C., Priewasser, R., Agostinelli, M., Marsili, S. and Huemer, M. (2012) Comparative study and improvement of battery open-circuit voltage estimation methods. *IEEE 55th International Midwest Symposium on Circuits and Systems (MWSCAS)*, Boise, ID, USA, pp. 1076–1079.

Wang, L., Zhao, J.S., He, X.M., et al. (2012) Electrochemical impedance spectroscopy (EIS) study of $LiNi_{1/3}Co_{1/3}Mn_{1/3}O_2$ for Li-ion batteries, *International Journal of Electrochemical Science*, 7, 345–353.

Wei, X.Z., Bing Zhu, B. and Xu, W. (2009) Internal resistance identification in vehicle power lithium-ion battery and application in lifetime evaluation. *International Conference on Measuring Technology and Mechatronics Automation*, Zhangjiajie, China, Vol. 3, 388–392.

Ye, Y.M., and Cheng, K.W.E. (2012) Zero-current switching switched-capacitor zero-voltage-gap automatic equalization system for series battery string, *IEEE Transactions on Power Electronics*, 27(7), 3234–3242.

Ye, Y.M., and Cheng, K.W.E. (2015) Modeling and analysis of series–parallel switched-capacitor voltage equalizer for battery/supercapacitor strings, *IEEE Journal of Emerging and Selected Topics in Power Electronics*, 3(4), 977–983.

# *Chapter 13*

# Integration of energy and information in electric vehicle systems

*C.C. Chan*[1], *Linni Jian*[2] *and Christopher H.T. Lee*[1]

In this chapter, the integration of energy and information in electric vehicle (EV) systems is discussed. The importance of open mind and philosophy of engineering are elaborated. Then, the correlation of energy and information in EV systems is explored. After the comprehensive study of the correlation between energy and information, the smart charging algorithms are discussed. In particular, the stimulated synergy by the integration of information and energy flows can definitely accomplish the win-win situation.

## 13.1 Introduction

With the increasing concerns of the energy crisis and the environmental protection, the commercialization of EVs has become very promising (Chan and Chau, 2001). Undoubtedly, the EV technology will definitely serve for a paradigm shift between the environment, energy and information, yet there are still obstacles to be resolved. Before reaching to the 'promised land' through the EV penetration, the development and research on this topic should never stop.

To boost up the market penetration for EV applications, there are four major factors need to be considered, namely, the initial cost, the convenience of use, the energy consumption and the exhaust emission. In particular, up to recent development, only the latter two factors are fulfiled while the former two are not satisfied yet. Hence, to improve the current situations, extra efforts should be made upon the three basic factors as follows:

- Excellent performance products with reasonable cost;
- Good infrastructures upon effectively and friendly application channels;
- Satisfactory business model serves as the leverage for the battery cost.

[1]Department of Electrical and Electronic Engineering, The University of Hong Kong, Hong Kong, China
[2]Department of Electrical and Electronic Engineering, Southern University of Science and Technology, Shenzhen, China

To achieve the ultimate goal, compromises between the major roles in the automobile industry, the electric power industry and all the stakeholders, must be accomplished. Because the functionalities, production processes and commercial chain of EVs are very different as compared with the conventional vehicles, the EV industry has been regarded as the disruptive industry.

Regarding the functionalities, the EV serves not only as a transportation for human being, but also as an electric utility with function of energy storage. Hence, the implementation, or collaboration, between the EVs and the smart grid is very important. In particular, four zeros, namely the zero emission, the zero gasoline, the zero traffic accident and the zero traffic jam are set as the common goals for the aforementioned integration and collaboration (Zhu and Howe, 2007). Undoubtedly, achieving the long-term goal is a long and sophisticated march, yet the smart charging has been confirmed as the very promising step close to the existing stage.

## 13.2 Renaissance scientists and engineers

From the fourteenth to the seventeenth century, the renaissance, which activated a cultural movement, was launched (Wikipedia, website(b)). Instead of merely understanding why and how things work, the renaissance scientists and engineers are also keen on why and how the world works (Chan, 2013). The corresponding characters of these pioneers can be summarized as follows:

- Global thinking rather than local thinking;
- Circle thinking rather than linear thinking;
- Closed loop thinking rather open loop thinking;
- Life cycle thinking rather than partial life thinking.

'Yi-Jing', namely the principles of changing, are well known as its nature of balance, holistic, dialectical, periodicity and unity of opposites. These concepts have also drawn many attentions from scientists and engineers to represent the spirit of Chinese ancient philosophy.

## 13.3 Engineering philosophy of electric vehicles

The integration of automobile engineering and electrical engineering generally can be regarded as the core value of the overall EV engineering philosophy. To provide excellent EV performances with reasonable cost, the system integration and optimization are the first task to be tackled (Burke, 2015).

Since the characteristics of electric propulsion and engine propulsion are very different in nature, a new design approach is needed for EV engineering. In the meantime, to allow the EVs to compete with the common internal combustion engine vehicles (ICEVs), the advanced energy sources and intelligent energy management have both served as the key roles. Undoubtedly, the overall cost effectiveness should be one of the major concern for the EV market penetration.

To elaborate the EV engineering philosophy, the analogue of the marriage can be employed. EV engineering philosophy can be regarded as the marriage between automotive engineering and electrical engineering that includes motor, power electronic converter, controller, battery or other energy storage device and energy management system. According to the general perception, marriage means the bride and the groom have truly understand the characters among their partner so that they are able to cope with each other harmoniously (Chan, 2007). Consequently, when it comes to the EV application, the required drivability at maximum energy efficiency and minimum emission can be accomplished. To emphasize, the correlation between energy and information should serve as a critical role.

To make it short, the system integration and optimization have served as the core part for engineering philosophy. The integrated system design can be summarized into the six fundamental principles as follows:

1. Debate, define, revise and pursue the purpose and objective
    - Before the statement of a requirement is launched, the methods used to test the statement should be defined. In the meantime, the requirements should be able to reflect the constraints of technology and budgets.
2. Think holistically
    - The synergy of the individual parts should be considered so that the whole system can achieve more than the sum of the individual parts, while each part is more than a fraction of the whole system.
3. Be creative
    - Practice to learn to see the wood before the trees.
4. Follow a disciplined procedure
    - Try to follow the procedure as: divide and conquer, combine and rule.
5. Focus on the people
    - Focus on ergonomics, ethics and trust.
6. Manage the project and the relationships
    - Remember the adage: all for one, and one for all.

## 13.4 Rapid adoption of new electric vehicles

The environmental problems, including the energy consumption, carbon dioxide ($CO_2$) emissions and air pollutants by transportation, have become more severe in the past few decades, while the situations are even worse in the emerging and developing economies. About 23% of the global $CO_2$ emissions are produced from fuel combustion by the transport sector (PRTM Management Consultants, Inc, 2011). As suggested by a recent study, the main contributor to urban ambient particulate matter with a size up to 2.5 μm (PM2.5) in several regions, ranging from 12% to 37%, and a global average of 25% were caused by the traffic (Karaguliana *et al.*, 2015). As indicated in the context of China's accelerating industrialization and urbanization, the adoption of the new EV systems that are defined as the vehicles that are partially or fully powered by electricity has served as one of the key solutions to minimize the environmental impacts from the transportation (State Council, 2012).

On owning over 35% of the global sales of the EVs, China has recently overtaken the leading position from the United States and been ranked No. 1 in the production and sales of EVs all over the world (China Association of Automobile Manufacturers, website). In particular, in 2015, China has manufactured and sold 340,471 units and 331,092 units of EVs, respectively, which has recorded 3.3 times and 3.4 times higher than those happened in 2014. Consequently, the sales are expected to be 700,000 units in 2016, as estimated by the China Association of Automobile Manufacturers (Chinese Central Government. Made in China 2025, website).

According to the development blueprint, China expects to see more than 5 million EVs with 4.5 million charging stations running within the country by 2020. As there is an annually compound average growth rate of 11.4% for the China's automotive industry between 2004 and 2014 (EU SME Centre, website), it can deduce the expected growth of 8–10% for the production and sales of EVs in the next decade (Wang *et al.*, website). All these intriguing figures somehow have shown that the world is confident with the EVs to improve the energy utilization as well as the environmental problems.

## 13.5 Comparison of information and energy

Same as different data formats, including text, sound, picture, video, movie and the others, employed in information technology (IT) industry, different energy formats, such as power, heat, cold, fuel, gas, water and the others, can be found in energy technology (ET) industry. Because of the prevalence of data operating system (DOS) which manages the computer resources efficiently with organized arrangement, IT industry has effectively solved generic data conversion methodology among different formats.

However, the concept of ET industry with the network based is still in the beginning stage, especially regarding to the distributed energy resource (DER) application perspective. Hence, there is unfortunately no genetic energy conversion methodology available between different energy formats at the current stage. It should be noted that many sophisticated details at the sub-system levels rely on different physical or chemical characteristics so that all these details are somehow difficult to be generalized. Therefore, managing the energy system with different formats on a single platform is a very challenging task. To manage the energy system with different formats has undoubtedly become a fundamental barrier to hinder the development of DER-based energy system.

The energy coupling and optimization with the aid of internet of things (IOT) is shown in Figure 13.1. By borrowing the concept of the DOS in computer architecture, the energy operating system may also be developed to manage the energies.

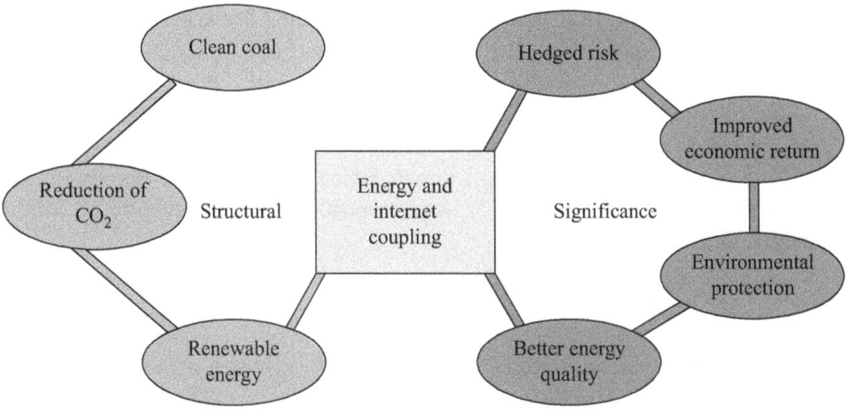


*Figure 13.1 Energy coupling and optimization with the aid of internet of things*

## 13.6 Relationship between energy and information

The duty cycle and efficiency of steam engine were analyzed by Lazare Carnot in 1803, while the accelerator and motion devices have been proposed to represent losses of moment of activity in any machine. In 1824, Lazare's son Sadi Carnot, developed from his father's work to publish the reflections on the motive power of fire. It showed that all heat engines utilized the difference between temperatures, so that work or kinetic power could be generated by the reactions among the hot and cold objects. To improve the comprehensiveness, the analogy of the falling water in a water wheel has also been proposed (Wikipedia, website(a)). Count Rumford has utilized the cannon bores as an example to show that the heat could be generated by the frictional force (McCulloch, 1876). The first law of thermodynamics suggests that the internal energy within an isolated system is always constant. Hence, the amount of energy and matter in the universe are kept constant, while energy can only be converted from one form to another.

Unfortunately, the effects of friction and dissipation could not be quantitatively explained by the first law of thermodynamics, based on the heat-friction experiments by James Joule in 1843. To elaborate the idea, Rudolf Clausius in the 1850s, objected to the supposition that heat could not pass from a colder to a hotter body of itself, unless there was any change occurred in the working body. The entropy was described by Clausius as the transformation-content, namely, the dissipative energy used within a thermodynamic system or working body of chemical species during a change of state (Clausius, 1850). However, this new concept violated the earlier views of the theories of Isaac Newton, which suggested heat, as the object with mass, was an indestructible particle. In the meantime, all spontaneous changes are governed by the second law of thermodynamics.

Maxwell's demon was an imaginary experiment proposed by James Clerk Maxwell in 1867. This experiment was first suggested in a letter that Maxwell wrote to Peter Guthrie Tait, while it was then publicly presented in his book on thermodynamics titled as *Theory of Heat* in 1872 (Leff and Rex, 2002). Throughout the demon experiment, Maxwell planned to suggest that the second law of thermodynamics could only show a statistical certainty.

According to the general theory of statistics, the relationship between energy and information was suggested by Ludwig Boltzmann in 1877. The expression between the information and energy was proposed by Chan recently as follows (Chan and Jian, 2013):

$$\sum I \Leftrightarrow \sum E \tag{13.1}$$

where $\sum I$ and $\sum E$ involve various information and various energies, respectively.

Previous results of molecular motions were observed based on the Boltzmann's research. 'On the reduction of entropy in a thermodynamic system by the intervention of intelligent beings', introduced the thought experiment by Leó Szilárd (1929), while it was called as Szilárd's engine. In particular, this concept has become very important regarding the trials to understand the Maxwell's demon. According to the point of view of Szilárd, Maxwell's demon, who could delete and save the information, was very intelligent. With the implementation of the second law of thermodynamics, the consideration of entropy should be included when Maxwell's demon gained the information.

Based on the inspiration of Szilárd, the correlation between entropy in information and entropy in thermodynamics was suggested by Léon Nicolas Brillouin in 1956. The formula governing the relationships was then suggested as:

$$S = K_B \ln 2 \tag{13.2}$$

According to (13.2), when 1 bit of information is obtained, the entropy $S$ value of $K_B\ln 2$ in (J/K) in the system must be decreased, where $K_B = 1.38 \times 10^{-23}$ is the Boltzmann constant. Based on the information theory, the information entropy $H$ of a discrete random variable $X$ with possible values $\{x_1, x_2, \ldots, x_n\}$ and probability mass function $p(x)$ was proposed by Claude Shannon (1948), where the relationships can be described as:

$$H(X) = -\sum_{i=1}^{n} p(x_i)\log_b p(x_i) \tag{13.3}$$

where $b$ is the logarithm base of 2.

The problem of quantitative metrics in information can be effectively solved by Shannon's theory. Hypothetically, the thermodynamic entropy of a system can be minimized by using information of the individual molecules by the Maxwell's demon. Yet, according to the arguments by Landauer *et al.* (1961), the Maxwell's demon can only operate upon the increase of thermodynamic entropy in the process by at least the amount of Shannon information which first acquired and stored. Consequently, the total thermodynamic entropy will not decrease so that the paradox is resolved. Upon the implication of Landauer's principle, a computer must dissipate certain amount of heat in order to process a given amount of information, such that modern computers are far from the efficiency limit.

The laboratory demon was recently produced by the Japanese researchers (Toyabe *et al.*, 2010). The produced demon stayed along with Bennett, who originally believed that Maxwell's demon could manage the door so that the molecules were controlled to flow from one room to another. However, the phenomenon appeared if and only if the demon deleted its memory of the judgment on the molecular movements (Bennett, 1987). Based on the proposed experiment by the Japanese, the second law of thermodynamics can be described as:

$$\langle \Delta F - W \rangle \leq K_B T I \tag{13.4}$$

where $\Delta F$ is the difference between free energies among various states, $W$ is the work done on the system, $T$ is the environment temperature and $I$ is the manual information content obtained by measurements.

The association between the decrease of the entropy and the information access was investigated by Szilárd. Based on the developed model by Szilárd, the new concept, which suggested 1 bit of information in the system was equivalent to $K_B T\ln 2$, was proposed by Rolf Launder. Consequently, the entropy earned by the Maxwell's demon from the control of the gate would be reduced accordingly.

According to the Japanese experiment and Bennett's paper in 1987, the Maxwell's demon was suggested to be able to abandon or restore the information, upon limited of memory. In the other word, the entropy of the system would be increased, when the demon abandoned any information. Just when the occasion that the demon has used up its memory, it would have to delete some information. Consequently, the entropy of the system was also increased. By combining all the proposed concept, the correlation between information and energy can be expressed effectively by the Szilárd's work as:

$$\{\langle \Delta F - W \rangle / 1 \text{ bit}\} \leq K_b T \ln 2 \tag{13.5}$$

As suggested by the Japanese experiment, enough energy can be obtained by the microscopic particle through the process of measurement so that the exact amount of energy that is transformed from the information during the process (or measurement) can be figured out, provided that no external force is exerted (such as injecting energy). This phenomenon somehow can provide evidence to support that the energy in fact comes from information.

Entropy consists of two different forms, namely the macro and the microscopic forms. The macro form, which is associated with thermal energy, is also known as the Clausius form (thermodynamic entropy). In the meantime, the microscopic form, which correlates to quantum energy, is instead known as the Boltzmann form (information entropy). Entropy has also been employed in computer science and communication networking. The fundamental meanings of both thermodynamic and information entropies are the same, where both of them are regarded as energy to the described system.

The existence of energy is governed by the 'field', while the presence of field is governed by the material of field. In other words, without the material, no field can be generated. Due to the conservation of energy and mass, energy can never separate from the existence of mass, yet any particular mass can only be generated by another type of mass. According to the concept of the mass-energy conservation, the mass basically is a type of energy state or carrier, while energy is an adjunct of mass. To describe the correlation between the energy and the mass, the famous equation can be utilized as:

$$E = mc^2 \tag{13.6}$$

The mentioned mass-energy conservation tried to describe the relationship between mass and energy in amount, yet the calculation of the quantity of transformation from mass to energy has never been explored.

According to the common understanding in physics, 'equals sign' implies the same meaning of physical quantity, or they consist of independent physical properties with the same nature, such as the quantities '$m$' and '$E$' in the equation of mass-energy conservation. Information can be used to describe the interchange between energy and mass, in terms of the change of state, the change of time and the change of dimension. According to (13.5), it suggests that the information has the direct relationship with the energy, such that more energy is needed if more information is wanted. In other words, better control capability can be achieved if more information of one kind of mass is obtained.

The quantum state of a physical system changes with time, which can be effectively described by the Schrödinger equation. The time-dependent Schrödinger equation (single non-relativistic particle) can be governed as:

$$ih\frac{\partial}{\partial t}\psi(r,t) = \left[-\frac{h^2}{2\mu}\nabla^2 + V(r,t)\right]\psi(r,t) \tag{13.7}$$

where $i$ is the imaginary unit, $h$ is the Planck constant divided by $2\pi$, $\mu$ is the particle's 'reduced mass', $V$ is its potential energy, $\nabla^2$ is the Laplacian operator and $\psi$ is the wave function. When it comes to a real case, all of the focused matters, such as cost, time and risks, can be described by $\psi$. The most important part with the largest weight can be calculated based on the Schrödinger equation and per differential or integral methods.

## 13.7 Utilization of energy and information for continuous development

Upon the development of the analysis among energy and information, the penetration of novel energy form can be invented, so that the most effective blueprint for the renewable energy can be obtained. The most important incidents can never be obtained merely based on the consideration of the upper and lower limits of information entropy. Yet, the physical or thermodynamic limits of energy transmission and the decision making on social, economic and political issues are the complex interdependence correlation among this topic.

The energy and material resources are scarce resources that are limited. Fortunately, the solar energy, as one of the renewable energies, can be instead harnessed in various forms. The correlation between the following three parts, namely the energy remains in the system, the degrees of freedom and its total content of information, can be quantitatively described. In addition, the energy and information management is another critical task to be tackled. Undoubtedly, the proper method in handling the information and energy distribution within the grid and storage systems, such as solar panels, wind turbines, coal power plants and EVs as mobile storage, is important. However, it is hindered by the total content of information stored, the technological choices for energy production and the social choices on the issues of arrangements.

The scarce energy and matter are the basic concern, while the values, justice and equality should be fairly allocated and distributed with various factors. In particular, the shared information can be spread to everyone effectively to improve its accessibility to the potential readers. Merely producing more energy with more derived information is not enough, unless there is any change on the social perception of the energy usage. Consequently, the new business models and opportunities for innovation can be provided so that the more effective understandings of local and global societies toward the scarce resources on Earth can be accomplished.

For the applications of generation of energy, transmission and distribution, the employment of the control and energy utilization by the information sharing has been proposed in the past few decades. Upon the advancement of computer technologies, better control algorithm as well as more effective operation can be accomplished. The voltages and currents can be regarded as the derived information for the electrical systems, while the synchronized vector approach can be utilized to result the quicker and more accurate real-time responses.

Upon any fault or malfunction within the electrical system, the change among the energy or information can be shown through the voltage drop or an increased current magnitudes. The fault can then be detected by the protection system and the circuit breakers are activated to remedy the fault in the short period of time as requested. The spontaneous information can be shown by the change in energy so that the current status of the system can be handled in a proper way. The control and reaction with the energy can be achieved by monitoring the change of the information entropy. The information entropy in the functional system can be utilized to reach the optimization status in a healthy system.

Various energy information, such as transportation, working environment, preparation of food and civic services, have been employed as energy consumption information for a metropolitan city. To be specific, the big data can be utilized to serve as an important role to manage the whole city in a better way. For instances, optimization of the daily trips can achieve minimized energy consumption with the help of a global positioning system (GPS) on any vehicles. The improved platform and services can be achieved upon the analysis of the behavior of citizens. Various aspects, such as sustainable energy employment, social harmony and integration, can be accomplished upon the upcoming development in big data analysis and utilization.

*Table 13.1 Fast charging, rapid charging and quick charging*

| Options | Power rating (kW) | | |
|---|---|---|---|
| | **Heavy duty** | **SUV/sedan** | **Small sedan** |
| Fast charging, 10 min, 100% SoC | 500 | 250 | 125 |
| Rapid charging, 15 min, 80% SoC | 250 | 125 | 60 |
| Quick charging, 60 min, 70% SoC | 75 | 35 | 20 |

## 13.8 Smart charging

### *13.8.1 Background information*

There are two traditional methods, namely the fast charging and the slow charging (Botsford and Szczepanek, 2009), for the plug-in EVs (PEVs). Generally speaking, the plug-in EVs can be classified as the plug-in hybrid EVs (PHEVs) and the battery EVs (BEVs). The two charging methods can be distinguished by the charging power rating, as well as the charging time. At slow charging circumstance, normally 6–8 h are needed to fully charge the battery with a charging power of around 3 kW. Hence, slow charging method is suitable for the overnight charging. On the other hand, the fast charging method means quicker and faster charging mechanism, while there is no definite specification for this technology. Therefore, any charging methods, which can provide charging rate faster than the slow charging method, can be considered as fast charging. In particular, the fast charging method is further categorized as three sub-classes, namely the fast charging, the rapid charging and the quick charging as shown in Table 13.1 in which they adopt different values of charging time with different levels of state-of-charge (SoC).

Regardless of the status of the utility grid, both charging methods, namely slow charging and fast charging, can be operated normally. These new electric loads, upon the increasing penetration of plug-in EVs, will undoubtedly initiate the great surges as demand in peak hours. Hence, there may be negative effects toward the stability and security of the power grid. In the meantime, by carrying information into power, the smart charging has been confirmed as one of the best solution to the mentioned challenge. Upon the employment of smart charging, the real-time status of power grid and the demand of the PEV owners are monitored and analyzed. Consequently, the charging power can be adjusted with online algorithm under the smart charging operation so that the conventional peak load problems can be suppressed. The daily average power load curves within a regional grid is shown in Figure 13.2. It can be found that with the adoption of the smart charging algorithm, the power load curve can be flattened and hence the peak load problems can be relieved (Department of Economic and Social Affairs, 2006). After that, the stability and efficiency as well as the average operating cost of the grid can be improved (Dyke *et al.*, 2010; Hajimiragha *et al.*, 2010; Kiviluoma and Meibom, 2011).

By adopting the electric power generated from renewable energy sources, such as wind power, solar power and so forth, the EV batteries can be charged more

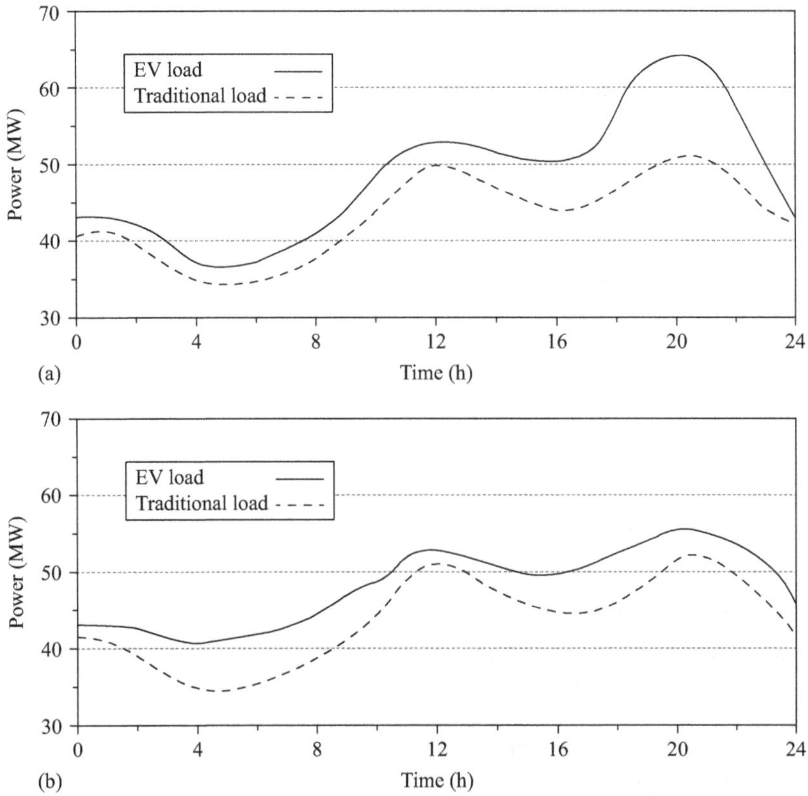


*Figure 13.2 Average power load curves within a regional grid with different charging schemes: (a) Traditional charging. (b) Smart charging*

effectively with the smart charging operation. The renewable energy sources, which rely on climate conditions, are fundamentally different from the conventional power sources. The integration between these intermittent electrical energy sources into the conventional power grids has become a hot issue in the past few decades (Wang *et al.*, 2012). With the support by the smart grid and smart charging algorithms, the EV batteries can serve as the bridge between the renewable energy and the power loads (Judd and Overbye, 2008; Saber and Venayagamoorthy, 2011).

The blueprint of smart charging within the smart grids is depicted in Figure 13.3 (Fang *et al.*, 2012; Galli *et al.*, 2011; Li *et al.*, 2010; Slootweg, 2009). The key of the smart charging is to communicate the information between all the parties, such as the PEV, EV supply equipment (EVSE), regional power grid and the control center, within the group. In the current smart charging algorithm, the EV batteries cannot return the excessive electric power back to the power grid. Yet, the concept of transferring the electric power from the EV batteries to the power grid has been proposed and known as an essential feature of the vehicle-to-grid (V2G) technology (Clement-Nyns *et al.*, 2010; Clement-Nyns *et al.*, 2011; Guille and Gross, 2009; Gungor *et al.*, 2011; Jian *et al.*, 2013; Sortomme *et al.*, 2011).

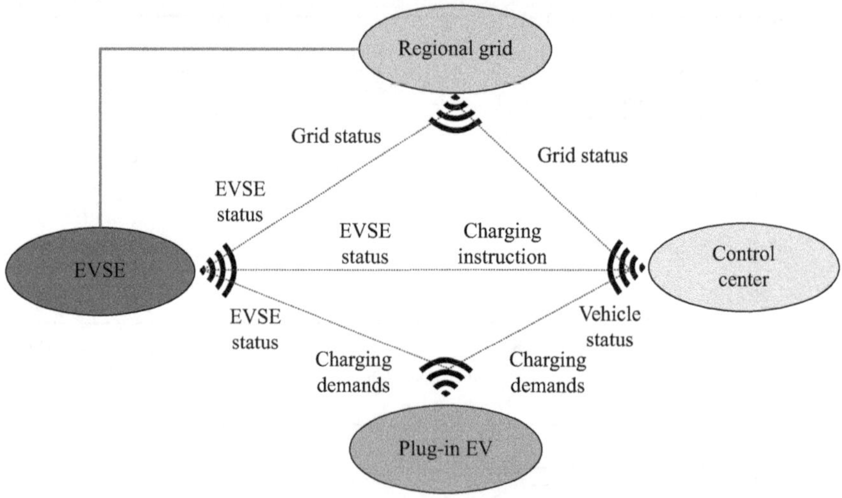


*Figure 13.3 Blueprint of smart charging within the smart grid*

### 13.8.2 Stakeholders

The active cooperation, collaboration and coordination among the key stakeholders, namely, the automakers, electricity providers, vehicle owners and charging service providers, are the key to success of market penetration for PEVs. Yet, as expected, different stakeholders show different interests as well as different demands (Schurr, 2012).

According to the perspective from the automakers, the convenient and friendly charging facilities must be the major concern. Meanwhile, the charging cost should be reasonable and affordable. Then, the EVSE or the charging posts should be supported by different connector standards. Finally, a long-term relationship should be established among the customers. As expected, the fulfilment of the needs of the customers can allow the product to become more competitive. The existing connector standards for EV charging methods vary with different regions globally. In addition, the non-contact charging or wireless charging, which is based on the technology of electromagnetic coupling, can be regarded as a potential charging option in the near future (Jang *et al.*, 2012; Kobayashi *et al.*, 2011; Matsuda *et al.*, 2006; Mohrehkesh and Nadeem, 2011). Obviously, the automakers as well as the customers always expect that their EVs can be recharged conveniently at anytime and anywhere.

According to the perspective from the electricity providers, the charging patterns must be the major concern. If possible, the electricity providers aim to shift the charging loads away from the peak hours, so that the overall power load curves can be leveled and flattened. In the meantime, the grid problems or faults, which might be arisen from the EV charging, such as extreme surges in power loadings, fluctuations in power flows, power losses and charging profile patterns, should be considered and analyzed. Finally, the adoption rate of the renewable energy should be enhanced (Lopes *et al.*, 2011).

According to the perspectives from the vehicle-owner, a safe charging system must be the major concern. The second criteria they may pay attention should then be the convenient and friendly charging facilities. In other words, the EVSE or charging posts should be widely available at the parking slots so that they can be easily accessed. In the third place, the charging cost should be payable by different payment options, such as by cash, debit card, credit card or even mobile phone. Finally, the PEVs should be charged up by electric power from the renewable energy sources, whenever possible.

According to the perspective from the charging service providers, the utilization frequency of their charging facilities must be the major concern. Meanwhile, the price should be set within the range that is profitable. Finally, the low operating cost and long equipment life are favorable.

It should be noted that the battery swapping scheme has been provided by some power utility companies and charging service providers in the past decades (Jian *et al.*, 2011; Li *et al.*, 2011; Worley and Klabjan, 2011). The battery swapping scheme serves as one type of the infrastructure as well as business model for EV systems. In general, this type of scheme can offer the definite benefits as follows:

- Because the battery pack is an external device and managed by the power utility companies or charging service providers, the initial manufacturing cost of EVs can be minimized.
- Since the battery swapping can be completed within few minutes of time so that quick refueling can be achieved. According to the latest report, the battery swapping processes can be completed in less than 1 min. This battery swapping time is even quicker than the traditional fuel vehicles which rely on the oil refueling mechanism (Squatriglia, 2009). Not surprisingly, the reported time should be the optimum case while the whole processes should generally take a few minutes to complete.
- The power utility companies can choose the best time slots to recharge the batteries, such that the efficiency and stability of the power system, especially for the system with the integration of renewable energy, can be improved. Consequently, the overall performance of power grid and renewable energy can be optimized.
- It should be noted that the battery can be charged and discharged with the better managed schedule, so that the life of the battery can be increased.
- Since the relatively obsolete battery packs can be employed as the energy storage on other applications, the battery packs can be fully utilized until they have become valueless.

However, the battery swapping scheme is still an ambiguity issue among the automakers, while some of the automakers tend to design the battery pack as an integrated part for the EV system. In other words, the battery packs should not be separated from the EV so that the swapping shame is not achievable. Otherwise, the overall design considerations, such as the vehicular ventilating system, the controlling strategy, the vehicular body structure and strength, have to be reconfigured. Apart from the technical issues, the legal aspect of the battery pack and the potential loss of core businesses are the other important factors that the automakers are concerned with. Hence, compromises between the automakers, the power utility companies and the charging service providers should be accomplished in order to achieve a win-win battery swapping scheme. To be specific, the battery swapping scheme should be favorable for the applications requiring the fast refueling characteristics.

In addition to the battery swapping, the battery leasing scheme is another promising infrastructure and business model. Because the battery is owned by the service provider but not the vehicle owner, the battery leasing scheme consists of the definite advantages of reduced initial selling cost of EVs. Moreover, the scheme consists of the advantage of optimized charging schedule, which is beneficial to the grid as well as the battery health. Yet, some compromises have to confirm between the automakers and the service providers in advance.

As suggested, the fundamental principles of the integration of information and energy should be adoptable to any types of infrastructures and business models. The functional financial settlement system that serves the smart charging is shown in Figure 13.4. To provide the charging services, the charging service providers can purchase the electricity from the power retailers or the power grid company. In the meantime, the charging service providers should pay rent to the real estate owners for the accommodation of the charging facilities. It is not surprising that the vehicle owners have to pay the charging fee to the charging service providers. In general, to support the infrastructure development, the government would impose tax on electricity providers, real estate owners and charging service providers. To promote the market penetration of the BEVs and PHEVs, the government, in the beginning stage, may offer subsidies to the electricity providers, charging service providers and vehicle owners.

The time-based variable price or smart price of electrical energy should definitely be one of the key incentives to attract all the major players to take further

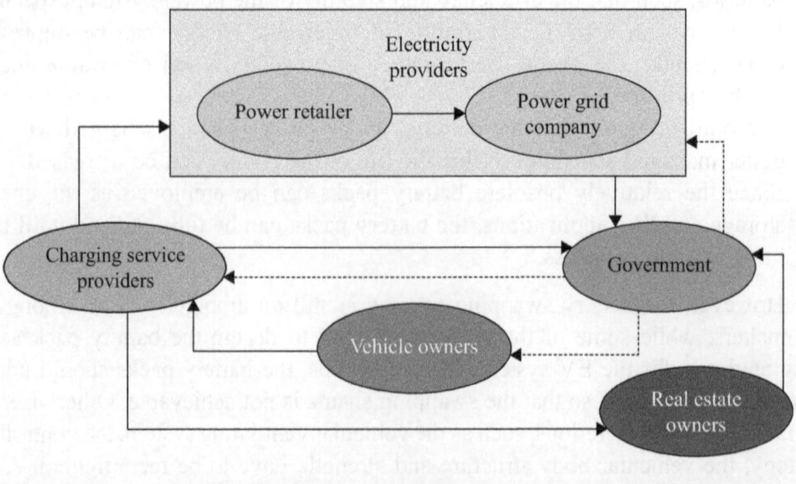


*Figure 13.4 Functional financial settlement system for smart charging*

steps to maximize the overall energy efficiency and minimize the greenhouse gases emission (Al-Awami and Sortomme, 2012; Cao *et al.*, 2012; Faruqui *et al.*, website; Vojdani, 2008). It should be noted that the understanding of the PEV connectivity conditions among the stakeholder groups is limited while huge gaps are remained in the knowledge base. These gaps are essential signal to the early market penetration, in the situation when most of the stakeholders are still on a steep learning curve (APEC Consultant Report, 2012). Hence, the knowledge base for mutual understanding among stakeholders should be improved so that consensus can be compromised.

### *13.8.3 Energy flow and information flow for smart charging*

The integration of the flows between energy and information is the key issue of EV smart charging (O'Doherty, 2012). The energy and information flows among renewable energy generation, smart grid, advanced metering infrastructure, smart home, distributed network, micro-grid and EVs should be seamlessly correlated, as shown in Figure 13.5. Some key targets, such as 10% penetration of PEVs by 2020 and 40% utilization of renewable energy for EV charging, are expected to be accomplished with the integration system.

The planning of the optimal charging schedule, as shown in Figure 13.6, is another important task to be handled. The status of power grid, the parameters of on-board EV batteries, the traffic conditions, the weather conditions and the EV owners' demands or personal behaviors, are the key factors to be considered. Huge computing resource should be used upon the development of the cloud computing technology (Baliga *et al.*, 2011; Grobauer *et al.*, 2011; Patidar *et al.*, 2012). In addition, the optimal charging references may need to be implemented with the EVSEs. The cooperation, collaboration and coordination among several key players/stakeholders, namely, the automakers, electricity providers, vehicle owners and charging service providers, are the key of success toward the market penetration for PEVs.

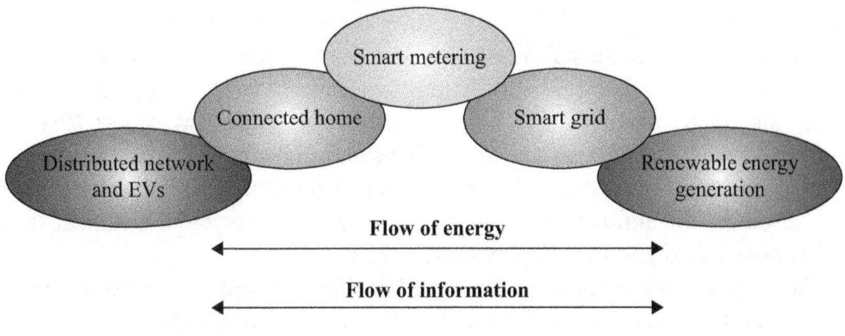


*Figure 13.5 Integration of flows between energy and information in smart charging*

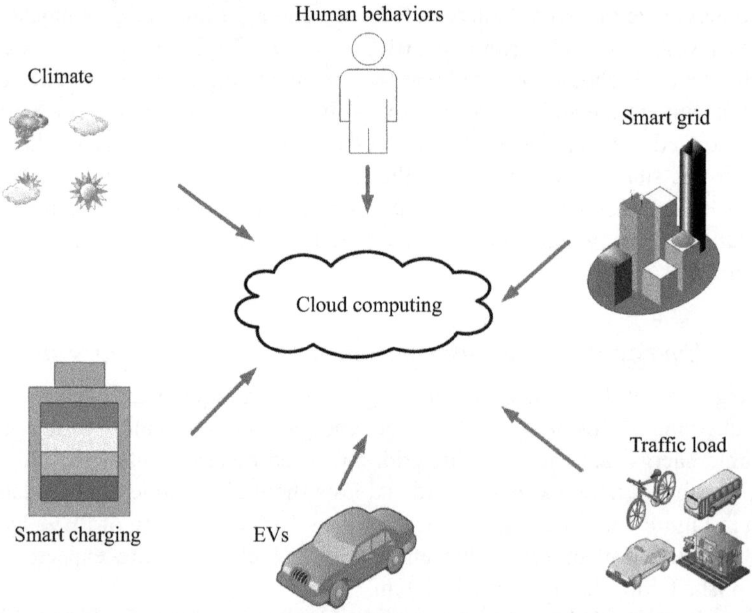


*Figure 13.6 Optimal charging schedule in smart charging*

### 13.8.4 Challenges for smart charging

The determination of which information type should be integrated into the energy flow is the major part of smart charging. This challenge can be assorted as follows:

- The impacts on the energy interchange among power grid and PEVs.
- The effects toward the overall energy efficiency and exhaust emission.
- The method of acquisition, transmission and process of these information.
- The effectiveness and accuracy of the information utilization.

The problem existing in the smart charging can be regarded into three major levels, as shown in Figure 13.7. The top level is concerned with the basic business model issue. The fundamental questions exist in the top level includes: (i) Which charging method, such as on-board battery charging, battery swapping or battery leasing, should be adopted? (ii) Who should be the major investors? (iii) Who should be beneficial via the systems? All these questions involve many issues, such as the living styles of citizens, the environmental impact, the urban planning, the travel distance, the infrastructure location, the parking issues, the impact to the existing power grid and the support from government.

The way to handle the aforementioned problem should be the considerations within the middle level. The fundamental questions exist in the middle level includes: (i) What or who are the control objects? (ii) What are the major variables? (iii) What are the key constraints? (iv) What are the solutions to the charging instructions?

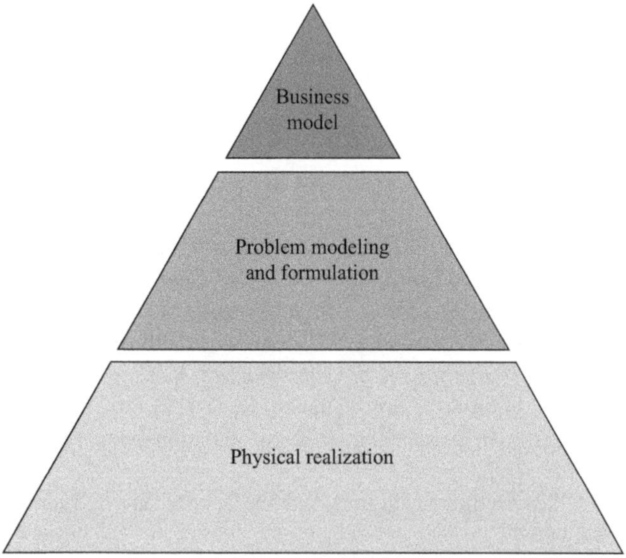


*Figure 13.7 Key considerations for smart charging*

The realization of the designed business model and optimal charging strategy is the major concern at the bottom level. The fundamental questions exist in the bottom level includes: (i) What are the basic layouts of the charging posts or charging stations? (ii) What are the process to build the sensor network? (iii) What are the configurations of the computing resources?

## 13.9 Conclusions

In the coming generation, the implementation among sustainable renewable energy and electrified transportation will become the key to success for the sustainable transportation. The overall sense of electro-mobility has been discussed in this chapter. To accelerate the development of electro-mobility, the good output, good infrastructure as well as good business model are essential. In particular, the concerns and correlations among the major stakeholders, namely the automakers, electricity providers, vehicle owners and charging service providers, have to be carefully handled. The integration between the flows of information and energy has high potential to achieve the desirable goal. Upon the flow integration, different needs from various stakeholders can be accomplished harmoniously. To be specific, the dynamic smart charging schedule can be achieved upon the analysis of the information from automobile, power grid and drivers' needs. The exploration of the basic theory between information and energy will become a hot research topic in the coming future. Yet, this particular topic can practically result the win-win ecosystem.

The essence of renaissance scientists and engineers, as well as the inspiration of innovative ideas, have become the indispensable part to cope with the complex engineering challenges in the new century. The word 'Why and How Things work' and 'Why and How the World works' are the two typical questions the renaissance scientists and engineers often ask. To work for a brighter and better future, we have to learn from the spirit of our predecessors.

## References

Al-Awami, A. and Sortomme, E. (2012) Coordinating vehicle-to-grid services with energy trading. *IEEE Transactions on Smart Grid*, **3**, 453–462.

APEC Consultant Report (2012) Stock-take of electric vehicle interface with electricity and smart grids across APEC economies and the potential for harmonization. *APEC Workshop on EV Connectivity, Wellington, New Zealand.*

Baliga, J., Ayre, R., Hinton, K. and Tucker, R. (2011) Green cloud computing: balancing energy in processing, storage, and transport, *Proceedings of the IEEE*, **99**, 149–167.

Bennett, C.H. (1987) Demons, engines and the second law. *Scientific American*, **257**, 108–116.

Botsford, C. and Szczepanek, A. (2009) Fast charging vs. slow charging: pros and cons for the new age of electric vehicles. *Proceeding of EVS24, Stavanger, Norway.*

Burke, A.F. (2015) Advanced batteries for vehicle applications, *Encyclopedia of Automotive Engineering, Hybrid and Electric Powertrain, Energy Source Batteries,* New York, USA, Wiley.

Cao, Y., Tang, S., Li, C., *et al.* (2012) An optimized EV charging model considering TOU price and SOC curve. *IEEE Transactions on Smart Grid*, **3**, 388–393.

Chan, C.C (2007) The state of the art of electric, hybrid, and fuel cell vehicles. *Proceedings of IEEE*, **95**, 704–718.

Chan, C.C. (2013) Renaissance scientists and engineers – mass, Energy and information. *Studies in Science and Technology*, **2**, 101–106.

Chan, C.C. and Chau, K.T. (2001) *Modern Electric Vehicle Technology*, Oxford: Oxford University Press.

Chan, C.C. and Jian, L. (2013) Correlation between energy and information. *Journal of Asian Electric Vehicles*, **11**, 1625–1634.

China Association of Automobile Manufacturers (2015). (website) Available at: http://www.caam.org.cn/

Chinese Central Government. Made in China 2025. (website) Available at: http://www.gov.cn

Clausius, R. (1850) On the motive power of heat, and on the laws which can be deduced from it for the theory of heat. *Poggendorff's Annalender Physick*, **79**, 368–500.

Clement-Nyns, K., Haesen. E. and Driesen, J. (2010) The impact of charging plug-in hybrid electric vehicles on a residential distribution grid. *IEEE Transactions on Power Systems*, **25**, 371–380.

Clement-Nyns, K., Haesen, E. and Driesen, J. (2011) The impact of vehicle-to-grid on the distribution grid. *Electric Power Systems Research,* **81**, 185–192.

Department of Economic and Social Affairs. (2006) *Multi dimensional issues in international electric power grid interconnections*, United Nations, New York, USA.

Dyke, K.J., Schofield, N. and Barnes, M. (2010) The impact of transport electrification on electrical networks. *IEEE Transactions on Industrial Electronics*, **57**, 3917–3926.

EU SME Centre. *The Automotive Market in China.* (website) Available at: http://www.ccilc.pt/

Fang, X., Misra, S., Xue, G. and Yang, D. (2012) Smart grid – the new and improved power grid: a survey. *IEEE Communications Surveys & Tutorials*, **14**, 944–980.

Faruqui, A., Hledik, R., Levy, A. and Madian, A. (website) *Will smart prices induce smart charging of electric vehicles?* Available at: http://ssrn.com/abstrat=1915658

Galli, S., Scaglione, A. and Wang, Z. (2011) For the grid and through the grid: the role of power line communications in the smart grid, *Proceedings of the IEEE*, **99**, 998–1027.

Grobauer, B., Walloschek, T. and Stocker, E. (2011) Understanding cloud computing vulnerabilities. *IEEE Security & Privacy*, **9**, 50–57.

Guille, C. and Gross, G. (2009) A conceptual framework for the vehicle-to-grid implementation. *Energy Policy*, **37**, 4379–4390.

Gungor, V.C., Sahin, D., Kocak, T., *et al.* (2011) Smart grid technologies: communication technologies and standards. *IEEE Transactions on Industrial Informatics*, **7**, 529–539.

Hajimiragha, A., Canizares, C.A., Fowler, M.W. and Elkamel, A. (2010) Optimal transition to plug-in hybrid electric vehicles in Ontario, Canada, considering the electricity-grid limitations. *IEEE Transactions on Industrial Electronics*, **57**, 690–701.

Jang, Y., Ko, Y. and Jeong, S. (2012) Optimal design of the wireless charging electric vehicle. *2012 IEEE International Electric Vehicle Conference*, Greenville, USA.

Jian, L., Xu, G., Xue, H. and Chang, M. (2011) Intelligent multi-mode energy-refreshing station for electric vehicles within the framework of smart grid, *IEEE International Conference on Information and Automation*, Shenzhen, China.

Jian, L., Xue, H., Xu, G., Zhu, X., Zhao, D. and Shao, Z. (2013) Regulated charging of plug-in hybrid electric vehicles for minimizing load variance in household smart micro-grid. *IEEE Transactions on Industrial Informatics*, **60**, 3218–3226.

Judd, S.L. and Overbye, T.J. (2008) An evaluation of PHEV contributions to power system disturbances and economics. *The 40th North American Power Symposium*, Calgary, North America.

Karaguliana, F., Belisd, C.A., Francisco, C., *et al.* (2015) Contributions to cities' ambient particulate matter (PM): a systematic review of local source contributions at global level. *Atmospheric Environment*, **120**, 475–483.

Kiviluoma, J. and Meibom, P. (2011) Methodology for modeling plug-in electric vehicles in the power system and cost estimates for a system with either smart or dumb electric vehicles. *Energy*, **6**, 1758–1767.

Kobayashi, K., Pontefract, T., Kamiya, Y. and Daisho, Y. (2011) Development and performance evaluation of a non-contact rapid charging inductive power supply system for electric micro-bus. *IEEE Vehicle Power and Propulsion Conference*, Chicago, USA.

Landauer, R. (1961) Irreversibility and heat generation in the computing process. *IBM Journal of Research and Development*, **5**, 183–191.

Leff, H. and Rex, A.F. (2002) *Maxwell's Demon 2: Entropy, Classical and Quantum Information, Computing*, CRC Press, Boca Raton, FL, USA.

Li, F., Qiao, W., Sun, H., *et al.* (2010) Smart transmission grid: vision and framework. *IEEE Transactions on Smart Grid*, **1**, 168–177.

Li, S., Kolmanovsky, I.V. and Ulsoy, A.G. (2011) Battery swapping modularity design for plug-in HEVs using the augmented Lagrangian decomposition. *American Control Conference*, San Francisco, USA.

Lopes, J.A.P., Soares, F.J. and Almeida, P.M.R. (2011) Integration of electric vehicles in the electric power system. *Proceedings of the IEEE*, **99**, 168–183.

Matsuda, Y., Sakamoto, H., Shibuya, H. and Murata, S. (2006) A non-contact energy transferring system for an electric vehicle – charging system based on recycled products. *Journal of Applied Physics*, 99.

McCulloch, R.S. (1876) *Treatise on the Mechanical Theory of Heat and Its Applications to the Steam-Engine, etc*., New York: D. Van Nostrand.

Mohrehkesh, S. and Nadeem, T. (2011) Toward a wireless charging for battery electric vehicles at traffic intersections. *14th International IEEE Conference on Intelligent Transportation Systems*, Washington, DC, USA.

O'Doherty, P. (2012) Charging ahead – key role of the vertically integrated utility in smart EV charging and smart grid. *Fully Charged 2012, International Electric Vehicle Submit*, Dublin, Ireland.

Patidar, S., Rane, D. and Jain, P. (2012) A survey paper on cloud computing. *International Conference on Advanced Computing & Communication Technologies*, Rohtak, India.

PRTM Management Consultants, Inc. (2011) *The China New Energy Vehicles Program – Challenges and Opportunities*, Waltham, MA, USA, World Bank. Available at: http://siteresources.worldbank.org/EXTNEWSCHINESE/Resources/3196537 1202098669693/EV_Report_en.pdf

Saber, A.Y. and Venayagamoorthy, G.K. (2011) Plug-in vehicles and renewable energy sources for cost and emission reductions. *IEEE Transactions on Industrial Electronics*, **58**, 1229–1238.

Schurr, A. (2012) Smarter cities need smarter charging for plug-in EVs. *Fully Charged 2012, International Electric Vehicle Submit*, Dublin, Ireland.

Shannon, C.E. (1948) A mathematical theory of communication. *Berll System Technical Journal*, **27**, 379–423.

Slootweg, H. (2009) Smart grids – the future or fantasy. *2009 IET Conference: Smart Metering – Making It Happen,* London, United Kingdom.

Sortomme, E., Hindi, M.M., Macpherson, S.D.J. and Venkata, S.S. (2011) Coordinated charging of plug-in hybrid electric vehicles to minimize distribution system losses. *IEEE Transactions on Smart Grid*, **2**, 198–205.

State Council. (2012) Program of the state council on issuing the planning for the development of the energy-saving and new energy automobile industry (2012–2020) (file number: State Council[2012]22). Available at: http://www.miit.gov.cn/n11293472/n11505629/n11506426/n11515200/n11515446/n11926400/15116596.html

Squatriglia, C. (2009) Better place unveils and electric car battery swap station. Available at: http://www.wired.com/autopia/2009/05/better-place/

Szilárd, L. (1929) Über die Entropieverminderung in einem thermodynamischen System bei Eingriffen intelligenter Wesen. *Zeitschrift für Physik*, **53**, 840–856.

Toyabe, S., Sagawa, T., Ueda, M. Muneyuki, E. and Sano, M. (2010) Experimental demonstration of information-to-energy conversion and validation of the generalized Jarzynski equality. *Nature Physics*, **6**, 988–992.

Vojdani, A. (2008) Smart integration. *IEEE Power & Energy Magazine*, **6**, 71–79.

Wang, L., Singh, C. and Kusiak, A. (2012) Guest editorial: special issue on integration of intermittent renewable energy resources into power grid. *IEEE System Journal*, **6**, 2–3.

Wang, A., Liao, W. and Hein, A. (website) Bigger, better, broader: a perspective on China's auto market in 2020. Available at: http://www.mckinsey.com

Wikipedia. (website(a)) Entropy. Available at: http://en.wikipedia.org/wiki/Entropy#cite_note-2

Wikipedia (2014). (website(b)) History of science in the Renaissance. Available at: http://en.wikipedia.org/wiki/History_of_science_in_the_Renaissance

Worley, O. and Klabjan, D. (2011) Optimization of battery charging and purchasing at electric vehicle battery swap stations. *IEEE Vehicle Power and Propulsion Conference*, Chicago, USA.

Zhu, Z.Q. and Howe, D. (2007) Electrical machines and drives for electric, hybrid, and fuel cell vehicles. *Proceedings of IEEE*, **95**, 746–765.

## Chapter 14

# Optimal scheduling with vehicle-to-grid ancillary services

*Junhao Lin[1], James J.Q. Yu[1], Ka-Cheong Leung[1] and Victor O.K. Li[1]*

## 14.1 Overview

In this chapter, we propose a method for the utility grid operator and aggregators to coordinate multiple grid-connected electric vehicles (EVs) so as to provide electric-power ancillary services accounting for quality-of-service (QoS) guarantees for EV charging in a multilevel vehicle-to-grid (V2G) system. This method includes applying consensus-algorithm-based distributed control, designing the control objective functions and constraints for providing ancillary services, and designing the constraints for charging/discharging power schedules and the individual QoS requirements of EVs. The consensus-algorithm-based distributed control consists of operation protocols for control and communication among the utility grid operator, aggregators, and EVs, calculating the control signals in the utility grid operator and aggregators, determining the optimal charging/discharging power schedules of EVs so as to provide a required type of ancillary services as well as satisfy the QoS requirements of EVs.

### *14.1.1 Electric vehicles and ancillary services*

With the worldwide concerns on climate change due to global warming, many countries have established environmental policies to control the greenhouse gas (GHG) emissions, the primary cause of global warming. For example, California has adopted the Global Warming Solutions Act of 2006 to reduce its GHG emissions to 80% of the 1990 levels by 2050. One way to achieve GHG reduction is wide deployment of EVs. According to Electric Power Research Institute (2013), it is estimated that EVs would account for over 5% of new vehicle sales by 2020. Annual carbon dioxide ($CO_2$) reductions contributed by transportation electrification are expected to reach beyond 100 million tons by 2025 and 500 million tons by 2050.

The adoption of EVs will bring environmental and economic benefits including reducing GHG emissions, reducing dependence on fossil fuels, improving the

[1]Department of Electrical and Electronic Engineering, The University of Hong Kong, Hong Kong, China

stability and flexibility of the power system by utilizing the batteries installed in EVs as energy storage. The use of the batteries of grid-connected EVs for the provision of electric-power ancillary services is in the focus of this chapter.

The Federal Energy Regulatory Commission (FERC) defines ancillary services as those "necessary to support the transmission of electric power from seller to purchaser given the obligations of control areas and transmitting utilities within those control areas to maintain reliable operations of the interconnected transmission system." Hirst and Kriby (1996) outlined seven main types of ancillary services:

- scheduling and dispatch,
- load following,
- reliability,
- supplemental operating,
- energy imbalance,
- real-power loss replacement, and
- voltage control.

The EV fleets are able to provide ancillary services to the power grid by utilizing their batteries as distributed energy storage. Grid-connected EVs can charge or discharge their batteries to absorb or deliver power to the grid. The concept of V2G has been applied to describe such operation mode of EVs. V2G can be both unidirectional when EVs provide ancillary services by modulating their charging rates, and bidirectional when EVs are also allowed to discharge their batteries to inject energy back to the grid. V2G ancillary services can be used to refer to the ancillary services provided by V2G power. The economic benefits of V2G are compelling (Kempton and Tomić, 2005a). V2G can provide fast-response service because of the fast-response characteristics of battery power. It can smooth out the power fluctuations between generation and load, and is important for supporting the integration of renewable energy generation (Kempton and Tomić, 2005b).

### *14.1.2 Current research*

In order to provide V2G ancillary services, the charging/discharging power schedules of EVs should be coordinated or controlled properly. Research on charging/discharging control of EVs can be categorized as centralized control and distributed control. The aggregator of EVs acts as the central controller in centralized control (Han *et al.*, 2010; Sortomme and El-Sharkawi, 2012). The scalability issue is the main drawback of centralized control; the high computational complexity of the centralized control algorithms becomes a burden against efficient and prompt operations of the system as the number of EVs scales up.

Distributed control enjoys the advantage of high scalability over centralized control. It can be classified as fully distributed control and consensus-algorithm-based distributed control. In fully distributed control, no central node is involved in the decision-making process of the distributed EV nodes. For provision of frequency regulation, fully distributed control strategies based on local frequency measurement have been proposed (Ota *et al.*, 2012; Yang *et al.*, 2013). However, those fully distributed strategies cannot achieve the global optimum among the EV fleet since the control is for a single EV and the discharging/charging schedules of EVs are not coordinated.

In consensus-algorithm-based distributed control, a central node acts as the consensus manager to coordinate the distributed nodes in the decision-making process. Decisions are made locally by the distributed nodes. Consensus has been studied in system control area (Olfati-Saber and Murray, 2004). Consensus-algorithm-based distributed control has been applied to coordinated EV charging (Gan *et al.*, 2013; Ma *et al.*, 2010). However, only a few proposals apply consensus-algorithm-based distributed control to solve the control problem of V2G ancillary services.

The system for providing V2G ancillary services has been structured as a hierarchical or multilevel architecture. However, most existing work on charging/discharging control of V2G ancillary services are designed for a system with only one aggregator to coordinate the schedules of EVs. In existing solutions that consider multiple aggregators, the control algorithms of the solutions can only support one level of aggregators (Yao *et al.*, 2013). Therefore, a method that can support the charging/discharging control for a multilevel V2G system would be of value.

Although EVs can earn rewards for their EV owners/users by providing ancillary services, the fundamental function of EVs is for transportation. The charging needs of EVs must be met when the EVs are plugged in and provide ancillary services. The existing work models the charging needs as a desired value of the state-of-charge (SOC) of the batteries (Ota *et al.*, 2012). However, it would be more realistic to model the charging needs as a desired range of SOC. In addition, other QoS guarantees of the EVs should be taken into account, such as control of the battery degradation due to providing ancillary services.

Based on the discussion above, it is noted that none of the existing solutions are designed for the distributed charging/discharging control based on the consensus algorithm that supports a multilevel V2G system. Our goal is to develop such a system.

## 14.2 System architecture

In this section, we propose a multilevel system architecture, as shown in Figure 14.1, for the provision and operation of bidirectional V2G frequency regulation service. A preliminary proposal of the architecture without the operation protocol design can be found in our prior work (Lin *et al.*, 2013).

The system, consists of three key components: the utility grid operator, aggregators, and many EVs. The V2G system allows bidirectional power flow, namely, EVs can not only consume electric energy from the power grid by charging their batteries but also deliver electric energy back to the grid by discharging their batteries. Allowing discharging of the batteries would increase the number of charging/discharging cycles and therefore shorten the longevity of the batteries. Hence, in practice, some of the EV owners/users may not allow their EVs to provide bidirectional V2G ancillary services. To take this situation into consideration,

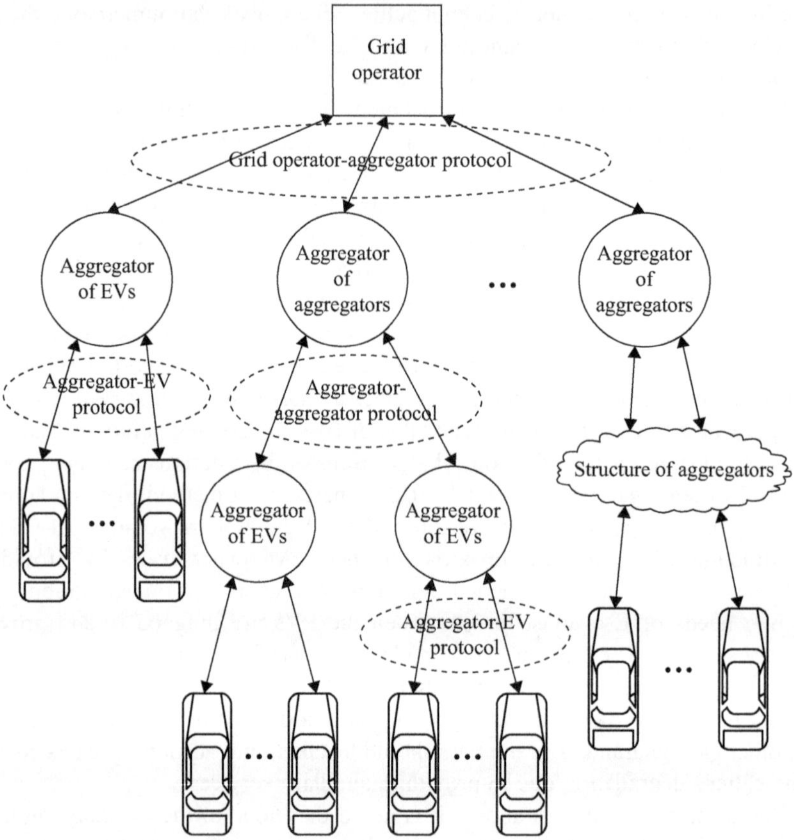


*Figure 14.1 System architecture. © 2014 IEEE. Reprinted with permission from Lin et al. (2014)*

a participation ratio of bidirectional V2G, $\theta_{bi} \in \{0, 1\}$, is used to denote the proportion of the EVs that participate in bidirectional V2G in the system. The subscript "bi" of $\theta_{bi}$ is short for "bidirectional." The system has a hierarchical structure with multiple levels of nodes. The utility grid operator is the root node. The aggregators directly connected to the grid operator are called level-1 aggregators. An aggregator directly connected to a group of level-$(l+1)$ aggregators is called a level-l aggregator, where $l = 1,\ 2, \cdots, N_L$ and $N_L$ is the number of aggregator levels in the system. For convenience, the aggregators directly connected to EVs are called aggregators of EVs. Correspondingly, all the other aggregators are called aggregators of aggregators. Each aggregator node can be viewed as the "root node" of a subtree of aggregators and EVs. Each aggregator node is uniquely numbered according to its location in the multilevel architecture. As shown in Figure 14.1, a level-$i$ aggregator that is the $j$-th aggregator of all level-$i$ aggregator (numbered from left to right in Figure 14.1) is labeled as $A_{ij}$, where $1 \le i \le N_L$. The aggregators at the same level may have different numbers of sub-levels. The size of a subtree is determined by the size of its subordinate EV fleets and other geographical, economic, and/or technical factors, such as the communication radius, delay, and cost between nodes at different levels. For instance, a parking lot or a certain area of a large parking lot can install an aggregator of EVs. A number of such parking areas can be controlled by an aggregator of aggregators.

Three types of operation protocols for control and communication between neighboring levels, namely, operator-aggregator protocol, aggregator-aggregator protocol, and aggregator-EV protocol are designed. Each protocol operates between a node and its immediate subordinate nodes as illustrated in Figure 14.1. The operator-aggregator protocol governs how the grid operator assigns the requests for ancillary services to and coordinates the level-1 aggregators to meet the demand of ancillary services. The requests for ancillary services are the aggregators' shares for the demand of ancillary services according to their signed contracts for providing ancillary services. The aggregator-aggregator protocol governs how an aggregator of aggregators coordinates its immediate subordinate aggregator to meet requests for ancillary services. The aggregator-EV protocol specifies the process and algorithm for an aggregator of EVs to coordinate its connected EVs to decide their charging/discharging schedules. In other words, the aggregators act as the interface between the utility grid and EV fleets so that the grid operator does not need to care about the individual charging/discharging profiles of EVs. These EVs can collectively form a massive energy storage system to provide ancillary services.

Consensus-algorithm-based distributed control is applied in the three types of operation protocols. Consensus algorithm requires an iterative process to reach the final solution. Figure 14.2 shows a simple example of a consensus-algorithm-based distributed control. In each round of iteration, the consensus manager calculates a consensus result according to the information states or decisions received from the local controllers and broadcasts the result to the local controllers as the control signal. Then, every local controller adjusts its information state or decision based on the consensus result and reports its decision to the consensus manager. The iterative process ends when the predetermined stopping criteria are met.

The three types of operation protocols have a nested relation as shown in Figure 14.3. The operator-aggregator includes processes of aggregator-aggregator

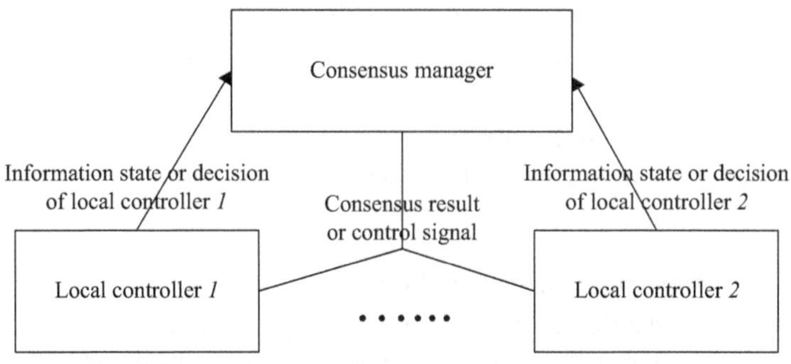


*Figure 14.2 A simple example of a consensus-algorithm-based distributed control*

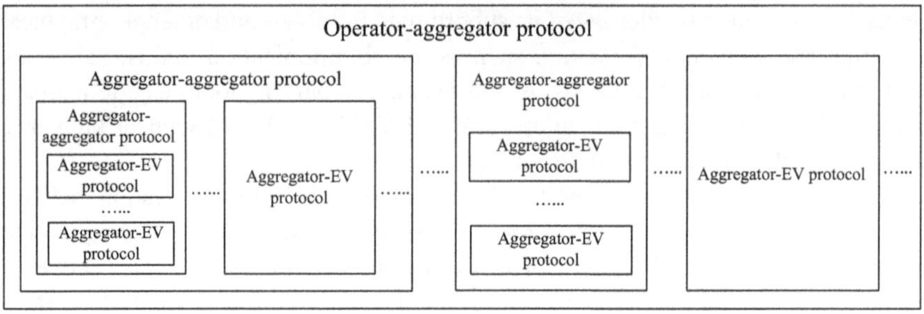


*Figure 14.3 The nested relation of three types of operation protocols*

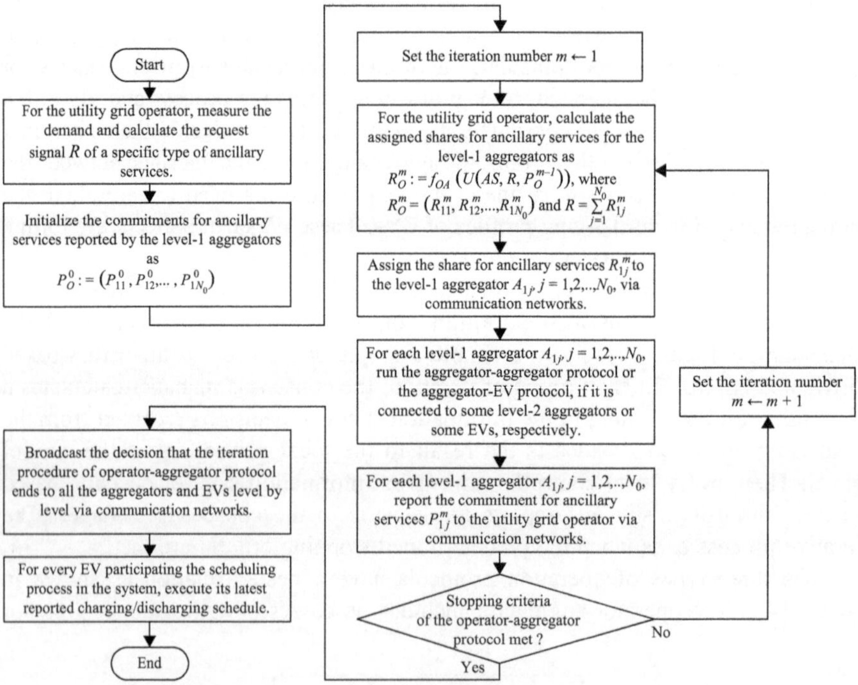


*Figure 14.4 Operator-aggregator protocol*

protocol. The aggregator-aggregator protocol includes processes of aggregator-aggregator protocol or aggregator-EV protocol, which are illustrated in the following subsections.

## *14.2.1 Operator-aggregator protocol*

Figure 14.4 shows the flowchart of the operator-aggregator protocol. The operator-aggregator protocol operates between the utility grid operator and the level-1 aggregators. The overall scheduling process of the multilevel V2G system starts from the operator-aggregator protocol. In the iterative part of the operator-aggregator protocol, the grid operator should wait for its subordinate node – the level-1 aggregators to run the aggregator-aggregator protocol or the aggregator-EV protocol. The notation of parameters and variables in Figure 14.4 is explained as follows. $N_0$ is the number of the level-1 aggregators. The subscript $j = 1, 2, \cdots, N_0$ in the assigned share $R_{1j}$ and the reported commitment $P_{1j}$ for ancillary services of aggregator $A_{1j}$ indicates that $A_{1j}$ is the $j$-th level-1 aggregator (numbered from left to right in Figure 14.1). The request signal $R$, the arrays of the assigned shares $R_O$, and the reported commitments $P_O$ are generic formulations, i.e., their actual forms and dimensions are determined by the specific type of ancillary services the system is currently concerned. The function $f_{OA}$ is for calculating the consensus result or control signal, i.e., the shares $R_O$ for ancillary services, according to the latest information states or decisions, i.e., the commitments $P_O$ for ancillary services, reported by the level-1 aggregators. The stopping criteria can be based on the number of iterations performed, and/or the convergence of the control signal $R_O$ within the convergence tolerance. When the operator-aggregator protocol finishes, the overall scheduling process of the V2G system ends and all the EVs participating in the scheduling process should execute their latest reported charging/discharging schedules so as to provide the specified type of ancillary services AS to the power grid.

### *14.2.2 Aggregator-aggregator protocol*

Figure 14.5 shows the flowchart of the aggregator-aggregator protocol. The aggregator-aggregator protocol operates between an aggregator of aggregators and its immediate subordinate aggregators. Figure 14.5 presents the operation protocol for a level-$i$ aggregator $A_{ij}$ and its subordinate level-$(i+1)$ aggregators, where $1 \leq i \leq N_L$. As indicated in Figures 14.2–14.5, a (level one) aggregator and its subordinate aggregators start operating the aggregator-aggregator protocol when the grid operator assigns the share for ancillary services to the level-1 aggregator. A level-$i$ aggregator, where $2 \leq i \leq N_L$, and its subordinate aggregators start operating the aggregator-aggregator protocol when the upper level aggregator of the level-$i$ aggregator assigns the share for ancillary services to the level-$i$ aggregator in the upper level aggregator-aggregator protocol. For the convenience of notation, the subordinate level-$(i+1)$ aggregators of the level-$i$ aggregator $A_{ij}$ are denoted as $a_{(i+1)k}$, $k = 1, 2, \cdots, N_{ij}$, in the aggregator-aggregator protocol, where $N_{ij}$ is the number of the subordinate level-$(i+1)$ aggregators of $A_{ij}$. The assigned share and the reported commitment for ancillary services of aggregator $a_{(i+1)k}$, $k = 1, 2, \cdots, N_{ij}$, are denoted as $R_{(i+1)k}$ and $P_{(i+1)k}$, respectively. The function $f_{AA}$ is for calculating the shares $R_{A_{ij}}$ according to the latest reported commitment $P_{A_{ij}}$ for ancillary services of the subordinate level-$(i+1)$ aggregators of $A_{ij}$. The stopping criteria can be based on the number of iterations performed, and/or the convergence of the control signal $R_{A_{ij}}$ within the convergence tolerance. By the end of the aggregator-aggregator protocol, $A_{ij}$ should calculate and report its commitment for ancillary services $P_{ij}$ to its upper level node (the grid operator or a level-$(i-1)$ aggregator) based on the commitments $P_{A_{ij}}$ reported by its subordinate level-$(i+1)$ aggregators.

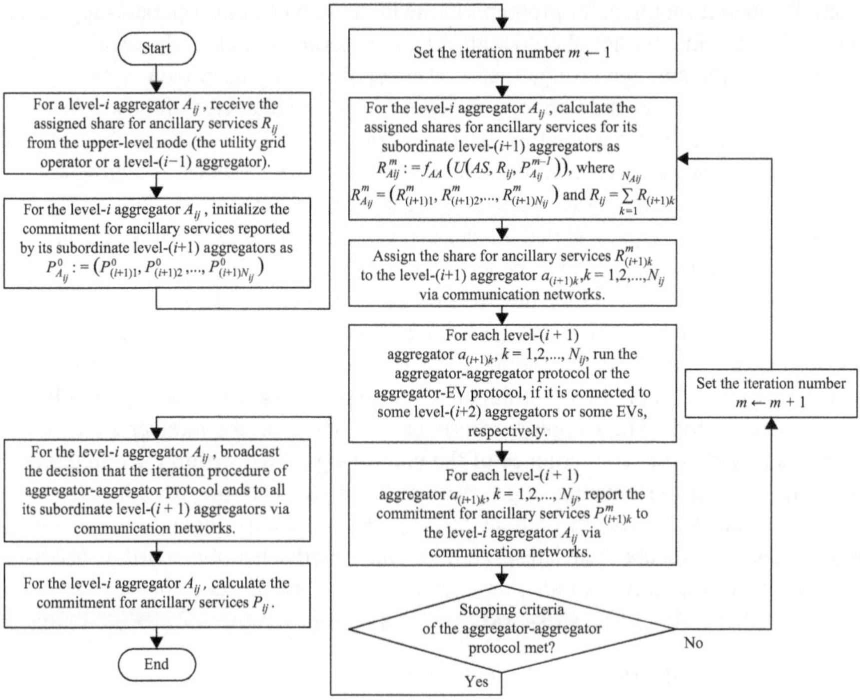


*Figure 14.5 Aggregator-aggregator protocol*

## 14.2.3 Aggregator-EV protocol

Figure 14.6 shows the flowchart of the aggregator-EV protocol. The aggregator-EV protocol operates between an aggregator of EVs and its subordinate EVs. Figure 14.6 presents the operation protocol for a level-$i$ aggregator $A_{ij}$ and its subordinate EVs, where $1 \leq i \leq N_L$. As indicated in Figures 14.2–14.6, the aggregator-EV protocol is triggered by either the operator-aggregator protocol or the aggregator-aggregator protocol. The aggregator-EV protocol is different from the other two types of operation protocols in that it directly tackles the problem of determining the charging/discharging schedules of the EVs. The subordinate EVs of the aggregator $A_{ij}$ are denoted as $EV^{(ij)}_n n = 1, 2, \cdots, N_{ij}$, where $N_{ij}$ is the number of $A_{ij}$'s subordinate EVs. $EV^{(ij)}_n$'s charging/discharging schedule $Q_n$ refers to the specific values of power, time, and duration that EV will charge or discharge its battery at. The function $f_{AE}$ is for calculation of the consensus result or control signals in the aggregator-EV protocol. The function $f_{AE}$ is a function of the EVs' aggregate schedule $Q_A$ rather than every individual EV's schedule, since the aggregator only cares about the collective charging/discharging behaviors of the EVs when it coordinates them to meet the assigned share $R_{ij}$ for ancillary services. The individual QoS requirement for battery charging/discharging of EVs should be taken into account when the EVs decide their own schedules. The stopping criteria

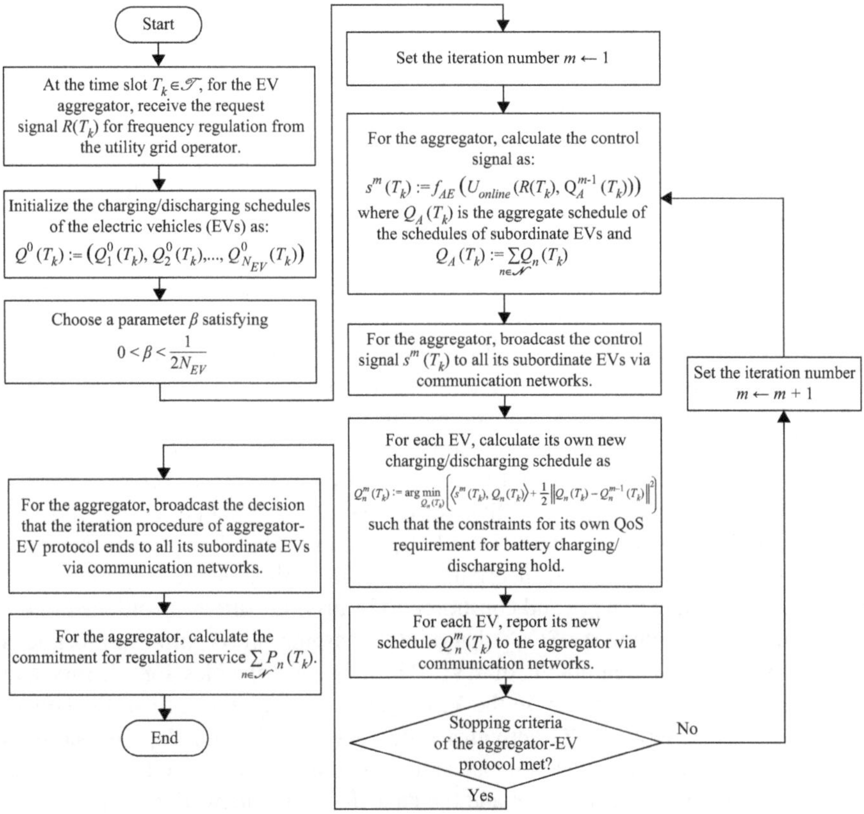


*Figure 14.6 Aggregator-EV protocol*

can be based on the number of iteration performed, and/or the convergence of the control signals within the convergence tolerance. By employing the aggregator-EV protocol, $A_{ij}$ can calculate and report its commitment for ancillary services $P_{ij}$ to its upper level node (the grid operator or a level-$(i-1)$ aggregator) based on the schedules $Q$ reported by its subordinate EVs.

### 14.2.4 EV requirements

When an EV is plugged in and operating in the V2G mode, meeting the QoS requirement for battery charging/discharging set by its user, rather than the requests for ancillary services from the utility grid operator, should be the first priority. The QoS requirements should include but are not limited to:

- Charging needs: An EV must get charged with adequate amount of electricity specified by its user before it is plugged out.
- Smart charging/discharging: An EV's charging/discharging strategy should consider the SOC of its battery. It should avoid deep discharging and over-charging of

the battery and should reduce the number of battery charging/discharging cycles while providing ancillary services so as to prolong the battery life.
- Security and privacy: The driving patterns and records of rewards related to the provision of ancillary services by EV users should remain private.

The first requirement in the list above is the primary purpose of EV charging. The provision of ancillary services should never impair the fundamental function, i.e., transporting people or goods, of EVs. Therefore, EV users should be guaranteed that their EVs will be adequately charged before being plugged out. The first requirement can be incorporated into the optimal scheduling process as a constraint. A smart charging/discharging strategy mentioned in the second requirement should be able to achieve better utilization of the battery storage in EVs, e.g., storing electricity in advance to cope with the variability and uncertainty of the demand of ancillary services, and protect the battery life from being severely harmed by charging/discharging frequently. The smart charging/discharging strategy should synthesize the EVs' capability and availability of power and energy, i.e., the output power limits, capacity, and SOC of the battery, the amount of electricity to charge and planned plug-out time of the EVs, etc., to decide the optimal schedules. The second requirement can be incorporated into the optimal scheduling process by the design of the function $f_{AE}$ and constraints. The driving patterns mentioned in the third requirement include the SOC, charging/discharging power, time and locations of plugging-in and plugging-out of EVs. Three basic principles for security and protection of EV users' privacy in the V2G system are suggested. First, the collection of user data during V2G activities should ask for the users' permissions in advance. Second, an authentication mechanism should be established to ensure legal and safe access to the V2G system and the data. Third, the network security system should be robust against cyber-attacks and prevent the data from being intercepted during transmission.

## 14.3 System model and problem formulation

### *14.3.1 Control objective for V2G regulation service*

The existing control schemes (Donadee and Ilic, 2014; Han *et al.*, 2011; Ota *et al.*, 2009, 2012; Shi and Wong, 2011; Wang *et al.*, 2013; Yang *et al.*, 2013) for the V2G regulation service seek to offset the regulation demand with the charging/discharging power of EVs so that the sum of the regulation demand and the aggregate power of EVs is zero. However, since frequency regulation is a zero-energy service (Kirby, 2004), the charging needs of EVs, which consume energy, is unlikely to be met merely by offsetting the regulation demand.

This section generalizes the optimal scheduling method for the aggregator-EV protocol proposed in our prior work (Lin *et al.*, 2014). The imbalance between generation and load originates from their respective uncertainties. Given that such imbalance cannot be fully compensated by EVs, we propose to schedule the charging/discharging power of EVs to absorb the uncertainties of generation and load.

In other words, EVs are employed to smooth out the power imbalance fluctuations of the grid by minimizing the variance of the profile of the total power, which is the sum of the regulation demand and the aggregate power of EVs.

From an economic perspective, the proposed control objective is to minimize the costs for the system to meet the regulation demand and the charging needs of EVs. Without V2G or other storage technologies, conventional generators have to be used to supply frequency regulation at a great expense, including the high ramping costs and the lost opportunity costs of the underutilized regulation capacity (Kirby, 2004). By implementing the proposed control objective, the random fluctuations of the regulation demand are absorbed by EVs. Therefore, the resultant total power of the regulation demand and the powers of EVs can be met by conventional generators with minimal ramping costs, and the required generation reserve for the regulation service is minimized as well.

## *14.3.2 Models and constraints*

The coordinated scheduling, which determines the charging/discharging schedule of EVs, is the core of the aggregator-EV protocol. Consider a scenario in which an aggregator of EVs coordinates $N_{EV}$ EVs to schedule their charging/discharging profiles so as to provide the V2G regulation service over a participation period $[T_{begin}, T_{end}]$, which is divided equally into $N_T$ time slots of length $\Delta t$. Let $\mathcal{T} = \{T_k | k = 1, 2, \ldots, N_T\}$ be the set of the slotted participation period, $R(T_k)$ be the assigned share of the actual regulation demand, i.e., the actual regulation request, at time slot $T_k$, and $P_n(T_k)$ be the charging/discharging power of EV $n$ at $T_k$, for $T_k \in \mathcal{T}$ and $n \in \mathcal{N} = \{1, 2, \ldots, N_{EV}\}$. We assume that the share of the regulation demand of an aggregator accounts for a fixed proportion of the total regulation demand in the grid during $\mathcal{T}$. $R(T_k) > 0$ means that the grid calls for regulation up due to generation shortfall. Similarly, when $R(T_k) < 0$, regulation down is required to absorb excessive power from the grid. When $P_n(T_k) > 0$, EV $n$ is charging. When $P_n(T_k) < 0$, it is discharging and delivering power back to the grid. Each EV will provide either unidirectional or bidirectional V2G regulation service according to its contract to the aggregator. Let $\theta_{bi} \in [0, 1]$ denote the proportion of the EVs that participate in bidirectional V2G. Note that those EVs participating in unidirectional V2G will not discharge their batteries to provide regulation up.

Denote the plug-in time and plug-out time of EV $n$ as $T_{n,in}$ and $T_{n,out}$, respectively. When EV $n$ is plugged in, its charging/discharging power should follow:

$$P_{n,discharge} \leq P_n(t) \leq P_{n,charge}, t \in [T_{n,in}, T_{n,out}], \tag{14.1}$$

where $P_{n,discharge} \leq 0$ and $P_{n,charge} > 0$ denote the limits of discharging power and charging power of EV $n$, respectively.

Denote the profile of the actual regulation requests as $R(\mathcal{T}) = (R(T_1), R(T_2), \ldots, R(T_{N_T}))^T$, where $(\cdot)^T$ denotes transposition. Denote the charging/discharging schedule of EV $n$ as $P_n(\mathcal{T}) = (P_n(T_1), P_n(T_2), \ldots, P_n(T_{N_T}))^T$, and the schedules of all EVs as $P_{\mathcal{N}}(\mathcal{T}) = (P_1(\mathcal{T}), P_2(\mathcal{T}), \ldots, P_{N_{EV}}(\mathcal{T}))$. Then, the profile of the total power, which is the sum of the regulation requests and the aggregate power of EVs, is defined as follows:

$$P_{total}(\mathscr{T}) = R(\mathscr{T}) + P_A(\mathscr{T}), \tag{14.2}$$

where $P_A(\mathscr{T})$ denotes the profile of the aggregate power of EVs defined as:

$$P_A(\mathscr{T}) = \sum_{n \in \mathscr{N}} P_n(\mathscr{T}). \tag{14.3}$$

Let $SOC_{n,0}$, $SOC_n(T_k)$, and $C_n$ be the initial SOC, SOC at the end of $T_k$, and capacity of the battery pack of EV $n$, respectively. Considering the energy conversion efficiency between the power grid and the batteries of EVs, $SOC_n(T_k)$ can be calculated as:

$$SOC_n(T_k) = SOC_{n,0} + \frac{\Delta t}{C_n} \sum_{i=1}^{k} \eta(P_n(T_i))P_n(T_i), \tag{14.4}$$

where $\eta(x)$ calculates the energy conversion efficiency of a given charging/discharging power $x$ of an EV. Assume that the charging and discharging efficiencies are, respectively, identical among the EVs. Then, $\eta(x)$ is defined as:

$$\eta(x) = \begin{cases} \eta_{ch} & \text{if } x \geq 0 \\ \dfrac{1}{\eta_{dch}} & \text{if } x < 0 \end{cases}, \tag{14.5}$$

where $\eta_{ch}$ and $\eta_{dch}$ are the charging and discharging efficiencies of the EVs, respectively, and therefore we have:

$$0 < \eta_{ch}, \eta_{dch} \leq 1. \tag{14.6}$$

Two constraints for the SOC of the battery pack during the plug-in period of EV $n$, where $n \in \mathscr{N}$ are proposed as:

$$SOC_n(T_{N_T}) \geq SOC_{n,MinCh}, \tag{14.7}$$

$$SOC_{n,min} \leq SOC_n(T_k) \leq SOC_{n,max}, T_k \in \mathscr{T}. \tag{14.8}$$

$SOC_{n,MinCh}$ denotes the minimum value of SOC that EV $n$ needs to reach before it is plugged out. The constraint represented by (14.7) ensures that EV $n$ will have been charged up with enough energy for the next trip when it is plugged out. $SOC_{n,min}$ and $SOC_{n,max}$ denote the lower and upper SOC limits, respectively, of EV $n$ for all $T_k \in \mathscr{T}$. The constraint in (14.8) prevents deep discharging or over-charging of the battery so as to prolong the battery life.

### *14.3.3 Formulation of forecast-based scheduling*

Assume that, before the participation period $\mathscr{T}$, the aggregator receives the forecasting profile $R_f(\mathscr{T}) = (R_f(T_1), R_f(T_2), \ldots, R_f(T_{N_T}))^T$ of its actual regulation requests $R(\mathscr{T})$, and is able to communicate with $N_{EV}$ EVs that are going to participate in the V2G regulation service during $\mathscr{T}$. Then, according to the control objective proposed in Section 14.3.1, it should coordinate these $N_{EV}$ EVs to determine their optimal charging/discharging schedules $P_{\mathscr{N}}(\mathscr{T})$ by the following optimization problem:

$$\min_{P_{\mathscr{N}}(\mathscr{T})} U_f(P_A(\mathscr{T})) \tag{14.9}$$

such that $\forall n \in \mathscr{N}$, (14.1), (14.7), and (14.8) hold, where $U_f(P_A(\mathscr{T}))$ calculates the variance of the profile of the total power $P_{total}(\mathscr{T})$. Therefore, we have:

$$\begin{aligned} & U_f(P_A(\mathscr{T})) \\ & \quad = \mathrm{Var}(P_{total}(\mathscr{T})) \\ & \quad = \frac{1}{N_T}\sum_{T_i\in\mathscr{T}}(R_f(T_i) + P_A(T_i) - \frac{1}{N_T}\left(\sum_{T_j\in\mathscr{T}}(R_f(T_j) + P_A(T_j))\right)^2, \end{aligned} \tag{14.10}$$

where Var(·) denotes the function for calculating variance.

**Lemma 1** *The objective function (14.10), $U_f : \mathbb{R}^{N_T} \rightarrow \mathbb{R}$ of the optimization problem (14.9) is convex.*

*Proof.* $\forall 1 \le i,j \le N_T$, according to the first-order partial derivatives of $U_f(P_A(\mathscr{T}))$, $\frac{\partial U_f(P_A(\mathscr{T}))}{\partial P_A(\mathrm{T}_i)}$, the second-order derivatives are as follows:

$$\frac{\partial^2 U_f(P_A(\mathscr{T}))}{\partial P_A(T_i)^2} = \frac{2}{N_T^2}(N_T - 1), \tag{14.11}$$

$$\frac{\partial^2 U_f(P_A(\mathscr{T}))}{\partial P_A(T_i)\partial P_A(T_j)} = -\frac{2}{N_T^2}. \tag{14.12}$$

Thus,

$$\nabla^2 U_f(P_A(\mathscr{T})) = \frac{2}{N_T^2}\begin{bmatrix} N_T - 1 & -1 & -1 & \cdots & -1 \\ -1 & N_T - 1 & -1 & \cdots & -1 \\ -1 & -1 & \ddots & \ddots & \vdots \\ \vdots & \vdots & \ddots & N_{T-1} & -1 \\ -1 & -1 & \dots & -1 & N_T - 1 \end{bmatrix}. \tag{14.13}$$

Due to the symmetry of $\nabla^2 U_f(P_A(\mathscr{T}))$, it is obvious that all the principal minors of order $i$ are equal to the leading principal minor of the same order $i$, $i = 1, 2, \ldots, N_T$. Further, by applying mathematical induction, it can be proved that all the leading principal minors of $\nabla^2 U_f(P_A(\mathcal{T}))$ are non-negative. Hence, $\nabla^2 U_f(P_A(\mathcal{T}))$ is positive semidefinite. In addition, since the domain of $U_f$, $\mathbf{dom}\{U_f\} = R^{N_T}$, is convex, it follows from the second-order conditions for convex function (Boyd and Vandenberghe, 2004) that the function $U_f : \mathbb{R}^{N_T} \to \mathbb{R}$ is convex.

**Lemma 2** *The feasible set of the optimization problem (14.9) is convex.*

*Proof.* The feasible set of (14.9) is defined by the constraints (14.1), (14.7), and (14.8). It is obvious that the region defined by (14.1) is convex. The constraint (14.7) can be transformed into (14.14). The constraint (14.8) can be transformed into (14.15) and (14.16). Thus, $\forall n \in \mathcal{N}$, $k \in \{1, 2, \ldots, N_T\}$ we have:

$$\sum_{i=1}^{N_T} \eta(P_n(T_i))P_n(T_i) \geq \frac{C_n}{\Delta t}(SOC_{n,MinCh} - SOC_{n,0}), \tag{14.14}$$

$$\sum_{i=1}^{k} \eta(P_n(T_i))P_n(T_i) \geq \frac{C_n}{\Delta t}(SOC_{n,min} - SOC_{n,0}), \tag{14.15}$$

$$\sum_{i=1}^{k} \eta(P_n(T_i))P_n(T_i) \leq \frac{C_n}{\Delta t}(SOC_{n,max} - SOC_{n,0}). \tag{14.16}$$

We prove that the region defined by (14.14), which equivalent to (14.7), is convex. Assume that $P_{n,1}(\mathcal{T}), P_{n,2}(\mathcal{T})$ are any two points that satisfy (14.14). $\forall 0 \leq \lambda_1, \lambda_2 \leq 1, \lambda_1 + \lambda_2 = 1$, we denote:

$$P_{n,3}(\mathcal{T}) = \lambda_1 P_{n,1}(\mathcal{T}) + \lambda_2 P_{n,2}(\mathcal{T}). \tag{14.17}$$

It can be shown that the following inequality holds $\forall T_i \in \mathcal{T}$,

$$\eta(P_{n,3}(T_i))P_{n,3}(T_i) \geq \lambda_1 \eta(P_{n,1}(T_i))P_{n,1}(T_i) + \lambda_2 \eta(P_{n,2}(T_i))P_{n,2}(T_i). \tag{14.18}$$

It follows from (14.18) that $P_{n,3}(\mathcal{T})$ also satisfies (14.14). Hence, the feasible set defined by (14.14) is convex. Similarly, it can be shown that, $\forall k \in \{1, 2, \ldots, N_T\}$, the feasible set defined by (14.15) is also convex.

When $k = 1$, it is obvious that the region defined by (14.16) is convex. Then, by applying mathematical induction, it can be shown that $\forall k \in \{1, 2, \ldots, N_T\}$, the region defined by (14.16) is also convex.

To conclude, the region defined by each of the constraints (14.1), (14.7), (14.8) is convex. Therefore, the feasible set of (14.9) is convex.

**Theorem 1** *The optimization problem in* (14.9) *is a convex optimization problem.*

*Proof.* According to Lemmas 1 and 2, it follows from the definition of convex optimization problem (Boyd and Vandenberghe, 2004) that (14.9) is a convex optimization problem. □

The solution of the forecast-based scheduling problem in (14.9) provides the best possible schedules $P_{\mathcal{N}}(\mathcal{T})$ if and only if the forecasting profile of regulation requests $R_f(\mathcal{T})$ is accurate, i.e. $R_f(\mathcal{T}) = R(\mathcal{T})$. However, in reality, the forecast of the regulation demand is highly inaccurate and vulnerable to forecasting errors of generation and load. Hence, the forecast-based scheduling in (14.9) is not practical.

### *14.3.4 Formulation of online scheduling*

In practice, the regulation demand is derived from the regulation signals measured in real time. Hence, it is more realistic to adopt online scheduling, which schedules the charging/discharging power of EVs in response to the real-time input of a regulation request. Consider a scenario of online scheduling where at each time slot $\mathrm{T}_k \in \mathcal{T}$, the aggregator receives the real-time signal of a regulation request $R(T_k)$ and then coordinates the EVs to update their charging/discharging schedules from $T_k$ to $T_{N_T}$, i.e. $\{P_n(T_j)|n \in \mathcal{N}, k \leq j \leq N_T\}$, so that $\mathrm{Var}(P_{total}(\mathcal{T}))$ is minimized. We have:

$$\begin{aligned}\mathrm{Var}(P_{total}(\mathcal{T})) = &\frac{1}{N_T}\sum_{i=1}^{k}\left(R(T_i) + \sum_{n\in\mathcal{N}} P_n(T_i) - \mathrm{A}(P_{total}(\mathcal{T}))\right)^2 \\ &+\frac{1}{N_T}\sum_{j=k+1}^{N_T}(R_f(T_j) + \sum_{n\in\mathcal{N}} P_n(T_j) - \mathrm{A}(P_{total}(\mathcal{T})))^2,\end{aligned} \tag{14.19}$$

where $\mathrm{A}(P_{total}(\mathcal{T}))$ denotes the average of $P_{total}(\mathcal{T})$ as:

$$\mathrm{A}(P_{total}(\mathcal{T})) = \frac{1}{N_T}\left(\sum_{i=1}^{k} R(T_i) + \sum_{j=k+1}^{N_T} R_f(T_j) + \sum_{T_l\in\mathcal{T}}\sum_{n\in\mathcal{N}} P_n(T_l)\right). \tag{14.20}$$

From (14.19), the forecasts of the future regulation requests $R_f(T_j), j = k+1, k+2, \ldots, T_k$, can be approximated by:

$$R_f(T_j) = \mathbb{E}(R(T_j)|\{R(T_i)|1 \leq i \leq k\}). \tag{14.21}$$

However, the calculation of the conditional expectation in (14.21) requires the distribution of the regulation demand which is not known a priori. Nonetheless, since frequency regulation is a zero-energy service, the expectation of the total energy that the regulation service requires is zero over a long period of time. Therefore, we can make the following assumption:

$$\mathbb{E}(R_S(\mathcal{T})) = 0, \tag{14.22}$$

where

$$R_S(\mathcal{T}) = \sum_{T_i\in\mathcal{T}} R(T_i). \tag{14.23}$$

From (14.21) and (14.22), we have:

$$\sum_{j=k+1}^{N_T} R_f(T_j) = \mathbb{E}\left( \sum_{j=k+1}^{N_T} R(T_j) | \{R(T_i)|1 \le i \le k\} \right) = -\sum_{i=1}^{k} R(T_i). \tag{14.24}$$

By applying the Cauchy-Schwarz Inequality, we can derive a lower bound of the second summation in (14.19) when $k \le N_T - 1$ as follows:

$$\frac{1}{N_T}\sum_{j=k+1}^{N_T}\left( R_f(T_j) + \sum_{n\in\mathcal{N}} P_n(T_j) - \mathrm{A}(P_{total}(\mathcal{T})) \right)^2$$
$$\ge \frac{1}{N_T(N_T-k)}\left( \sum_{j=k+1}^{N_T} R_f(T_j) + \sum_{n\in\mathcal{N}}\sum_{j=k+1}^{N_T} P_n(T_j) - (N_T-k)\mathrm{A}(P_{total}(\mathcal{T})) \right)^2. \tag{14.25}$$

The equality of (14.25) holds if and only if the following condition is satisfied:

$$\forall k+1 \le j, l \le N_T, R_f(T_j) + P_A(T_j) = R_f(T_l) + P_A(T_l). \tag{14.26}$$

The condition (14.26) also minimizes (14.19). Therefore, the lower bound derived in (14.25) can be used to approximate (14.19) since we seek to minimize the variance of $P_{total}(\mathcal{T})$.

Based on (14.24) and (14.25), an approximation of $\mathrm{Var}(P_{total}(\mathcal{T}))$ in (14.19) for $k \le N_T - 1$ is derived and used as the objective function for the online scheduling problem as follows:

$$\begin{aligned} U_o(Q_A(T_k)) =& \frac{1}{N_T}\sum_{i=1}^{k}\left( R(T_i) + \sum_{n\in\mathcal{N}} P_n(T_i) \right)^2 \\ &+ \frac{\alpha(k)}{N_T}\left( -\sum_{i=1}^{k} R(T_i) + \sum_{n\in\mathcal{N}} FP_n(T_k) \right)^2 \\ &- \frac{1}{N_T^2}\left( \sum_{n\in\mathcal{N}}\left( \sum_{i=1}^{k} P_n(T_i) + FP(T_k) \right) \right)^2, \end{aligned} \tag{14.27}$$

where

$$Q_A(T_k) = (P_A(T_k), FP_A(T_k))^T = \sum_{n\in\mathcal{N}} Q_n(T_k), \tag{14.28}$$

and

$$\alpha(k) = \begin{cases} \dfrac{1}{N_T - k} & \text{if } 1 \le k \le N_T - 1 \\ 0 & \text{if } k = N_T \end{cases}. \tag{14.29}$$

In (14.28), $Q_n(T_k) = (P_n(T_k), FP_n(T_k))^T$ denotes the control variables or schedule of EV $n, \forall n \in \mathcal{N}$, at $T_k$, where $FP_n(T_k)$ is the sum of the future charging/discharging powers of EV $n$ as follows:

$$FP_n(T_k) = \sum_{i=k+1}^{N_T} P_n(T_i). \tag{14.30}$$

Note that the charging/discharging powers of the EVs before $T_k$, i.e., $\{P_n(T_i)|n \in \mathcal{N}, 1 \leq i \leq N_T\}$, are not included in the control variables since they are already historical data.

Denote the schedules of all EVs at $T_k$ as $Q_{\mathcal{N}}(T_k) = (Q_1(T_k), Q_2(T_k), \ldots, Q_{N_{EV}}(T_k))$. The online scheduling problem for the V2G regulation service is formulated as follows:

At each $T_k \in \mathcal{T}$,

$$\min_{Q_{\mathcal{N}}(T_k)} U_o(Q_A(T_k)) \tag{14.31}$$

such that for all $n \in \mathcal{N}$,

$$\eta(P_n(T_k))P_n(T_k) + \eta(FP_n(T_k))FP_n(T_k)$$

$$\geq \frac{C_n}{\Delta t}(SOC_{n,MinCh} + SOC_{n,MOS}(T_k) - SOC_n(T_{k-1})) \tag{14.32}$$

and (14.1), (14.8) hold.

From (14.32), $SOC_{n,MOS}(T_k)$ denotes the SOC "margin of safety" of EV $n \in \mathcal{N}$ defined as follows:

$$SOC_{n,MOS}(T_k) = \begin{cases} \mu(SOC_{n,max} - SOC_{n,MinCh}) & T_k \in [T_{n,in}, (1-\tau)T_{n,in} + \tau T_{n,out})] \\ 0 & \text{otherwise} \end{cases}, \tag{14.33}$$

where $\mu \in [0, 1]$ quantifies the relative amount of the safety margin, and $\tau \in [0, 1]$ determines the ratio of the time that the safety margin is in effect. The parameters $\mu$ and $\tau$ are identical among the EVs.

The constraint (14.32) is derived from (14.7) with $SOC_{n,MOS}(T_k)$ added to the charging requirement of EV $n$, where $n = 1, 2, \ldots, N_{EV}$, in (14.7). The purpose of introducing the margin of safety for charging is to cope with the uncertainty of the regulation requests. Because the objective function (14.27) for the online scheduling problem in (14.31) approximates $\text{Var}(P_{total}(\mathcal{T}))$ in (14.19) based on the zero-sum assumption in (14.22) of the regulation requests $R(\mathcal{T})$, such approximation may be inaccurate when the assumption stated in (14.22) does not hold in some cases. By introducing the margin of safety for charging in (14.32), the EVs would buffer some more energy on top of their minimum charging requirements to meet the extra energy needs for regulation up, i.e., $R_S(\mathcal{T}) > 0$.

**Lemma 3** *For all $T_k \in \mathcal{T}$, the objective function (14.27), $U_o : \mathbb{R}^2 \to \mathbb{R}$ of the optimization problem (14.31) is convex.*

*Proof.* $\forall 1 \leq k \leq N_T - 1$, according to the gradient of $U_o(Q_A(T_k))$, $\nabla U_o(Q_A(T_k))$, the Hessian of $U_o(Q_A(T_k))$ is as follows:

$$\nabla^2 U_o(Q_A(T_k)) = \frac{2}{N_T^2} \begin{bmatrix} N_T - 1 & -1 \\ -1 & \dfrac{N_T}{N_T - k} - 1 \end{bmatrix}. \tag{14.34}$$

When $k = N_T$, the Hessian of $U_o(Q_A(T_{N_T}))$ is as follows:

$$\nabla^2 U_o(Q_A(T_{N_T})) = \frac{2}{N_T^2} \begin{bmatrix} N_T - 1 & 0 \\ 0 & 0 \end{bmatrix}. \tag{14.35}$$

It can be checked that both (14.34) and (14.35) are positive semidefinite. In addition, since the domain of $U_o$, $\mathbf{dom}\{U_o\} = R^2$, is convex, it follows from the second-order conditions for convex function (Boyd and Vandenberghe, 2004) that the function $U_o : \mathbb{R}^2 \to \mathbb{R}$ is convex. □

**Lemma 4** *For all $T_k \in \mathcal{T}$, the feasible set of the optimization problem (14.31) is convex.*

*Proof.* The feasible set of (14.31) is defined by the constraints (14.1), (14.8), and (14.32). Similar to the proof of the convexity of the region defined by (14.14) in Lemma 2, it can be shown that, for all $T_k \in \mathcal{T}$, the region defined by (14.32) is convex. In addition, it is obvious that the region defined by each of the constraints (14.1) and (14.8) is convex. Therefore, for all $T_k \in \mathcal{T}$, the feasible set of (14.31) is convex. □

**Theorem 2** *For all $T_k \in \mathcal{T}$, the optimization problem in (14.31) is a convex optimization problem.*

*Proof.* According to Lemmas 3 and 4, it follows from the definition of convex optimization problem (Boyd and Vandenberghe, 2004) that (14.31) is a convex optimization problem. □

Although the proposed online scheduling problem in (14.31) only optimizes an approximation of (14.19), it is more practical than the forecast-based scheduling problem in (14.9) because it does not require the forecasts of the regulation requests. In addition, (14.31) incurs much lower computational complexity than (14.9) since it reduces the number of control variables significantly.

## 14.4 Decentralized scheduling algorithm

In this section, two classes decentralized algorithms, namely, Algorithms 14.1 and 14.2, and Algorithms 14.3 and 14.4, are proposed to solve the forecast-based scheduling problem in (14.9) and the online scheduling problem in (14.31), respectively. They are inspired by the decentralized algorithm proposed for optimal EV charging control (Gan *et al.*, 2013) and based on the gradient projection method (Bertsekas and Tsitsiklis, 1989).

Since Algorithms 14.1 and 14.2 distribute the computational efforts to EVs, the aggregator only needs to perform simple arithmetic for calculating the control signals with a computational complexity equal to $O(N_{EV})$. In each round of the iterations, every EV should update its own schedule by solving an optimization problem. The computational complexity of an EV is $O(1)$ in terms of the scale of the set of EVs. Therefore, Algorithms 14.1 and 14.2 are highly scalable.

**Lemma 5** $\forall m \geq 1$, *the following inequality holds:*

$$\begin{aligned} &\langle \nabla U_f(P_A^{m-1}(\mathscr{T})) - \nabla U_f(P_A^m(\mathscr{T})), P_A^{m-1}(\mathscr{T}) - P_A^m(\mathscr{T})\rangle \\ &\leq \frac{2}{N_T}\|P_A^{m-1}(\mathscr{T}) - P_A^m(\mathscr{T})\|^2. \end{aligned} \tag{14.36}$$

*Proof.* According to the first-order partial derivatives of $U_f(P_A(\mathscr{T}))$ derived in (14.44), we have,

$$\begin{aligned} &\langle \nabla U_f(P_A^{m-1}(\mathscr{T})) - \nabla U_f(P_A^m(\mathscr{T})), P_A^{m-1}(\mathscr{T}) - P_A^m(\mathscr{T})\rangle \\ &= \frac{2}{N_T}\sum_{T_k\in\mathscr{T}} (P_A^{m-1}(T_k) - P_A^m(T_k))^2 - \frac{2}{N_T^2}\left(\sum_{T_j\in\mathscr{T}} (P_A^{m-1}(T_j) - P_A^m(T_j))\right)^2 \\ &\leq \frac{2}{N_T}\sum_{T_k\in\mathscr{T}} (P_A^{m-1}(T_k) - P_A^m(T_k))^2 \\ &= \frac{2}{N_T}\|P_A^{m-1}(\mathscr{T}) - P_A^m(\mathscr{T})\|^2. \qquad \square \end{aligned} \tag{14.37}$$

**Lemma 6** $\forall n \in \mathscr{N}, m \geq 1$, *the following inequality holds:*

$$\langle s_f^m(\mathscr{T}), P_n^m(\mathscr{T}) - P_n^{m-1}(\mathscr{T})\rangle \geq -\|P_n^m(\mathscr{T}) - P_n^{m-1}(\mathscr{T})\|^2. \tag{14.38}$$

*Proof.* See the proof of Lemma 1 of Gan *et al.* (2013). □

**Theorem 3** *In Algorithms 14.1 and 14.2, the schedules $P^m_{\mathcal{N}}(\mathcal{T})$ converge to one of the optimal solutions for the forecast-based scheduling problem in (14.9) as $m \to \infty$.*

*Proof.* $\forall m \geq 1$,

$$\begin{aligned}
&U_f(P^m_A(\mathcal{T})) \\
&\leq U_f(P^{m-1}_A(\mathcal{T})) - \langle \nabla U_f(P^m_A(\mathcal{T})), P^{m-1}_A(\mathcal{T}) - P^m_A(\mathcal{T})\rangle \\
&\leq U_f(P^{m-1}_A(\mathcal{T})) - \langle \nabla U_f(P^{m-1}_A(\mathcal{T})), P^{m-1}_A(\mathcal{T}) - P^m_A(\mathcal{T})\rangle \\
&\quad + \frac{2}{N_T}\|P^{m-1}_A(\mathcal{T}) - P^m_A(\mathcal{T})\|^2 \\
&= U_f(P^{m-1}_A(\mathcal{T})) - \frac{1}{\beta}\sum_{n\in\mathcal{N}}\langle s^m_f(\mathcal{T}), P^{m-1}_n(\mathcal{T}) - P^m_n(\mathcal{T})\rangle \\
&\quad + \frac{2}{N_T}\|P^{m-1}_A(\mathcal{T}) - P^m_A(\mathcal{T})\|^2 \\
&\leq U_f(P^{m-1}_A(\mathcal{T})) - \frac{1}{\beta}\sum_{n\in\mathcal{N}}\|P^{m-1}_n(\mathcal{T}) - P^m_n(\mathcal{T})\|^2 \\
&\quad + \frac{2}{N_T}\|P^{m-1}_A(\mathcal{T}) - P^m_A(\mathcal{T})\|^2 \\
&\leq U_f(P^{m-1}_A(\mathcal{T})) + \left(\frac{2}{N_T} - \frac{1}{\beta N_{EV}}\right)\|P^{m-1}_A(\mathcal{T}) - P^m_A(\mathcal{T})\|^2 \\
&\leq U_f(P^{m-1}_A(\mathcal{T})).
\end{aligned} \tag{14.39}$$

The first inequality holds due to the first-order condition (Boyd and Vandenberghe, 2004) of the convex function $U_f$. The second inequality is due to Lemma 5, the third inequality is due to Lemma 6, the fourth inequality is due to the Cauchy-Schwarz inequality, and the fifth inequality is due to $0 < \beta < \frac{N_T}{2N_{EV}}$.

According to (14.39), $U_f(P^m_A(\mathcal{T}))$ is nonincreasing as $m$ increases. Further, it is easy to check that $U_f(P^m_A(\mathcal{T})) = U_f(P^{m-1}_A(\mathcal{T}))$ if and only if $P^m_{\mathcal{N}}(\mathcal{T}) = P^{m-1}_{\mathcal{N}}(\mathcal{T})$. If $P^m_{\mathcal{N}}(\mathcal{T}) = P^{m-1}_{\mathcal{N}}(\mathcal{T})$, it follows from the proof of Theorem 3 of Gan *et al.*, (2013) that $P^m_{\mathcal{N}}(\mathcal{T})$ minimizes $U_f$. To conclude, $P^m_{\mathcal{N}}(\mathcal{T})$ minimizes $U_f$ as $m \to \infty$. □

**Lemma 7** $\forall m \geq 1, T_k \in \mathcal{T}$, *the following inequality holds:*

$$\begin{aligned}
&\langle \nabla U_o(Q^{m-1}_A(T_k)) - \nabla U_o(Q^m_A(T_k)), Q^{m-1}_A(T_k) - Q^m_A(T_k)\rangle \\
&\leq \frac{2}{N_T}\|Q^{m-1}_A(T_k) - Q^m_A(T_k)\|^2.
\end{aligned} \tag{14.40}$$

*Proof.* Similar to the proof of Lemma 5, the inequality (14.40) can be proved by simple derivation.

$\forall m \geq 1, T_k \in \mathscr{T}$, according to the gradient of $U_o(Q_A(T_k))$, $\nabla U_o(Q_A(T_k))$, derived in (14.48), we have,

$$
\begin{aligned}
&\langle \nabla U_o(Q_A^{m-1}(T_k)) - \nabla U_o(Q_A^m(T_k)), Q_A^{m-1}(T_k) - Q_A^m(T_k)\rangle \\
&\quad = \frac{2}{N_T}(P_A^{m-1}(T_k) - P_A^m(T_k))^2 + \alpha(k)\frac{2}{N_T}(FP_A^{m-1}(T_k) \\
&\qquad - FP_A^m(T_k))^2 - \frac{2}{N_T^2}(P_A^{m-1}(T_k) - P_A^m(T_k) + FP_A^{m-1}(T_k) - FP_A^m(T_k))^2 \\
&\quad \leq \frac{2}{N_T}((P_A^{m-1}(T_k) - P_A^m(T_k))^2 + (FP_A^{m-1}(T_k) - FP_A^m(T_k))^2) \\
&\qquad - (1-\alpha(k))\frac{2}{N_T}(FP_A^{m-1}(T_k) - FP_A^m(T_k))^2 \\
&\quad \leq \frac{2}{N_T}\|Q_A^{m-1}(T_k) - Q_A^m(T_k)\|^2.
\end{aligned}
\tag{14.41}
$$

The second equality is due to $\alpha(k) \leq 1$. □

**Lemma 8** *$\forall n \in \mathscr{N}, m \geq 1, T_k \in \mathscr{T}$, the following inequality holds:*

$$\langle s_o^m(T_k), Q_n^m(T_k) - Q_n^{m-1}(T_k)\rangle \geq -\|Q_n^m(T_k) - Q_n^{m-1}(T_k)\|^2 \tag{14.42}$$

*Proof.* See the proof of Lemma 1 of Gan *et al.* (2013). □

**Theorem 4** *In Algorithms 14.3 and 14.4, at any time slot $T_k \in \mathscr{T}$, the schedules $Q_{\mathscr{N}}^m(T_k)$ converge to one of the optimal solutions for the online scheduling problem in (14.31) as $m \to \infty$.*

*Proof.* Similar to the proof of Theorem 3, by applying the first-order condition of the convex function $U_o$, Lemmas 7 and 8, and the Cauchy-Schwarz inequality, successively, it can be shown that, $\forall T_k \in \mathscr{T}$, $U_o(Q_A^m(T_k))$ is nonincreasing as $m$ increases. It can be checked that $U_o(Q_A^m(T_k)) = U_o(Q_A^{m-1}(T_k))$ if and only if $Q_A^m(T_k) = Q_A^{m-1}(T_k)$, and such $Q_A^m(T_k)$ minimizes $U_o$. To conclude, $Q_A^m(T_k)$ minimizes $U_o$ as $m \to \infty$. □

In Algorithms 14.1 and 14.2, $\langle\cdot,\cdot\rangle$ represents the dot product operation and $\|\cdot\|$ denotes the Euclidean norm.

**Algorithm 14.1** *Forecast-based scheduling for aggregator*
***Input:*** *The participation period $\mathscr{T}$. Before the participation period $\mathscr{T}$ starts, the aggregator receives the forecasting profile of the regulation requests $R_f(\mathscr{T})$ and knows the number of EVs $N_{EV}$.*
***Output:*** *The charging/discharging schedules of EVs $P_{\mathscr{N}}(\mathscr{T}) = (P_1(\mathscr{T}), P_2(\mathscr{T}), \ldots, P_{N_{EV}}(\mathscr{T}))$.*

*Choose a parameter* $\beta$ *such that* $0 < \beta < \frac{N_T}{2N_{EV}}$.
*Wait for the initial schedule* $P_n^0(\mathcal{T})$ *of every EV* $n \in \mathcal{N}$.
*Set the iteration number* $m \leftarrow 1$, *repeat Steps 1–3.*

1. *Calculate the control signal* $s_f^m(\mathcal{T}) = (s_f^m(T_1), s_f^m(T_2), \ldots, s_f^m(T_{N_T}))^T$ *as follows:*

$$s_f^m(\mathcal{T}) = \beta \nabla U_f(P_A^{m-1}(\mathcal{T})) \tag{14.43}$$

*Therefore* $\forall T_k \in \mathcal{T}$,

$$s_f^m(T_k) = \beta \frac{\partial U_f(P_A^{m-1}(\mathcal{T}))}{\partial P_A(T_k)} = \frac{2\beta}{N_T}(R_f(T_k) + \sum_{n \in \mathcal{N}} P_n^{m-1}(T_k)) - \frac{2\beta}{N_T^2}\left(\sum_{T_j \in \mathcal{T}} (R_f(T_j) + \sum_{n \in \mathcal{N}} P_n^{m-1}(T_j))\right) \tag{14.44}$$

*Broadcast the control signal* $s_f^m(\mathcal{T})$ *to all EVs.*

2. *Wait for the updated schedule* $P_n^m(\mathcal{T})$ *reported by every EV* $n \in \mathcal{N}$.
3. *If the stopping criteria are not met, set* $m \leftarrow m+1$ *and go to Step 1. Otherwise, broadcast the message that the iteration process ends to all EVs.*

***Return*** $P_{\mathcal{N}}(\mathcal{T}) = P_{\mathcal{N}}^m(\mathcal{T})$.

**Algorithm 14.2** *Forecast-based scheduling for each EV* $n \in \mathcal{N}$
***Input:*** *The participation period* $\mathcal{T}$. *EV* $n \in \mathcal{N}$ *knows its own constraints (14.1), (14.7), and (14.8).*
***Output:*** *The charging/discharging schedule of EV n,* $P_n(\mathcal{T})$.

*Initialize the schedule* $P_n^0(\mathcal{T})$ *such that* $P_n^0(\mathcal{T})$ *lies in the boundary of the region defined by the constraint (14.7). Then report* $P_n^0(\mathcal{T})$ *to the aggregator. Set the iteration number* $m \leftarrow 1$, *repeat Steps 1–3.*

1. *Wait for the updated control signal* $s_f^m(\mathcal{T})$ *broadcast by the aggregator.*
2. *Calculate a new schedule* $P_n^m(\mathcal{T})$ *as:*

$$P_n^m(\mathcal{T}) = \arg \min_{P_n(\mathcal{T})} \left( \langle s_f^m(\mathcal{T}), P_n(\mathcal{T}) \rangle + \frac{1}{2} \|P_n(\mathcal{T}) - P_n^{m-1}(\mathcal{T})\|^2 \right), \tag{14.45}$$

*such that (14.1), (14.7), and (14.8) hold.*
*Report* $P_n^m(\mathcal{T})$ *to the aggregator.*

3. *If the aggregator has not announced that the iteration process has ended, set* $m \leftarrow m+1$ *and go to Step 1.*

***Return*** $P_n(\mathcal{T}) = P_n^m(\mathcal{T})$.

### 14.4.1 Forecast-based scheduling

For the forecast-based scheduling, it is assumed that all $N_{EV}$ EVs are available to run Algorithms 14.1 and 14.2 under the coordination of the aggregator before the participation period $\mathscr{T}$ starts. Since the forecasting inputs of the regulation requests over such a long time horizon $\mathscr{T}$ (a span of hours in our context), are highly unreliable, forecast-based scheduling is not practical in the real world. Nonetheless, Algorithms 14.1 and 14.2 can still be useful to obtain the best possible scheduling results as the performance bound when we assume that $R_f(\mathscr{T})$ is accurate in the simulation.

The stopping criteria of Algorithms 14.1–14.4 can be based on the number of iterations performed, i.e., $m = M_f$, where $M_f$ is the maximum number of iterations, and/or the convergence of the control signal $s_f^m(\mathscr{T})$ within the convergence tolerance, i.e. $\| s_f^m(\mathscr{T}) - s_f^{m-1}(\mathscr{T})\| \leq \varepsilon_f$, where $\varepsilon_f > 0$ is the convergence tolerance.

The forecast-based scheduling problem in (14.9) minimizes the variance of $P_{total}(\mathscr{T})$. Therefore, without considering the constraints of (14.9), the optimal solutions of (14.9) should follow:

$$\forall T_i \in \mathscr{T}, P_{total}(\mathrm{T}_i) = \mathrm{A}(P_{total}(\mathscr{T})) \tag{14.46}$$

The value of $\mathrm{A}(P_{total}(\mathscr{T}))$ does not affect the optimality of (14.9) as long as (14.46) is satisfied. Since the constraints (14.7) and (14.8) allow the final SOC of an EV $n \in \mathscr{N}$ to be within a given range, i.e., $SOC_n(T_k) \in \left[SOC_{n,MinCh}, SOC_{n,max}\right]$, the total energy consumption of EVs, $\sum_{T_k \in \mathscr{T}} P_A(T_k)$, is not fixed. The proposed Algorithms 14.1 and 14.2 does not determine the value of $\mathrm{A}(P_{total}(\mathscr{T}))$. Nevertheless, the value of $\mathrm{A}(P_{total}(\mathscr{T}))$ in the optimization result is related to the schedules $P^0_{\mathscr{N}}(\mathscr{T})$ in the initialization step, since the searching for the optimal schedules $P_{\mathscr{N}}(\mathscr{T})$ starts from $P^0_{\mathscr{N}}(\mathscr{T})$.

In each round of the iterations of Algorithms 14.1 and 14.2, the aggregator calculates and broadcasts the control signal $s_f^m(\mathscr{T}) \in \mathbb{R}^{N_T \times 1}$ from the schedules $P^{m-1}_{\mathscr{N}}(\mathscr{T}) \in \mathbb{R}^{N_T \times N_{EV}}$ received from the EVs. Every EV $n \in \mathscr{N}$ needs to solve the optimization problem (14.46) to obtain its updated schedule $P^m_n(\mathscr{T}) \in \mathbb{R}^{N_T \times 1}$ and reports $P^m_n(\mathscr{T})$ to the aggregator. Therefore, the total communication overheads $CO_f$ of Algorithms 14.1 and 14.2 are calculated as:

$$CO_f = D \cdot m_f \cdot N_T \cdot (N_{EV} + 1) \tag{14.47}$$

where $D$ and $m_f$ denote the size of a one-dimensional control variable, e.g. $P_n(T_k)$, and the number of iterations performed, respectively.

**Algorithm 14.3** *Online scheduling for aggregator*
***Input:*** *At any time slot $T_k \in \mathscr{T}$, the aggregator knows the total number of time slots $N_T$, and the number of EVs $N_{EV}$, and has received the regulation requests $\{R(T_i)|1 \leq i \leq k\}$.*
***Output:*** *The charging/discharging schedules of EVs at $T_k$, $Q_{\mathscr{N}}(T_k) = (Q_1(T_k), Q_2(T_k), \ldots, Q_{N_{EV}}(T_k))$.*

*Choose a parameter* $\beta$ *such that* $0 < \beta < \frac{N_T}{2N_{EV}}$.
*Wait for the initial schedule* $Q_n^0(T_k)$ *of every EV* $n \in \mathcal{N}$.
*Set the iteration number* $m \leftarrow 1$, *repeat Steps 1 3.*

1. *Calculate the control signal* $s_o^m(T_k) \in \mathbb{R}^{2\times 1}$ *as follows:*

$$\begin{aligned} s_o^m(T_k) &= \beta \nabla U_o(Q_A^{m-1}(T_k)) \\ &= \beta \left( \frac{\partial U_o(Q_A^{m-1}(T_k))}{\partial P_A(T_k)}, \frac{\partial U_o(Q_A^{m-1}(T_k))}{\partial FP_A(T_k)} \right)^T \\ &= \frac{2\beta}{N_T} \begin{pmatrix} R(T_k) + P_A^{m-1}(T_k) \\ \alpha(k)\left( -\sum_{i=1}^{k} R(T_i) + FP_A^{m-1}(T_k) \right) \end{pmatrix} \\ &\quad - \begin{pmatrix} \frac{2\beta}{N_T^2} \\ \frac{2\beta}{N_T^2} \end{pmatrix} \left( \sum_{i=1}^{k-1} P_A(T_i) + P_A^{m-1}(T_k) + FP_A^{m-1}(T_k) \right) \end{aligned} \tag{14.48}$$

*Broadcast the control signal* $s_o^m(T_k)$ *to all EVs.*

2. *Wait for the updated schedule* $Q_n^m(T_k)$ *reported by every EV* $n \in \mathcal{N}$
3. *If the stopping criteria are not met, set* $m \leftarrow m+1$ *and go to Step 1. Otherwise, broadcast the message that the iteration process ends to all EVs.*

***Return*** $Q_{\mathcal{N}}(T_k) = Q_{\mathcal{N}}^m(T_k)$.

**Algorithm 14.4** *Online scheduling for each EV* $n \in \mathcal{N}$
***Input:*** *At any time slot* $T_k \in \mathcal{T}$, *EV* $n \in \mathcal{N}$ *knows its own constraints (14.1), (14.8), and (14.32).*
***Output:*** *The charging/discharging schedules of EV n at* $T_k$, $Q_n(T_k)$.
*Initialize the schedule* $Q_n^0(T_k)$ *as:*

$$Q_n^0(T_k) = \begin{cases} \text{a point at the boundary of the region defined by (14.32)} & k = 1 \\ \dfrac{FP(T_{k-1})}{T_{n,out} - T_{k-1}} (\Delta t, T_{n,out} - T_k)^T & \text{otherwise} \end{cases} \tag{14.49}$$

*and report* $Q_n^0(T_k)$ *to the aggregator.*
*Set the iteration number* $m \leftarrow 1$, *repeat Steps 1–3.*

1. *Wait for the updated control signal* $s_o^m(T_k)$ *broadcast by the aggregator.*

2. *Calculate a new schedule $Q_n^m(T_k)$ as:*

$$Q_n^m(T_k) = arg \min_{Q_n(T_k)} \left( \langle s_o^m(T_k), Q_n(T_k) \rangle + \frac{1}{2} \| Q_n(T_k) - Q_n^{m-1}(T_k) \|^2 \right), \tag{14.50}$$

*such that (14.1),(14.8), and (14.32) hold. Report $Q_n^m(T_k)$ to the aggregator.*

*If the aggregator has not announced that the iteration process has ended, set $m \leftarrow m+1$ and go to Step 1).*

***Return*** $Q_n(T_k) = Q_n^m(T_k)$.

## *14.4.2 Online scheduling*

For the online scheduling, Algorithms 14.3 and 14.4 are performed at every slot $T_k \in \mathcal{T}$ to update the schedules of EVs according to the newly received regulation request $R(T_k)$.

Similar to those of Algorithms 14.1 and 14.2, the stopping criteria of Algorithms 14.3 and 14.4 can be $m = M_o$ where $M_o$ is the maximum number of iterations, and/or $\| s_o^m(\mathcal{T}) - s_o^{m-1}(\mathcal{T}) \| \le \varepsilon_o$, where $\varepsilon_o > 0$ is the convergence tolerance.

At every slot $T_k \in \mathcal{T}$, the optimization results $Q_{\mathcal{N}}(T_k)$ are influenced by the profiles of the current and past regulation requests $\{R(T_i) | 1 \le i \le k\}$ and the historical charging/discharging power of the EVs $\{P_n(T_k) | \forall n \in \mathcal{N}, 1 \le i < k\}$.

In each round of the iterations of Algorithms 14.3 and 14.4, the aggregator calculates and broadcasts the control signal $s_o^m(T_k) \in \mathbb{R}^{2\times 1}$ from the schedules $Q_{\mathcal{N}}^{m-1}(T_k) \in \mathbb{R}^{2\times N_{EV}}$ received from the EVs. Every EV $n \in \mathcal{N}$ needs to solve the optimization problem (14.45) to obtain its updated schedule $Q_n^m(T_k) \in \mathbb{R}^{2\times 1}$ and reports $Q_n^m(T_k)$ to the aggregator. Therefore, the total communication overhead $CO_o$ of Algorithms 14.3 and 14.4 at each time slot is calculated as:

$$CO_o = D \cdot m_o \cdot 2 \cdot (N_{EV} + 1) \tag{14.51}$$

where $m_o$ denotes the number of iterations performed.

# 14.5 Case studies

This section synthesizes and presents the simulation results reported in our prior works (Lin *et al.*, 2013, 2014).

## *14.5.1 V2G scheduling algorithms*

The scheduling algorithms, namely, Algorithms 14.1 and 14.2 for the forecast-based scheduling problem in (14.9), Algorithms 14.3 and 14.4 for the online scheduling problem in (14.31), and an extended version, which is introduced below, of the optimal decentralized charging (ODC) algorithm proposed in Gan *et al.* (2013), will be investigated by computer simulation.

Since the ODC algorithm proposed in Gan *et al.* (2013) does not consider discharging of EV batteries, we extend ODC by enabling discharging to fit in our context. Hence, the optimization problem of ODC with discharging (ODCD) is discussed as follows:

$$\min_{P_{\mathcal{N}}(\mathcal{T})} \sum_{T_i \in \mathcal{T}} u\left( R_f(T_i) + \sum_{n \in \mathcal{N}} P_n(T_i) \right)$$

such that for all $n \in \mathcal{N}$, (14.1) holds, and

$$SOC_n(T_{N_T}) = SOC_{n,MinCh}, \tag{14.52}$$

where $u : \mathbb{R} \to \mathbb{R}$ is strictly convex. According to Theorem 2 of Gan *et al.* (2013), the optimal total power profile obtained by (14.52) is independent on the choice of $u$.

According to Property 1 of Gan *et al.* (2013), ODC is able to obtain a flat total power profile by scheduling the charging activities of EVs. However, when discharging is introduced and ODC is extended to ODCD, such valley-filling property of ODC may not be inherited by ODCD. It will be studied in the simulation.

### 14.5.2 Performance metric

According to the proposed control objective for the V2G regulation service, the variance of the profile of the total power, $\text{Var}(P_{total}(\mathcal{T}))$, is used as the performance metric.

A smaller $\text{Var}(P_{total}(\mathcal{T}))$ implies a more flattened profile of the total power, indicating that the fluctuations of the regulation requests are better absorbed by the aggregated EV power, and therefore a better performance.

### 14.5.3 Simulation setup

The simulation scenario is an aggregator coordinating $N_{EV} = 1,000$ EVs to decide their charging/discharging schedules from 19:00 in the evening to 7:00 on the next morning. This 12-hour period of time is divided equally into $N_T = 144$ slots of length $\Delta t = 5$ minutes.

The hypothetic EV group consists of four models of EVs currently on the market: Chevrolet Volt with a 16.5-kWh battery pack (Che, 2014), Ford C-MAX Energi with a 7.6-kWh battery pack (For, 2014), Nissan Leaf SV with a 24-kWh battery pack (Nis, 2014), and Tesla Model S with a 60-kWh battery pack (Tes, 2014). Each of the four models accounts for 25% of the 1000 EVs. All EVs are assumed to have been contracted to provide the V2G regulation service, either unidirectionally or bidirectionally. According to the standard Level 2 charging in the US, we assume that the charging power of Chevrolet Volt and C-MAX Energi can vary from 0 to 4.0 kW. Nissan Leaf SV and Tesla Model S have their dedicated 240-volt chargers with charging powers of 6.6 kW (Nis, 2014) and 10 kW (Tes, 2014), respectively. Thus, we assume that the charging power of Leaf SV and Model S can vary from 0 to 6.6 kW and from 0 to 10 kW, respectively. Furthermore, the discharging power of all of the four models are assumed to vary from −4 kW to 0 for bidirectional V2G. According to Shao *et al.* (2012), the distribution of plug-in

*Table 14.1 Values or distributions of the parameters for all EVs*

| Parameter | Value | Parameter | Distribution |
|---|---|---|---|
| $\mu$ | 0.4 | $SOC_{n,0}$ | $U[0.45,0.55]$ |
| $\tau$ | 0.7 | $SOC_{n,MinCh}$ | $U[0.75,0.85]$ |
| $\eta_{ch}$ | 0.9 | $SOC_{n,max}$ | $U[0.9,0.99]$ |
| $\eta_{dch}$ | 0.9 | $SOC_{n,min}$ | $U[0.3,0.4]$ |

time of EVs is close to a normal distribution. Hence, in the simulation, the plug-in times of EVs are generated based on a normal distribution with the mean at 19:00 and the standard deviation is equal to 1 hour first, and then any plug-in time before 19:00 is set to be 19:00. Similarly, the plug-out times are also generated based on a normal distribution with the mean at 7:00 and the standard deviation is equal to 1 hour first. Then, any plug-out time after 7:00 is set to be 7:00. The values or distributions of the parameters for EVs are summarized in Table 14.1.

The fast response regulation signals of the PJM market (Reg, 2013) from 18 December 2012 to 18 January 2013 are used in the simulation. A total of 31 sets of the 12-hour profiles of regulation signals are extracted from the 32-day period. According to Rebours *et al.* (2007), the regulation signal is normalized to be within the range of [−1,1], and related to the regulation demand linearly. The ratio between the regulation demand and regulation signal is set by the balancing authority of a specific control area. Since most of the 1,000 EVs in the hypothetic EV group have charging and discharging power limits of 4 kW and −4 kW, respectively, and the regulation service is usually bid on an MW basis (Kempton and Tomić, 2005b), it is reasonable to assume that the aggregator would receive the assigned regulation requests within the range of $[-2, 2]$ MW in this case. Hence, the ratio between the regulation requests and regulation signal would be 2 MW in the simulation. According to (14.22), the expectation of the sum of the regulation requests over the 12-hour participation period is assumed to be zero. Figure 14.7 shows the histogram of the sum of the 12-hour regulation requests. The minimum and maximum values of the sum, $R_S(\mathcal{T})$, are −13.442 MW and 8.161 MW. It can be observed that the total energy needs of frequency regulation over the specified 12-hour period may be non-zero although in most cases the total energy needs are close to zero.

Both Algorithms 14.1 and 14.2 for forecast-based scheduling and ODCD require the forecasting profile of the regulation requests. In the simulation, it is assumed that the forecasts of the regulation requests are accurate, i.e., $R_f(T_i) = R(T_i), \forall T_i \in \mathcal{T}$.

For the stopping criteria of Algorithms 14.1 and 14.2, $\varepsilon_f = 10^{-5}$ and $M_f = 100$. As for Algorithms 14.3 and 14.4, $\varepsilon_o = 10^{-5}$ and $M_o = 50$.

### 14.5.4 Simulation results

The scheduling results of the three algorithms with various values of the participation ratio of the bidirectional V2G, $\theta_{bi}$, and the sum of the regulation requests, $R_S(\mathcal{T})$, are studied. Table 14.2 lists the performances of the three algorithms with $\theta_{bi}$ equal to 1, 0.5, and 0, respectively, in three special cases of $R_S(\mathcal{T})$, namely, the

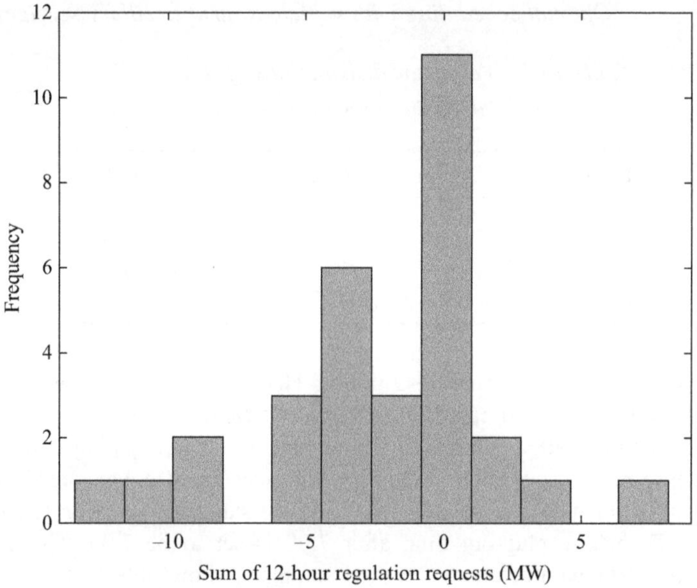


*Figure 14.7 Histogram of the sum of the 12-hour regulation signals with 31 samples. © 2014 IEEE. Reprinted with permission from Lin* et al. *(2014)*

*Table 14.2 Performances*[2] *of the three scheduling algorithms with various values of $R_S(\mathcal{T})$ and $\theta_{bi}$: (a) $R_S(\mathcal{T}) = -24.09$ kW; (b) $R_S(\mathcal{T}) = 8.161 \times 10^3$ kW; (c) $R_S(\mathcal{T}) = -1.344 \times 10^4$ kW*

(a)

| $\theta_{bi}$ \ Algorithm | Forecast-based scheduling | Online scheduling | ODCD |
|---|---|---|---|
| 1 | 0.023 | 9.988 | $1.858 \times 10^3$ |
| 0.5 | 0.172 | 11.903 | $1.789 \times 10^3$ |
| 0 | $2.638 \times 10^3$ | $4.065 \times 10^3$ | $1.268 \times 10^4$ |

(b)

| $\theta_{bi}$ \ Algorithm | Forecast-based scheduling | Online scheduling | ODCD |
|---|---|---|---|
| 1 | 0.0038 | 45.888 | $2.095 \times 10^3$ |
| 0.5 | 0.272 | 54.215 | $2.090 \times 10^3$ |
| 0 | $4.807 \times 10^3$ | $1.395 \times 10^4$ | $2.540 \times 10^4$ |

(c)

| $\theta_{bi}$ \ Algorithm | Forecast-based scheduling | Online scheduling | ODCD |
|---|---|---|---|
| 1 | 0.114 | $4.418 \times 10^2$ | $5.139 \times 10^2$ |
| 0.5 | 0.257 | $4.716 \times 10^2$ | $5.125 \times 10^2$ |
| 0 | $2.126 \times 10^3$ | $1.642 \times 10^3$ | $3.967 \times 10^3$ |

[2]The unit of the performance metric $\mathrm{Var}(P_{total}(\mathcal{T}))$ is $(\mathrm{kW})^2$.

minimum absolute, maximum, and minimum values of $R_S(\mathscr{T})$, respectively, among the 31 sets of the profiles of the regulation requests.

First, we compare the scheduling results in the ideal case when the sum of the regulation requests is close to zero. Among the 31 sets of profiles of regulation requests, the one with its sum closest to zero has $R_S(\mathscr{T}) = -24.09$ kW. Figure 14.8

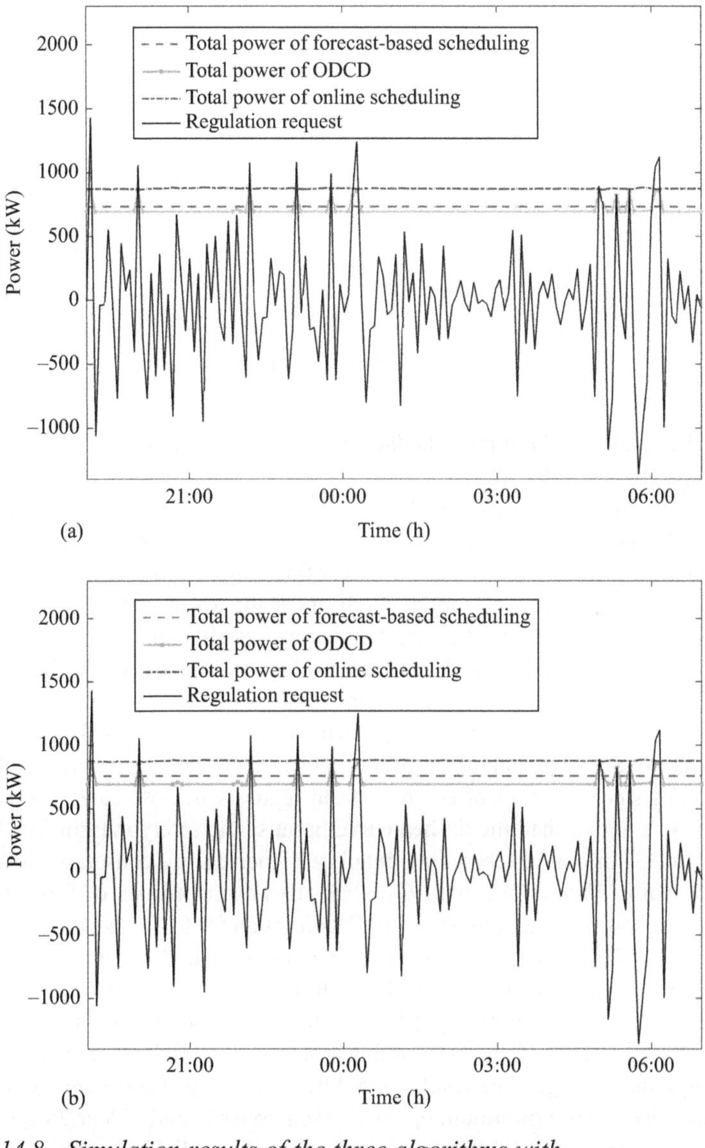


*Figure 14.8 Simulation results of the three algorithms with $\sum_{T_k \in \mathscr{T}} R(T_k) = -24.09$ kW and various configurations of $\theta_{bi}$: (a) $\theta_{bi} = 1$; (b) $\theta_{bi} = 0.5$; (c) $\theta_{bi} = 0$. ©2014 IEEE. Reprinted with permission from Lin* et al. *(2014)*

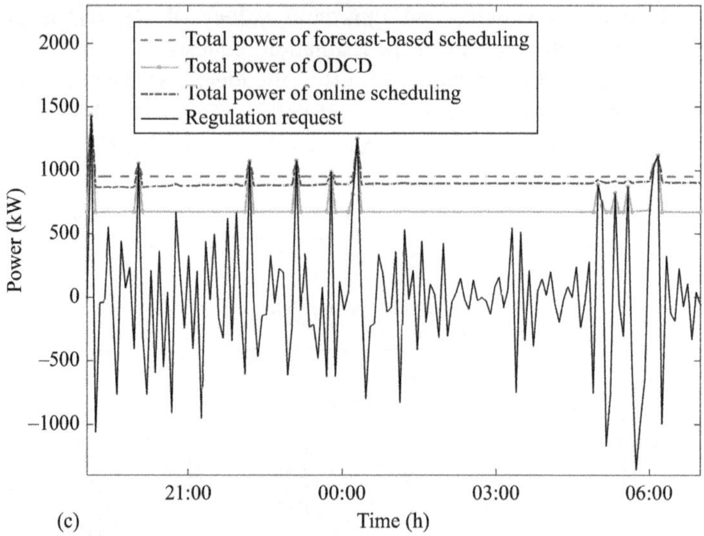


*Figure 14.8 (Continued)*

presents the results of the three scheduling algorithms when $\theta_{bi}$, is equal to $1, 0.5$, and 0, respectively.

As shown in Figure 14.8(a) with $\theta_{bi} = 1$, i.e., all EVs participate in the bidirectional V2G, the total power profile of the proposed forecast-based scheduling (the dashed curve) is flat, indicating that the power fluctuations represented by the profile of the regulation requests are smoothed out under the assumption of accurate forecasts of the regulation requests. The total power profile of the proposed online scheduling (the dash-dotted curve) is almost as flat as the dashed curve. Note that the dashed curve and the dash-dotted curve are close to two constant positive loads of about 731 kW and 874 kW, respectively. As indicated in Section 14.3.1, the constant loads are approximately equal to the power consumptions for satisfying the charging needs of EVs since the sum of the regulation requests is close to zero. The dash-dotted curve is higher than the dashed curve because the safety margin for charging stated in (14.33) of online scheduling entails a higher energy consumption of EVs.

As shown in Figure 14.8 with $\theta_{bi}= 0.5$, the power profiles of forecast-based scheduling and online scheduling are still flat although 50% of EVs participate in the unidirectional V2G for which discharging is not allowed. However, when all EVs participate in unidirectional V2G, as shown in Figure 14.8 with $\theta_{bi}= 0$, the performances of the two scheduling algorithms deteriorate since the EVs are unable to discharge their batteries to maintain a flat profile of the total power when the regulation requests are large. The results of online scheduling at different values of $\theta_{bi}$ imply that there exists a minimum $\theta_{bi}$ for a given upper bound of $\mathrm{Var}(P_{total}(\mathscr{T}))$. In other words, it may not always be necessary to have all EVs enabled with the bidirectional V2G.

The reliability and resilience of the proposed online scheduling in some extreme cases when $R_S(\mathscr{T})$ is far beyond zero are also investigated. Figure 14.9 presents the

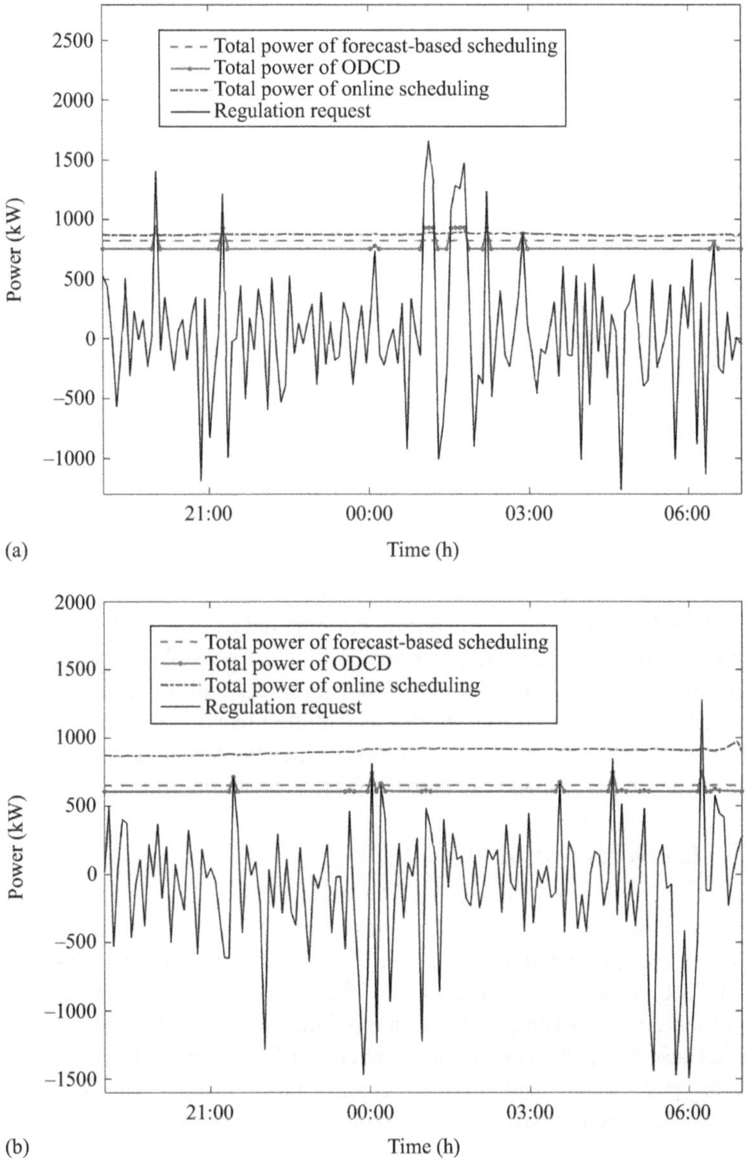


*Figure 14.9 Simulation results of the three algorithms when* $\theta_{bi}$= 0.5 *in two extreme cases of* $R_S(\mathscr{T})$*: (a)* $R_S(\mathscr{T}) = 8.161 \times 10^3$ *kW; (b)* $R_S(\mathscr{T}) = -1.344 \times 10^4$ *kW. ©2014 IEEE. Reprinted with permission from Lin* et al.*(2014)*

simulation results of the three scheduling algorithms when the sum of the regulation requests $R_S(\mathscr{T})$ is the maximum and the minimum, respectively, among the 31 sets of the profiles of the regulation requests. The total power profiles of online scheduling (the dash-dotted curves) are still nearly as flat as those of forecast-based scheduling (the dash curves) even when the zero-energy assumption specified in (14.22) is seriously violated and $R_S(\mathscr{T})$ takes large absolute values. Although the safety margin of online scheduling results in a total load a bit higher than that of forecast-based scheduling, it makes online scheduling resilient to different regulation requests. In addition, since the total power profile is flattened by online scheduling, such total load can be economically met by load following or even generation dispatch. Therefore, online scheduling will not impair the stability of the grid.

The proposed forecast-based scheduling and online scheduling outperform ODCD. As shown in Figures 14.8 and 14.9, ODCD fails to produce flat total power profiles even though the forecasts of the regulation requests have been assumed to be accurate. There are obvious spikes in the total power profiles, shown by the dots and solid curves, corresponding to ODCD when the power demands of the regulation-up requests are so high that the EV fleet should collectively provide discharging power to the grid. This is because ODCD does not account for the effect of the round-trip efficiency of a battery. It indicates that the method proposed in Gan *et al.* (2013) is not suitable to deal with the discharging control of EVs. Moreover, ODCD is impractical for the scheduling control of the regulation service since it requires the forecasting profile of regulation requests.

To summarize, the simulation results further corroborate our statements in Sections 14.3.3 and 14.3.4 that the optimization result of forecast-based scheduling can serve as a performance bound since it is the best possible schedules when the forecasts of the regulation requests are accurate, but online scheduling is more suitable and practical for real-world implementation because it does not depend on the forecasts of the regulation requests, determines the schedules in real time, and has a comparable performance to forecast-based scheduling.

### *14.5.5 Convergence rates*

The convergence rates of Algorithms 14.1 and 14.2 for forecast-based scheduling for various values of the ratio of the bidirectional V2G $\theta_{bi}$ and the convergence tolerance $\varepsilon_f$ are presented in Table 14.3. The numbers of iterations shown in Table 14.3 are averaged over the 31 sets of the profiles of the regulation requests. It can be observed that given $\varepsilon_f$, the number of iterations increases as $\theta_{bi}$ decreases. This is because as $\theta_{bi}$ decreases, the number of EVs that are allowed to discharge their batteries decreases, and hence it becomes more difficult for EVs to suppress the peaks of the regulation requests and flatten the total power profile. According to

*Table 14.3 Mean numbers of iterations of Algorithms 14.1 and 14.2 in various scenarios*

| $\theta_{bi}$ \ $\varepsilon_f$ | $10^{-4}$ | $10^{-5}$ | $10^{-6}$ |
|---|---|---|---|
| 1 | 6.0 | 7.3 | 8.3 |
| 0.5 | 9.0 | 11.0 | 13.7 |
| 0 | 18.0 | 42.3 | 88.7 |

*Table 14.4 Mean numbers of iterations of algorithms 14.3 and 14.4 in various scenarios*

| $\theta_{bi}$ \ $\varepsilon_o$ | $10^{-4}$ | $10^{-5}$ | $10^{-6}$ |
|---|---|---|---|
| 1 | 3.9 | 5.2 | 7.4 |
| 0.5 | 4.1 | 5.8 | 8.2 |
| 0 | 4.1 | 5.7 | 7.8 |

(14.47) and Table 14.3, the communication overheads of Algorithms 14.1 and 14.2 can be obtained. For instance, when $\theta_{bi} = 1$ and $\varepsilon_f = 10^{-5}$, the average number of iterations is 7.3. Suppose the size of a one-dimensional control variable $D = 8$ bytes. Then, the average communication overhead $CO_f$ in this case is about 8.028 MB according to (14.47). Forecast-based scheduling is performed one time to determine the schedules of all EVs during the participation period $\mathscr{T}$.

The convergence rates of Algorithms 14.3 and 14.4 for online scheduling in various values of $\theta_{bi}$ and $\varepsilon_o$ are presented in Table 14.4. The numbers of iterations in Table 14.4 are averaged over the $N_T = 144$ time slots and the 31 sets of regulation requests. It can be observed that, different from the convergence rate of Algorithms 14.1 and 14.2, the ratio of the bidirectional V2G $\theta_{bi}$ does not have a significant effect on the number of iterations performed in Algorithms 14.3 and 14.4. This can be explained by the much lower dimension of the control variables for online scheduling compared to that for forecast-based scheduling. In online scheduling, at each time slot, an EV should only determine a two-dimensional schedule. Therefore, the schedules of EVs can soon converge to the best-effort ones to minimize the variance of the total power profile. According to (14.51) and Table 14.4, the communication overheads of Algorithms 14.3 and 14.4 can be obtained. For instance, when $\theta_{bi} = 1$ and $\varepsilon_o = 10^{-5}$, the average number of iterations at each time slot is 5.2. Thus, the average communication overhead per slot $CO_o$ in this case is about 81.331 kB according to (14.51). Since Algorithms 14.3 and 14.4 should be performed $N_T = 144$ times during the participation period $\mathscr{T}$, the total overhead throughout $\mathscr{T}$ is about 11.437 MB.

## 14.6 Conclusion

The optimal scheduling for an aggregator coordinating its EVs to provide the V2G regulation service is studied. Based on the zero-energy characteristics of frequency regulation, we propose an online scheduling method which depends on the forecast of the regulation demand and allows each EV to determine its own schedule in real time. Our method jointly guarantees adequate charging of EVs and optimizes the quality of the regulation service. A simulation study of 1,000 hypothetical EVs shows that the proposed online scheduling algorithm performs nearly as well as the forecast-based scheduling algorithm, demonstrating the practicability of online charging/discharging control for the provision of the V2G regulation service. Future work will extend the methods and models proposed in this chapter to solve the inter-level control in a multilevel V2G system.

## Acknowledgements

This research is supported in part by the Research Grants Council, Hong Kong Special Administrative Region, China (Project No. 17261416).

## References

Fast response regulation signal, Feb 2013. URL http://www.pjm.com/markets-and-operations/ancillary-services/mkt-based-regulation/fast-response-regulation-signal.aspx.

2014 volt models & specifications, May 2014. URL http://www.chevrolet.com/volt-electric-car/specs/options.html.

Model: C-MAX energi, May 2014. URL http://www.ford.com/cars/cmax/trim/energi/.

2014 leaf compare specifications, May 2014. URL http://www.nissanusa.com/electric-cars/leaf/versions-specs/.

Model S specifications, May 2014. URL http://www.teslamotors.com/models/specs.

D. Bertsekas and J. Tsitsiklis. *Parallel and distributed computation*. Prentice Hall, Englewood Cliffs, NJ, 1989.

S. Boyd and L. Vandenberghe. *Convex Optimization*. Cambridge University Press, *Cambridge, UK*, 2004.

J. Donadee and M. Ilic. Stochastic optimization of grid to vehicle frequency regulation capacity bids. *IEEE Transactions on Smart Grid*, 5(2):1061–1069, Mar 2014.

Electric Power Research Institute. Electric transportation – program 18. Technical report, 2013.

L. Gan, U. Topcu, and S. H. Low. Optimal decentralized protocol for electric vehicle charging. *IEEE Transactions on Power Systems*, 28(2):940–951, May 2013.

S. Han, S. Han, and K. Sezaki. Development of an optimal vehicle-to-grid aggregator for frequency regulation. *IEEE Transactions on Smart Grid*, 1(1):65–72, Jun 2010.

S. Han, S. Han, and K. Sezaki. Optimal control of the plug-in electric vehicles for V2G frequency regulation using quadratic programming. In *Proceedings of IEEE PES ISGT 2011*, Jan 2011.

E. Hirst and B. Kriby. Electric-power ancillary service. Technical report, Oak Ridge National Laboratory, 1996.

W. Kempton and J. Tomić. Vehicle-to-grid power fundamentals: calculating capacity and net revenue. *Journal of Power Sources*, 144(1):268–279, Jun 2005a.

W. Kempton and J. Tomić. Vehicle-to-grid power implementation: from stabilizing the grid to supporting large-scale renewable energy. *Journal of Power Sources*, 144(1):280–294, Jun 2005b.

B.J. Kirby. Frequency regulation basics and trends. Technical Report ORNL/TM-2004/291, Oak Ridge National Laboratory, Dec 2004.

J. Lin, K.-C. Leung, and V.O.K. Li. Online scheduling for vehicle-to-grid regulation service. In *Proceedings of IEEE SmartGridComm 2013*, pp. 43–48, Oct 2013.

J. Lin, K.-C. Leung, and V.O.K. Li. Optimal scheduling with vehicle-to-grid regulation service. *Internet of Things Journal, IEEE*, 1(6):556–569, Dec 2014.

Z. Ma, D. Callaway, and I. Hiskens. Decentralized charging control for large populations of plug-in electric vehicles: Application of the nash certainty equivalence principle. In *Proceedings of IEEE CCA 2010*, pp.191–195, Sept 2010.

R. Olfati-Saber and R.M. Murray. Consensus problems in networks of agents with switching topology and time-delays. *IEEE Transaction on Automatic Control*, 49 (9):1520–1533, Sept 2004.

Y. Ota, H. Taniguchi, T. Nakajima, K.M. Liyanage, and A. Yokoyama. An autonomous distributed vehicle-to-grid control of grid-connected electric vehicle. In *Proceedings of ICIIS 2009*, pp. 414–418, Dec 2009.

Y. Ota, H. Taniguchi, T. Nakajima, K.M. Liyanage, J. Baba, and A. Yokoyama. Autonomous distributed V2G (vehicle-to-grid) satisfying scheduled charging. *IEEE Transactions on Smart Grid*, 3(1):559–564, Mar 2012.

Y. Rebours, D.S. Kirschen, M. Trotignon, and S. Rossignol. A survey of frequency and voltage control ancillary services — part I: technical features. *IEEE Transactions on Power Systems*, 22(1):350–357, Feb 2007.

S. Shao, M. Pipattanasomporn, and S. Rahman. Grid integration of electric vehicles and demand response with customer choice. *IEEE Transactions on Smart Grid*, 3(1):543–550, Mar 2012.

W. Shi and V.W.S. Wong. Real-time vehicle-to-grid control algorithm under price uncertainty. In *Proceedings of IEEE SmartGridComm* 2011, pp. 261–266, Oct 2011.

E. Sortomme and M.A. El-Sharkawi. Optimal scheduling of vehicle-to-grid energy and ancillary services. *IEEE Transactions on Smart Grid*, 3(1):351–359, Mar 2012.

R. Wang, Y. Li, P. Wang, and D. Niyato. Design of a V2G aggregator to optimize PHEV charging and frequency regulation control. In *Proceedings of IEEE SmartGridComm 2013*, pp. 127–132, Oct 2013.

H. Yang, C.Y. Chung, and J. Zhao. Application of plug-in electric vehicles to frequency regulation based on distributed signal acquisition via limited communication. *IEEE Transactions on Power Systems*, 28(2):1017–1026, May 2013.

W. Yao, J. Zhao, F. Wen, Y. Xue, and G. Ledwich. A hierarchical decomposition approach for coordinated dispatch of plug-in electric vehicles. *IEEE Transactions on Power Systems*, 28(3):2768–2778, Aug 2013.

# Chapter 15

# Vehicle-to-home, vehicle-to-vehicle, and vehicle-to-grid energy systems

*Shuang Gao*[1]*, Zhen Zhang*[1] *and Chunhua Liu*[2]

As a large market share of electric vehicles (EVs) is expected in the transportation section, the high penetration of EVs in the power grid has placed new challenges on the power system planning and operation. Their negative impacts on energy supply, transformer deterioration, and peak demand increment are revealed. Therefore, control methods for integrating EVs into the power grid are developed to mitigate the negative effects and exploit the aggregated EVs as energy storage devices to provide ancillary services.

In this chapter, the vehicle-to-grid (V2G) operation in modern power system and its control framework are first introduced, which combines communication and EV charging infrastructures. Then, three different operation modes of EVs integrated to the power grid are described, which consists of vehicle-to-home (V2H), vehicle-to-vehicle (V2V), and V2G. The modeling and the control methods for these three operation modes are given. The planning of EV charging infrastructure and the ancillary services provided by V2G are discussed.

## 15.1 Introduction

As the production and commercialization of EVs are supported by the government policy in most of the countries in the world, a variety of EV models including the grid-connected EVs such as battery electric vehicles (BEVs) and plug-in hybrid electric vehicles (PHEVs) are offered by automotive manufacturers (Chau, 2014). A larger market share of EVs is expected in the near future. According to a recent study on its potential in the transportation section, up to 86% of the light-vehicles are predicted to be replaced by EVs over the next two decades (Becker and Sidhu, 2009). Also, the tax reduction and economic incentives are provided by the government to realize their ambitious target to put one million EVs on the road by 2020, aiming to position the country as a market leader for electric mobility. However, a large penetration of EVs in the power grid place new challenges on the power system

[1]School of Electrical Engineering and Automation, Tianjin University, China
[2]School of Energy and Environment, City University of Hong Kong, Hong Kong, China

operation (Shafie-Khah and Catalao, 2015). The negative impacts on energy supply, transformer deterioration, and peak demand increment are revealed. Therefore, while the extensive use of EVs is highly desirable, both globally and locally, the control method to facilitate the integration of EV into the power grid needs to be developed.

Among various renewable energy sources, the wind power, hydro power, and solar power have been well developed as an integral part of the emerging power grid. One common challenge associated with renewable energy sources must be the intermittent power generation. In order to accommodate the large-scale installation of renewable energy generation units, the energy storage system is required to compensate for the surplus or insufficient power generation so as to balance the power supply and power demand at all times. The widespread EVs in the power grid have good prospects of acting as distributed energy storage because of the sufficient power capacity from a large number of EV on-board batteries and the flexible control allowed by the advance power electronic chargers. Namely, smart charging of EV is implemented as controllable load to consume the excessive electricity production of renewable energy like wind power usually in the midnight of low load demand, and batteries are discharged to feed power back to the power grid when the renewable power suddenly falls off, becoming insufficient to supply the load. The energy management of grid-connected EVs in the corporation with renewable energies is of high relevance for the countries, where a large portion of electricity is produced by renewable energy sources. Thus, V2G operation associated with the intermittent power generation is becoming attractive and is studied in the scenarios of electric mobility development and a generation mix dominated by renewable energy sources.

Control methods for managing the EV battery energy have been developed depending on the structure of power network where EVs are connected to and the scope of EV interaction with power grid, electric appliances, and other EVs. If the energy management in a single EV is taken into account, the control objective is chosen to be the minimization of total cost from the amount of gasoline consumed by combustion engine and the amount of electricity bought from the power grid (Ansari *et al.*, 2015). An emerging concept of V2H describes the interaction of a single EV with the other electric appliances in a household usually containing small-scale renewable power generation like one wind turbine or several photovoltaic (PV) panels. The EV charging and discharging are controlled to smooth the household daily profile (DLP) through the on-board or off-board bi-directional charger connected to the home-grid (Jian *et al.*, 2013). The other common concepts such as vehicle-to-building (V2B) and vehicle-to-infrastructure (V2I) refer to the power network covering a building or certain region where the power exchange is enabled for EVs, generation resources, and electricity consumers (Vasiladiotis and Rufer, 2015). For instance, an optimization scheme is derived for a designated EV charging station composed of fast charging facility for a number of EVs and an array of PV panels on the roof. The aggregated charging power of EVs is coordinated with the electricity produced by PV panels to minimize the power being supplied by the distribution utility (Chukwu and Mahajan, 2014). These emerging concepts in V2G interaction allow the flexible use of EVs in particular case and take the merits of simplicity in mathematical model, but cannot offer a sizeable energy storage to interact with renewable generation, for instance a small PV farm typically ranging from 2 to 5 MW at medium voltage levels. Due to the limited storage capacity of a single EV, the intelligent control must be able to govern a large number of EVs so that the aggregated V2G power can be sufficient to interact with fluctuating generation as well as regulation for system stabilization. Therefore, energy management for a single EV and EVs in particular applications is not the main concern, and the scope of research should be expanded to the optimal control of V2G power for grid supports. Considering the complexity of a V2G optimization problem, a hierarchical structure is usually formed to deal with the control of each single EV's charging, EV aggregator, and interaction with distribution grid operator. As depicted in Figure 15.1, the EV aggregator as the new entity in the V2G control framework is engaged in power system operation.

Various power electronic converters are employed to implement the optimal control of V2G power. For the EVs connected to the power grid at the residential buildings, the converter functions to regulate the charging/discharging rate of an individual EV, thus usually being low-voltage and low-power bi-directional converter (Yilmaz and Krein, 2013). EV charging stations are designed to accommodate the simultaneous charging load from a fleet of EVs, resulting in a considerable total charging demand being supplied by distribution utility. A main AC-DC converter is often installed to control the power flow between the EV charging station and the

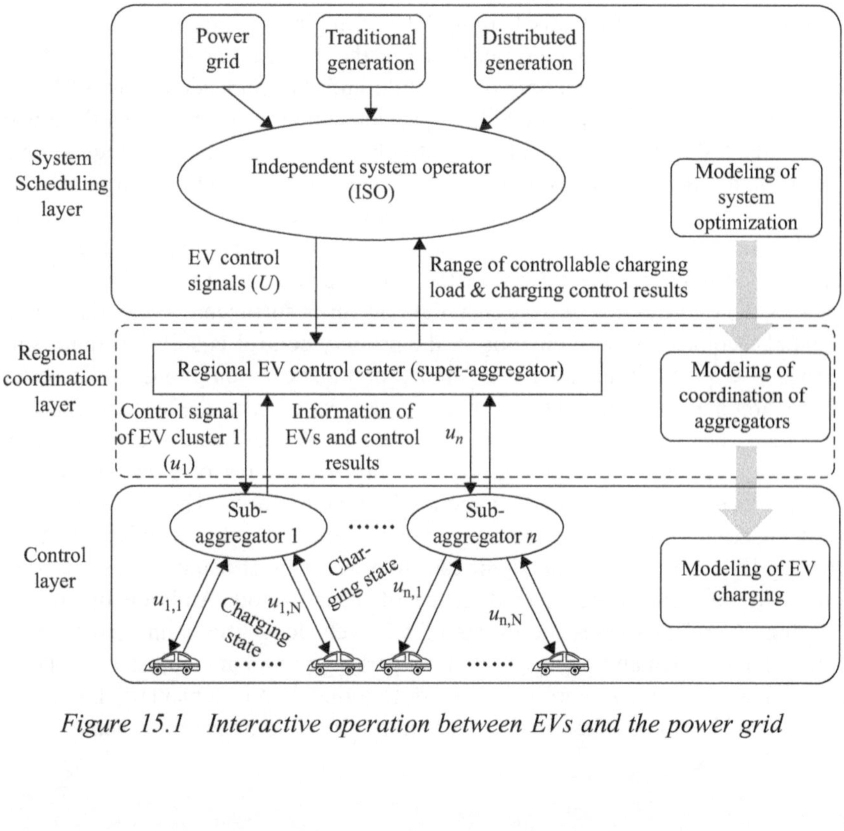


*Figure 15.1 Interactive operation between EVs and the power grid*

medium-voltage distribution grid, and also improve the power quality at the AC side by adjusting the power factor and reducing the harmonics. The EV converters can be designed as AC-DC or DC-DC type depending on the structure of power network normally comprising DC links such as EV charging station, microgrid, and the more complex distribution grid. The analysis of the operating characteristics of the EV converters is required to build an accurate mathematical model of V2G system. On the other hand, the information and communication technologies (ICTs) and metering network are also assumed to be perfectly established in the study of V2G system. Much of the previous work has been targeted at overly ideal V2G model for the sake of convenience or simplicity, regardless of the limits of power electronic devices and ICT network. Recent studies on ICT supported V2G model attempt to include as many practical components as possible in order to formulate the V2G optimization problem that is directly relevant to practical EV applications. In other words, the commonly used methodology of studying over-simplified ideal cases should be refrained. A V2G model and optimal control under the constraints of charging infrastructure and information technologies should be developed to reflect the realistic operating limits of power converters and the communication network customized to V2G applications.

In recent years, the application of EVs in optimal scheduling of the power system has attracted great attention, mainly containing load shifting and generation reserve, as EVs are considered as a controllable load and an available resource. From the viewpoint of power system, EVs are regarded as virtual power plants to participate in the unit commitment planning. In reality, mobility needs must be fulfilled and the use of EV discharging for peaking power plants is hardly to be a solution to reduce the generation cost due to the limited storage capacity of EVs compared with the total energy consumption and the tremendous cost of battery degradation. The application of EVs to provide spinning reserve takes the advantages of the EV fast response and the relatively high revenue for both regulation up and regulation down services, which means both charging and discharging EVs at the specified time intervals can get paid. Among the three control steps, namely primary, secondary, and tertiary control, the primary control requires the shortest duration for injecting the power to the grid typically up to 15 min, and thus becomes the only reserve service that V2G power is sufficient to provide. But the potential income from participating in the primary control is still uncertain, since the negative profit is counted even under optimistic assumptions about battery degradation and regulation reserve requests associated with the system frequency deviation. The main purpose of load shifting is to coordinate the EV charging demand with the daily energy consumption profile, and perform peak shaving by discharging the EVs under certain economic circumstances. The necessity of EV smart charging is justified according to the analysis on the negative impact of arbitrary EV charging and the significant improvement of using delayed charging control to shift the charging load to the off-peak period. Additionally, the EV charging control possesses the merits of relatively long grid-connected time and highly shiftable demand owing to the long parking period and adjustable charging rate of EV chargers. Accordingly, smart charging is an elementary EV energy control that should be taken into account in the V2G optimization scheme.

Other than optimal scheduling of V2G energy, ancillary services that provide dynamic regulation power to compensate for the continuous power imbalance between the demand and the supply are more suitable for EVs to perform in the advanced distribution grid with a large share of renewable generation. On the one hand, the dynamic regulation is performed on the time scale of seconds, which requires alternating charging and discharging of EV batteries to provide positive and negative regulations. Consequently, EVs characterized by fast power response and bi-directional four-quadrant power flow via the power electronic charger are perfect candidate for dynamic power regulations aiming to stabilize the grid frequency and voltage. Moreover, the reactive power can be independently injected at AC terminal even without the EV connected to the charging facility, by the virtue of the capacitor positioned at the DC link of the converter. On the other hand, the positive regulations over a long time must be equivalent to negative regulations, which makes the net energy to be roughly zero. This is a significant advantage of dynamic regulation performed by EVs as no energy change occurs on the battery storage. Also, small energy decline only involves shallow discharging that has the minimal impact on the battery cycle, and therefore prevents the battery from wearing out which is one key challenge of V2G application.

Multi-objective optimization for V2G operation has been developed, which utilizes the V2G power to achieve the multiple goals such as minimizing the cost and emission, minimizing the electricity price and voltage deviation, and minimizing the cost with the maximum renewable energy consumed (Li *et al.*, 2014). In order to fully utilize the aforementioned merits of EVs for dynamic regulations while the necessary V2G energy scheduling is implemented in corporation with the renewable generation and daily load profile, the idea of an integrated control scheme has been developed to perform the dynamic regulation and energy scheduling simultaneously. The frequency and voltage stabilization achieved by regulating the instant power exchange with EVs is viewed as complementary to energy scheduling that is targeted at varying the battery energy storage to fulfil the mobility and grid needs. Meanwhile, the ancillary services provided in the control scheme prove to have the minimum impact on the battery service lifetime, thus being more economic and practical V2G applications. As this multi-functional V2G power regulation inevitably increases the control complexity, an EV dispatching method for the EV aggregator must be included in the control scheme so that the scheduled power is properly allocated to each available EV.

The objective of V2G is to integrate EVs in the distribution grid with a large share of wind power generation and provide V2G ancillary services via modern power electronic converters. The key issues are:

1. to develop a V2G mathematical model and optimal control method consisting of the EV model and power system optimization to fulfil the mobility needs and improve the grid reliability and stability in both dynamic and steady states;
2. to develop a control approach for V2G ancillary services with the consideration of the active and reactive power regulations through power electronic converters and the communication infrastructure available for V2G;

3. to develop a multi-functional control scheme for V2G energy scheduling and dynamic regulation services, containing an EV dispatching method for V2G power allocation in each EV aggregation;
4. to evaluate the V2G control scheme by integrating software simulation and hardware experiment; also, the practical application of V2G control scheme with the existing communication technologies is analyzed.

The mathematical model of V2G operation can be formulated by solving an optimization problem which combines the models of EVs, charging facility, and the power grid. A wide variety of optimization problems are formed to manage the energy storage provided by EV fleets. The objective functions usually consider the investment of EV charging infrastructure, the operating cost of EV smart charging or V2G, satisfaction degree of EV users, and the improvement on the reliability and stability of the power grid while V2G control methods are applied to provide grid support. In a typical V2G optimization problem, the objective function for minimizing the investment of V2G infrastructure can be expressed as:

$$\min f_1 = \sum_{i=1}^{m} C_i^{SC} + \sum_{i=1}^{n} C_i^{FC} + \sum_{i=1}^{m} C_i^{SC}(t) + \sum_{i=1}^{n} C_i^{FC}(t) \tag{15.1}$$

where $C_i^{SC}$ and $C^{FC}$ are the one-time investment of slow charging and fast charging facilities, respectively. The fast charging facility represents the centralized charging mode, where a number of EVs are connected to a charging station to fill the battery in less than 1 h, for example, the charging station installed on highways or urban areas. The slow charging facility contains the home chargers widespread in the distribution grid, where EVs can be parked at home and charged for a relatively long time. Other than the capital investment for constructing EV charging infrastructure and upgrading the transformer and transmission lines, the cost for maintaining the operation of charging facilities is taken into account. $C_i^{SC}(t)$ and $C_i^{FC}(t)$ are the continued investment of slow charging and fast charging facilities, respectively; $m$ and $n$ represent the number of distributed generators (DGs) and charging stations, respectively.

The objective functions can be various in the modeling of the optimization problem. Other criteria for optimizing the V2G operation include the improvement on the power quality and economy of power grid operation and maximize the profits and satisfaction for the EV users. Since EVs are mainly connected to the distribution grid, a typical objective function is formed to minimize the total power losses of the distribution network, which are impacted by the distribution of the charging or discharging power over the operating period:

$$f_3 = \sum_{t \in T} \sum_{i,j \in l} I_{ij}^2(t) R_{ij} \tag{15.2}$$

where $f_3$ is the power loss of the system; $I_{ij}$ represents the current between node $i$ and $j$; $R_{ij}$ is the resistance between node $i$ and $j$.

The objective functions with respect to user's benefits and the cost or reliability issues on the grid side can be put together to formulate a multi-objective optimization problem subject to the following constraints. The constraints can be categorized into three groups: limitations of traffic condition, power system operation, and electric devices.

1. Limitations of transport network

$$\sum_{j=1}^{n} X_{SC,ij} = 1 \tag{15.3}$$

$$\sum_{j=1}^{n} X_{FC,ij} = 1 \tag{15.4}$$

$$X_{SC,ij} \times X_{FC,ij} = 0 \tag{15.5}$$

where $X_{SC,ij}$ and $X_{FC,ij}$ are the locations of the facility in the power grid that is integrated with the transport network in the optimization problem. For locating the EV infrastructure and charging battery at the charging facility, transport network and traffic condition are associated with the layout of the feeders that supply the electric power. This equation indicates that the fast charging facility and slow charging facility are at different node of the power grid and transport network.

2. Limitations of power grid operation

$$P_i = U_i \sum_{j=1}^{n} U_j (G_{ij} \cos \delta_{ij} + B_{ij} \sin \delta_{ij}) \tag{15.6}$$

$$Q_i = U_i \sum_{j=1}^{n} U_j (G_{ij} \sin \delta_{ij} - B_{ij} \cos \delta_{ij}) \tag{15.7}$$

where $P_i$ is the active power injected into node $i$; $Q_i$ is the reactive power injected into node $i$; $U_j$ is the voltage magnitude of node $j$; $G_{ij}$ and $B_{ij}$ are the real part and imaginary part of nodal admittance matrix, respectively; $\theta_{ij}$ is the phase angle between node $i$ and $j$.

$$\sum_{j=1}^{N_i} P^i_{ev,j,\max} \leq P^i_{\max} \tag{15.8}$$

where $N_i$ is number of charging stations connected to node $i$; $P^i_{ev,j,max}$ is the maximum EV charging power of station $j$ which is connected to node $i$; $P^i_{max}$ is the maximum power of node $i$ .

$$V_{i,\min} \leq V_i(t) \leq V_{i,\max}, i \in N \tag{15.9}$$

where $V_i(t)$ is the voltage of node $i$ at time $t$; $V_{i,min}$ and $V_{i,max}$ are the upper and lower limits of the voltage of node $i$, respectively:

$$S_j \leq S_{j,\max}, j \in N_{branch} \tag{15.10}$$

where $S_j$ is the power flow of branch $j$; $S_{j,max}$ is the maximum transmission capability of branch $j$; $N_{branch}$ is the set of branches.

3. Limitations of electric devices

$$M_{DG} \leq M_{DG,\max} \tag{15.11}$$

$$G_{i,DG} \leq G_{DG,\max} \tag{15.12}$$

$$M_{CS,\min} \leq M_{CS} \leq M_{CS,\max} \tag{15.13}$$

$$G_{j,CS} \leq G_{CS,\max} \tag{15.14}$$

where $M_{DG}$ is the number of DGs in power system, and the limit of it is represented by $M_{DG,max}$. $G_{i,DG}$ is the capacity of $DG\ i$; $M_{CS}$ means the number of charging stations; $M_{CS,min}$ and $M_{CS,max}$ are the minimum and maximum numbers of charging stations, respectively; $G_{j,CS}$ is the capacity of charging station connected to the grid and $G_{CS,max}$ is its upper limit.

## 15.2 Vehicle-to-home

One important emerging concept in the integration of EVs is the V2H that is derived in the background of smart home (Xu and Chung, 2013). The EV converter is also needed to enable the two-way power flow but it does not have to be AC-DC power electronic converter which directly connects an EV to the single-phase AC grid (Erdinc *et al.*, 2015). The power network of a smart home can be designed to contain an internal low-voltage DC grid to facilitate the use of household appliances with DC input such as computer, fluorescent lights, and wind power or solar power generation units. In the smart home, AC power is converted into DC when entering the household network, using a high-efficiency rectifier which distributes the power directly to the DC load and DC renewable generation sources as well as energy storage and EV (Hegazy *et al.*, 2013). Referring to Figure 15.2, it can be seen that the number of power conversion stages of DC equipment can be reduced compared to the direct connection to the AC grid, and thus minimizing the energy losses caused by additional AC/DC and DC/AC power conversion. A home control unit is responsible for the overall scheduling of power output in generation, energy distribution, and load. In the V2H scenario, the home controller can monitor the EV state and coordinate the EV charging/discharging rate to follow the generation pattern of renewable energies connected to the same DC link.

### *15.2.1 PHEV control strategy for V2H applications*

The power exchange between EV and the grid can be optimized under the energy management structure of a single EV. The price signal is broadcasted by the grid operator and used by single EV to manage the energy stored in the battery. The financial benefits can be earned from shifting the EV charging load to the period of low electricity price. Other than using the battery energy to drive the vehicle electrically, the fuel cost for internal combustion engine (ICE) is also considered in the total energy cost which must be minimized by the optimal control method. In addition

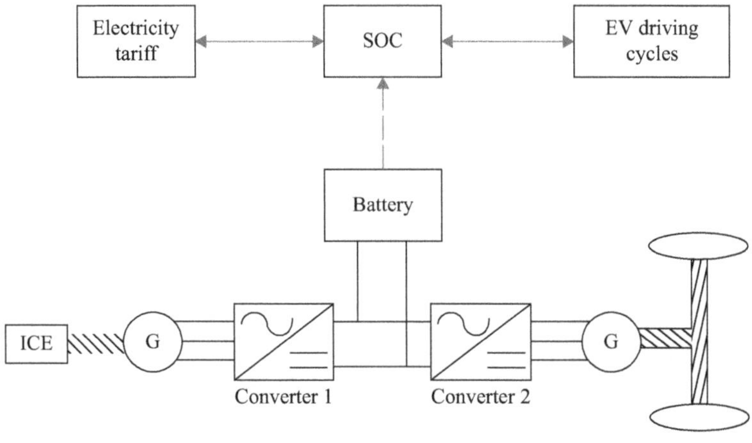


*Figure 15.2 Control of battery energy in an EV according to the changing electricity price*

to reducing the electricity bill, the revenue for selling V2G power to the grid can be included in the cost calculation. In this case, the EV charger must be capable of two-way power transfer and an AC-DC power electronic converter must be installed to connect the EV battery to the AC grid. As shown in Figure 15.2, the battery state of charge (SoC), as the indicator of energy storage, is usually chosen as decision variable and the optimal charging period is calculated and implemented by an EV converter.

V2H is taken as a good example of smart home application in which the rooftop PV panels or small wind turbines are installed. An optimal control method is often derived to maximize the portion of the EV charging demand served by the renewable energy. An internal DC grid is constructed within smart home shown in Figure 15.3, so that the rooftop PV, micro-turbines, and EV battery can be incorporated in its native DC form. Since the main load and distributed resource are DC in nature, the renewable power generation can be fed directly to EV battery without AC-DC conversion step, resulting in higher energy efficiency of the entire charging infrastructure. Another significant advantage of introducing the DC link into the charging infrastructure is that the well-established primary AC-DC converter is capable of regulating the power flows and its power quality in terms of power factor, voltage magnitude, and harmonic distortion.

### *15.2.2 V2H with demand response*

Several V2H control algorithms have been incorporated with the demand response strategies and the microgrid management models. Other than covering the load in the smart home, the remaining energy of EVs can be used as energy storage and feeding energy back to the grid outside the smart home. Various demand response strategies can be implemented for the EV coordination, such as the operating cost minimization, peak shaving/valley filling, frequency support, and volt-var control. The majority of V2H with demand response strategies investigate EVs for load shaping associated

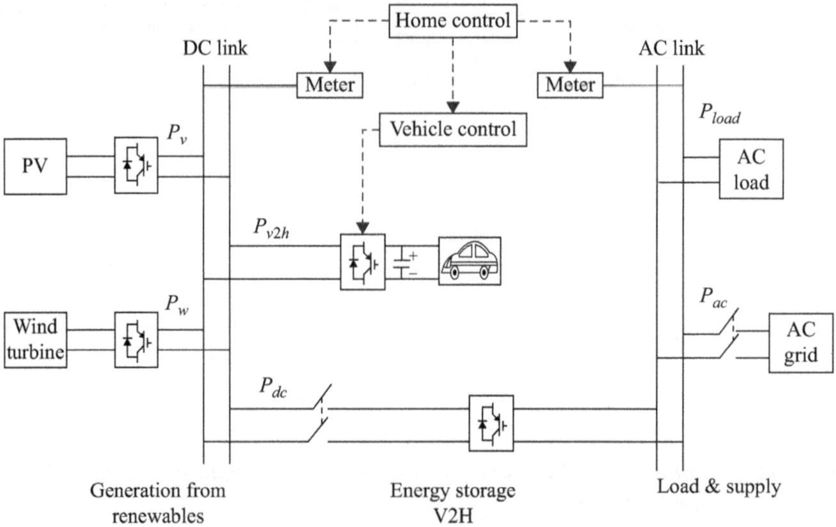


*Figure 15.3 Home control and vehicle control in V2H*

with the other smart electric appliances in a household, which can be grouped into three main types based on their features and behaviors, namely the even-smooth type, stochastic-behavior type, and fixed-program type. It can be modeled as daily load profiles (DLPs) to express their functions to the grid (Liu *et al.*, 2013). As depicted in Figure 15.4, it shows the typical DLPs of a household with three types of loads, which illustrates that the fixed-program type and the stochastic-behavior type dominate the power consumption. The DLP model can be obtained based on a huge measuring data collection. Figure 15.5 shows the flowchart of DLP generators according to the usage probability of household electric devices.

Thus, the objective can be set to minimize the total operating cost of a household or smooth the DLP. The EV modeling can be included in the constraints of a demand response problem. Similar to the energy storage installed at the customer side, EV is modeled based on the state of energy in a battery. Equations are added to prevent the deep discharge of EVs by imposing a least SoC limit. Moreover, the energy state of EV battery at the arrival time and the departure time should satisfy the driving need and the predefined parking time period.

## 15.3 Vehicle-to-vehicle

The control methods for EV charging and discharging can be categorized as central control and distributed control. Central control refers to the control structure where one control center manages the charging plan of all the EVs in the whole region. The control signal computed by the control center for each individual EV is transferred by the communication network and gives the selection of charging spot, charging, and departure time. Distributed control refers to the control framework, which includes multiple layers and EV aggregators are inserted to the middle layer

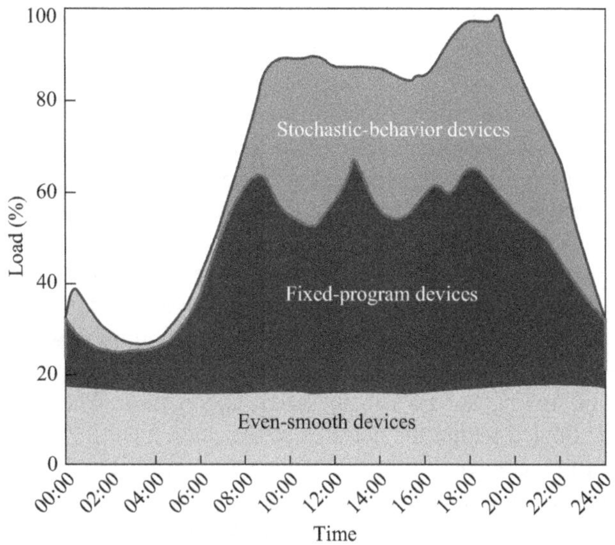


*Figure 15.4 Household DLP curves*

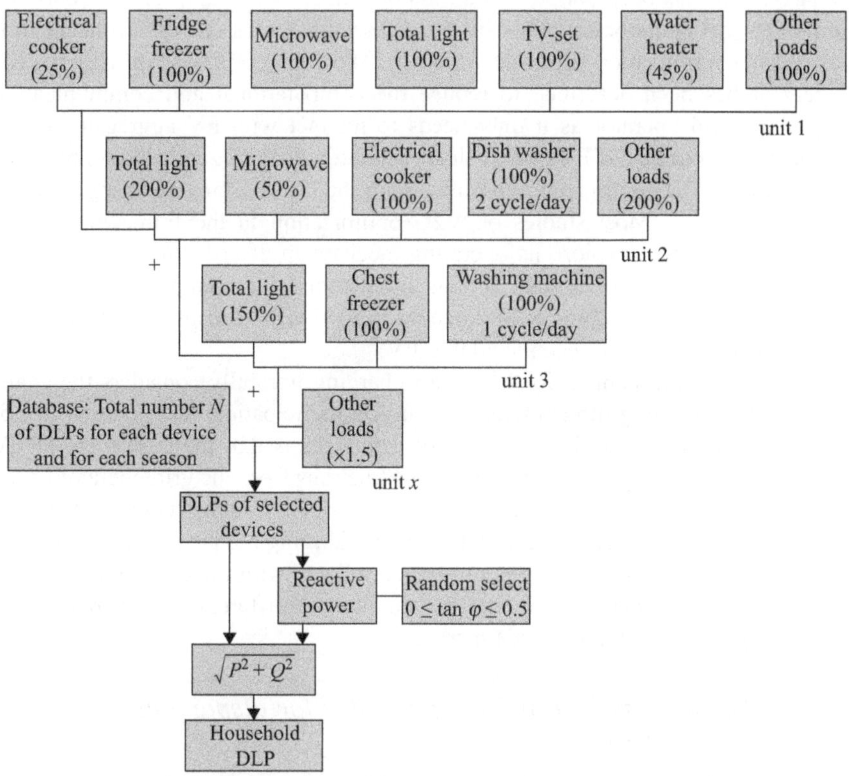


*Figure 15.5 Household DLP generator*

to control a group of EVs in a certain region. In this way, the control center only needs to deal with the several EV aggregators. The charging plan of each individual EV in the group supervised by the EV aggregator is given by the new entities in the distribution grid. The autonomous control under the EV aggregator decreases the complexity of the optimization problem formulated for smart charging of EVs, and thus the computation time and communication cost can be minimized.

The V2G control can be planned in a local power network such as an islanded microgrid, an EV charging station and a building or household power network (Shao *et al.*, 2015). Unlike V2H which only consider an EV's interaction with the other smart household appliances in the power network inside a home, a fleet of EVs are engaged in the optimal control scheme of the entire power network. The control strategies of V2V are needed to coordinate the charging/discharging and power exchange among the EVs in the same aggregation. The power exchange among EVs is allowed within the network so that the urgent EV charging demand can be satisfied immediately by the EVs which are intended to feed their power to the grid. In this way, the double energy conversion losses induced by AC/DC and DC/AC power conversion between individual EV and AC grid can be eliminated.

### *15.3.1 Concept and structure of EV aggregator*

EV aggregator has been adopted as an intermediate control unit inserted between the power grid operator and EVs in many research works on V2G concept, framework and optimization (Bessa *et al.*, 2014). The main purpose of introducing EV aggregator has been described to reduce the computational and communication load of the grid operator as it only needs to interact with EV aggregator representing a large number of EVs rather than each individual EV, and also simplify the formulation of V2G optimization problem with the EV charging/discharging being decision variables. Most studies on V2G optimization in the framework of the power grid operation explore its operating mechanism of how the optimal aggregated power is realized through timing and varying the charging of every EV underneath. Accordingly, an EV dispatching method is used to allocate the targeted V2G regulation power for each available EV.

The market mechanism for EV smart charging normally considers the availability of EV charging infrastructure and user's participation. The prerequisite of the aforementioned control and regulation strategies is that the customers of EV charging facilities fully accept the control instructions from the grid operator or the EV aggregator. The profit for the customers that sign a contract with the charging station manager is also guaranteed in this smart charging mode. The role played by the new entity such as EV aggregator is to coordinate with the distribution system operator (DSO) and manage the charging facilities of different sizes to participate in the energy and ancillary service market.

### *15.3.2 Control method of EV aggregator for dispatching a fleet of EVs*

Owing to the limited power rating of individual EV, a number of EVs are aggregated to form a sizable ancillary service provider. Consequently, an intermediate control unit named EV aggregator is designed to take charge of vehicles underneath and interact with the grid operator. The aggregator is responsible for assembling EV power capacities to bid into the ancillary service market and provide more reliable service against the stochastic availability of individual vehicle by modulating the charging/discharging of a fleet of EVs. Unlike the battery energy storage device, the V2G power is composed of a certain number of EVs, the aggregated regulation power has to be allocated to each individual EV.

The V2G power at each time interval is optimized by the aforementioned control method, and the signals are given out to each EV aggregation. To implement V2G energy scheduling, EVs are required to charge or discharge at different rates depending on the driving demand and battery storage capacity of each individual vehicle. An EV dispatching method is designed for EV aggregator to allocate the scheduled power to individual EV. In the framework, the EVs are split into five operating modes to perform the multi-objective control scheme. EVs that are not engaged in the optimal energy scheduling are tagged as in the idle mode and deployed for dynamic power regulation to stabilize the system frequency and bus voltage. EVs are ranked in each mode mainly depending on the driving pattern, current SoC departure time, and charging preference, and then the charging/discharging rate is calculated by the dispatching method. It should be noted that there will be internal energy exchange among the EVs which are electrically connected within a single aggregation, which is defined as V2V operation, since in this dispatching method different EVs are selected to perform charging and discharging in the same time interval. The major merit of V2V is that the charging demand is locally supplied so that power transfer to the external power grid is minimized.

As shown in Figure 15.6, at the bottom level, EV charging facilities execute the control signal sent by the aggregator. As depicted in Figure 15.7, the SoC of the EV

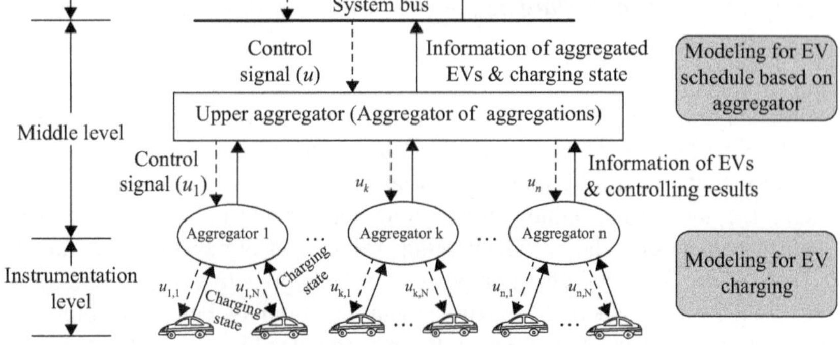


*Figure 15.6 Definition of various EVs*

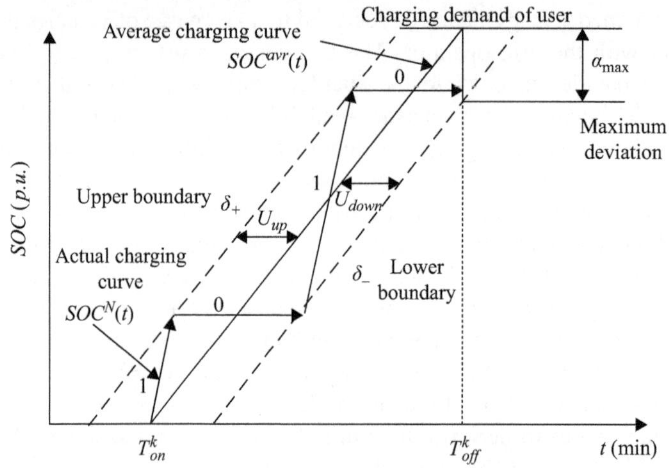


*Figure 15.7 Control strategy of EV SoC*

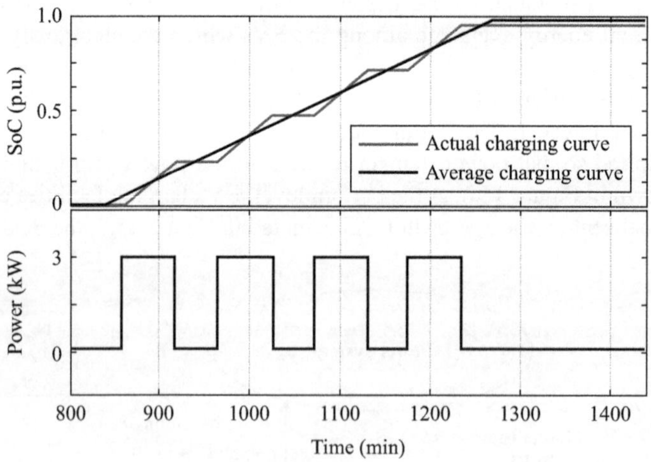


*Figure 15.8 SoC and charging power of EV under smart charging scheme*

battery is varying while charging EVs according to the control signal. Figure 15.8 shows the charging state of the EV and the changes on SoC over the operating period. The control strategies in the middle level are mainly designed for the aggregator, including the autonomous control for each aggregator and the coordinated control among the aggregators. The overall charging profile of an aggregator is the sum of all EVs in its realm. The total charging power for the aggregators of different sizes is shown in Figure 15.9, where the maximum and minimum load profiles represent the adjustable capacity of an EV aggregator. It can be found in Figure 15.10 that the actual

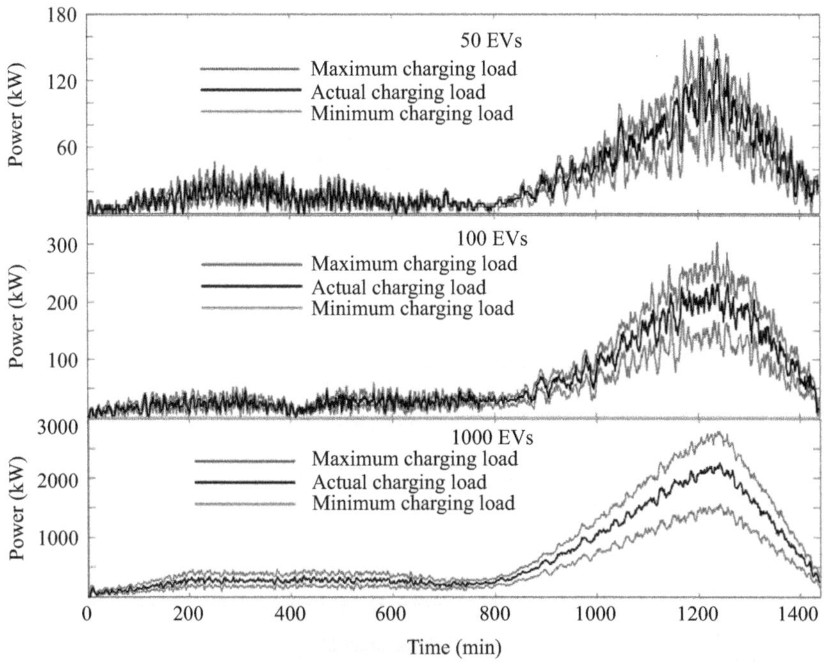


*Figure 15.9 Total charging loads of EV groups with different numbers of EVs*

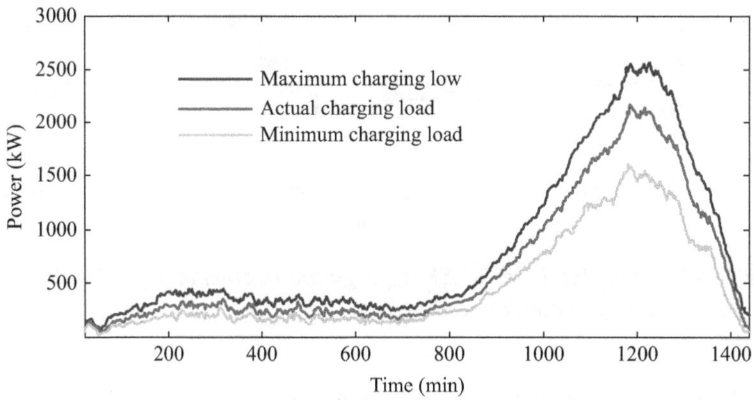


*Figure 15.10 Charging loads of EV aggregation within boundaries*

charging profile of EV aggregation is between the upper and lower limits. The control target is given by the optimization in the system level. EV aggregations with their considerable battery energy storage can participate in various ancillary services in the power system. The EV aggregator aims to track the target by allocating the power to each EV. Figures 15.11 and 15.12 show a comparison between the actual charging

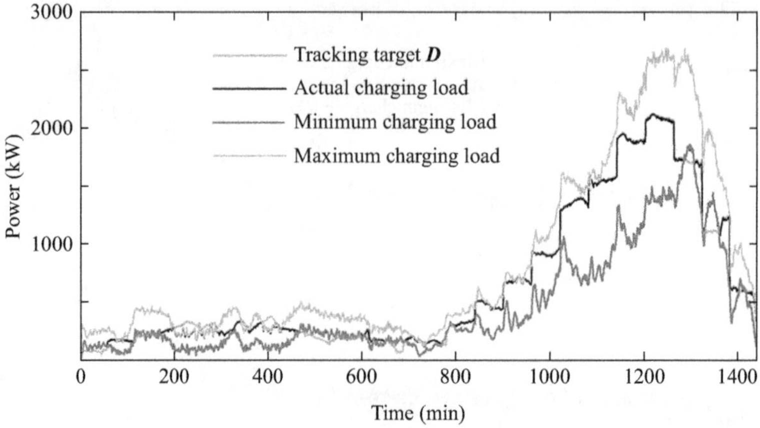


*Figure 15.11 Charging loads of EV aggregation in comparison with the target (smaller boundary of SoC)*

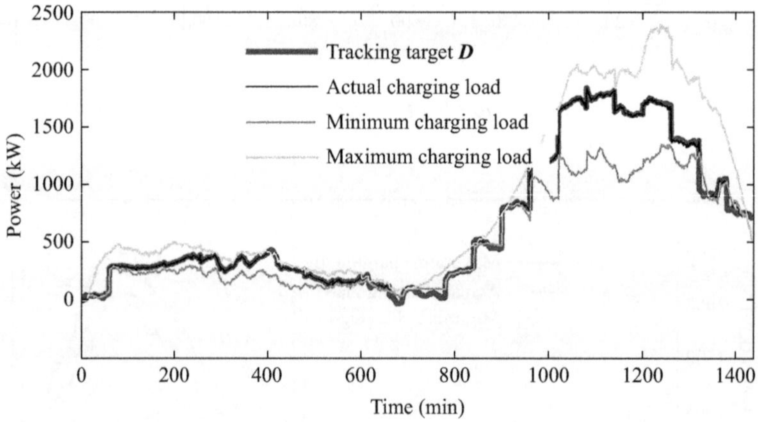


*Figure 15.12 Charging loads of EV aggregation in comparison with the target (larger boundary of SoC)*

curve of EV aggregation and the control target. The overall charging power is consistent with the target as long as the target power is within the adjustable capacity. Moreover, the limits of the battery SoC when the EV has to depart from the charging facility can be set as constraints to measure the user satisfaction level. The boundary of SoC at the departure time can be adjusted by each EV owner in the control algorithm. The upper and lower boundaries of SoC in Figure 15.12 are set to be larger than those boundaries in Figure 15.11, which changes the charging profile and the calculated adjustable capacity of EV aggregation. The target can be more precisely tracked if the larger SoC deviation from the desired value is allowed.

## 15.4 Vehicle-to-grid

As mentioned earlier, the objectives of V2G are to integrate EVs into the power grid and unleash full potential of smart charging and V2G operation to support the grid operation in terms of power quality, reliability, cost-efficiency, and sustainability, which offers the following advantages: (1) enhance the power supply in accordance with the power quality standards; (2) encourage the participation of EVs into the intelligent energy management; (3) provide ancillary services using EVs while satisfying the needs for driving and parking of the vehicles; (4) integrate energy flow and information flow to improve the interaction between EVs and power grid; (5) facilitate the large penetration of renewable energy sources.

Figure 15.13 depicts the modeling of V2G operation. The major components can be categorized into three layers. The upper layer provides the ancillary services to the power grid. Energy scheduling, spinning reserve (energy storage), and frequency and voltage regulations are performed on different time scales, ranging from hours to seconds. EV aggregator as the middle layer interacts with the grid operator and EVs are dispatched to achieve the instructions given by the upper optimization model. Communication and electric infrastructures are at the lower layer where two-way information flow and bi-directional power flow are enabled by the current ICT and power electronic technologies. Particularly, EV charging facilities are integrated into a cyber-physical system in such a way that the charging and discharging of EVs can be controlled to serve the grid as distributed energy storage. From the perspective of power system, the optimal control of V2G power should be designed considering operating limits of power system, charging infrastructure, and the limited information obtained from the communication and metering units.

### *15.4.1 Planning of V2G infrastructure in the smart grid*

The planning of the EV charging infrastructure in the distribution grid is an integral part of building an EV smart charging network. The main consideration of EV

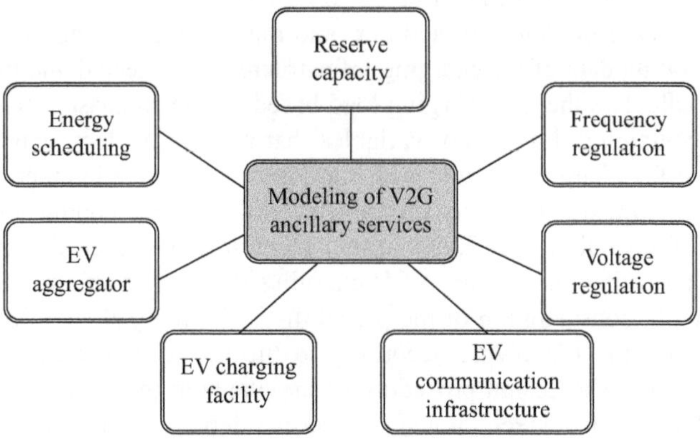


*Figure 15.13 Modeling of V2G operation*

infrastructure planning in the recent studies is given to determine the number and distribution of different types of EV infrastructure so as to meet the charging demand for mobile need and at the same time minimize the installation and operation cost of both EV charging infrastructure and the power grid. The application of energy internet has considerable impact on the existing planning strategies for EV charging network in the distribution grid. Since a large amount of renewable energy resources are integrated into the energy internet such as wind power and solar panels, the objective function must be formed to accommodate and consume the clean energy to the maximum. Furthermore, energy resources other than electricity, for example, thermal energy and load, natural gas, in the energy internet provides new opportunities for building integrated energy system where multiple energy resources can be complementary with each other to achieve high energy efficiency.

The EV charging and discharging can be controlled in accordance with the optimization algorithm to mitigate the power fluctuation caused by the intermittent power generation from renewable energy resources such as wind and solar power plants. The EVs connected to the charging infrastructure, which enable bi-directional power and information flow can be used as flexible electric load and battery energy storage to maintain the power balance so that a greater amount of renewable power generation units can be accommodated in the distribution grid without any new investment to improve the power system reliability. Meanwhile, the construction of EV charging infrastructure is associated to the model of energy internet. The capacity and location of each individual EV charging facility need to match with the structure of energy internet as well as the monitoring condition and management ability. Based on the existing planning and construction method for integrating distributed power generation units in the energy internet, the theory for integrating EV charging infrastructure can be derived to plan the EV smart charging network according to the real need of EV charging demand at each node of the charging network which combines the information from the transport network and drive behavior of each individual EV.

An integrated planning model needs to be built considering load forecasting, locating, and sizing of charging facilities, and multi-objective optimization methods. The operating data of the charging infrastructure is collected and processed in real time to calculate the EV charging load based on the stochastic model of load forecasting containing the random variables that reflects the temporal and spatial uncertainty of EV charging. In this way, further studies on the interaction between EV charging infrastructure and the power grid can be conducted with the assumption of several typical EV charging scenarios. Based on the precise load forecasting, the siting and sizing of EV charging facilities is able to meet the EV charging need in consideration of the probabilistic driving path and driving direction selected by the EV owner according to the traffic condition in the urban transport network. The design procedure of the planning of EV charging network blended with the data and technologies of energy internet is shown in Figure 15.14. New concepts must be considered into the optimization problem formulated for planning the EV charging network as the energy internet is introduced.

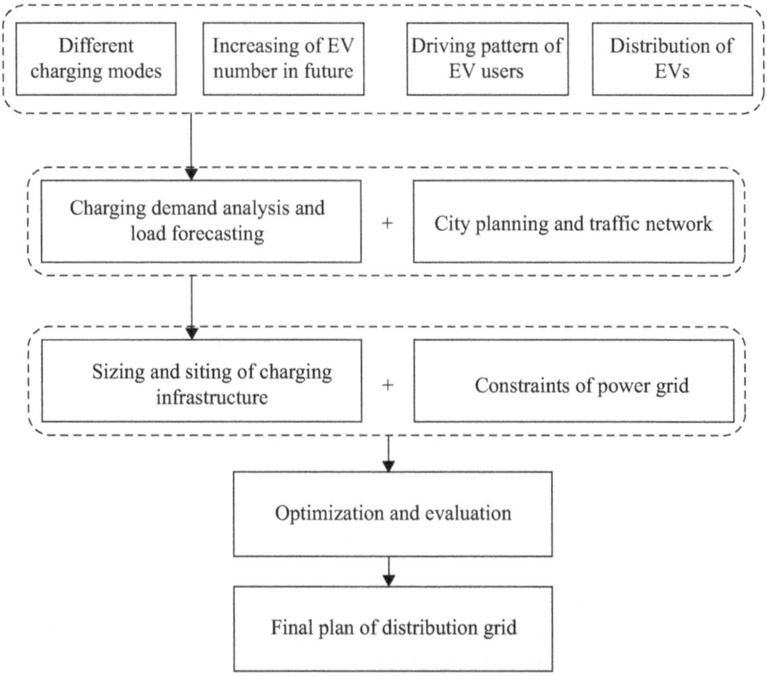


*Figure 15.14 Flowchart for planning of EV charging infrastructure*

The research focus is directed to the supply and consumption of clean energy and the complementary utilization of diverse energy systems other than electricity grid. The EV charging network can perform as controllable load or functions as backup energy storage so that a larger scale of clean energy can be accommodated to the power grid and the energy efficiency can be improved due to the comprehensive utilization of different energy resources, such as cooling and heating, electricity, and natural gas.

## *15.4.2 Ancillary services provided by V2G*

Ancillary services refer to those functions performed by the power system (generation, transmission, and distribution) to support energy transmission from primary resources to the consumers while maintaining reliable operation of the integrated transmission system. In Europe, the Union of the Electricity Industry (EURELECTRIC) has defined ancillary services as "all services required by the transmission or distribution system operator to enable them to maintain the integrity and stability of the transmission or distribution system as well as the power quality." In the United States, the Federal Energy Regulatory Commission (FERC) has defined ancillary services as "those services necessary to support the transmission of electric power from seller to purchaser given the obligations of control areas and transmitting utilities within those control area to maintain reliable operations of the interconnected transmission system." From the viewpoint of electric power market, the ancillary services market usually contains regulation and frequency response, operating reserve, energy imbalance, and sometimes volt-var operation and black start functionalities (Lin *et al.*, 2014). There are several ancillary services where EVs with its V2G capability can provide, including load shifting/peak shaving, spinning reserve, frequency regulation, reactive power, and voltage supports. Those V2G services can also facilitate the renewable integration in the setting of future distribution grid where abundant renewable resources are used for power generation.

The modeling of V2G system and its control algorithm are firstly developed. The impact of large-scale EV penetration on the power quality is quantitatively analyzed in terms of voltage profile, transmission line loading, and power losses. In the coordinated charging/discharging EVs (Vaya and Andersson, 2013), a V2G optimization problem is formulated to regulate battery storage energy for mobility needs and power grid supports such as acting as shiftable load and spinning reserve. In the area of dynamic power regulation of EVs (Sun *et al.*, 2014), the battery energy storage has been modeled to repeatedly absorb and feed back the power to the local grid to compensate the instant power imbalance caused by fluctuating wind energy. The work can be further extended to manage the dynamic power of battery storage in V2G regulation services. There are several key technologies to solve the issues of V2G ancillary services.

#### 15.4.2.1 V2G mathematical model and system configuration

A mathematical model to investigate the integration of EVs into the power grid must be built. The V2G model should include power system operation, characteristics of EVs and EV charging facilities that enable the bi-directional power flow for smart charging, and V2G operation (Gao *et al.*, 2010). The system configuration is carried out to determine the initial values of the parameters and variables in the EV model and power system model by performing power flow calculation and detecting the states of the EV as well as the key factor of the on-board battery like the SoC. Since there are no available design equations to form the EV model in the simulation environment of power system optimization, the design process is based on several considerations: specifications of different EV models, driving pattern of EV owner (availability of grid-connected EVs), spatial distribution of EVs in the power grid, power capacity of charging infrastructure, and charging scenarios. In block 1 of Figure 15.15, the V2G model and system configuration are shown in the block named "control center."

In this research area, the V2G concept can be introduced to microgrid or distribution grid with high penetration of DGs to support the grid operation (Liu *et al.*, 2013). Furthermore, the V2G concept is applied to develop a V2G model that combines the EV model with the power system model. Namely, the key characteristics of different types of EVs is extracted to formulate an EV mathematical model and turned into the power system optimization problem to participate in various grid supports, including load shift (peak load shaving and valley filling), operating cost minimization, and frequency and voltage stabilization.

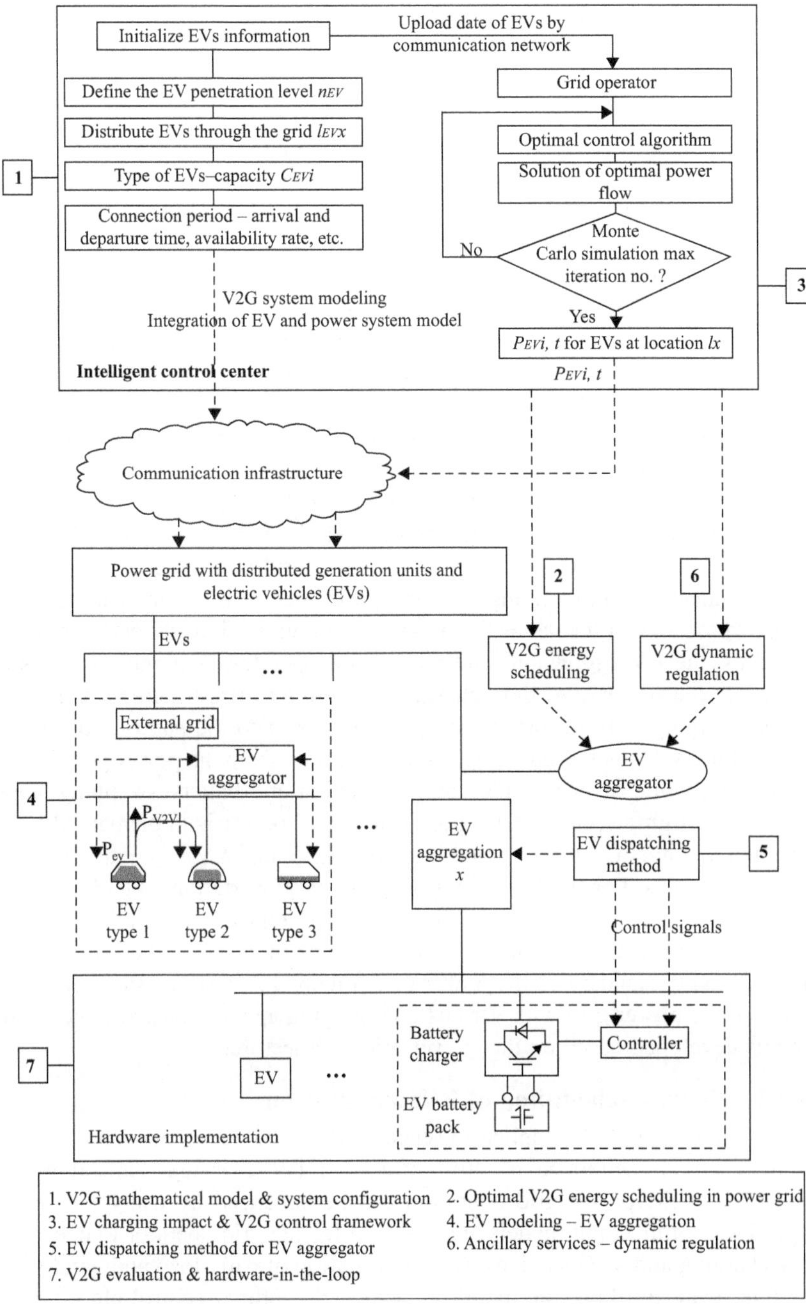


*Figure 15.15 Key technologies in providing V2G ancillary services*

#### 15.4.2.2 Impact evaluation of EV integration and V2G control framework

Although the above modeling approaches can provide the general idea about the potential benefits and impacts of EVs on the power system, quantitative analysis must be conducted by means of power flow computation while the EV charging load over the planning time period is added on the top of the daily load demand (Gao *et al.*, 2012). The power quality in terms of bus voltage and transmission line loading is assessed in different types of distribution grid. Voltage variation and transmission line loading are important factors concerned by the distribution network operator (DNO) because the line loading limits the energy supply while voltage variations disrupt the operation of electric appliances. In order to restrict the voltage variation within a tolerable range and reduce the line loading, control methods are needed to adjust the EV charging rate and charging time to coordinate the EV charging load with the changes in energy consumption and power generation from renewable energy sources.

The conventional resource-centric control model of the electric power system is no longer sustainable, thus a control framework for V2G operation is developed on the basis of the existing structure of the power grid with its diverse resources. The communication and metering network is assumed to be established for smart grid application so that it enables the two-way communication and data acquisition among EVs, the EV aggregator, and power grid operator. A number of EVs are plugged into the power grid through various charging infrastructures such as the large DC charging stations, battery swapping station, and the parking lots at residential buildings. Domestic charging draws electricity from the power grid at a low charging rate, while dedicated EV charging station and battery swapping station provide fast charging options. The coordinated V2G power is implemented by the bi-directional converters that connect individual EVs and EV aggregations at different power levels. For the EV aggregation at charging station or parking lots, a high-power AC-DC converter is usually adopted to adjust the overall power flow for fulfilling the charging requirements while maintaining the power factor within the power quality standard of the power grid. Figure 15.15 shows the hierarchical structure of the power grid with various EV charging applications, and the control scheme is developed based on this power grid architecture.

#### 15.4.2.3 Energy scheduling of EVs integration

Control schemes should be applied to schedule the V2G power in such a way that not only the aforementioned negative effects of EV charging demand will be eliminated, but the power grid operation can be enhanced by using EV battery storage. Two optimization models are designed for the grid operator to coordinate the EV charging and V2G services with the variation of daily load and wind power generation while fulfilling the mobility needs. The optimal control algorithm for V2G energy scheduling is targeted to: (1) minimize the total power losses of the test distribution grid; and (2) minimize the operating cost subject to the constraints from both power system model and EV model (Gao *et al.*, 2011). As a result of power losses minimization, the EV charging load is redistributed over the time and also the improvement on the power quality is evaluated in terms of voltage deviation and transmission line loading. In the cost minimization, a V2G revenue is defined to offer incentives to V2G participants while a penalty will be imposed on the unserved EV charging demand. The participation in V2G is significantly related to the value of V2G revenue, so the utility needs to choose appropriate revenue to encourage the participants engaged in the V2G program without being a too expensive option for the power grid operator. All the EVs are required to be fully charged by the end of the parking period although the parking periods, initial and targeted SoCs as well as the participation in V2G services are subject to the choice of EV owner. A Monte Carlo simulation approach is adopted to reflect the uncertain availability of grid-connected EVs and thus illustrate the effectiveness of the optimal control method regardless of the variation of numerical results in any specific circumstances. Intuitively, the EV charging load should be shifted to the off-peak hours and the energy can be fed back to the grid to supply the peak load of the day. The optimal solutions can be found by the simulation model so that the charging/discharging rates for each EV aggregation over the whole parking period are properly determined. This energy scheduling method for EVs in the grid with wind power generation has been developed.

#### 15.4.2.4 V2G dynamic regulation services

One way to include EVs in active power market is for them to provide regulation reserve to the utility. Due to the relatively limited energy capacity and fast response speed, EVs are expected to participate in the primary control market in load frequency control (LFC) which requires the positive spinning reserve up to 15 min in case of equipment or power supplier failures. Despite the fact that the primary control market requires the least duration for regulation reserve, this V2G service induces the deep discharging of EV battery and thus resulting in large battery degradation costs. Rather than providing regulation reserve in which battery storage energy is significantly reduced, the depth of discharge in dynamic frequency regulation does not seriously affect the battery cycle life, alleviating the battery degradation. Moreover, owing to the alternating charging and discharging in response to the negative and positive regulation demands, the net variation of energy can be zero. Namely, there is no energy difference in the battery storage while providing dynamic frequency regulation on the time scale of seconds. It is a remarkable advantage of frequency regulation over the V2G load regulation, since the approximate mean of long-term energy is zero and there is no increase on the battery degradation cost. In the microgrid, EV aggregations close to the terminal of the wind plant can be employed to flatten the power output of wind power generator. The dynamic V2G power regulation is conducted to compensate for the real-time power imbalance caused by the continuous variation of wind power generation, keeping the frequency and voltage stable. As tagged as block 6 in Figure 15.15, a simulation model of the V2G frequency regulation in the power grid with wind energy has been developed to analyze the system dynamics in the process of V2G frequency stabilization.

Additionally, with the introduction of high-performance power electronics, EVs are capable of providing reactive power regulation. Since the bi-directional four-quadrant power transfer is allowed by the EV charger, reactive power regulation can be accomplished without any additional investment. If the capacitor positioned at the DC link of the converter is sufficient to the source of reactive power, the charging facility alone can maintain the regulation service without an EV connected. EV aggregations scattered over the feeders are employed to inject the reactive power so as to mitigate the voltage drop at the connected bus in the test distribution grid. Similar to the aforementioned energy scheduling, an optimal control algorithm in couple with AC power flow calculation is devised to minimize the total power losses while the power limits of charging facilities are taken into account.

#### 15.4.2.5 Evaluation of V2G control scheme and hardware-in-the-loop design process

In order to investigate the practicality of the V2G control scheme, a small power network is constructed to interact with the simulation software running on the computer. Battery packs, programmable load, and power supply are connected together by AC-DC and DC-DC converter modules associated with its electronic controllers to form a down-scale experimental setup so as to simulate the operation of a microgrid with EVs and renewable power generation units. An experimental platform is firstly set up. The power electronic converter and its controller are tested for smart power delivery in a simplified power network (Gao *et al.*, 2014). The topology of the EV battery converter and the control block diagram are depicted in the bottom block of Figure 15.15. The power flow of the battery packs are regulated by the bi-directional converter in accordance with the control signal sent by simulation software and meanwhile responds to the varying power supply from renewable energy sources. Both steady-state and dynamic performances can be measured to evaluate whether the control of battery storage can effectively mitigate the power fluctuation caused by intermittent power supply or varying electric load. Additionally, based on the features that are extracted from experimental data of the hardware-in-the-loop system for V2G power regulation, the mathematical model and the corresponding optimal control algorithm can be further improved.

#### 15.4.2.6 Cyber-physical system for V2G applications

The control of EV charging relies on the successful integration of communication network and EV smart charging network. V2G applications are supported by cyber-physical system where advanced control method is applied to EV charging infrastructure by using ICT (Cao *et al.*, 2014). The concept of energy internet and its technological development provide a solution to the proper coordinated control between EVs and the power grid. As shown in Figure 15.16, a cyber-physical system equipped with communication and metering devices is established to support the aforementioned V2G applications.

Energy internet plays an important role in the smart transport system, smart city, and smart grid. All the electric appliances in the grid if needed are connected to the Internet, which allows the communication among the devices and the data transfer. In the energy internet, the smart identification, location, tracking, monitoring, and management are realized by using a variety of technologies and the corresponding

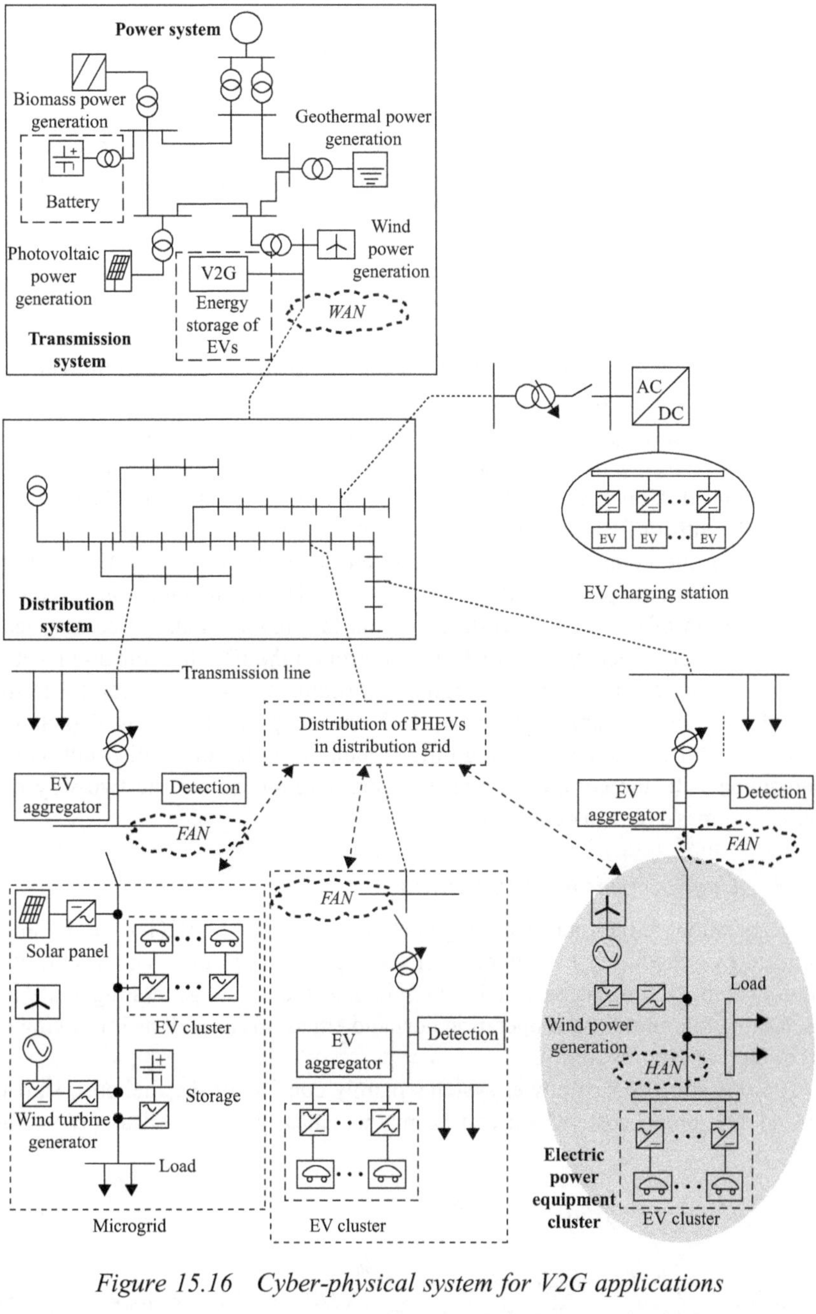


*Figure 15.16 Cyber-physical system for V2G applications*

protocol, consisting of radio frequency identification (RFID), infrared sensor, global positioning system (GPS), and information sensing devices like laser scanner. The development of energy internet provides new challenges and opportunities to the charging infrastructure investment, intelligent control, and market operation of EV charging network. More importantly, EV charging facilities, if properly managed, can be viewed as the virtual battery energy storage, and gives one of the best solutions to the supply and consumption of clean energy resources.

The communication interface must be unified for connecting EV charging infrastructure into the power grid. At present, the international standardization organization (ISO) has already set ISO standard for communication between EVs and the charging facilities. The Society of Automotive Engineers (SAE) is committed to develop the standard J2836 and J2847, which can be applied to the communication network covering EVs, charging infrastructure, power grid, and customers. The bi-directional power flow and information are taken into account to perform V2G operations. A joint research has been carried out by the Institute of Electrical and Electronics Engineers (IEEE) and SAE on interoperability standard that supports the integration of smart charging network into the energy internet. Another on-going project is to add information exchange mode for the interface between EV charging facilities and the power grid in the system of International Electrotechnical Commission (IEC) 61850, which is specifically designed for distributed control of EVs in the framework of energy internet.

Based on the communication interface, the interaction between EV charging infrastructure and the power grid can be achieved. The communication network is built on the top of electrical infrastructure and the massive data processing technology is adopted to achieve the real-time control on the EV charging and locating according to the results of smart charging algorithm. A multi-layer control framework of EV smart charging network with the support of ICT is depicted in Figure 15.16, where the EV charging infrastructure installed in the distribution grid is integrated with the communication infrastructure sustaining the two-way data transfer and massive data processing.

### 15.4.3 Cost-emission optimization

As aforementioned, the total operating cost and emissions of V2G systems can be optimized by scheduling EV aggregations and generating units. The key to cost-emission optimizations is to efficiently schedule the small generating units and EVs, aiming to minimize the operation cost and emissions as well as improving the system reserve and reliability.

The operating cost of power systems mainly consists of the fuel cost, start-up cost and shut-down cost, which can be mathematically expressed as (Wu *et al.*, 2012):

$$FC_i(P_i(t)) = a_i + b_iP_i(t) + c_iP_i^2(t) \tag{15.15}$$

$$SC_i(t) = \begin{cases} hcost_i & MD_i \le X_i^{off}(t) \le H_i^{off} \\ ccost_i & X_i^{off}(t) > H_i^{off} \end{cases} \qquad H_i^{off} = MD_i + cshour \tag{15.16}$$

where $FC_i()$ and $SC_i()$ denote the fuel cost and start-up cost, respectively. Due to the independence on other parameters, the shut-down cost is normally taken into account as a constant, like zero, for standard power systems. In terms of the $i$th unit, $P_i(t)$ is the output power, $hcost_i$ and $ccost_i$ are the hot start-up cost and cold start-up cost, $cshour_i$ is the cold start hour, $MD_i$ is the minimum down time, and $X_i^{off}(t)$ is the duration of continuously off. In addition, $a_i$, $b_i$, $c_i$ are positive fuel cost coefficients.

The fuel cost function is denoted as $EC_i$, which can be given by (Wu *et al.*, 2012):

$$EC_i(P_i(t)) = \alpha_i + \beta_i P_i(t) + \gamma_i P_i^2(t) \tag{15.17}$$

where $\alpha_i$, $\beta_i$, $\gamma_i$ are the emission coefficients of the $i$th unit. Thus, the fitness function of the V2G cost-emission optimization can be obtained as (Wu *et al.*, 2010):

$$\begin{aligned} \min TC = \sum_{i=1}^{N}\sum_{t=1}^{H} \big[ & w_c(FC_i(P_i(t)) + SC_i(1 - I_i(t-1))) \\ & + w_e(\psi_i EC_i(P_i(t)))\big] I_i(t) \end{aligned} \tag{15.18}$$

where $N$ is the number of units, $H$ is the scheduling hours, $I_i(t)$ is the status of the $i$th unit, and $\psi_i$ is the emission penalty factor. Additionally, $w_c$, $w_e$ are the weight factors, which make the fitness function take into account the cost or emission by setting as one and ignore the cost or emission by setting as zero.

Besides, the cost-emission optimization essentially takes into account practical constraints as follows.

- The scheduling period is set to 24 h, which means that the scheduling plan of the generating units and EVs is daily updated.
- The maximum capacity of the power grid is fixed for EVs, which means a constant total number of EVs for simultaneously connecting to the grid. It can be expressed as (Wu *et al.*, 2012):

$$\sum_{t=1}^{H} N_{EV}(t) = N_{EV\max} \tag{15.19}$$

$$N_{EV}(t) \le N_{EV\max} \tag{15.20}$$

  where $N_{EV}(t)$ is the number of EVs connected to the grid, and $N_{EGVmax}$ is the total number of EVs involved in the V2G system.
- When EVs are plugged out, the corresponding SoC should be maintained at a certain desirable level. Actually, the owner can set the desirable SoC and lowest SoC limitation based on their requirements. In order to avoid unexpected unplugs of EVs, the lowest SoC limitation is normally set to 60%. It means that the EV is not allowed to feed the grid once below the lowest SoC limitation. Besides, in order to extend the battery lifetime, EVs are assumed to perform the discharge once a day.
- The output power of the generating units and EVs has to satisfy the load demand. Meanwhile, the spinning reserves should be maintained for the

system reliability. It can be given by (Wu *et al.*, 2012):

$$\sum_{i=1}^{N} I_i(t)P_i(t) + P_{EV}N_{EV}(t) = D(t) \tag{15.21}$$

$$\sum_{i=1}^{N} I_i(t)P_{i\max}(t) + P_{EV\max}N_{EV}(t) \geq D(t) + R(t) \tag{15.22}$$

where $P_{EV}$ is the available capacity of each EV, $P_{EVmax}$ is the maximum capacity of each EV, $P_{imax}(t)$ is the maximum output power, $D(t)$ is the load demand, and $R(t)$ is the spinning reserve.

- The output power of each generating unit cannot be beyond its extreme boundaries. Meanwhile, the durations of committed and uncommitted units should not less than their minimum start-up and shut-down times, respectively. It can be expressed as (Wu *et al.*, 2012):

$$P_{i\min} \leq P_i(t) \leq P_{i\max} \tag{15.23}$$

$$\begin{cases} MU_i \leq X_i^{on}(t) \\ MD_i \leq X_i^{off}(t) \end{cases} \tag{15.24}$$

where $P_{imin}$ and $P_{imax}$ represent the minimum and maximum output limits of the $i$th unit, respectively. $MU_i$ represents the minimum up time, and $X_i^{on}(t)$ represents the duration of continuously on.

## 15.5 Conclusion

In this chapter, the interactions between EVs and the power grid are categorized as V2H, V2V, and V2G energy systems. The methodologies typically used for modeling EVs with the electric power network in these three operating modes are presented respectively in the above sections. Two-way energy exchange is allowed by applying power electronic devices as the EV charging facility. Communication network is established on the top of the electric power network which enables flexible interaction between EVs and the control entities in the power grid. The EV as one of the household electric appliances can participate in the control scheme of smart home, which is referred to as V2H. Although the battery energy in each EV is small, the energy stored in a group number of EVs which are widespread in the distribution grid can be considerable for providing grid supports. Thus, an EV aggregator as the new entity in the power system needs to be introduced to manage the battery energy in each individual EV and interacts with the power grid operator. The autonomous control of an aggregator regulates the charging state of each EV in its realm, including the power exchange among EVs, namely V2V. The coordinated control of aggregator deals with the interaction between aggregators. EVs aggregated at different locations in the distribution grid can be utilized to achieve an overall objective. A variety of

V2G ancillary services can be provided by this aggregated EV energy storage system. By applying V2H, V2V, and V2G operations in the power grid, the energy stored in the on-board battery of EVs can be coordinated to offer a more economically efficient use of electricity while satisfying the mobility needs. The new concepts, such as energy internet, that will emerge in the future will provide more flexibility and economic possibilities for the interaction of EVs with the smart grid as the technical advances of ICT and metering in the cyber-physical system.

## References

Ansari, M., Al-Awami, A.T., Sortomme, E., and Abido, M.A. (2015) Coordinated bidding of ancillary services for vehicle-to-grid using fuzzy optimization. *IEEE Transactions on Smart Grid*, 6(1), 261–270.

Becker, T.A. and Sidhu, I. (2009) *Electric Vehicles in the United States, a New Model with Forecasts to 2030*, Center for Entrepreneurship & Technology (CET), Berkeley, CA, USA: University of California.

Bessa, R.J., Matos, M.A., and Soares, F.J. (2014) Framework for the participation of EV aggregators in the electricity market. *Proceedings of IEEE Conference on Electric Vehicle*, pp. 1–8.

Cao, Y., Wang, N., Kamel, G., and Kim, Y.J. (2014) An electric vehicle charging management scheme based on publish/subscribe communication framework. *IEEE System Journal*, 1(1), 1–14.

Chau, K.T. (2014) Pure Electric Vehicles. In *Alternative Fuels and Advanced Vehicle Technologies for Improved Environmental Performance*, ed. R. Folkson, Amsterdam: Woodhead Publishing.

Chukwu, U.C. and Mahajan, S.M. (2014) V2G parking lot with PV rooftop for capacity enhancement of a distribution system. *IEEE Transactions on Sustainable Energy*, 5(1), 119–127.

Erdinc, O., Paterakis, N.G., Mendes, T.D.P., Bakirtzis, A.G., and Catalao, J.P.S. (2015) Smart household operation considering bi-directional EV and ESS utilization by real-time pricing-based DR. *IEEE Transactions on Smart Grid*, 6(3), 1281–1291.

Gao, S., Chau, K.T., Chan C.C., and Wu, D. (2010) Modelling, evaluation and optimization of vehicle-to-grid operation. *World Electric Vehicle Journal*, 4, 809–817.

Gao, S., Chau, K.T., Chan C.C., Liu C., and Wu, D. (2011) Optimal control framework and scheme for integrating plug-in hybrid electric vehicles into grid. *Journal of Asian Electric Vehicles*, 9(1), 1473–1481.

Gao, S., Chau, K.T., Liu C., Wu, D., and Li, J. (2012) SMES control for power grid integrating renewable generation and electric vehicles. *IEEE Transactions on Applied Superconductivity*, 22(3), 1–4.

Gao, S., Chau, K.T., Liu C., Wu, D., and Chan C.C. (2014) Integrated energy management of plug-in electric vehicles in power grid with renewables. *IEEE Transactions on Vehicular Technology*, 63(7), 3019–3027.

Hegazy, O., Barrero, R., Van Mierlo, J., and Coosemans, T. (2013) An advanced power electronics interface for electric vehicles applications. *IEEE Transactions on Power Electronics*, 28(12), 5508–5521.

Jian, L., Xue, H., Xu, G., Zhu, X., Zhao, D., and Shao, Z.Y. (2013) Regulated charging of plug-in hybrid electric vehicles for minimizing load variance in household smart microgrid. *IEEE Transactions on Industrial Electronics*, 60 (8), 3218–3226.

Li, R., Wu, Q., and Oren, S.S. (2014) Distribution locational marginal pricing for optimal electric vehicle charging management. *IEEE Transactions on Power Systems*, 29(1), 203–211.

Lin, J., Leung, K., and Li, V.O.K. (2014) Optimal scheduling with vehicle-to-grid regulation service. *IEEE Internet of Things Journal*, 1(6), 556–569.

Liu, C., Chau, K.T., Wu, D., and Gao, S. (2013) Opportunities and challenges of vehicle-to-home, vehicle-to-vehicle, and vehicle-to-grid technologies. *Proceedings of the IEEE*, 101(11), 2409–2427.

Liu, H., Hu, Z., Song, Y., and Wang, J. (2013) Vehicle-to-grid control for supplementary frequency regulation considering charging demands. *IEEE Transactions on Power Systems*, 1(1), 1–10.

Shafie-Khah, M. and Catalao, J.P.S. (2015) A stochastic multi-layer agent-based model to study electricity market participants behavior. *IEEE Transactions on Power System*, 30(2), 867–881.

Shao, C., Wang, X., Wang, X., and Du., C. (2015) Layered and distributed charge load dispatch of considerable electric vehicles. *IEEE Transactions on Power Systems*, 30(4), 1858–1867.

Sun, S., Dong, M., and Liang, B. (2014) Real-time welfare-maximizing regulation allocation in dynamic aggregator-EVs system. *IEEE Transactions on Smart Grid*, 5(3), 1397–1409.

Vasiladiotis, M. and Rufer, A. (2015) A modular multiport power electronic transformer with integrated split battery energy storage for versatile ultrafast EV charging stations. *IEEE Transactions on Industrial Electronics*, 62(5), 3213–3222.

Vaya, M.G. and Andersson, G. (2013) Optimal bidding strategy of a plug-in electric vehicle aggregator in day-ahead electricity markets. *IEEE Transactions on Power Systems*, 30(4), 2375–2385.

Xu, N.Z. and Chung, C.Y. (2013) Reliability evaluation of distribution systems including vehicle-to-home and vehicle-to-grid. *IEEE Transactions on Power Systems*, 1(1), 1–10.

Yilmaz, M. and Krein, P.T. (2013) Review of battery charger topologies, charging power levels and infrastructure for plug-in electric and hybrid vehicles. *IEEE Transactions on Power Electronics*, 28(5), 2151–2169.

Wu, D., Chau, K.T., and Gao, S. (2010) Cost-emission analysis of vehicle-to-grid system. *World Electric Vehicle Journal*, 4, 767–773.

Wu, D., Chau, K.T., Liu, C., and Gao, S. (2012) Genetic algorithm based cost-emission optimization of unit commitment integrating with griddable vehicles. *Journal of Asian Electric Vehicles*, 10(1), 1567–1573.

# Chapter 16
# Vehicle-to-grid power interface

*Zheng Wang*[1] *and Yue Zhang*[1]

The power interface plays an important role in efficient energy conversion between the grid and the energy storage elements such as the battery pack of electric vehicles (EVs) and hybrid electric vehicles (HEVs). On the basis of infrastructure and size, the power interface for EVs can be categorized into on-board chargers and off-board chargers. Between them, the on-board chargers enable the convenient charging with only suitable outlets. At the same time, the bidirectional power converters become the emerging techniques for EV chargers with the requirements of vehicle-to-grid (V2G) operation for smart grid as well as the increasingly strict limits of power quality. This chapter will take an overview of the power converters for power interface of EVs, including not only unidirectional topologies but also bidirectional topologies.

## 16.1 Introduction

High-efficiency charging is one of the technical challenges for battery management in EVs. The efficiency and performance of chargers are very important to the operation and life time of battery packs. EV battery chargers can be classified into on-board or off-board chargers. Between them, the on-board chargers decrease the requirements of charging infrastructure and simplify interconnection issues. The on-board charging can be implemented wherever a suitable power outlet is available. On the other hand, the off-board chargers can achieve fast charging by using high-power converters. Figure 16.1 shows the configuration of energy conversion systems for EVs, where the battery charger is used as the power interface between the EVs and the grid (Yilmaz and Krein, 2013). Normally, the topologies of battery charger for EVs consist of two-stages: one AC/DC rectifier with power factor correction (PFC) and a unidirectional or bidirectional DC/DC converter to interface with battery. The unidirectional EV chargers transmit power from the grid to the battery pack and operate with almost the fixed input power factor. In contrast, the bidirectional topologies generally transfer the power in two ways between the EVs and the grid. Unidirectional charging has the advantages of limited hardware requirements and simple control strategies

[1]School of Electrical Engineering, Southeast University, Nanjing, China

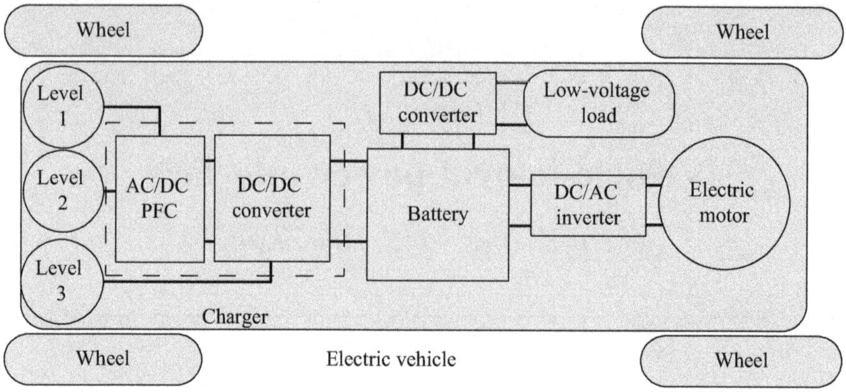


*Figure 16.1 Configuration of energy conversion systems in EVs*

compared to bidirectional charging. Furthermore, the battery in the unidirectional charging mode does not need to charge or discharge frequently. Thus, the life of the battery will be extended. Unidirectional chargers also simplify EVs feeder management in heavy load. The techniques of bidirectional chargers are emerging nowadays due to the benefits of V2G applications in the smart grid. In addition, the good harmonic performance and flexible power factor can be provided on grid side with the bidirectional EV chargers. The bidirectional EV chargers need precise sensors, smart distribution system and flexible communication. It will increase cost of power devices and the base installation.

It should be noted that V2G power interface denotes the power interface in general between EVs and the grid, which includes the unidirectional EV chargers for grid-to-vehicle (G2V) power flow and the bidirectional EV chargers for both G2V and V2G power flows.

Based on power level, the EV chargers are classified into the level-1, level-2 and level-3 charging. The level-1 and level-3 are normally single-phase and three-phase solutions, respectively, whereas the level-2 can be applied in both solutions (Yilmaz and Krein, 2013). The level-1 charging (120 V/240 V) is slow and convenient charging which can be used at home. Due to the power limits, most of on-board chargers use the level-1 charging (Kim *et al.*, 2010). The level-2 charging (240 V) is a compromise between time and convenience, which can be used through both private and public facilities. For fast charging, the level-3 (at least 480 V) charging technique is developed, which is a kind of DC charging. The level-3 charging is applied in commercial and public site such as market parks, operating as a charging station. The off-board chargers are used for the level-3 charging. But the high-power rapid charging may generate on-peak impacts on grid and power equipments compared to lower power charging (Williamson *et al.*, 2015). To standardize the EV chargers and limit their deleterious effects on the grid, several organizations, such as IEEE, the Society of Automotive Engineers (SAE) and the Infrastructure Working Council (IWC), plan to establish charging standards of EVs. Besides, the harmonic and DC current injection into the grid and EV chargers have been governed by the international standards such as IEEE-1547, SAE-J2894, IEC1000-3-2, IEC 61851-1 and the U.S. National Electric Code (NEC) 690 (Khaligh and Dusmez, 2012).

## 16.2 Two-stage power interface

The typical power conversion of the two-stage power interface contains an AC/DC rectifier on grid side and a DC/DC converter on battery side. The configuration and operation of AC/DC rectifiers on grid side should comply the requirements of grid codes, whilst the DC/DC converters ought to realize the efficient charging and discharging operation for battery packs.

### *16.2.1 AC/DC rectifiers*

Battery chargers may produce deleterious harmonic effects on grid because of discrete operation of power switches in rectifiers. Thus, the rectifiers need to shape the grid currents for low total harmonic distortion (THD) and high power factor while converting AC supply into an intermediate DC link. Depending on the direction of power flow, the AC/DC rectifiers are classified into the unidirectional rectifiers and the bidirectional rectifiers. The diode rectifiers with PFC stage and the front-end pulse width modulation (PWM) rectifiers are the two main categories, accordingly.

#### 16.2.1.1 Conventional PFC rectifiers

The topology shown in Figure 16.2 is the most popular topology for diode rectifier with PFC stage because of its simple structure and control strategy. It consists of a diode bridge rectifier converting AC input current into DC current, and a boost converter for PFC. The switch $S_o$ can be driven by an integrated control chip which generates PWM signals to control the boost converter for near unity power factor on the grid side. Both the average current mode control and the peak current mode control can be used to maintain stable and low-distortion sinusoidal line current of the conventional PFC rectifiers (Darly *et al.*, 2011).

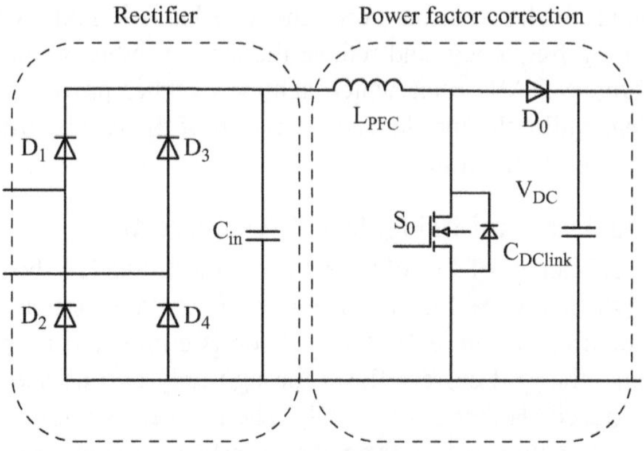


*Figure 16.2 Conventional PFC rectifiers*

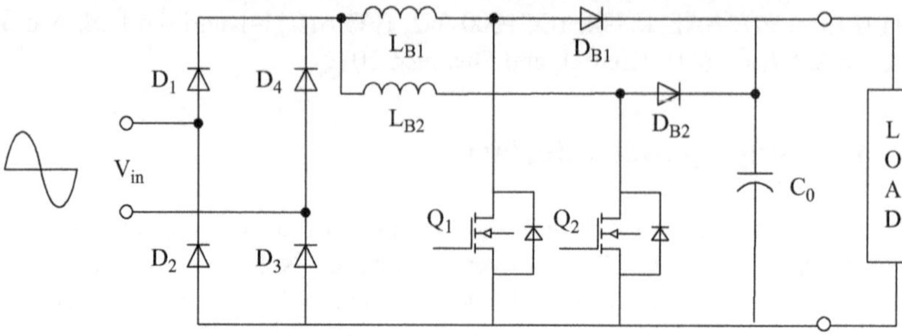


*Figure 16.3 Interleaved PFC rectifier*

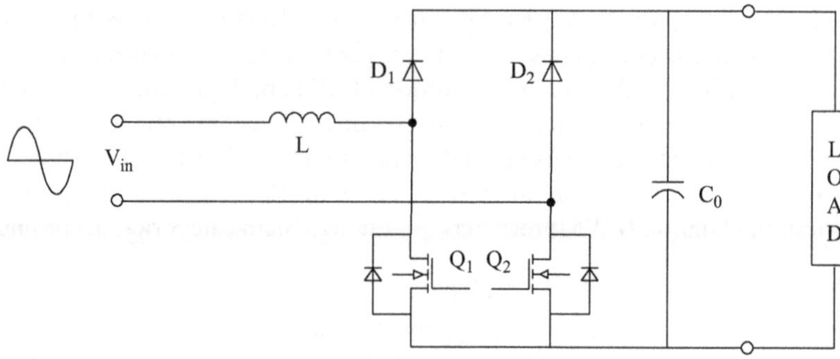


*Figure 16.4 Bridgeless PFC boost converter*

The interleaved PFC rectifier is described in Figure 16.3, which is suitable for high power-level EV chargers. It combines two conventional boost converters in parallel, operating 180° out of phase. The currents through one of the two inductors $L_{B1}$ and $L_{B2}$ are equal to half of the input current. Because the inductors' ripple currents are out of phase, they cancel with each other and the total input ripple current is reduced finally. Furthermore, the interleaved mode will double the equivalent switching frequency, and reduce the size of inductor and input electromagnetic interference (EMI) filter which achieves higher power density (Musavi *et al.*, 2011). Meanwhile, the interleaved operation of the two rectifiers can cancel out the ripples in the output voltage.

#### 16.2.1.2 Bridgeless/semi-bridgeless PFC converters

To improve the efficiency of the PFC rectifiers, the bridgeless boost topology is invented as shown in Figure 16.4 (Sudheer *et al.*, 2015). Compared to the conventional PFC rectifier in Figure 16.2, one diode is eliminated in the current path on grid side. Thus, the grid current flows through only two devices, and the conduction loss is reduced (Huber *et al.*, 2008). The bridgeless boost converters have higher efficiency, but they suffer from the significant common-mode noise. As shown in Figure 16.2, the output ground of the conventional PFC rectifier is always

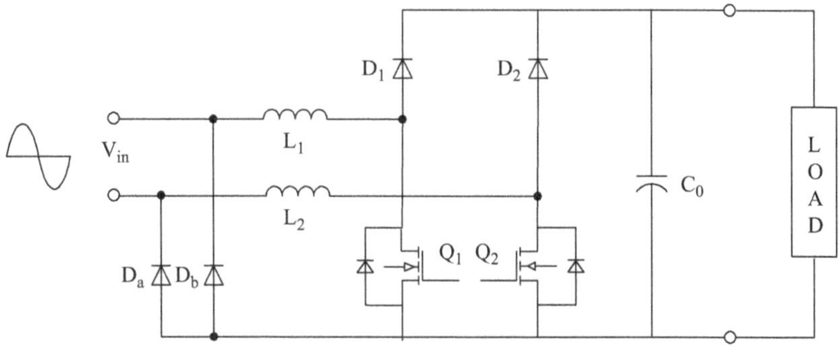


*Figure 16.5 Semi-bridgeless PFC boost converter*

connected to the grid through the diodes of full-bridge rectifier. However, in the bridgeless PFC boost rectifier of Figure 16.4, the output ground of rectifier is kept being connected to the grid through the free-wheeling diode of one power switch during the positive-half fundamental cycle. On the other hand, the relative voltage between the output ground of rectifier and the grid is high-frequency pulsating during the negative-half fundamental cycle. This high-frequency pulsating voltage charges and discharges the equivalent parasitic capacitance between the output ground and the AC line ground. Thus, a large leakage current is produced in this bridgeless PFC boost rectifier.

To maintain the low conduction loss while avoiding the large leakage current, a semi-bridgeless PFC rectifier is presented in Figure 16.5. Two additional diodes, $D_a$ and $D_b$, and a second inductor are added to the bridgeless PFC rectifier in Figure 16.5. By this way, two boost DC/DC converters are formed. During the positive-half fundamental cycle of grid voltage, the first boost converter $L_1$, $Q_1$ and $D_1$ work and the diode $D_a$ connects the output ground of rectifier to the grid. During the negative-half fundamental cycle of grid voltage, the second boost converter $L_2$, $Q_2$ and $D_2$ work and the diode $D_b$ connects the output ground of rectifier to the grid. So, the relative voltage between the output ground of rectifier and the grid varies slowly, and the common-mode noise can be eliminated (Huber *et al.*, 2008; Musavi *et al.*, 2013).

Figure 16.6 shows the configuration of the interleaved bridgeless PFC converter. This topology is developed for high power level EV chargers. Compared to the interleaved boost PFC mentioned earlier, the bridgeless PFC converter uses four active switches and four fast diodes. The gating signals for two groups of MOSFETs are 180° out of phase, similar to the interleaved boost converter. A detailed converter description and steady state operation analysis are given in Musavi *et al.* (2013). This converter topology can achieve high input power factor and high efficiency on wide load range and low input current harmonics.

#### 16.2.1.3 Half-bridge/full-bridge AC/DC rectifiers

The half-bridge and full-bridge configurations are widely used for the front-end PWM rectifiers. Figure 16.7 shows the single-phase half-bridge and full-bridge

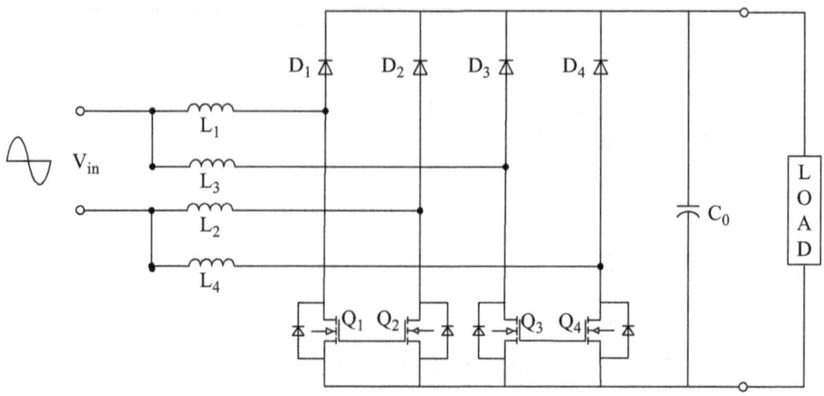


*Figure 16.6 Interleaved bridgeless PFC boost converter*

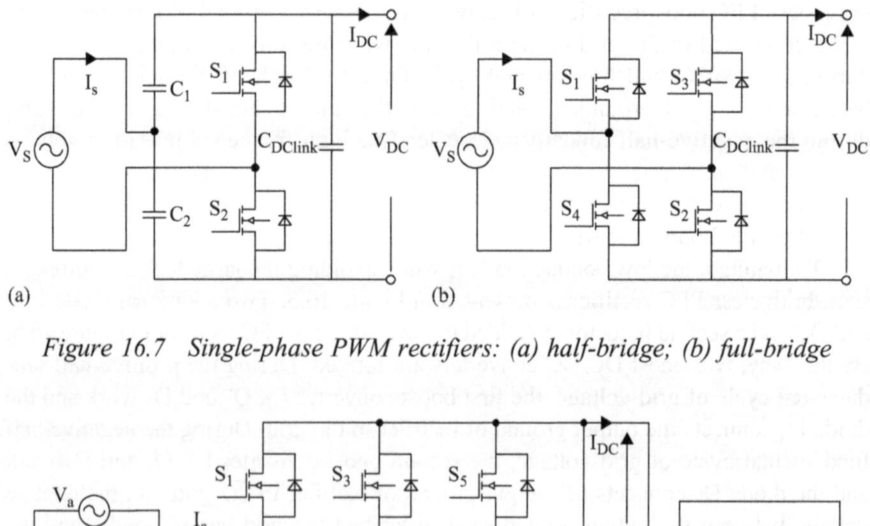


*Figure 16.7 Single-phase PWM rectifiers: (a) half-bridge; (b) full-bridge*

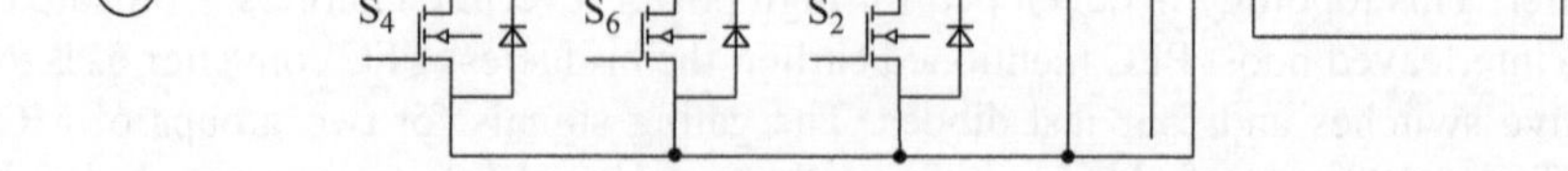


*Figure 16.8 Three-phase PWM rectifier*

PWM rectifiers, and Figure 16.8 shows the three-phase full-bridge PWM rectifier. The half-bridge topologies can achieve voltage doubling with the split capacitors in DC link. But the semiconductor devices with higher voltage ratings are required and the imbalance in the split capacitors should be avoided for the half-bridge AC/DC rectifiers. The full-bridge PWM converters can avoid the issue of capacitor

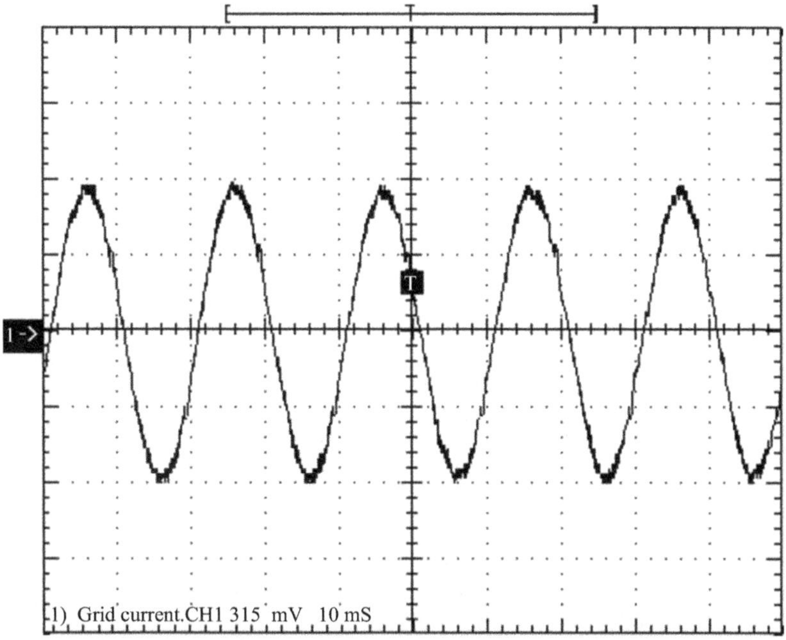


*Figure 16.9 Measured grid current of single-phase PWM rectifier (X: 10 ms/div; Y: 3 A/div)*

voltage imbalances and allow power switches with lower voltage rating, but they need more semiconductor devices, which increases the cost and complexity of the control. The PWM technique is widely used to trigger the power switches and modulate the voltage reference for the half-bridge and full-bridge rectifiers. Both the unipolar and the bipolar carrier based modulation strategies can be used. For the three-phase PWM rectifiers, the sinusoidal PWM (SPWM) and the space vector modulation (SVM) schemes are applied for generating the switching signals (Pahlevaninezhad *et al.*, 2012). By using the closed-loop control of the output DC voltage and the PWM strategies, the sinusoidal grid current waveforms are available for both the single-phase PWM rectifier and the three-phase PWM rectifier as shown in Figures 16.9 and 16.10, respectively (Wang *et al.*, 2013; Zou *et al.*, 2014).

#### 16.2.1.4 Multilevel AC/DC rectifiers

Figure 16.11 shows a single-phase unidirectional multilevel rectifier used in on-board battery chargers. This topology has four operation modes: mode1 ($S_1$ and $S_2$ ON), mode2 ($S_1$ ON and $S_2$ OFF), mode3 ($S_1$ OFF $S_2$ ON) and mode4 ($S_1$ and $S_2$ OFF). The blocking voltage of each power device is clamping to half of the DC link voltage $V_{DC}$. The single-phase unidirectional multilevel EV chargers can achieve low THD, high power factor and reduced EMI (Singh *et al.*, 2003). Voltage regulating function of this multilevel topology can make DC output voltage robust to load variation (Carlton and Dunford, 2001).

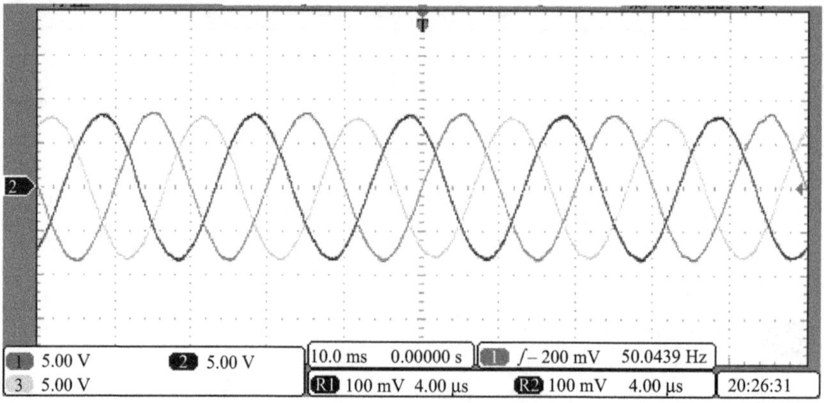


*Figure 16.10 Measured grid current of three-phase PWM rectifier (X: 10 ms/div; Y: 5 A/div)*

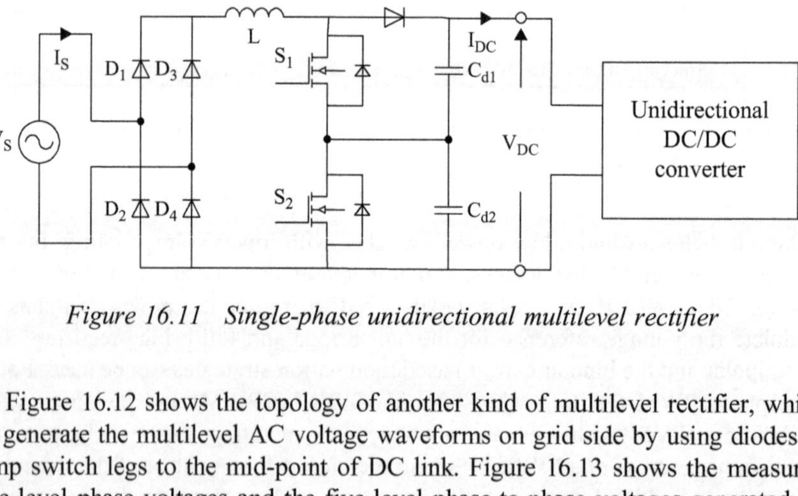


*Figure 16.11 Single-phase unidirectional multilevel rectifier*

Figure 16.12 shows the topology of another kind of multilevel rectifier, which can generate the multilevel AC voltage waveforms on grid side by using diodes to clamp switch legs to the mid-point of DC link. Figure 16.13 shows the measured three-level phase voltages and the five-level phase-to-phase voltages generated by the multilevel rectifier in Figure 16.13. Since the low current ripples are achieved in line currents of multilevel AC/DC rectifiers, the filter size and switching frequency can be reduced for the converter. Besides, the voltage stresses of power devices are reduced. However, it has disadvantages including more components and complex modulation strategies compared to two-level rectifiers (Wang *et al.*, 2015a).

#### 16.2.1.5 Current-source AC/DC rectifiers

Compared to the voltage-source rectifiers, the current-source rectifiers offers the advantages of good output voltage waveforms, low dv/dt and high short-circuit fault tolerant capability (Wang *et al.*, 2011; Wang *et al.*, 2012). Figure 16.14 (a) shows the topology of the current-source rectifier. Instead of the electrolytic capacitors, a DC choke is used as the energy storage element in DC link. The AC capacitors are needed for both phase commutation and harmonic filtering. The reverse blocking ability is

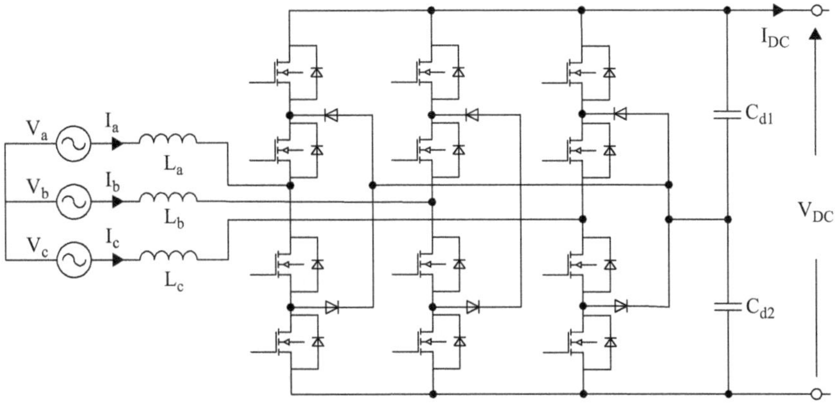


*Figure 16.12 Neural-point-clamping three-level rectifier*

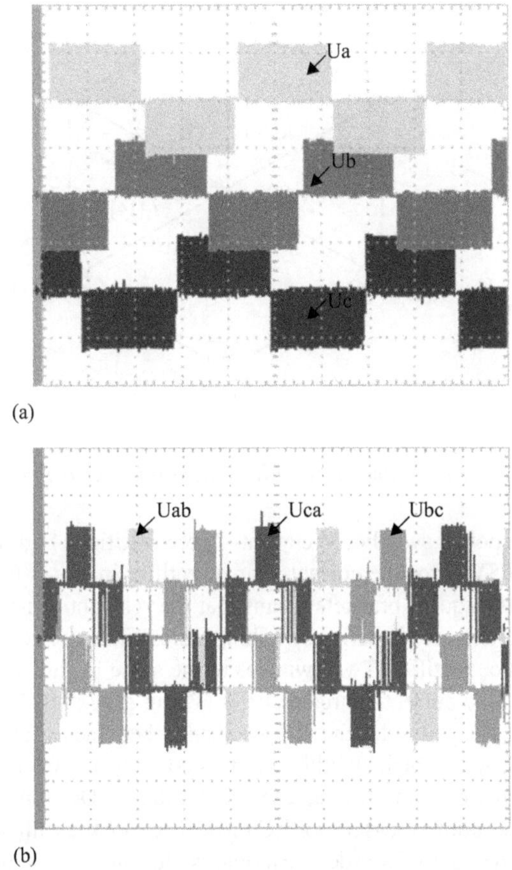


*Figure 16.13 Measured voltage waveforms of neutral-point-clamping three-level rectifier (X: 5 ms/div; Y: 50 V/div): (a) phase voltage; (b) phase-to-phase voltage*

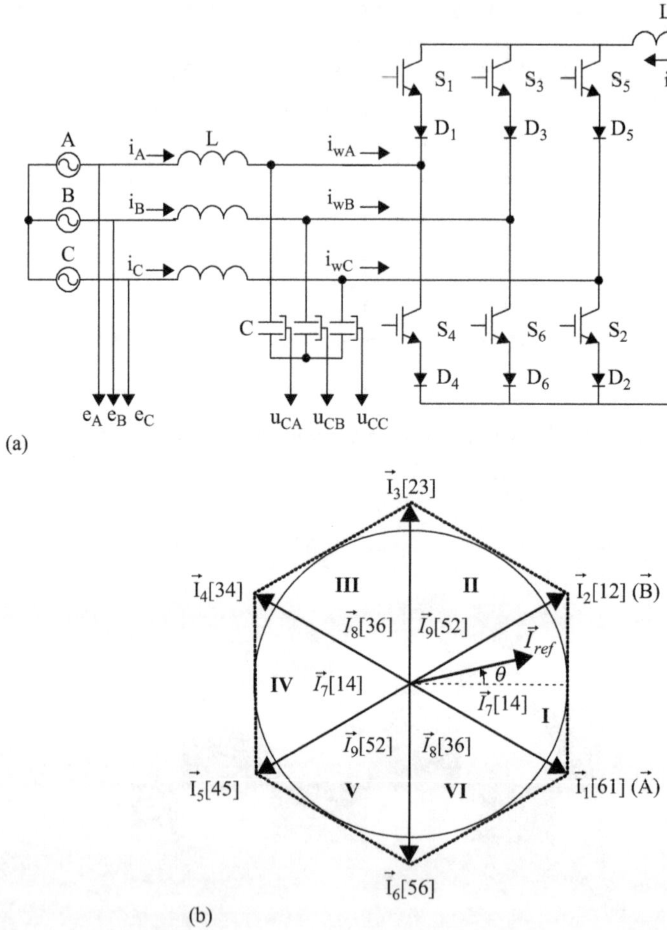


*Figure 16.14 Current-source rectifier: (a) topology; (b) vector diagram*

also in need for the power switches in current-source rectifiers. Figure 16.14(b) shows the diagram of the SVM for the current-source rectifier in which the significance of numbers given in the square brackets means that the relevant switches are turned on at the same time. There are six active switching states and three zero switching states for the current-source rectifier. Two switches in the same leg are allowed to conduct at the same time in the current-source rectifier (Wang *et al.*, 2012). Due to the existence of LC filter on grid side, the LC resonance becomes a challenge issue for current-source rectifiers, which should be mitigated well. Figure 16.15(a) shows the measured waveforms of grid current, capacitor voltage, DC link current and grid voltage without any countermeasure for LC resonance. So, the fifth-order harmonic is observed clearly since the fifth-order harmonic is close to the resonant frequency of LC filter. On the other hand, the LC resonance can be mitigated by using control means on grid side as shown in Figure 16.15(b).

(a)
(b)

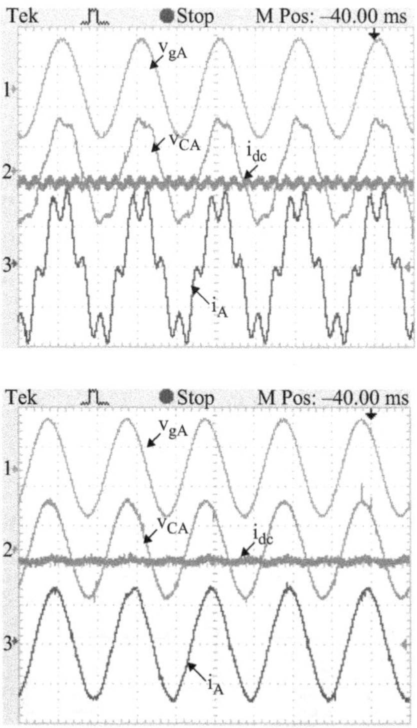


*Figure 16.15 Measured waveforms of current-source rectifier (X: 10 ms/div; Y: $v_{gA}$ 100 V/div, $v_{CA}$ 100 V/div, $i_A$ 2 A/div, $i_{dc}$ 2.5 A/div): (a) without LC resonance damping; (b) with LC resonance damping*

## *16.2.2 DC/DC converters*

The DC/DC converters function to match DC voltage levels between the output DC side of AC/DC rectifiers and DC side of battery packs, and also exchange power between the two DC sides. According to the direction of power flow, the DC/DC converters of battery chargers can be classified into unidirectional and bidirectional topologies. According to the galvanic isolation ability, the DC/DC converters of battery chargers can be categorized into non-isolated and isolated topologies.

### **16.2.2.1 Non-isolated DC/DC converters**

The non-isolated unidirectional converters usually have simple topologies and they normally use fewer components. But there exist challenges in high-gain voltage conversion and electrical isolation with non-isolated converters. Most of non-isolated DC/DC converters are derivatives from the basic topologies such as buck, boost, buck-boost, cuk, sepic, zeta. Applying the three-level technology to buck-boost converter, the three-level DC/DC converter can be realized (Gautam *et al.*, 2011; Mohammadpour *et al.*, 2015). By using the three-level DC/DC converter, the

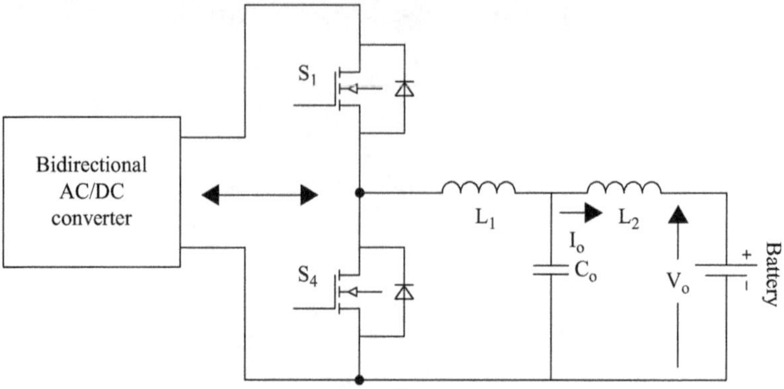


*Figure 16.16 Non-isolated bidirectional half-bridge DC/DC converter*

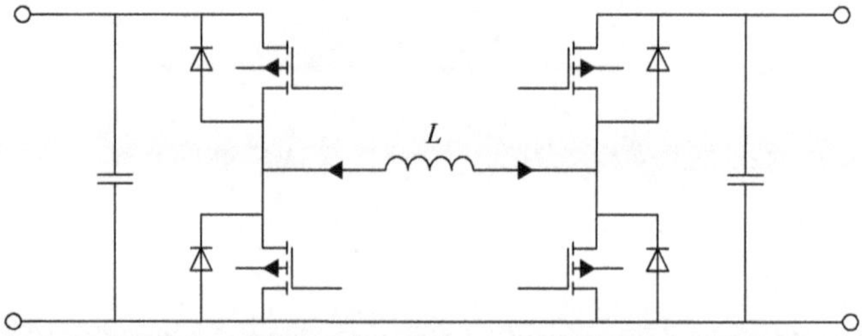


*Figure 16.17 Non-isolated bidirectional buck-boost converter*

sizes of the filter inductors and voltage stresses of power switches are reduced. To increase the power level of DC/DC conversion, several power legs can be connected in parallel, and thus extended to the multi-phase structure (Gautam *et al.*, 2012). The multi-phase DC/DC converter can reduce the current stress of switches and increase the delivered power of the converter.

Different from the unidirectional DC/DC converter, the non-isolated bidirectional DC/DC converter can achieve the bidirectional power transfer capability. Figure 16.16 shows a typical non-isolated bidirectional half-bridge DC/DC converter for EV chargers. The high-voltage DC side is from output side of AC/DC rectifier, and the low-voltage DC side is from battery side. Furthermore, two bidirectional half-bridge DC/DC converters can be cascaded to form the non-isolated bidirectional buck-boost converter, which is shown in Figure 16.17. In this topology, the buck and the boost capabilities can be provided for both the two power flow directions.

Figure 16.18 presents another DC/DC converter that can be utilized in EV chargers. It is capable of bidirectional operation in buck or boost modes. Under any operating modes, only one of the power switches is controlled by PWM, one switch

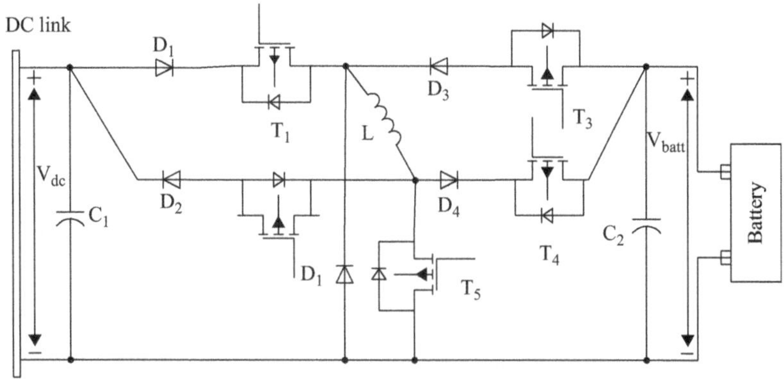


*Figure 16.18 Fully bidirectional universal DC/DC converter*

keeps ON and other switches keep OFF. Therefore, the switching losses are less than those of other buck or boost converters. For instance, when the power is transferred from $V_{dc}$ to $V_{batt}$ in buck mode, $T_1$ operates under PWM modulation, when $T_4$ keeps on. The buck circuit consists of $T_1$, L, $C_2$ and $D_5$. In addition, this converter shares only one inductor in all modes, decreasing the overall size and cost (Onar *et al.*, 2013).

#### 16.2.2.2 Isolated DC/DC converters

The isolated DC/DC converters for battery chargers are based on the high-frequency transformers which provide the galvanic isolation between the input sides and the output sides. The flyback and forward converters are two unidirectional isolated DC/DC converters with simple configuration and low costs. But their power ratings are limited. The half-bridge and full-bridge configurations are two other widely used bidirectional isolated DC/DC converters for EV chargers. The half-bridge DC/DC converter uses less switching devices, and the transformer's leakage inductance and switching devices' parasitic capacitance could realize soft-switching for the power switches. But the issue of magnetic deflection should be considered for the half-bridge DC/DC converters. For full-bridge DC/DC converters, the switching devices bear low voltage and current stresses. The zero-voltage switching (ZVS) operation can be realized. Thus, this topology is preferable to high power applications.

##### *16.2.2.2.1 Phase-shifted ZVS full-bridge converters*

Figure 16.19 presents the configuration of a phase-shifted ZVS full-bridge converter. Because the driving signals of $Q_1/Q_3$ are ahead of $Q_2/Q_4$, the $Q_1/Q_3$ bridge is called leading leg when the $Q_2/Q_4$ bridge is called lagging leg. The switches in the same bridge operate complementary. $C_1$~$C_4$ are the parasitic capacitors of $Q_1$~$Q_4$, respectively. The converter regulates its output voltage by changing the phase-shifted angle between the leading leg and the lagging leg. The resonant inductor $L_r$ contains leakage inductor of the high-frequency transformer. The converter can

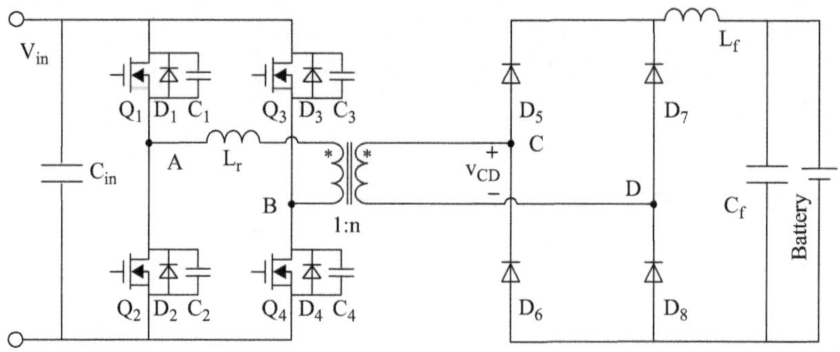


*Figure 16.19 Phase-shifted full-bridge DC/DC converter*

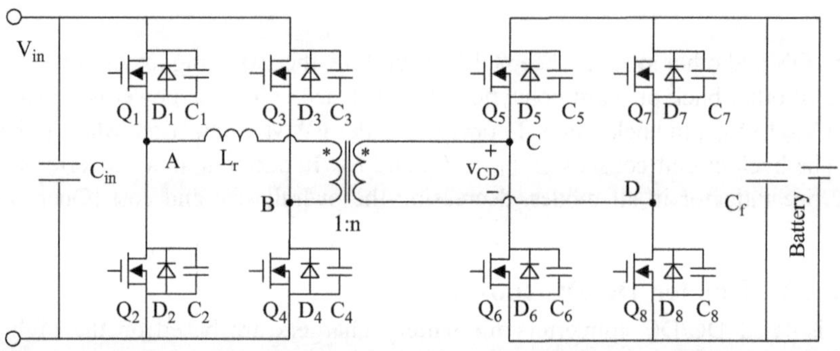


*Figure 16.20 Dual active full-bridge DC/DC converter*

realize ZVS operation by the resonance of $L_r$ and parasitic capacitors of switches. For example, after $Q_1$ turns off, the primary current of transformer charges $C_1$ while discharging $C_2$. In the next operation interval, due to natural conduction of diode $D_2$, the output voltage of $Q_2$ is clamped to zero and $Q_2$ achieves ZVS on (Gautam *et al.*, 2013b).

By using active full-bridge instead of diode-bridge on the output side, the dual active full-bridge DC/DC converter is presented in Figure 16.20. The dual active full-bridge DC/DC converter is widely used due to bidirectional power flow and soft switching capability (Kim *et al.*, 2015). The phase-shifted angle between the voltages of primary side and secondary side can be used to tune the transferred power. In initial state, the switches $Q_2$ and $Q_3$ are turned on. When the switches $Q_2$ and $Q_3$ turn off, the resonance occurs between the parasitic capacitors ($C_1$, $C_2$, $C_3$, $C_4$) and the winding inductance $L_r$. Thus, the parasitic capacitors paralleled to switches $Q_2$ and $Q_3$ are charging while the parasitic capacitors paralleled to switches $Q_1$ and $Q_4$ are discharging. When the voltages on parasitic capacitors ($C_1$, $C_4$) become zero, the anti-diodes of $Q_1$ and $Q_4$ are freewheeling. Thus, the ZVS condition is achieved during turn-on operation of $Q_1$ and $Q_4$. When the direction of leakage inductance current is changed to the opposite direction, $Q_1$ and $Q_4$ carry the input current of DC converter.

The commutation of switches on primary side within the positive half cycle is completed. The commutation processes of power switches on secondary side are similar. Figure 16.21 shows the waveforms of the dual active full-bridge DC/DC converter, where the phase-shifted angle between the voltages on the primary side and the secondary side is used to control the power exchanged between the two sides.

*16.2.2.2.2 Resonant converters*

One difficulty of the conventional unidirectional full-bridge phase-shift DC/DC converter is the voltage spikes across the output diodes on secondary side, which is caused by the interaction between the leakage inductance of transformer and the output inductor. Resonant converters are proposed to mitigate the aforementioned difficulty. Figure 16.22 shows the configuration of LLC resonant DC/DC converter, which is one variation of multi-resonant converter. The LLC resonant converter is able to perform ZVS conditions for power switches on the primary side and zero-current switching (ZCS) conditions for diodes on the secondary side. Thus, its switching frequency and power density can be increased. The resonant tank of LLC resonant converter consists of the resonant inductance $L_r$, the resonant capacitor $C_r$

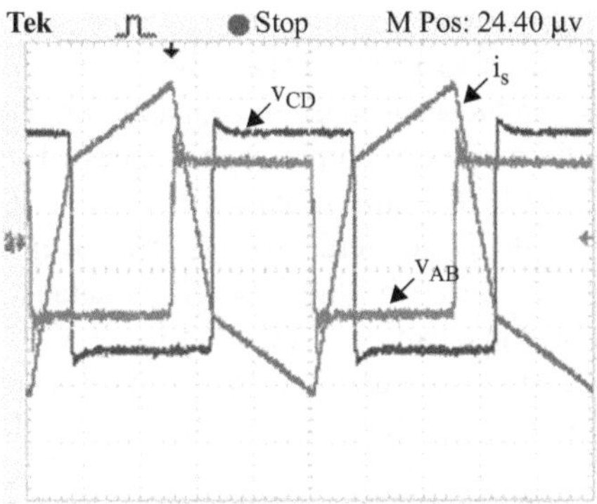


*Figure 16.21 Measured waveforms of dual active full-bridge DC/DC converter (X: 10 μs/div; Y: $v_{CD}$ 100 V/div, $v_{AB}$ 50 V/div, $i_s$ 1A/div).*

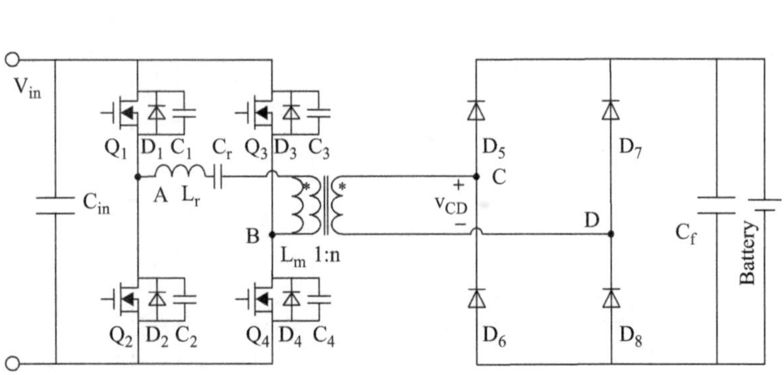


*Figure 16.22 LLC resonant DC/DC converter*

and the paralleled inductance $L_m$ of the transformer. $L_m$ can be integrated into the magnetizing inductance of transformer and $L_r$ can be fully or partially integrated into the transformer's leakage inductance. So, the overall volume of magnetic components can be further minimized (Gang *et al.*, 2015). There are two resonant modes in LLC resonant converter: one is the resonance between the resonant inductance $L_r$ and the resonant capacitor $C_r$, and the resonant frequency is $f_r$. This condition occurs when the resonant current on primary side is larger than the magnetizing current in $L_m$. The voltage on $L_m$ will be clamped to the reflected voltage from secondary side. The second resonance occurs among $L_r$, $C_r$ and $L_m$. The corresponding resonant frequency is $f_m$. It happens when the resonant current on primary side is equal to the magnetizing current and the diodes on secondary side turn off (Kim and Moon, 2014). When the switching frequency $f_s$ is designed to be $f_m < f_s < f_r$, both ZVS conditions for switches on primary side and ZCS conditions for diodes on secondary side can be achieved. However, under some light load conditions, the soft-switching ability may be lost for LLC resonant converters.

Figure 16.23 presents another resonant converter, called series-parallel current-driven (SPCD) full-bridge DC/DC converter. The SPCD full-bridge topology consists of a series branch $L_s$ and a parallel branch including $L_p$ and $C_p$. The series inductor $L_s$ is used to convert voltage pulses into current pulses. The parallel branch including $L_p$ and $C_p$ is used to produce the reactive current for ZVS condition. Since both the leakage inductance and the magnetizing inductance provide the energy to charge and discharge the output capacitances of the power switches on the primary side, the small load current can still achieve ZVS operation for the power switches. So the SPCD converter can operate over a wide range of load variations. The resonance caused by the leakage inductance $L_r$, the magnetizing inductance $L_p$, the output capacitances of the diodes ($C_5$, $C_6$, $C_7$, $C_8$) and the output capacitance $C_P$ provide smooth transition for the diodes on the secondary side. Thus, the voltage spikes are suppressed effectively on the secondary side of transformer without clamping circuits. However, the SPCD full-bridge DC/DC converter has higher current with the triangular waveform, which restricts its use in high current applications (Gautam *et al.*, 2011; Gautam *et al.*, 2012; Pahlevani *et al.*, 2015).

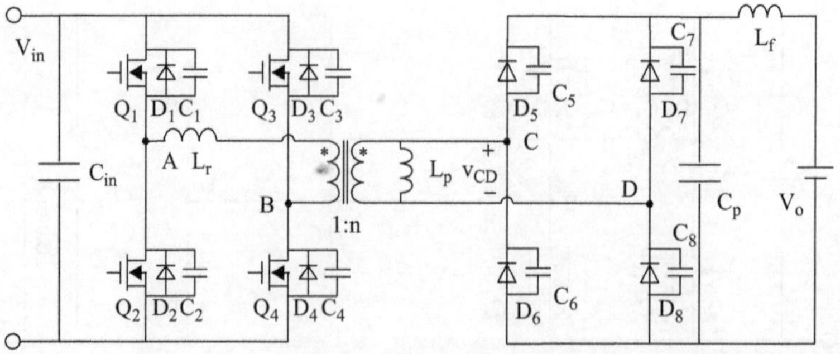


*Figure 16.23 Series-parallel current-driven full-bridge DC/DC converter*

#### *16.2.2.2.3 Interleaved full-bridge/half-bridge DC/DC converters*

Interleaving the DC/DC converters is an effective method to increase the power level of EV chargers. When the converters operate in parallel, the input current ripple, current stress of the semiconductor devices and conducting loss will be reduced. Figure 16.24 shows an isolated interleaved ZVS full-bridge converter with output voltage doubler rectifier. It is with input-parallel and output-series configuration. The interleaved DC/DC converter operates efficiently, shares equal output power and uniformly distributes thermal losses among the individual cells. By using the output voltage doubler rectifier and the interleaved operation, this converter configuration significantly reduces the number of output rectifier diodes and the input filtering requirements. Besides, the reverse recovery losses of secondary side diodes are mitigated due to the natural commutation (Gautam *et al.*, 2013a).

Figure 16.25 shows an interleaved three-phase ZCS isolated DC/DC converter which is with input-series and output-parallel connection. This method can increase input voltage blocking capability and decrease output current ripple. The current-source ZCS full-bridge phase-shifted DC/DC converter is the basic structure element of this topology. The interleaved angle among gating signals is decided by the number of converters (Mohammadpour *et al.*, 2015). For each ZCS full-bridge phase-shifted DC/DC converter, the two upper switches $S_{1A}$ and $S_{1B}$ conduct complementarily while the two lower switches $S_{2A}$ and $S_{2B}$ conduct complementarily. But there ought to be an overlapping time between the complementary switches. The resonance between the inductance $L_1$ and the capacitor $C_1$ makes the zero-current condition for the power switches on primary side to turn off. Besides, the diodes on the secondary side turn on and turn off with zero-voltage conditions. Such interleaved ZCS DC/DC converter enables high power density and reliability.

Another interleaved current-source full-bridge converter is shown in Figure 16.26. It is input-parallel and output-series connection. Both the input current

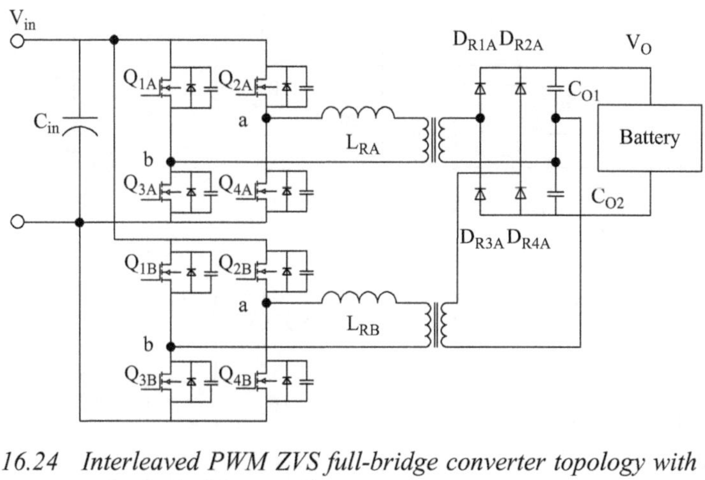


*Figure 16.24 Interleaved PWM ZVS full-bridge converter topology with output voltage doubler rectifier*

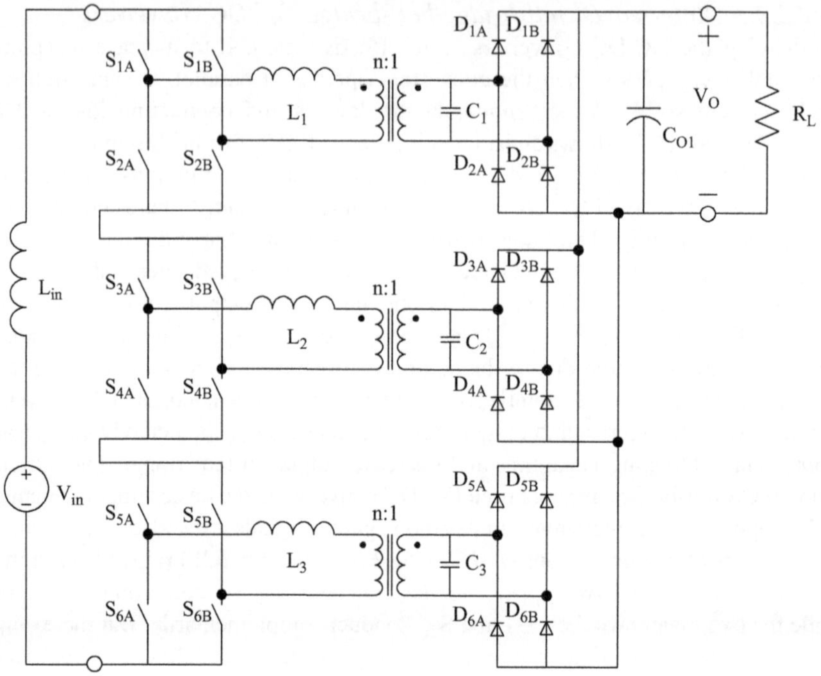


*Figure 16.25 Interleaved three-phase current-source ZCS DC/DC converter*

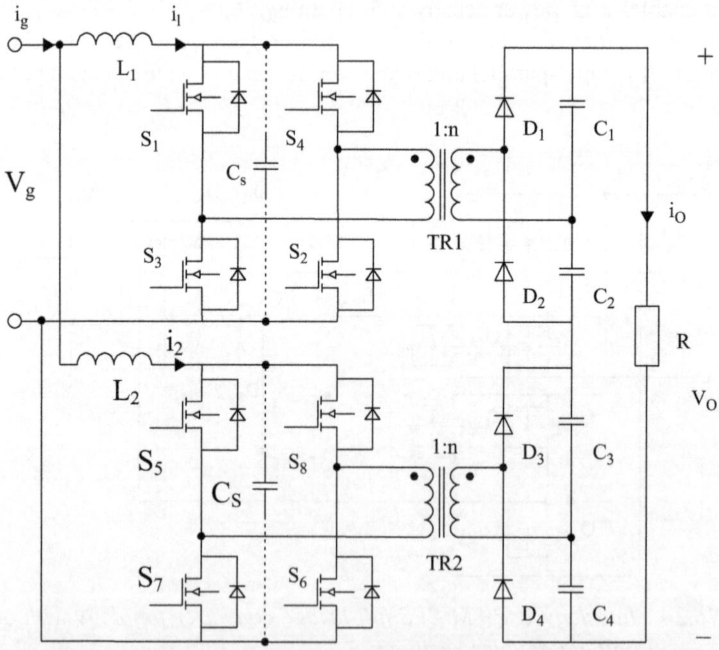


*Figure 16.26 Interleaved current-source full-bridge converter*

ripple and the output voltage ripple can be reduced. In addition, the size of the magnetic components and current stress of the semiconductor devices on the input side are reduced. Similarly, smaller voltage rating components can be used on the output side (Kong and Khambadkone, 2007).

## 16.3 Three-stage power interface

Usually, the EVs are equipped with two different DC voltage levels: one is to provide the high DC voltage for the traction motor, and the other is to feed the low DC voltage link for the low-voltage loads in EVs. The demand for two DC voltage levels can be met by two bidirectional DC/DC converters sharing the same battery stack or by using two battery packs, namely, the propulsion battery and the auxiliary battery. For both of the two solutions, additional DC/DC conversion stage is needed to match the difference between the high DC voltage side and the low DC voltage side. Therefore, the EV chargers are called as the three-stage power interface. Various three-stage converters and control methods have been developed. Figure 16.27 shows a block diagram of conventional three-stage converter. It consists of a bidirectional AC/DC rectifier and two bidirectional DC/DC converters for low-voltage side and high-voltage side, respectively.

Figure 16.28 presents a multifunctional on-board battery charger which connects auxiliary battery and propulsion battery. The battery charger consists of three full-bridge modules. The first function is G2V mode where the grid charges the

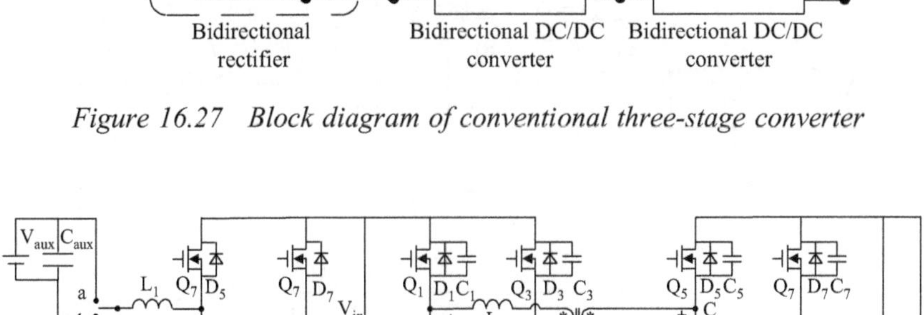


*Figure 16.27 Block diagram of conventional three-stage converter*

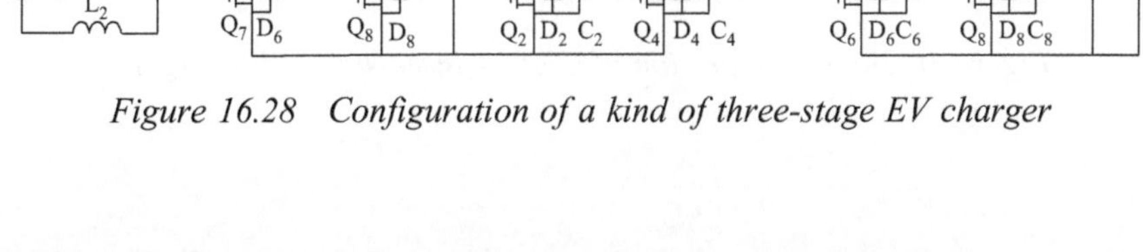


*Figure 16.28 Configuration of a kind of three-stage EV charger*

propulsion battery when EV connects to the grid. Under this case, the charger converts AC power to DC form. Since the charger is bidirectional, in V2G operation, the propulsion battery can supply surplus energy to the grid. For the case of V2G operation, the front-end full-bridge converter acts as a grid inverter. When the charger is disconnected from the grid, the propulsion battery charges the auxiliary battery through the three full-bridges in the driving state. Under this case, the charger acts as a step-down converter. It is obvious that the power switches and inductors are shared in the three functions. Thus, the power density of the EV charger can be enhanced (Kim and Kang, 2015).

Figure 16.29 shows another single-phase three-stage battery charger for EVs that also operates in three different modes. Compared to Figure 16.28, the non-isolated bidirectional two-stage charger is used to connect the propulsion battery to the grid in the configuration of Figure 16.29. An isolated unidirectional diode-rectifier is used to deliver power from the AC side to the auxiliary battery (Pinto *et al.*, 2014).

Recently, a three-port isolated (TPI) bidirectional DC/DC converter is studied for three energy ports by integrating three windings in high-frequency transformer magnetically as shown in Figure 16.30 (Wang *et al.*, 2015b). The zero-voltage transition condition is provided by the leakage inductances of transformer and parasitic capacitance of switching devices. The switching frequency can thus be increased. Due to the advantages of compact structure and high power density, the TPI bidirectional DC/DC converters are suitable to connect multiple energy ports of EVs. The three energy ports are connected to the propulsion battery, auxiliary battery and the grid-side DC link for the three-stage battery charging. Figure 16.31 (a) and (b) shows the measured steady-state voltage waveforms and current waveforms in the three windings of the TPI bidirectional DC/DC converter using the phase-shifted control.

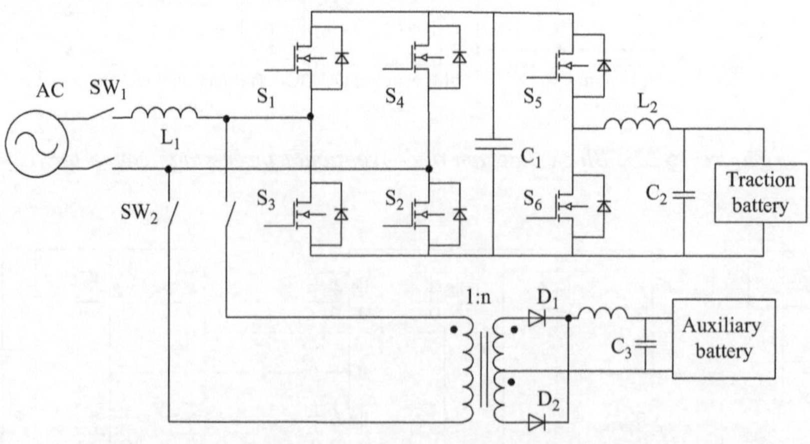


*Figure 16.29 Configuration of another three-stage EV charger*

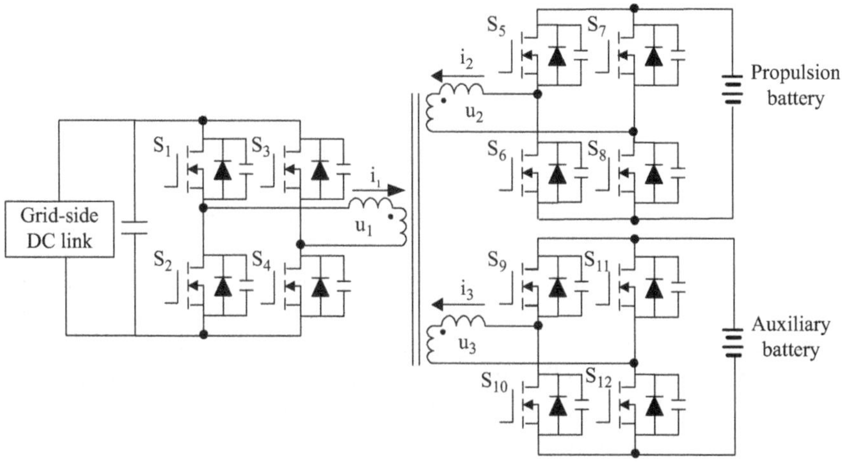


*Figure 16.30 Three-port isolated (TPI) bidirectional DC/DC converter*

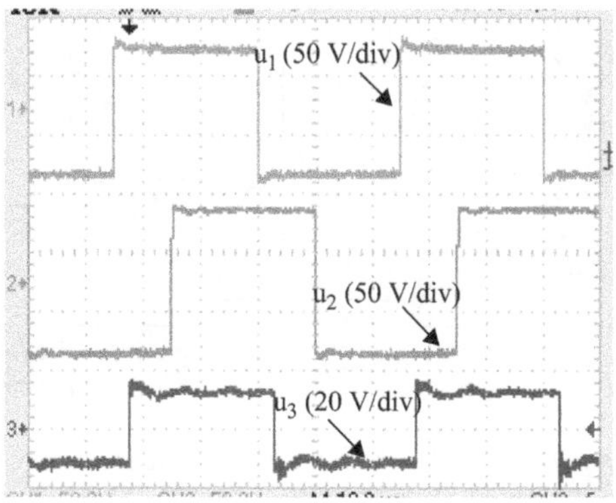


(a)

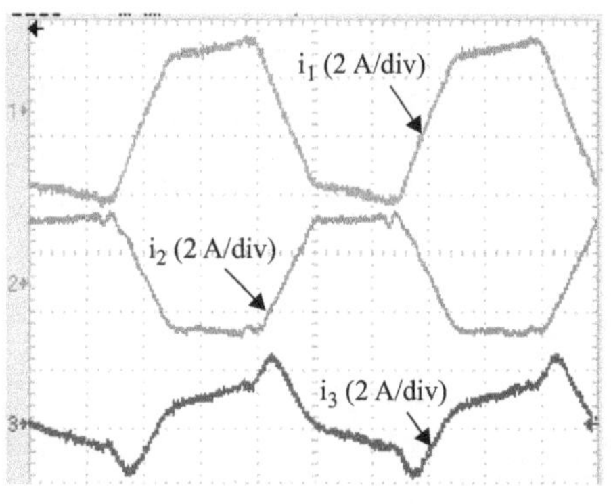


(b)

*Figure 16.31 Measured waveforms in three windings of TPI bidirectional DC/DC converter (10 μs/div): (a) voltage; (b) current*

## 16.4 Integrated power interface for multiple DC levels

In the EV chargers with multiple DC levels, the AC/DC rectifiers and different DC/DC converters can be integrated for compact design. Figure 16.32 presents the configuration of an integrated charger by integrating DC/DC converters with DC levels.

In Figure 16.33, a non-isolated buck-boost diode rectifier whose inductor is shared with an integrated bidirectional DC/DC converter is presented. With the buck (Bridge–T1–D5) and boost (Bridge–T2–D6) capability, this integrated charger can work with wide input voltage ranges for charging mode. It is also able to step-up (T4–D7–T2–D8–T5) the voltage in driving mode and step-down the voltage (T6–D9–D6–T3–D5) in regenerative braking of the propulsion drive. With the integrated configuration, the driving mode, the regenerative mode and the charging mode could be realized. However, due to the power losses caused by additional semiconductor devices in current passing path, the efficiency is slightly lower than traditional topologies (Lee *et al.*, 2009).

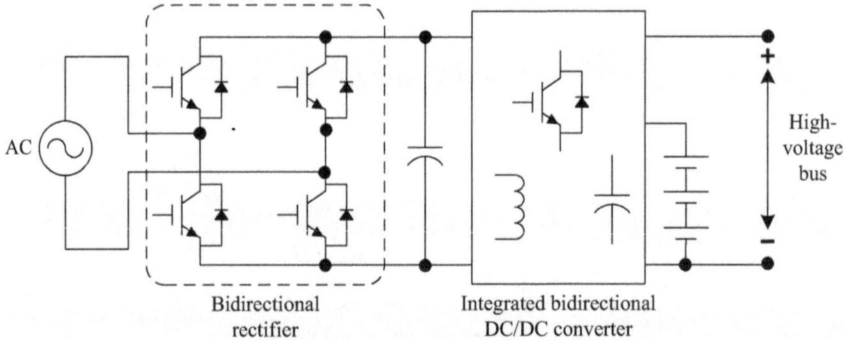


*Figure 16.32 Block diagram of integrated converter*

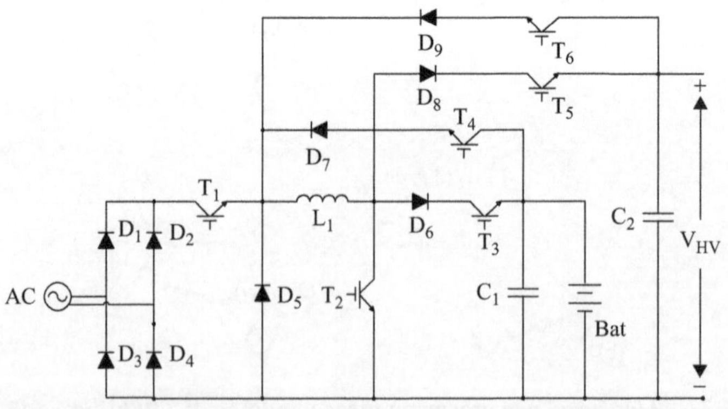


*Figure 16.33 Battery charger with integrated DC/DC converter*

Figure 16.34 shows another integrated topology offering further reduction in the number of components using one inductor, seven diodes and four switches (Dusmez and Khaligh, 2012). In the charging mode, the DC/DC converter in the second stage works in buck-boost mode (T1–D5). In the driving mode, this converter works in boost mode (T2–T3–D7). In the regenerative braking mode, this converter works in buck mode (T4–D5–D6). This integrated on-board charger topology shares one inductor for all operation modes.

To achieve the low input current THD, an integrated charger with three-level AC/DC rectifiers is presented in Figure 16.35 (Erb *et al.*, 2010a; Erb *et al.*, 2010b). A single-phase neutral-point-clamping three-level rectifier is connected to the integrated DC/DC converter for different DC voltage levels.

The AC/DC rectifiers and the DC/DC bidirectional converters can be further integrated into one converter, which is described in Figure 16.36. This combination reduces the number of components of the charger significantly. It is named integrated direct converter. Therefore, the topology in Figure 16.37 is presented where a direct AC-DC converter is used (Chen *et al.*, 2011). It has four operation modes, the driving mode (T6–T3–D3), the regenerative braking mode (T9–D2–T6), the

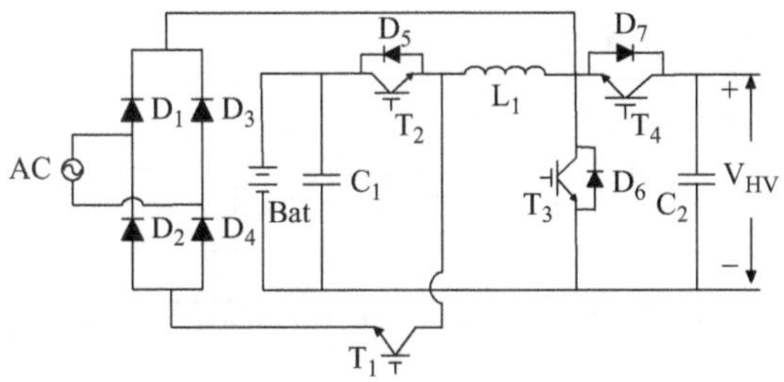


*Figure 16.34 Battery charger with integrated DC/DC converter with reduced hardware*

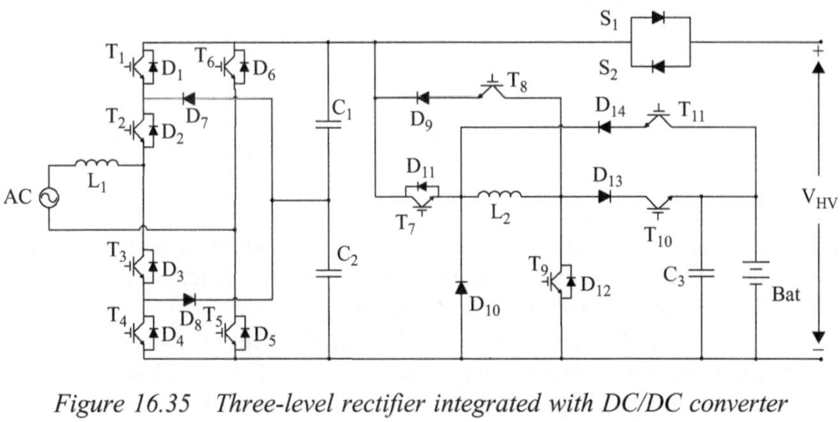


*Figure 16.35 Three-level rectifier integrated with DC/DC converter*

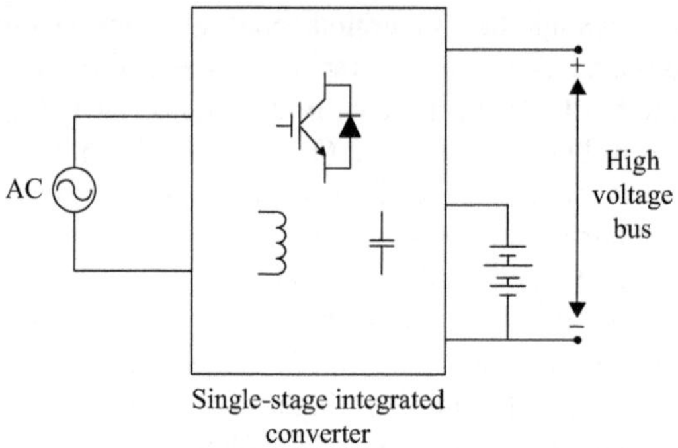


*Figure 16.36 Block diagram of integrated direct converter*

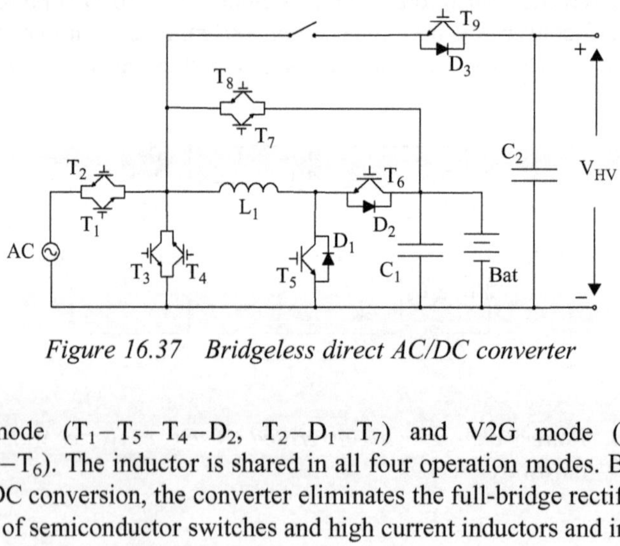


*Figure 16.37 Bridgeless direct AC/DC converter*

charging mode ($T_1$–$T_5$–$T_4$–$D_2$, $T_2$–$D_1$–$T_7$) and V2G mode ($T_8$–$T_5$–$T_1$, $T_2$–$D_1$–$T_3$–$T_6$). The inductor is shared in all four operation modes. Based on the direct AC/DC conversion, the converter eliminates the full-bridge rectifier, reduces the number of semiconductor switches and high current inductors and improves the conversion efficiency (Chen *et al.*, 2011).

## References

Carlton, D., and Dunford, W.G. (2001) Multilevel, unidirectional AC-DC converters, a cost effective alternative to bi-directional converters, *Proceedings of IEEE PESC'01*, pp. 1911–1917.

Chen, H., Wang, X., and Khaligh, A. (2011) A single stage integrated bidirectional ac/dc and dc/dc converter for plug-in hybrid electric vehicles, *Proceedings of IEEE Vehicle Power Propulsion Conference*, pp. 1–6.

Darly, S.S., Ranjan, P.V, Bindu, K.V., and Rabi, B.J. (2011) A novel dual boost rectifier for power factor improvement, *Proceedings of Electrical Energy System Conference*, pp. 122–127.

Dusmez, S., and Khaligh, A. (2012) A novel low cost integrated on-board charger topology for electric vehicles and plug-in hybrid electric vehicles, *Proceedings of Application Power Electronics Conference*, pp. 2611–2616.

Erb, D.C., Onar, O.C., and Khaligh, A. (2010a) An integrated bidirectional power electronic converter with multi-level AC–DC/DC–AC converter and non-inverted buck-boost converter for PHEVs with minimal grid level disruptions, *Proceedings of IEEE Vehicle Power Propulsion Conference*, pp. 1–6.

Erb, D.C., Onar, O.C., and Khaligh, A. (2010b) Bi-directional charging topologies for plug-in hybrid electric vehicles, *Proceedings of IEEE Application Power Electronics Conference Exposition*, pp. 2066–2072.

Gang, Y., Patrick, D., and Daniel, S. (2015) Double-phase high-efficiency, wide load range high-voltage/low-voltage LLC DC/DC converter for electric/hybrid vehicles, *IEEE Transactions on Power Electronics*, 30, pp. 1876–1886.

Gautam, D., Musavi, F., Edington, M., Eberle, W., and Dunford, W.G. (2011) An automotive on-board 3.3 kW battery charger for PHEV application, *Proceedings of IEEE Vehicle Power Propulsion Conference*, pp. 1–6.

Gautam, D., Musavi, F., Edington, M., Eberle, W., and Dunford, W.G. (2012) A zero voltage switching full-bridge DC–DC converter with capacitive output filter for a plug-in-hybrid electric vehicle battery charger, *Proceedings of IEEE Application Power Electronics Conference Exposition*, pp. 1381–1386.

Gautam, D., Musavi, F., Edington, M., Eberle, W., and Dunford, W.G. (2013a) An isolated interleaved DC-DC converter with voltage doubler rectifier for PHEV battery charger, *Proceedings of IEEE Application Power Electronics Conference Exposition*, pp. 3067–3072.

Gautam, D., Musavi, F., Eberle, W., and Dunford, W.G. (2013b) A zero-voltage switching full-bridge DC-DC converter with capacitive output filter for plug-in hybrid electric vehicle battery charging, *IEEE Transactions on Power Electronics*, 28, pp. 5728–5735.

Huber, L., Yungtaek, J., and Jovanovic, M.M. (2008) Performance evaluation of bridgeless PFC boost rectifiers, *IEEE Transactions on Power Electronics*, 23, pp. 1381–1390.

Khaligh, A., and Dusmez, S. (2012) Comprehensive topological analysis of conductive and inductive charging solutions for plug-in electric vehicles, *IEEE Transactions on Vehicle Technology*, 61, pp. 3475–3489.

Kim, J.S., Choe, G.-Y., Jung, H.-M., Lee, B.-K., Cho, Y.-J., and Han, K.-B. (2010) Design and implementation of a high-efficiency on-board battery charger for electric vehicles with frequency control strategy, *Proceedings of IEEE Vehicle Power Propulsion Conference*, pp. 1–6.

Kim, J.W., and Moon, G.W. (2014) A new LLC series resonant converter with a narrow switching frequency variation and reduced conduction losses, *IEEE Transactions on Power Electronics*, 29, pp. 4278–4287.

Kim, S., and Kang, F. (2015) Multifunctional onboard battery charger for plug-in electric vehicles, *IEEE Transactions on Industrial Electronics*, 62, pp. 3460–3472.

Kim, J.H., Lee, I.O., and Moon, G.W. (2015) Integrated dual full-bridge converter with current-doubler rectifier for EV charger, *IEEE Transactions on Power Electronics*, 31, pp. 942–951.

Kong, X., and Khambadkone, A.M. (2007) Analysis and implementation of a high efficiency, interleaved current-fed full bridge converter for fuel cell system, *IEEE Transactions on Power Electronics*, 22, pp. 543–550.

Lee, Y.J., Khaligh, A., and Emadi, A. (2009) Advanced integrated bidirectional ac/dc and dc/dc converter for plug-in hybrid electric vehicles, *IEEE Transactions on Vehicle Technology*, 58, pp. 3970–3980.

Mohammadpour, A., Harfman, T.M., Rixin, L., Rajib, D., and Luis, G. (2015) Series-input parallel-output modular-phase DC-DC converter with soft switching and high frequency isolation, *IEEE Transactions on Power Electronics*, 31, pp. 111–119.

Musavi, F., Eberle, W., and Dunford, W.G. (2011) A high-performance single phase bridgeless interleaved PFC converter for plug-in hybrid electric vehicle battery chargers, *IEEE Transactions on Industrial Electronics*, 47, pp. 1833–1843.

Musavi, F., Eberle, W., and Dunford, W.G. (2013) A phase-shifted gating technique with simplified current sensing for the semi-bridgeless ac–dc converter, *IEEE Transactions on Vehicle Technology*, 62, pp. 1568–1576.

Onar, O.C., Kobayashi, J., and Khaligh, A. (2013) A fully directional universal power electronic interface for EV, HEV, and PHEV applications, *IEEE Transactions on Power Electronics*, 28, pp. 5489–5498.

Pahlevaninezhad, M., Drobnik, P., Das, J., Jain, P.K., and Bakhshai, A. (2012) A novel ZVZCS full-bridge DC/DC converter used for electric vehicles, *IEEE Transactions on Power Electronics*, 27, pp. 2752–2769.

Pahlevani, M., Eren, S., Bakhshai, A., and Jain, P. (2015) A series-parallel current-driven full-bridge DC/DC converter, *IEEE Transactions on Power Electronics*, 31, pp. 1275–1293.

Pinto, J.G., Vítor, M., Henrique, G., and Joáo, L.A. (2014) Onboard reconfigurable battery charger for electric vehicles with traction-to-auxiliary mode, *IEEE Transactions on Vehicle Technology*, 63, pp. 1104–1116.

Singh, B., Singh, B.N., Chandra, A., Haddad, K.A., Pandey, A., and Kothari, D.P. (2003) A review of single-phase improved power quality AC–DC converters, *IEEE Transactions on Industrial Electronics*, 50, pp. 962–981.

Sudheer, L., Kanimozhi, G., and Sreedevi, V.T. (2015) Integrator controlled semi-bridgeless PFC boost converter, *Proceedings of International Conference on Circuit, Power and Computing Technologies (ICCPCT)*, pp. 1–6.

Wang, Z., Wu, B., Xu, D., and Zargari, N.R. (2011) Hybrid PWM for high power current source inverter fed drives with low switching frequency, *IEEE Transactions on Power Electronics*, 26, pp. 1754–1764.

Wang, Z., Zhang, B., and Chu, K. (2013) Design and experimentation of interleaved PWM and generalized control schemes for paralleled grid converters of wind energy systems, *Proceedings of 39th Annual Meeting of IEEE Industrial Electronics Society*, pp. 1581–1586.

Wang, Z., Wu, B., Xu, D., and Zargari N.R. (2012) A current-source-converter-based high-power high-speed PMSM drive with 420-Hz switching frequency, *IEEE Transactions on Industrial Electronics*, 59, pp. 2970–2981.

Wang, Z., Chen, J., and Cheng, M. (2015a) Modeling and control of neutral-point-clamping (NPC) three-level inverters fed dual-three phase PMSM drives, *Proceedings of Energy Conversion Congress and Exhibition*, pp. 6565–6572.

Wang, Z., Chu, K., Zhang, B., and Zhang, Y. (2015b) Three-port bidirectional DC/DC converter with duty cycle and phase-shifting control, *Electric Machines and Control*, 19, pp. 81–87.

Williamson, S.S., Rathore, A.K., and Musavi, F. (2015) Industrial electronics for electric transportation: current state-of-the-art and future challenges, *IEEE Transactions on Industrial Electronics*, 62, pp. 3021–3032.

Yilmaz, M., and Krein, P.T. (2013) Review of battery charger topologies, charging power levels, and infrastructure for plug-in electric and hybrid vehicles, *IEEE Transactions on Power Electronics*, 28, pp. 2151–2169.

Zou, Z., Wang, Z., and Cheng, M. (2014) Modeling, analysis, and design of multi-function grid-interfaced inverters with output LCL filters, *IEEE Transactions on Power Electronics*, 29, pp. 3830–3839.

# Index